Development and Recognition of the Transformed Cell

Development and Recognition of the Transformed Cell

Edited by

Mark I. Greene
University of Pennsylvania School of Medicine
Philadelphia, Pennsylvania

and

Toshiyuki Hamaoka
Institute for Cancer Research
Osaka University Medical School
Osaka, Japan

PLENUM PRESS • NEW YORK AND LONDON

Library of Congress Cataloging in Publication Data

Development and recognition of the transformed cell.

Includes bibliographies and index.
1. Carcinogenesis. 2. Cell transformation. 3. Cancer cells—Growth. 4. Oncogenes. I. Greene, Mark I. II. Hamaoka, Toshiyuki. [DNLM: 1. Cell Transformation, Neoplastic. 2. Neoplasm Proteins—genetics. 3. Neoplasms—immunology. 4. Oncogenes. QZ 202 D489]
RC268.5.D48   1987   616.99'4071   87-17175
ISBN 0-306-42636-6

© 1987 Plenum Press, New York
A Division of Plenum Publishing Corporation
233 Spring Street, New York, N.Y. 10013

All rights reserved

No part of this book may be reproduced, stored in a retrieval system, or transmitted in any form or by any means, electronic, mechanical, photocopying, microfilming, recording, or otherwise, without written permission from the Publisher

Printed in the United States of America

Contributors

ANTHONY P. ALBINO, Department of Medicine and Divison of Immunology, Memorial Sloan–Kettering Cancer Center, New York, New York 10021

PETER M. BLUMBERG, Molecular Mechanisms of Tumor Promotion Section, Laboratory of Cellular Carcinogenesis and Tumor Promotion, National Cancer Institute, Bethesda, Maryland 20892

JOAN S. BRUGGE, Department of Microbiology, State University of New York, Stony Brook, New York 11794-8621

STEVEN J. BURAKOFF, Division of Pediatric Oncology, Dana–Farber Cancer Institute and Harvard Medical School, Boston, Massachusetts 02115

JANET S. BUTEL, Department of Virology and Epidemiology, Baylor College of Medicine, Houston, Texas 77030

GRAHAM CARPENTER, Departments of Medicine and Biochemistry, Vanderbilt University School of Medicine, Nashville, Tennessee 37232

AGNES CHAN, Department of Microbiology, University of British Columbia, Vancouver, British Columbia V6T 1W5, Canada

MARTIN A. CHEEVER, Department of Medicine, University of Washington, and Division of Oncology, Fred Hutchinson Cancer Research Center, Seattle, Washington 98195

DAVID A. CHERESH, Department of Immunology, Scripps Clinic and Research Foundation, La Jolla, California 92037

PAUL COUSSENS, Department of Microbiology, State University of New York, Stony Brook, New York 11794-8621. *Present address*: Department of Microbiology and Public Health, Michigan State University, East Lansing, Michigan 48824

CARLO M. CROCE, Wistar Institute of Anatomy and Biology, Philadelphia, Pennsylvania 19104-4268

LAURA J. DAVIS, Department of Medicine and Division of Immunology, Memorial Sloan–Kettering Cancer Center, New York, New York 10021

MARIE L. DELL'AQUILA, Molecular Mechanisms of Tumor Promotion Section, Laboratory of Cellular Carcinogenesis and Tumor Promotion, National Cancer Institute, Bethesda, Maryland 20892

DAVID J. DE VRIES, Molecular Mechanisms of Tumor Promotion Section, Laboratory of Cellular Carcinogenesis and Tumor Promotion, National Cancer Institute, Bethesda, Maryland 20892

ANTONIO DiGIACOMO, Trudeau Institute, Inc., Saranac Lake, New York 12983

NICOLAS C. DRACOPOLI, Department of Medicine and Division of Immunology, Memorial Sloan–Kettering Cancer Center, New York, New York 10021

JEFFREY A. DREBIN, Department of Pathology, Harvard Medical School, Boston, Massachusetts 02115

EARL S. DYE, Trudeau Institute, Inc., Saranac Lake, New York 12983

DOUGLAS V. FALLER, Division of Pediatric Oncology, Dana–Farber Cancer Institute and Harvard Medical School, Boston, Massachusetts 02115

DAVID C. FLYER, Division of Pediatric Oncology, Dana–Farber Cancer Institute and Harvard Medical School, Boston, Massachusetts 02115

JUNICHI FUJISAWA, Department of Viral Oncology, Japanese Foundation for Cancer Research, Tokyo 170, Japan

HIROMI FUJIWARA, Department of Oncogenesis, Institute for Cancer Research, Osaka University Medical School, Osaka 553, Japan

MASAHIRO FUKUZAWA, Department of Oncogenesis, Institute for Cancer Research, Osaka University Medical School, Osaka 553, Japan

KENNETH H. GRABSTEIN, Immunex Corporation, Seattle, Washington 98101

PHILIP D. GREENBERG, Departments of Medicine and Microbiology/Immunology, University of Washington, and Division of Oncology, Fred Hutchinson Cancer Research Center, Seattle, Washington 98195

MARK I. GREENE, Division of Immunology, Department of Pathology and Laboratory Medicine, University of Pennsylvania School of Medicine, Philadelphia, Pennsylvania 19104-6082

TOSHIYUKI HAMAOKA, Department of Oncogenesis, Institute for Cancer Research, Osaka University Medical School, Osaka 553, Japan

LISA W. HOSTETLER, Department of Immunology, University of Texas System Cancer Center, M. D. Anderson Hospital and Tumor Institute, Houston, Texas 77030

ALAN N. HOUGHTON, Department of Medicine and Division of Immunology, Memorial Sloan–Kettering Cancer Center, New York, New York 10021

NOBUMICHI HOZUMI, Division of Molecular Immunology and Neurobiology, Mount Sinai Hospital Research Institute, Toronto, Ontario M5G 1X5, Canada

MINORU IGARASHI, Department of Microbiology, Tohoku University School of Dentistry, Sendai 980, Japan

WENDELYN H. INMAN, Departments of Medicine and Biochemistry, Vanderbilt University School of Medicine, Nashville, Tennessee 37232

JUNICHIRO INOUE, Department of Viral Oncology, Japanese Foundation for Cancer Research, Tokyo 170, Japan

MICHAEL C. V. JENSEN, Department of Microbiology/Immunology, University of Washington, Seattle, Washington 98195

PAUL W. JOHNSON, Division of Molecular Immunology and Neurobiology, Mount Sinai Hospital Research Institute, Toronto, Ontario M5G 1X5, Canada

DONALD E. KERN, Department of Microbiology/Immunology, University of Washington, Seattle, Washington 98195

CHIHARU KIYOTAKI, Department of Oncogenesis, Institute for Cancer Research, Osaka University Medical School, Osaka 553, Japan

JAY P. KLARNET, Department of Medicine, University of Washington, and Division of Oncology, Fred Hutchinson Cancer Research Center, Seattle, Washington 98195

AKIRA KOBATA, Department of Biochemistry, Institute of Medical Science, University of Tokyo, Shirokanedai, Minato-ku, Tokyo 108, Japan

ATSUSHI KOSUGI, Department of Oncogenesis, Institute for Cancer Research, Osaka University Medical School, Osaka 553, Japan

MARGARET L. KRIPKE, Department of Immunology, University of Texas System Cancer Center, M. D. Anderson Hospital and Tumor Institute, Houston, Texas 77030

KATSUO KUMAGAI, Department of Microbiology, Tohoku University School of Dentistry, Sendai 980, Japan

JULIA G. LEVY, Department of Microbiology, University of British Columbia, Vancouver, British Columbia V6T 1W5, Canada

VICTORIA C. LINK, Department of Pathology, Harvard Medical School, Boston, Massachusetts 02115

GAIL R. MASSEY, Department of Pathology and Laboratory Medicine, University of Pennsylvania School of Medicine, Philadelphia, Pennsylvania 19104-6082

DANIEL MERUELO, Department of Pathology and Kaplan Cancer Center, New York University Medical Center, New York, New York 10016

YUMIKO MIZUSHIMA, Department of Oncogenesis, Institute for Cancer Research, Osaka University Medical School, Osaka 553, Japan

HIROTO NAKAJIMA, Department of Oncogenesis, Institute for Cancer Research, Osaka University Medical School, Osaka 553, Japan

ROBERT J. NORTH, Trudeau Institute, Inc., Saranac Lake, New York 12983

PETER C. NOWELL, Department of Pathology and Laboratory Medicine, University of Pennsylvania School of Medicine, Philadelphia, Pennsylvania 19104-6082

MASATO OGATA, Department of Oncogenesis, Institute for Cancer Research, Osaka University Medical School, Osaka 553, Japan

ANTHONY J. PAWSON, Division of Molecular and Developmental Biology, Mount Sinai Hospital Research Institute, Toronto, Ontario M5G 1X5, Canada

JOHN C. REED, Department of Pathology and Laboratory Medicine, University of Pennsylvania School of Medicine, Philadelphia, Pennsylvania 19104-6082

JOHN C. RODER, Division of Molecular Immunology and Neurobiology, Mount Sinai Hospital Research Institute, Toronto, Ontario M5G 1X5, Canada

NAOMI ROSENBERG, Departments of Pathology and Molecular Biology and Microbiology, Tufts University School of Medicine, Boston, Massachusetts 02111

KOHICHI SAKAMOTO, Department of Oncogenesis, Institute for Cancer Research, Osaka University Medical School, Osaka 553, Japan

SHIGETOSHI SANO, Department of Oncogenesis, Institute for Cancer Research, Osaka University Medical School, Osaka 553, Japan

SOICHIRO SATO, Department of Oncogenesis, Institute for Cancer Research, Osaka University Medical School, Osaka 553, Japan

MOTOHARU SEIKI, Department of Viral Oncology, Japanese Foundation for Cancer Research, Tokyo 170, Japan

CHARLES J. SHERR, Department of Tumor Cell Biology, St. Jude Children's Research Hospital, Memphis, Tennessee 38105

JUNKO SHIMA, Department of Oncogenesis, Institute for Cancer Research, Osaka University Medical School, Osaka 553, Japan

JUN SHIMIZU, Department of Oncogenesis, Institute for Cancer Research, Osaka University Medical School, Osaka 553, Japan

ERICA M. S. SIBINGA, Department of Pathology and Laboratory Medicine, University of Pennsylvania School of Medicine, Philadelphia, Pennsylvania 19104-6082

RAKESH SINGHAI, Department of Pathology, Tufts University, Boston, Massachusetts

ANTHEA T. STAMMERS, Department of Microbiology, University of British Columbia, Vancouver, British Columbia V6T 1W5, Canada

J. KEVIN STEELE, Department of Pathology, Harvard Medical School, Harvard University, Cambridge, Massachusetts

TAKASHI SUDA, Department of Oncogenesis, Institute for Cancer Research, Osaka University Medical School, Osaka 553, Japan

RYUJI SUZUKI, Department of Microbiology, Tohoku University School of Dentistry, Sendai 980, Japan

SATSUKI SUZUKI, Department of Microbiology, Tohoku University School of Dentistry, Sendai 980, Japan

TETSU TAKAHASHI, Department of Microbiology, Tohoku University, School of Dentistry, Sendai 980, Japan

YASUYUKI TAKAI, Department of Oncogenesis, Institute for Cancer Research, Osaka University Medical School, Osaka 553, Japan

REBECCA A. TAUB, Howard Hughes Medical Institute, Department of Human Genetics, University of Pennsylvania School of Medicine, Philadelphia, Pennsylvania 19104-6082

SATVIR S. TEVETHIA, Department of Microbiology, The Pennsylvania State University College of Medicine, Hershey, Pennsylvania 17033

KUMAO TOYOSHIMA, Institute of Medical Science, University of Tokyo, Tokyo 108, Japan

WILLIAM S. TRIMBLE, Division of Molecular Immunology and Neurobiology, Mount Sinai Hospital Research Institute, Toronto, Ontario M5G 1X5, Canada

BARBOUR S. WARREN, Molecular Mechanisms of Tumor Promotion Section, Laboratory of Cellular Carcinogenesis and Tumor Promotion, National Cancer Institute, Bethesda, Maryland 20892

TADASHI YAMAMOTO, Institute of Medical Science, University of Tokyo, Tokyo 108, Japan

MITSUAKI YOSHIDA, Department of Viral Oncology, Japanese Foundation for Cancer Research, Tokyo 170, Japan

TAKAYUKI YOSHIOKA, Department of Oncogenesis, Institute for Cancer Research, Osaka University Medical School, Osaka 553, Japan

Preface

The study of the phenotypic and genetic features that characterize the malignant cell is a rapidly growing and changing field. Clearly new insights into the processes involved in normal and abnormal cell growth will facilitate our understanding of events relevant to cancer and cellular differentiation. Early studies on genetic features associated with cancer focused on chromosomal abnormalities that were observable in several human malignancies. The more recent examination of oncogenes and the proteins they encode has helped pinpoint many steps in different processes that might be involved in cancer.

Immunologic studies of cancer have also developed from an imprecise series of investigations to a more detailed molecular examination of cell-surface structures that can be recognized immunologically. In the course of the development of modern tumor immunology, it has become clear that many of the antigens that can be recognized appear to be the products of genes involved in cell growth. Furthermore, changes in the cell surface of malignant cells have often been found to include alteration of nonprotein constituents.

The purpose of this volume—produced by contributions from major individual laboratories—is to provide a description of the current status of the field. It is organized into two general parts. The first deals with the molecular description of genes and proteins involved in normal and malignant cell growth. The second deals with alterations in the phenotype of cells that can be recognized by the immune system. Aspects of immune regulation relevant to tumor cell growth are also considered.

The development of this book was helped enormously by the assistance of Susanne Gallagher of the University of Pennsylvania and by support from the Otsuka Pharmaceutical Company and the Nippon Zoki Pharmaceutical Company.

The editors are also grateful to the editorial staff of Plenum Press, who made this book a relatively painless effort.

Mark I. Greene
Toshiyuki Hamaoka

Philadelphia and Osaka

Contents

CHAPTER 1
Cytogenetics of Neoplasia 1
  PETER C. NOWELL AND CARLO M. CROCE

CHAPTER 2
The Structure and Function of the Normal c-*myc* Gene and Its Alteration
in Malignant Cells 21
  REBECCA A. TAUB

CHAPTER 3
Protooncogene Expression in Lymphoid Cells: Implications for the
Regulation of Normal Cellular Growth 39
  JOHN C. REED

CHAPTER 4
Oncogene Products as Receptors 59
  ERICA M. S. SIBINGA, GAIL R. MASSEY, AND MARK I. GREENE

CHAPTER 5
Monoclonal Antibodies Reactive with the *neu* Oncogene Product Inhibit
the Neoplastic Properties of *neu*-Transformed Cells 69
  JEFFREY A. DREBIN, VICTORIA C. LINK, AND MARK I. GREENE

CHAPTER 6
Relationship of the c-*fms* Protooncogene Product to the CSF-1 Receptor ... 81
  CHARLES J. SHERR

Chapter 7
Two *erbB*-Related Protooncogenes Encoding Growth Factor Receptors .. 93
TADASHI YAMAMOTO AND KUMAO TOYOSHIMA

Chapter 8
The Receptor for Epidermal Growth Factor 111
WENDELYN H. INMAN AND GRAHAM CARPENTER

Chapter 9
The Role of the *abl* Gene in Transformation 123
NAOMI ROSENBERG

Chapter 10
Comparison of the Structural and Functional Properties of the Viral and Cellular *src* Gene Products 145
PAUL COUSSENS AND JOAN S. BRUGGE

Chapter 11
Protein Kinase C as the Site of Action of the Phorbol Ester Tumor Promoters 157
MARIE L. DELL'AQUILA, BARBOUR S. WARREN, DAVID J. DE VRIES, AND PETER M. BLUMBERG

Chapter 12
Involvement of Human Retrovirus in Specific T-Cell Leukemia 187
MITSUAKI YOSHIDA, MOTOHARU SEIKI, JUNICHI FUJISAWA, AND JUNICHIRO INOUE

Chapter 13
Mechanisms of Virus-Induced Alterations of Expression of Class I Genes and Their Role on Tumorigenesis 203
DANIEL MERUELO

Chapter 14
Antigenic Requirements for the Recognition of Moloney Murine Leukemia Virus-Induced Tumors by Cytotoxic T Lymphocytes 221
DAVID C. FLYER, DOUGLAS V. FALLER, AND STEVEN J. BURAKOFF

Chapter 15
SV40 Tumor Antigen: Importance of Cell Surface Localization in Transformation and Immunological Control of Neoplasia 231
SATVIR S. TEVETHIA AND JANET S. BUTEL

CONTENTS xv

CHAPTER 16
Correlation of Natural Killer Cell Recognition with *ras* Oncogene
Expression 243
  PAUL W. JOHNSON, WILLIAM S. TRIMBLE, NOBUMICHI HOZUMI,
  ANTHONY J. PAWSON, AND JOHN C. RODER

CHAPTER 17
A Regulatory Role of Natural Killer Cells (LGL) in T-Cell-Mediated
Immune Response 261
  KATSUO KUMAGAI, RYUJI SUZUKI, SATSUKI SUZUKI, TETSU TAKAHASHI, AND
  MINORU IGARASHI

CHAPTER 18
Immune Regulation in Neoplasia: Dominance of Suppressor Systems ... 279
  J. KEVIN STEELE, AGNES CHAN, ANTHEA T. STAMMERS, RAKESH SINGHAI,
  AND JULIA G. LEVY

CHAPTER 19
The Generation and Down-Regulation of the Immune Response to
Progressive Tumors 295
  ROBERT J. NORTH, ANTONIO DIGIACOMO, AND EARL S. DYE

CHAPTER 20
Origin and Significance of Transplantation Antigens Induced on Cells
Transformed by UV Radiation 307
  LISA W. HOSTETLER AND MARGARET L. KRIPKE

CHAPTER 21
Cellular and Molecular Mechanisms Involved in Tumor Eradication *in
Vivo* 331
  HIROMI FUJIWARA, TAKAYUKI YOSHIOKA, HIROTO NAKAJIMA,
  MASAHIRO FUKUZAWA, KOHICHI SAKAMOTO, MASATO OGATA,
  SHIGETOSHI SANO, JUN SHIMIZU, CHIHARU KIYOTAKI, AND
  TOSHIYUKI HAMAOKA

CHAPTER 22
Application of T Cell–T Cell Interaction to Enhanced Tumor-Specific
Immunity Capable of Eradicating Tumor Cells *in Vivo* 355
  TOSHIYUKI HAMAOKA, YASUYUKI TAKAI, ATSUSHI KOSUGI, JUNKO SHIMA,
  TAKASHI SUDA, YUMIKO MIZUSHIMA, SOICHIRO SATO, AND
  HIROMI FUJIWARA

Chapter 23

Antigens Expressed by Melanoma and Melanocytes: Studies of the Immunology, Biology, and Genetics of Melanoma 373

ALAN N. HOUGHTON, LAURA J. DAVIS, NICOLAS C. DRACOPOLI, AND ANTHONY P. ALBINO

Chapter 24

Malignant Transformational Changes of the Sugar Chains of Glycoproteins and Their Clinical Application 385

AKIRA KOBATA

Chapter 25

Ganglioside Involvement in Tumor Cell–Substratum Interactions 407

DAVID A. CHERESH

Chapter 26

Specific Adoptive Therapy of Disseminated Tumors: Requirements for Therapeutic Efficacy and Mechanisms by Which T Cells Mediate Tumor Eradication 429

PHILIP D. GREENBERG, DONALD E. KERN, JAY P. KLARNET, MICHAEL C. V. JENSEN, KENNETH H. GRABSTEIN, AND MARTIN A. CHEEVER

Index 447

1
Cytogenetics of Neoplasia

PETER C. NOWELL AND CARLO M. CROCE

1. Introduction

Based on investigations from many laboratories, four general statements can be made about karyotypic alterations in neoplasia: (1) most tumors have chromosome abnormalities, which are not present in other cells of the body; (2) in a given tumor, all the neoplastic cells often have the same cytogenetic change, or related changes; (3) chromosome abnormalities are more extensive in advanced tumors; and (4) although chromosome alterations often differ between tumors, there are nonrandom patterns. Each of the four observations has contributed to our understanding of the nature of the transformed cell and the development of malignant tumors.

The first two generalizations represent a significant portion of the evidence that somatic genetic changes are important in tumorigenesis. In addition, the second generalization, along with related biochemical, molecular genetic, and immunological data, has been the basis for the now generally accepted view that most neoplasms are unicellular in origin.[1-3] The fact that in a given tumor all the cells show the same chromosome abnormality (or related abnormalities) suggests the derivation of a tumor from a single altered cell. Presumably, the particular karyotypic change confers on the progenitor cell a selective growth advantage, allowing its progeny to expand as a neoplastic "clone."

1.1. Cytogenetics of Tumor Progression

The third observation listed above, that karyotypic abnormalities are more extensive in advanced tumors, has also contributed to fundamental concepts of tumor biology. It has led to the hypothesis that the phenomenon of clinical and

---

PETER C. NOWELL • Department of Pathology and Laboratory Medicine, University of Pennsylvania School of Medicine, Philadelphia, Pennslyvania 19104-6082. CARLO M. CROCE • Wistar Institute of Anatomy and Biology, Philadelphia, Pennsylvania 19104-4268.

biological tumor progression, the tendency of tumors to become more aggressive with time, may often reflect the sequential appearance in a neoplastic clone of subpopulations of cells with further alterations in genetic makeup.[1,4,5] Considerable evidence indicates that neoplastic cells show increased genetic instability, and are thus more likely than normal cells to generate genetic variants.[1,6-8] In this concept of "clonal evolution," most of the genetic variants that arise in a tumor cell population do not survive, but those few mutants that have an additional selective growth advantage, however, expand to become predominant subpopulations within the neoplasm and demonstrate the characteristics of more aggressive growth and increased "malignancy" that we recognize as tumor progression. The continued presence of multiple subpopulations within the neoplasm provides the basis for the heterogeneity that is also typically observed in malignant tumors.

Because it has been difficult to do serial investigations on the common human cancers, the most significant cytogenetic data supporting this view of tumor progression have been obtained from repeated studies on individual cases of leukemia and lymphoma. Best documented are the findings in chronic granulocytic leukemia. The cells in the early indolent stage of this disorder typically show only the t(9;22) chromosome translocation involving the c-*abl* oncogene that results in the Philadelphia chromosome, but the terminal accelerated phase of the disease apparently represents overgrowth of this leukemic population by one or more subclones having additional karyotypic change.[9,10] Similar associations between additional cytogenetic changes and tumor progression have been demonstrated in other human leukemias as well as some solid malignancies in man and experimental animals.[11] In several cases, the findings have begun to suggest particular genes (oncogenes) that may be associated with the chromosome alteration involved in the evolutionary process, and these will be discussed in subsequent sections.

1.2. Cytogenetics and Oncogenes

Currently, the most exciting developments in tumor cytogenetics relate to the fourth generalization listed in the opening paragraph: that specific alterations in particular chromosomes are associated, with different degrees of consistency, with specific types of tumors or with neoplasia in general.[11-13] Increasing recognition of these nonrandom karyotypic changes has suggested that they might indicate sites in the genome where particular genes important in carcinogenesis are located, and provide clues as to how the function of such "oncogenes" might be significantly altered. Much of the initial support for these hypotheses has come from the study of specific reciprocal chromosome translocations in various hematopoietic tumors such as Burkitt's lymphoma and chronic myelogenous leukemia (CML), in which alteration in structure and/or function of protooncogenes, the human homologues of retroviral oncogenes, has been demonstrated.[10-14] In addition, chromosome changes in other tumors, reflecting gain or loss of genetic material, are indicating a critical role for oncogene dosage in some instances of carcinogenesis. The remainder of this brief discussion will focus on this aspect of the cytogenetics and related molecular genetics of neoplasia, with emphasis on our own recent studies, and with no attempt to review exhaustively this rapidly developing field of cancer research.

2. Chromosome Translocations in Human Leukemia and Lymphoma

2.1. Lymphocytic Tumors

The first evidence for "activation" of a human protooncogene by chromosome translocation has come from studies of Burkitt's lymphoma. These studies are now being extended to other human B-cell and T-cell neoplasms. The Burkitt tumor will be considered first.

In approximately 75% of the cases of this tumor, there is a reciprocal translocation between chromosomes 8 and 14 (Fig. 1). Almost all of the remaining cases have "variant" translocations involving chromosomes 8 and 22 or chromosomes 2 and 8. In all instances, the breakpoint on chromosome 8 is the same, in the terminal portion of the long arm of the chromosome (band q24). The genes for the human immunoglobulin heavy and light chains have been mapped to the other chromosomal regions involved in these translocations: the heavy chain locus to band 14q32, the λ light chain genes to band 22q11, and the κ light chain genes to band 2p11. At the same time, the human homologue of the retroviral v-*myc* oncogene was mapped to the terminal portion of 8q, suggesting that this gene, as well as the immunoglobulin genes, might be significant in the pathogenesis of Burkitt's lymphoma.[14]

This hypothesis has now been investigated in several laboratories, using a combination of cytogenetic and molecular genetic techniques.[14–16] It has been demonstrated that in each of these translocations a transcriptionally active and rearranged immunoglobulin gene is brought into juxtaposition with the c-*myc* gene, resulting in deregulation of the oncogene. In the case of the common t(8;14) translocation, the immunoglobulin heavy chain locus is split, and the c-*myc* oncogene, with or without structural alteration, is brought into a "head-to-head" association with it. In the variant translocations, the κ or λ immunoglobulin gene is translocated to the 3' end of the c-*myc* oncogene, which remains on chromosome 8, usually without structural alteration (Fig. 1).

These various rearrangements are still being studied, but it appears that in each instance the c-*myc* protooncogene comes under the influence of enhancers in or adjacent to the immunoglobulin loci, resulting in deregulation of expression of the oncogene and a presumed critical role in the altered growth of the neoplastic B cells.[14–17] The c-*myc* gene product is a nuclear protein, apparently having a normal function in growth regulation of nonneoplastic lymphocytes and other cells,[17,18] as discussed in detail elsewhere in this volume. Our own studies of the translocated c-*myc* gene, on the $14q^+$ chromosome of Burkitt tumor cells, indicate that it can still be regulated in an appropriate cellular background (e.g., a fibro-

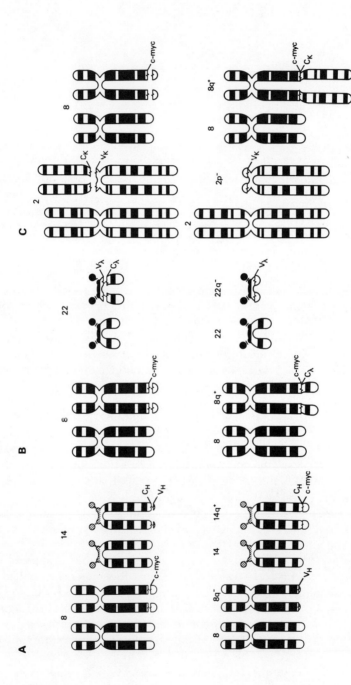

FIGURE 1. Diagram of the three chromosomal translocations (A, B, C) seen in Burkitt's lymphoma that result in the c-*myc* oncogene being brought into juxtaposition with an immunoglobulin gene locus (C_H, C_λ, C_κ). (From Ref. 14)

blast), but when under the influence of an enchancing element in a B cell, at the proper stage of differentiation, it does not respond to normal regulatory mechanisms.[14]

These findings concerning the c-*myc* gene in Burkitt's lymphoma, with similar observations in tumors or rats and mice,[5] have provided strong evidence for tumorigenic effects of a structurally unaltered, but deregulated, oncogene. The exact role of the *myc* gene product in the pathogenesis of the lymphoma remains uncertain, however, as well as the contribution, in endemic areas, of other factors such as chronic infection with malaria and with Epstein–Barr virus.

2.2. Translocations in Other B-Cell Lymphomas

Burkitt's lymphoma is a relatively rare neoplasm, particularly in this country, but it has recently been recognized that a significant proportion of other B-cell lymphomas and chronic B-cell leukemias have characteristic translocations that also involve the terminal portion of chromosome 14 (band q32).[13,19] In these tumors, the "donor" chromosomal site is different from the Burkitt tumor, being on either the long arm of chromosome 11 (q13) or the long arm of chromosome 18 (q21) (Fig. 2).

Because it seemed likely that these translocations might also involve the immunoglobulin heavy chain locus, we have used neoplastic cells from patients with B-cell tumors having either the t(11;14)(q13;q32) translocation or the t(14;18)(q32;q21) translocation to clone molecularly the breakpoints involved.[20,21] DNA probes flanking the breakpoints on chromosomes 11 and 18 were then used to detect rearrangement of the homologous sequences in B-cell lymphomas and leukemias carrying these translocations. The breakpoints in these tumors were found to be clustered within short segments, either on chromosome 11 or on chromosome 18, and the breakpoint on chromosome 14 was consistently 5' to the constant region of the immunoglobulin heavy chain gene. These findings suggested that unknown oncogenes, located on chromosomes 11 and 18, might be involved in these rearrangements in a manner analogous to the c-*myc* gene in the t(8;14) translocation of the Burkitt tumor (Fig. 2). We have suggested the names *bcl-1* and *bcl-2*, respectively, for these putative new oncogenes,[20,21] which thus far appear not to represent homologues of any of the retroviral oncogenes that have been described. A human counterpart of the v-*yes* oncogene, c-*yes*, has recently been mapped to the chromosome 18q21 region,[22] but it appears not to be homologous with the *bcl-2* gene that we have cloned (C. C. Croce, unpublished findings).

Our findings to date indicate that with both the *bcl-1* and *bcl-2* genes, the translocation is into the J region of the immunoglobulin locus on chromosome 14,[20,21] and this observation has recently been independently confirmed for *bcl-2*.[23] Furthermore, in the breakpoint regions of the translocated segments of chromosomes 11 and 18, we have identified several short signal sequences typically used in V–D–J joining during a normal rearrangement within the heavy chain locus. This indicated that following chromosome breakage, there might be an increased probability of the specific t(11;14) or t(14;18) translocations occurring when the recom-

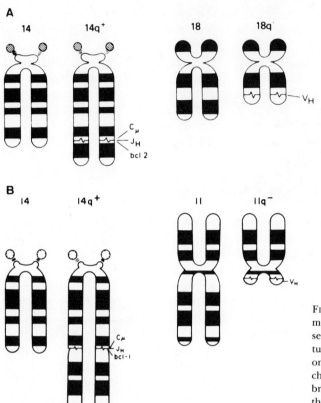

FIGURE 2. Diagram of two chromosomal translocations (A, B) seen in various lymphocytic tumors that result in two putative oncogenes, *bcl-1* and *bcl-2*, from chromosomes 11 and 18, being brought into juxtaposition with the immunoglobulin heavy chain locus (C_μ, J_H). (From Ref. 14)

binase system normally involved in immunoglobulin gene rearrangement erroneously utilized the chromosome 11 or chromosome 18 signal sequence.[20,21]

We have also recently used the *bcl-2* gene probe to identify a 6-kb RNA transcript, and determined that levels of this transcript are tenfold higher in leukemic B cells with the t(14;18) translocation than in neoplastic B cells without it, suggesting deregulation of the *bcl-2* gene.[21] These preliminary data are consistent with an oncogene activation mechanism similar to that of the Burkitt lymphoma, but more data must be obtained concerning the *bcl-1* and *bcl-2* genes before the nature of their presumed altered function can be definitely characterized.

2.3. Translocations in Acute Lymphocytic Leukemia (ALL)

In addition to these ongoing studies of chromosome rearrangements involving the immunoglobulin heavy chain locus as an "activating" site in various B-cell tumors, there are several other nonrandom translocations that have been described in ALL that are just beginning to be investigated at the molecular level. In some cases of ALL there is a t(9;22) translocation with chromosome breakpoints in the same regions as the typical t(9;22)(q34;q11) rearrangement that produces the Philadelphia chromosome in chronic myelogenous leukemia (CML; see Section 2.5).[9] We have recently obtained data on five cases of Ph-positive ALL, in both children and adults, which indicate that at the molecular level the chromosome 22 breakpoint is heterogeneous. In two adult cases, the translocation appeared to be identical with that in CML, which will be discussed subsequently. In two children and one young adult with Ph-positive ALL, however, the breakpoint in 22q was proximal to the so-called breakpoint cluster region (bcr) that characterizes CML (Fig. 3), and data from one patient indicated that the transcript from the translocated c-abl protooncogene was normal.[24] The breakpoint on 22q in these cases was also different from that of the variant t(8;22) translocation of the Burkitt lymphoma, with minimal, if any, involvement of the λ immunoglobulin light chain locus.

These data indicate that further investigation of chromosome translocations that appear to be identical at the level of the light microscope may provide molecular information that helps to explain differing phenotypes of the presenting neoplasms. The different breakpoints that have been recognized thus far, with molecular genetic techniques, in the q11 band of chromosome 22 are illustrated in Fig. 3, along with the specific tumors with which they have been associated.

With respect to other nonrandom translocations recognized in ALL, there is a group of cases of the childhood form of this disorder characterized by a

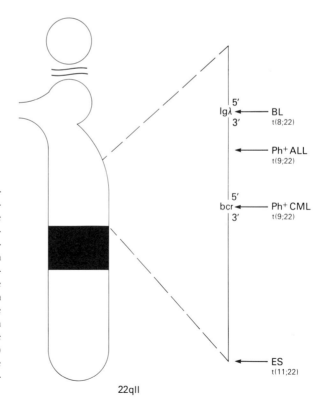

FIGURE 3. Schematic representation of 22q11 breakpoints in different human malignancies. The 22q11 breakpoint in Burkitt's lymphoma with the t(8;22) translocation involves the immunoglobulin λ locus. The breakpoint in Ph-positive ALL of childhood with the t(9;22) chromosome translocation is proximal to bcr and distal to the λ locus. The 22q11 breakpoint in Ph-positive CML is within bcr. The breakpoint in Ewing sarcoma (ES) is at 22q11–q12, distal to the breakpoints observed in hematopoietic disorders. (From Ref. 24)

t(4;11)(q21;q23) translocation involving the terminal portion of the long arm of chromosome 11, to which one of the c-*ets* oncogenes has been mapped.[25] Several other hemic and nonhemic neoplasms (e.g., acute monocytic leukemia, Ewing's sarcoma) have also been shown to have translocations involving this site,[25,26] including a t(11;22)(q23;q11) in Ewing's tumor (Fig. 3), and the possibility of altered structure or function of the c-*ets* oncogene is now being investigated.

It has been recently suggested that a characteristic t(1;19)(q21;p13) translocation associated with pre-B-cell acute leukemia might involve the insulin receptor gene that has recently been mapped to 19p13. Since this gene has homologies with both the c-*src* and c-*erbB* oncogenes, the possibility of a leukemogenic role for altered insulin receptor function has been suggested,[27] although to date there is no direct evidence for involvement of the gene.

2.4. Translocations in T-Cell Tumors

The relatively limited cytogenetic data on human T-cell neoplasms indicate that nonrandom chromosomal alterations are less frequent than in B-cell leukemias and lymphomas.[28-30] Recent findings, however, have suggested that approaches similar to those used for the study of oncogenes involved in B-cell tumors may be feasible for the investigation, at the molecular level, of T-cell neoplasia. Genes coding for subunits of the T-cell antigen receptor have now been mapped, and their possible role in leukemogenesis is being explored. The gene for the α subunit of the T-cell receptor is located at band 14q11, a site frequently involved in translocations and inversions in T-cell tumors.[31] The β chain gene has been mapped to the terminal portion of 7q (7q32-35), and the γ chain gene to the short arm of chromosome 7 (7p13).[32,33]

Studies using lymphocyte cultures from patients with chromosomal fragility syndromes such as ataxia telangiectasia, as well as from normal subjects, indicate that these three chromosomal sites are unusually subject to breakage in human T cells, as is, surprisingly, the 14q32 region, the site of the immunoglobulin heavy chain locus.[32,33] Presumably, the increased fragility at the T-cell receptor loci reflects the effect of somatic recombination that occurs normally at these sites during T-cell differentiation, and makes translocations involving these three chromosomal regions more likely. The situation is less clear with respect to the 14q32 region, where the various chromosomal rearrangements noted thus far in T-cell tumors[28,29] suggest that "activating" genes (immunoglobulin gene? T-cell genes?), as well as one or more protooncogenes, may all be located within this short chromosomal segment. It is clear that further molecular dissection of this portion of chromosome 14 is required as was described in the previous section for the q11 region of chromosome 22.

The most common translocations in T-cell tumors appear to be inversions and translocations between 14q11 and 14q32.[28,29] Rearrangements involving the T-cell receptor gene sites on chromosome 7 (p13 and q35) are relatively rare. A nonrandom translocation has been described in a small group of T-cell leukemias[30] that involves a translocation from the short arm of chromosome 11 (band p13) to the proximal region (q11) of chromosome 14 (Fig. 4). No known oncogene is impli-

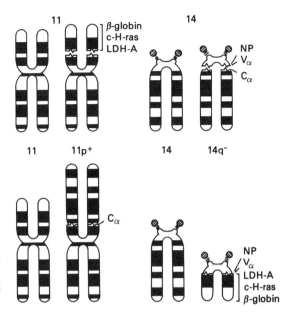

FIGURE 4. The t(11;14)(p13;q11) translocation in ALL. The translocation breakpoint on chromosome 14 splits the locus for the α chain of the T-cell receptor. The V_α genes remain on the 14q⁻ chromosome, while the C_α translocates to the involved chromosome 11 (11p⁺). The gene for human nucleoside phosphorylase remains on the involved chromosome 14 (14q⁻). The genes LDH-A, β-globin, and c-H-*ras* translocate to the involved chromosome 14 (14q⁻). (From Ref. 34)

cated, but analysis of this rearrangement in our laboratory and elsewhere indicates that it does split the gene for the α chain of the T-cell receptor, with the constant portion sequences translocated to chromosome 11 and the variable portion remaining on chromosome 14.[34,35] Several cases have also been reported in which there is a t(8;14)(q24;q11) translocation,[36,37] and we find that this rearrangement may bring a portion of the α T-cell receptor gene into juxtaposition with the c-*myc* protooncogene, resulting in deregulation of the oncogene.[37] Although still very preliminary, these data suggest that at least one of the T-cell receptor genes may have the same "activating" role in certain T-cell neoplasms as do the immunoglobulin genes in B-cell tumors.

2.5. Translocations in Nonlymphocytic Leukemias

The only cytogenetic abnormality that has been extensively studied in this group of neoplasms, with respect to oncogene involvement, is the characteristic t(9;22)(q34;q11) translocation that produces the Philadelphia chromosome in most typical cases of CML.[9,13] As noted above, mapping of the chromosome breakpoints in this rearrangement has indicated that the c-*abl* protooncogene is regularly translocated from its normal site on chromosome 9 to a very restricted region of chromosome 22, located between the λ immunoglobulin gene and the c-*sis* oncogene. The latter two genes do not appear to be significantly involved in the rearrangement, and the critical segment of chromosome 22 has been termed the "breakpoint cluster region" (bcr).[38]

Further investigation of the c-*abl* oncogene in the leukemic cells of CML has demonstrated the production of a novel 8-kb mRNA, larger than the normal transcript. Cloning of the joining region on chromosome 22 has shown that the altered

RNA is a hybrid molecule, containing both bcr and *abl* sequences.[10] Furthermore, this transcript appears to code for an abnormal protein of higher molecular weight and having tyrosine kinase activity, which has not been demonstrated with the normal gene product. Because some receptors for cellular growth factors are tyrosine kinases, it has been suggested that the altered bcr-*abl* product, if it proves to be a cell surface protein, might represent a hybrid receptor with altered growth regulatory effects.[10] Unlike the Burkitt tumor translocations, where the *myc* gene product appears unaltered, the data in CML indicate that the t(9;22) translocation in this disease results in a modified protein encoded by an oncogene. Interestingly, our initial findings in Ph-positive ALL, where the chromosome 22 breakpoint appears to be proximal to the bcr, suggest that the translocated *c-abl* oncogene in this circumstance is not transcribing the altered mRNA seen in CML.

2.6. Translocations in Acute Nonlymphocytic Leukemias

A number of specific cytogenetic rearrangements have been shown to be associated nonrandomly with different forms of acute nonlymphocytic leukemia. In several instances, the possibility of involvement of known oncogenes has been suggested, or possible activation mechanisms, but none has yet been adequately documented.

In nearly every case of acute promyelocytic leukemia (APL) there is a characteristic t(15;17)(q22;q21) translocation.[13,19] A family of related genes that includes the *c-erbA* and *neu* oncogenes and the gene for the nerve growth factor receptor (NGFR) have recently been mapped close to the breakpoint on chromosome 17, suggesting possible involvement.[39–41] If true, this would indicate that "activation" of an oncogene on 17q resulted from the juxtaposition with a critical sequence on chromosome 15, but currently there is no direct evidence relating to a specific gene on either of the two chromosomes that participate in this translocation.

The same can be said of the t(8;21)(q22;q22) translocation that characterizes a subgroup of cases of acute myelogenous leukemia (AML). The *c-mos* oncogene is proximal to the breakpoint on chromosome 8, and there is no evidence of its involvement.[19] Studies are under way to attempt to identify and characterize the relevant DNA sequence at band 8q22, as well as a possible activating sequence on chromosome 21. We have investigated a poorly differentiated case of acute leukemia characterized by an unusual t(17;21)(q21;q22) translocation in which the breakpoints on chromosomes 17 and 21 appeared to be identical, respectively, with those of the t(15;17) of APL and the t(8;21) of AML.[39,41] In this instance, as in APL, the *erbA* oncogene and the NGFR gene were close to the breakpoint on chromosome 17, suggesting that perhaps an "activating" sequence in this rearrangement was derived from chromosome 21, and that this might also be the "activating" gene in cases of AML with the t(8;21) translocation.

There is also a rare form of acute myelomonocytic leukemia (AMMOL) that is typically characterized by an inversion of chromosome 16 or a t(16;16) translocation, involving bands p13 and q22 in both instances. Interruption of the metallothionein gene complex in band 16q22 has been demonstrated in several of these cases, with the suggestion that these genes might activate a protooncogene from the 16p13 regions.[42] There have also been a few cases of myeloid leukemia reported with a t(6;9) translocation,[19] which might involve the site of the human *pim* oncogene at 6p21.[43] At present, however, all of these various possibilities concerning oncogene involvement in acute nonlymphocytic leukemia remain in the realm of speculation, but definitive results for at least some of them may soon be forthcoming.

3. Other Chromosome Alterations in Human Leukemia and Lymphoma

In addition to reciprocal translocations, there are other types of karyotypic abnormalities that have been identified as occurring nonrandomly in human hematopoietic tumors.[12] The most common of these are trisomies and monosomies, representing gain or loss of a whole chromosome. Other alterations involve deletions or duplications, in which only a segment of a chromosome has been lost or gained. Less frequent have been examples of the homogeneous staining regions (HSRs), double minutes (DMs), and abnormal banding regions (ABRs), which appear to represent gene amplification units.

In earlier studies of cell lines, it was demonstrated that these latter cytogenetic structures represented, in some cases, multiple copies of genes necessary for cell growth under specific culture conditions, and also that they might be relatively labile alternative forms of gene amplification.[44,45] Although these structures do not show the same consistent localization within the genome as the nonrandom translocations, deletions, and additions, it has been shown that they can involve human homologues of retroviral oncogenes.

HSRs, DMs, and ABRs are more common in solid tumors such as neuroblastoma[46,47] (see Section 4), but occasionally such amplification units have been recognized in hematopoietic neoplasms, both fresh specimens and derived cell lines. For example, it has been demonstrated that in cells of the HL60 cell line, originally established from a patient with APL, there are from 20 to 40 copies of the c-*myc* gene.[48] In some sublines of HL60 cells, these amplified copies appear to be present in the form of an ABR on chromosome 8, the normal location of c-*myc*,[49] and in other sublines the amplification has been in the form of DMs or of an HSR on another chromosome.[49,50] In a study involving another oncogene, we have demonstrated that multiple copies of the c-*abl* oncogene in the K562 cell line (from a case of CML), as well as similarly amplified copies of the C-λ immunoglobulin gene, are associated in an ABR located on what appears to be a modified Philadelphia chromosome.[51] There has also been a report of one case of AML in which amplified copies of the c-*myb* oncogene were associated with a karyotypic abnormality in the long arm of chromosome 6.[52]

It remains to be determined, however, how frequently this type of gene dosage modification plays a significant role in the development of human hematopoietic tumors *in vivo*. Present cytogenetic data indicate that gain or loss of a chromosome segment is the more usual mechanism by which oncogene dosage is significantly altered in leukemia and lymphoma. For example, trisomy 8 is the most common cytogenetic abnormality in human acute leukemia,[12,13] and trisomy 12 is the most frequent in the B-cell form of chronic lymphocytic leukemia.[28,53] Loss of all or part of chromosomes 5 and 7 is a common finding in myeloid preleukemia and leukemia.[12,13,19] Presumably, the selective advantage gained by tumor cells with such alterations results from gain or loss of a single copy of one or more "oncogenes," but there is also the possibility of significant concurrent structural genetic changes associated with the trisomies or of position effects in those deletions that are interstitial. In any event, no specific gene has yet been identified as definitely involved in any of these circumstances, although the c-*fms* oncogene (as well as a colony-stimulating factor gene) has been mapped to the region of 5q that is deleted in a number of myeloid disorders,[54,55] and the c-*met* oncogene is located in a similarly involved region of 7q.[56]

Chromosome alterations other than reciprocal translocations have also been observed to occur nonrandomly in a number of human solid tumors, and these will be considered in the next section.

4. Chromosome Alterations in Human Solid Tumors

There is much less information on the cytogenetics of human carcinomas and sarcomas than on hematopoietic tumors. It has been difficult to obtain adequate numbers of mitoses for study from early lesions, either through direct preparations or through tissue culture. Many advanced lesions have also been difficult to grow *in vitro,* and even when metaphases are obtained, the extensive heterogeneity and genetic instability within the tumor has tended to obscure nonrandom chromosome patterns. In the last few years, culture techniques have improved, and specific cytogenetic data have begun to emerge for a few solid tumors, with some molecular correlates. The findings thus far generally extend and confirm the more extensive observations with leukemias and lymphomas, and several illustrative examples will be summarized briefly below.

4.1. Malignant Melanoma

In a large collaborative study of melanocytic tumors, we have investigated the cytogenetics of a spectrum of lesions ranging from benign nevus to highly invasive and metastatic melanoma.[57] In addition to demonstrating a general pattern of greater karyotypic abnormality in parallel with clinical and biological tumor progression, we and others have also observed nonrandom alterations involving chromosomes 1, 6, and 7.[57-59] The latter is currently of greatest interest, since we have observed extra dosage of chromosome 7 in many advanced melanomas, and specifically of the short arm (7p). This extra dosage was consistently associated with expression on the tumor cells of the receptor for epidermal growth factor (EGFR).[60] Since the human protooncogene c-*erbB* is located on 7p and appears to code for a portion of the EGFR,[61] this may indicate that a single extra copy of an oncogene can result in significantly altered function, contributing to further selective growth advantage in already malignant cells.

Nonrandom involvement of chromosome 1 in malignant melanoma has been observed both in our series and by others, usually involving the proximal portion of the short arm (p11–p22).[57–59] Interestingly, the oncogene N-*ras*, as well as the gene for the β subunit for nerve growth factor (NGF), has been mapped to 1p22.[62] There is yet no direct evidence of involvement of the NGF gene, but an "activated" N-*ras* gene, with transforming ability in the NIH/3T3 assay, has been extracted from several melanoma cell lines.[63] Also, a similarly "activated" N-*ras* gene has been obtained from a neuroblastoma cell line,[64] but the proximal region of lp was not recognizably altered in the karyotype of these cells, and a nonrandom abnormality in lp reported for other neuroblastomas involves a more distal location (lp32).[65] More data are needed to determine the importance of functional changes in N-*ras* in the pathogenesis of human tumors of neural crest origin, such as melanoma and neuroblastoma, and how often such changes will be recognizable cytogenetically as alterations in the short arm of chromosome 1.

Abnormalities of chromosome 1 have also been reported as occurring with some frequency in other human solid tumors, including carcinoma of the testis, cervix, ovary, and breast.[12,66] In most of these circumstances, breakpoints have been described in the centromeric region, between p12 and q12, often in association with extra copies of all or part of the *long* arm. There have been similar observations, involving extra doses of 1q, in various hematopoietic neoplasms, both lymphoid and myeloid, suggesting that increased dosage of one or more genes in this region can provide a selective advantage to almost any neoplastic cell.[12,67] Conversely, nonrandom involvement of the proximal segment of the *short* arm of chromosome 1 has been rare except in melanoma, strongly suggesting that this chromosome region carries a gene specifically involved in the development of this particular neoplasm.

The findings with respect to chromosome 6 in human melanoma are even less clear at present. Various studies, including our own, have demonstrated nonrandom involvement of both the short and long arms, in some cases involving a region of 6q (q21–q22) also involved in deletions and translocations in a number of lymphocytic leukemias and in ovarian carcinoma.[12,13,19,57–59] Two protooncogenes have been mapped to this region, c-*myb* and mcf3. The latter is the human homologue of the retroviral v-*ros* oncogene, and its protein product appears to have functional similarities to the EGFR.[68] The findings to date indicate that c-*myb* is probably not involved in the observed rearrangements, but the role of mcf3 has not been evaluated.

4.2. Other Solid Tumors

In the previous section, we noted the possible involvement of the N-*ras* oncogene in neuroblastoma. Another interesting finding with respect to this tumor has been the recognition of an oncogene, N-*myc*, that is closely homologous to the c-*myc* gene, but is located on chromosome 2.[69] As mentioned earlier, gene amplification units, recognizable as chromosomal HSRs or DMs, are particularly common in neuroblastomas, and in a number of instances these have been shown to involve the N-*myc* oncogene. Furthermore, it has recently been demonstrated that multiple copies of the N-*myc* gene are characteristically associated with more aggressive stages of neuroblastoma, indicating a role in tumor progression of this disease.[70]

Other cytogenetic evidence indicating significant dosage changes for specific genes has been developed in connection with small cell carcinoma of the lung (SCLC) and the pediatric tumors retinoblastoma and Wilms's tumor. In SCLC, a small interstitial deletion in the short arm of chromosome 3 (3p14) has been demonstrated,[71] and this same site is involved in a constitutional t(3;8) translocation associated with familial renal carcinoma.[72] The c-*raf* oncogene has been localized to this general region of 3p,[73] but as yet no structural or functional alteration of this oncogene has been definitely demonstrated in either SCLC or renal tumors. It is interesting that a proportion of SCLC tumors, or cell lines derived from them, also show amplification of either N-*myc* or c-*myc*. As in neuroblastoma, such amplification may be involved in the more aggressive stages of the disease.[71]

The role of gene deletion has been most extensively studied in connection with retinoblastoma. In this disorder, either inherited or sporadic, a necessary step in the tumorigenic process appears to be deletion of a gene from the proximal portion of the long arm (band q14) in both No. 13 chromosomes.[74] In some cases these deletions may be recognizable cytogenetically; in other instances they may be submicroscopic. Several investigators have suggested that loss or inactivation of negative regulatory genes through such deletions may foster tumor development by derepressing other critical genes.[75,76] This process appears to require deletion of the genes from both chromosomes, and there is now evidence that a similar phenomenon may be involved in the pathogenesis of Wilms's tumor (with a deletion in 11p), SCLC, and occasional other solid malignancies.[76,77]

These limited examples indicate that the various types and patterns of nonrandom chromosome change that have been extensively studied in hematopoietic tumors can also be identified in solid malignancies. Recent improvements in culture techniques, and the potential for related molecular genetic studies, should provide the impetus for considerable additional work on the cytogenetics of solid tumors in the immediate future.

5. Conclusions

As indicated in the preceding sections, chromosome studies are contributing significantly to the most exciting development in fundamental cancer research in recent years: the recognition and investigation of individual genes important in tumorigenesis.[5,78,79] These "oncogenes" were first identified in mammalian cells through homology with retroviral oncogenes and through transfection assays. Karyotypic data are now helping to indicate the involvement of certain human protooncogenes in the pathogenesis of specific tumors, and are also indicating the location of putative new human oncogenes.[5,14,78]

The cytogenetic findings are also contributing to our understanding of the variety of mechanisms by which significant changes in the function of an oncogene can be brought about within the potentially neoplastic cell. Certain "activating" processes such as point mutation and promoter-insertion are not visible at the chromosome level. Karyotypic studies, however, have provided evidence on the importance of alterations in gene dosage, through amplification or deletion, as well as of chromosome translocation, with or without structural change in the oncogene.[5,14,70] In this regard, the most extensive work has been done on translocations, and the findings described above in Burkitt's lymphoma and in CML have clearly shown that this type of somatic genetic rearrangement represents one mechanism by which oncogene function can be critically altered and contribute to the development of human malignancies. The findings in these two neoplasms have also demonstrated that the mechanisms of activation of an oncogene, following chromosome translocation, may be different in different tumors. In Burkitt's lymphoma, a usually unaltered c-*myc* oncogene appears to be deregulated by its juxtaposition to a rearranged and transcriptionally active immunoglobulin gene.[14] In CML, altered function of the c-*abl* oncogene reflects a new gene product, resulting from a hybrid gene formed in the translocation event.[10]

These examples also indicate the complexity of the problems still remaining. Although the products of a number of oncogenes have been identified, as nuclear proteins, tyrosine kinases, growth factors, or their receptors,[5,78-80] we do not yet know the exact role of any oncogene in a mammalian tumor. Furthermore, we must continue to recognize the multiple steps and factors involved in the complex phenomenon of cellular transformation and subsequent tumor development. Some of these involve genetic alterations in the tumor cells, and some reflect changes in the host.[1,5] The neoplastic process will have to be carefully dissected in each circumstance, with the likelihood that the final clinical and biological endpoint in man may be reached by different pathways in different types of tumors, and even in different individual cases. Despite these difficulties, greatly improved understanding of the fundamental nature of neoplasia will certainly be forthcoming within the next decade, and cytogenetics, combined with molecular genetics, should make a major contribution.

References

1. Nowell, P., 1976, The clonal evolution of tumor cell populations, *Science* **194**:23–28.
2. Fialkow, P., 1979, Clonal origin of human tumors, *Annu. Rev. Med.* **30**:135–143.
3. Arnold, A., Cossman, J., Bakhshi, A., Jaffe, E. S., Waldmann, T. A., and Korsmeyer, S. J., 1983, Immunoglobulin gene rearrangements as unique clone markers in human lymphoid neoplasms, *N. Engl. J. Med.* **309**:1593–1599.
4. Cairns, J., 1981, The origin of human cancers, *Nature* **289**:353–357.
5. Klein, G., and Klein, E., 1985, Evolution of tumors and the impact of molecular oncology, *Nature* **315**:190–195.
6. Sager, R., 1985, Genetic instability, suppression, and human cancer, in: *Gene Regulation in the Expression of Malignancy* (L. Sachs, ed.), Oxford University Press, London, pp. 321–334.
7. German, J. (ed.), 1983, *Chromosome Mutation and Neoplasia*, Liss, New York.
8. Ling, V., Chambers, A. F., Harris, J. F., and Hill, R. P., 1985, Quantitative genetic analysis of tumor progression, *Cancer Metastasis Rev.* **4**:173–194.
9. Rowley, J. D., 1980, Ph-positive leukaemia, including chronic myelogenous leukaemia, *Clin. Haematol.* **9**:55–86.

10. Shtivelman, E., Lifshitz, B., Gale, R. P., and Cananni, E., 1985, Fused transcript of *abl* and *bcr* genes in chronic myelogenous leukaemia, *Nature* **315:**550–554.
11. Nowell, P., 1982, Cytogenetics, in: *Cancer: A Comprehensive Treatise*, 2nd ed. (F. Becker, ed.), Plenum Press, New York, pp. 3–46.
12. Sandberg, A. A., 1980, *The Chromosomes in Human Cancer and Leukemia*, Elsevier, Amsterdam.
13. Yunis, J. J., 1983, The chromosomal basis of human neoplasia, *Science* **221:**227–236.
14. Croce, C. M., and Nowell, P. C., 1985, Molecular basis of human B cell neoplasia, *Blood* **65:**1–7.
15. Leder, P., Battey, J., Lenoir, G., Moulding, C., Murphy, W., Potter, H., Stewart, T., and Taub, R., 1983, Translocations among antibody genes in human cancer, *Science* **222:**765.
16. Rabbitts, T. H., Foster, A., Hamlyn, P., and Baer, R., 1984, Effect of somatic mutation within translocated c-myc genes in Burkitt's lymphoma, *Nature* **309:**592.
17. Cole, M., 1985, Regulation and activation of *c-myc, Nature* **318:**510–511.
18. Kelly, K., Cochran, B. H., Stiles, C. D., and Leder, P., 1983, Cell-specific regulation of the *c-myc* gene by lymphocyte mitogens and platelet-derived growth factor, *Cell* **35:**603–610.
19. Rowley, J. D., 1984, Biological implications of consistent chromosome rearrangements in leukemia and lymphoma, *Cancer Res.* **44:**3159–3168.
20. Tsujimoto, Y., Jaffe, E., Cossman, J., Gorham, J., Nowell, P. C., and Croce, C. M., 1985, Clustering of breakpoints on chromosome 11 in human B-cell neoplasms with the t(11;14) chromosome translocation, *Nature* **315:**340–343.
21. Tsujimoto, Y., Cossman, J., Jaffe, E., and Croce, C. M., 1985, Involvement of the *bcl*-2 gene in human follicular lymphoma, *Science* **228:**1440–1443.
22. Yoshida, M. C., Sasaki, M., Mise, K., Semba, K., Nishizawa, M., Yamamoto, T., and Toyoshima, K., 1985, Regional mapping of the human proto-oncogene c-*yes* to chromosome 18 at band q21.3, *Jpn. J. Cancer Res.* **76:**559–562.
23. Bakhshi, A., Jensen, J. P., Goldman, P., Wright, J. J., McBride, O. W., Epstein, A. L., and Korsmeyer, S. J., 1985, Cloning the chromosomal breakpoint of t(14;18) human lymphomas: Clearing around J$_H$ on chromosome 14 and near a transcriptional unit on 18, *Cell* **41:**899–906.
24. Erikson, J., Griffin, C. G., ar-Rushdi, A., Valtieri, M., Hoxie, J., Finan, J., Emanuel, B. S., Rovera, G., Nowell, P. C., and Croce, C. M., 1986, Heterogeneity of chromosome 22 breakpoint in Ph-positive acute lymphocytic leukemia, *Proc. Natl. Acad. Sci. USA* **83:**1807–1811.
25. DeTaisne, C., Gegonne, A., Stehelin, D., Bernheim, A., and Berger, R., 1984, Chromosomal localization of the human proto-oncogene c-*ets, Nature* **310:**581–583.
26. Emanuel, B. S., Nowell, P. C., Croce, C. M., and Israel, M. I., 1986, Translocation breakpoint mapping: Molecular cytogenetic studies of chromosome 22, *Cancer Genet. Cytogenet.* **19:**81–92.
27. Yang-Feng, T.L., Francke, U., and Ullrich, A., 1985, Gene for human insulin receptor: Localization to site on chromosome 19 involved in pre-B-cell leukemia, *Science* **228:**728–731.
28. Nowell, P. C., Vonderheid, E. C., Besa, E., Hoxie, J., Moreau, L., and Finan, J., 1986, The commonest chromosome change in 86 chronic B-cell or T-cell tumors: A 14q32 translocation, *Cancer Genet. Cytogenet.* **19:**219–227.
29. Hecht, F., Morgan, R., Hecht, B. K.-M., and Smith, S. D., 1984, Common region on chromosome 14 in T-cell leukemia and lymphoma, *Science* **226:**1445–1446.
30. Williams, D. L., Look, A. T., Melvin, S. L., Roberson, P. K., Dahl, G., Flake, T., and Stass, S., 1984, New chromosomal translocations correlate with specific immunophenotypes of childhood acute lymphoblastic leukemia, *Cell* **36:**101–109.
31. Croce, C. M., Isobe, M., Palumbo, A., Puck, J., Ming, J., Tweardy, D., Erikson, J., Davis, M., and Rovera, G., 1985, Gene for alpha-chain of human T-cell receptor: Location on chromosome 14 region involved in T-cell neoplasms, *Science* **227:**1044–1047.
32. Isobe, M., Emanuel, B. S., Erikson, J., Nowell, P. C., and Croce, C. M., 1985, Location of gene for beta subunit of human T cell receptor at band 7q35, a region prone to rearrangements in T cells, *Science* **228:**580.
33. Morton, C. C., Duby, A. D., Eddy, R. L., Shows, T. B., and Seidman, J. G., 1985, Genes for the beta chain of human T-cell antigen receptor map to regions of chromosomal rearrangement in T cells, *Science* **228:**582–585.
34. Erikson, J., Williams, D. L., Finan, J., Nowell, P. C., and Croce, C. M., 1985, Locus of the alpha-chain of the T-cell receptor is split by chromosome translocation in T-cell leukemias, *Science* **229:**784–786.

35. Lewis, W. H., Michalopoulos, E. E., Williams, D. L., Minden, M. D., and Mak, T. W., 1985, Breakpoints in the human T-cell antigen receptor alpha-chain locus in two T-cell leukaemia patients with chromosomal translocations, *Nature* **317**:544–546.
36. Caubet, J. F., Mathieu-Mahul, D., Bernheim, A., Larsen, C. J., and Berger, R., 1985, Rearrangement of the c-myc proto-oncogene locus in a cell line of T-lymphoblastic origin, *C.R. Acad. Sci.* **300**:171–176.
37. Erikson, J., Finger, L., Showe, L., Nishizuka, K., Minowada, J., Finan, J., Emanuel, B. S., Nowell, P. C., and Croce, C. M., 1986, Deregulation of c-*myc* by translocation of the alpha-locus of T cell receptor in T cell leukemias, *Science* **232**:884–886.
38. Groffen, J., Stephenson, J. R., Heistercamp, N., DeKlein, A., Barton, C. R., and Grosvald, G., 1984, Philadelphia chromosomal breakpoints are clustered within a limited region, bcr, on chromosome 22, *Cell* **36**:93–99.
39. Dayton, A. I., Selden, J. R., Laws, G., Dorney, D. J., Finan, J., Tripputi, P., Emanuel, B. S., Rovera, G., Nowell, P. C., and Croce, C. M., 1984, A human c-*erbA* oncogene homologue is closely proximal to the chromosome 17 breakpoint in acute promyelocytic leukemia, *Proc. Natl. Acad. Sci. USA* **81**:4495.
40. Coussens, L., Yang-Feng, T. L., Liao, Y.-C., Chen, E., Gray, A., McGrath, J., Seeburg, P. H., Libermann, T. A., Schlessinger, J., Francke, U., Levinson, A., and Ullrich, A., 1985, Tyrosine kinase receptor with extensive homology to EGF receptor shares chromosomal location with *neu* oncogene, *Science* **230**:1132–1139.
41. Huebner, K., Isobe, M., Chao, M., Bothwell, M., Ross, A. H., Finan, J., Hoxie, J. A., Sehgal, A., Buck, C. R., Lanahan, A., Nowell, P. C., Koprowski, H., and Croce, C. M., 1986, The nerve growth factor receptor gene is at human chromosome region 17q12–17q22, distal to the chromosome 17 breakpoint in acute leukemias, *Proc. Natl. Acad. Sci. USA* **83**:1403–1407.
42. LeBeau, M. M., Diaz, M. O., Karin, M., and Rowley, J. D., 1985, Metallothionein gene cluster is split by chromosome 16 rearrangements in myelomonocytic leukaemia, *Nature* **313**:709–711.
43. Nagarajan, L., Louie, E., Yoshihide, T., ar-Rushdi, A., Huebner, K., and Croce, C. M., 1986, Localization of the human pim oncogene to a region of chromosome 6 involved in translocations in acute leukemias, *Proc. Natl. Acad. Sci. USA* **83**:2556–2560.
44. Biedler, J. L., Malera, P. W., and Spengler, B. A., 1983, Chromosome abnormalities and gene amplification: Comparison of antifolate-resistant and human neuroblastoma cell systems, in: *Chromosomes and Cancer: From Molecules to Man* (J. D. Rowley and J. E. Ultmann, eds.), Academic Press, New York, pp. 117–138.
45. Kaufman, R. J., Brown, P. C., and Schimke, R. T., 1979, Amplified dihydrofolate reductase genes in unstably methotrexate-resistant cells are associated with double minute chromosomes, *Proc. Natl. Acad. Sci. USA* **76**:5669–5673.
46. Schwab, M., Alitalo, K., Klempnauer, K.-H., Varmus, H. E., Bishop, J. M., Gilbert, F., Brodeur, G., Goldstein, M., and Trent, J., 1983, Amplified DNA with limited homology to myc cellular oncogene is shared by neuroblastoma cell lines and a neuroblastoma tumor, *Nature* **305**:245.
47. Alitalo, K., Schwab, M., Lin, C. C., Varmus, H. E., and Bishop, J. M., 1983, Homogeneously staining chromosomal regions contain amplified copies of an abundantly expressed cellular oncogene (c-myc) in malignant neuroendocrine cells from a human colon carcinoma, *Proc. Natl. Acad. Sci USA* **80**:1707–1711.
48. Dalla Favera, R., Wong-Staal, F., and Gallo, R. D., 1982, Oncogene amplification in promyelocytic leukaemia cell line HL-60 and primary leukaemic cells of the same patient, *Nature* **299**:61–63.
49. Nowell, P., Finan, J., Dalla Favera, R., Gallo, R., ar-Rushdi, A., Ramaczuk, P., Selden, J., Emanuel, B., Rovera, G., and Croce, C., 1983, Association of amplified oncogene c-*myc* with an abnormally banded chromosome 8 in a human leukemia cell line, *Nature* **306**:494–497.
50. Wolman, S. R., Lanfrancone, L., Dalla Favera, R., Ripley, S., and Anderson, A. S., 1985, Oncogene mobility in a human leukemia line HL-60, *Cancer Genet. Cytogenet.* **17**:133–141.
51. Selden, J., Emanuel, B., Wang, E., Cannizzaro, L., Palumbo, A., Erikson, J., Nowell, P., Rovera, G., and Croce, C., 1983, Amplified C-lambda and c-*abl* genes on the same market chromosome in K562 leukemia cells, *Proc. Natl. Acad. Sci. USA* **80**:7289–7292.
52. Pelicci, P.-G., Lanfrancone, L., Brathwaite, M. D., Wolman, S. R., and Dalla Favera, R., 1984, Amplification of the c-*myb* oncogene in a case of human acute myelogenous leukemia, *Science* **224**:1117–1121.

53. Juliusson, G., Robert, K.-H., Ost, A., Friberg, K., Biberfeld, P., Nilsson, B., Zech, L., and Gahrton, G., 1985, Prognostic information from cytogenetic analysis in chronic B-lymphocytic leukemia and leukemic immunocytoma, *Blood* **65**:134.
54. Neinhaus, A. W., Bunn, H. F., Turner, P. H., Gopal, T. V., Nash, W. G., O'Brien, S. J., and Sherr, C. J., 1985, Expression of the human c-*fms* proto-oncogene in hematopoietic cells and its deletion in the 5q⁻ syndrome, *Cell* **42**:421.
55. Huebner, K., Isobe, M., Croce, C. M., Golde, D. W., Kaufman, S. E., and Gasson, J. C., 1985, The human gene encoding GM-CSF is at 5q21–q32, the chromosome region deleted in the 5q⁻ anomaly, *Science* **230**:1282–1285.
56. Dean, M., Park, M., LeBeau, M. M., Robins, T. S., Diaz, M. O., Rowley, J. D., Blair, D. G., and Van de Woude, G. G., 1985, The human *met* oncogene is related to the tyrosine kinase oncogenes, *Nature* **318**:385–388.
57. Balaban, G., Herlyn, M., Clark, W. H., and Nowell, P. C., 1986, Karyotypic evolution in human malignant melanoma, *Cancer Genet. Cytogenet.* **19**:113–122.
58. Becher, R., Gibas, Z., Karakousis, C., and Sandberg, A. A., 1983, Nonrandom chromosome changes in malignant melanoma, *Cancer Res.* **43**:5010–5016.
59. Pathak, S., Drwinga, H. L., and Hsu, T. C., 1983, Inolvement of chromosome 6 in rearrangements in human malignant melanoma cell lines, *Cytogenet. Cell Genet.* **36**:573–579.
60. Koprowski, H., Herlyn, M., Balaban, G., Parmiter, A., Ross, A., and Nowell, P. C., 1985, Expression of the receptor for epidermal growth factor correlates with increased dosage of chromosome 7 in malignant melanoma, *Somat. Cell Mol. Genet.* **11**:297.
61. Downward, J., Yarden, Y., Mayes, E., Scrace, G., Totty, N., Stockwell, P., Ullrich, A., Schlessinger, J., and Waterfield, M. D., 1984, Close similarity of epidermal growth factor receptor and v-*erb*-B oncogene protein sequences, *Nature* **307**:521–527.
62. Francke, U., De Martinville, B., Coussens, L., and Ullrich, A., 1983, The human gene for the beta subunit of nerve growth factor is located on the proximal short arm of chromosome 1, *Science* **222**:1248–1251.
63. Albino, A. P., LeStrange, R., Oliff, A. I., Furth, M. E., and Old, L. J., 1984, Transforming *ras* genes from human melanoma: A manifestation of tumour heterogeneity? *Nature* **308**:69–72.
64. Taparowsky, E., Shimizu, K., Goldfarb, M., and Wigler, M., 1983, Structure and activation of the human N-*ras* gene, *Cell* **34**:581–586.
65. Gilbert, F., Balaban, G., Moorhead, P., Bianchi, D., and Schlesinger, H., 1982, Abnormalities in chromosome lp in human neuroblastoma tumors and cell lines, *Cancer Genet. Cytogenet.* **7**:33–42.
66. Brito-Babapulle, V., and Atkin, N. B., 1981, Breakpoints in chromosome 1 abnormalities of 218 human neoplasms, *Cancer Genet. Cytogenet.* **4**:215–225.
67. Rowley, J., 1977, Mapping of human chromosomal regions related to neoplasia, *Proc. Natl. Acad. Sci. USA* **74**:5729.
68. Rabin, M., Birnbaum, D., Wigler, M., and Ruddle, F. H., 1985, mcf3 oncogene mapped to region on chromosome 6 associated with malignant transformation, *Am J. Hum. Genet.* **37**:A36 (abstract).
69. Schwab, M., Alitalo, K., Klempnauer, K.-H., Varmus, H. E., Bishop, J. M., Gilbert, F., Brodeur, G., Goldstein, M., and Trent, J., 1983, Amplified DNA with limited homology to *myc* cellular oncogene is shared by human neuroblastoma cell lines and a neuroblastoma tumor, *Nature* **305**:245–248.
70. Seeger, R. C., Brodeur, G. M., Sather, H., Dalton, A., Siegel, S. E., Wong, K. Y., and Hammond, D., 1985, Association of multiple copies of the N-*myc* oncogene with rapid progression of neuroblastomas, *N. Engl. J. Med.* **313**:1111–1116.
71. Little, C. D., Nau, M. M., Carney, D. N., Gazdar, A. F., and Minna, J. D., 1983, Amplification and expression of the c-*myc* oncogene in human lung cancer cell lines, *Nature* **306**:194–196.
72. Cohen, A., Li, F., Berg, S., Marchetto, D., Tsai, S., Jacobs, S., and Brown, R., 1979, Hereditary renal-cell carcinoma associated with a chromosomal translocation, *N. Engl. J. Med.* **301**:592.
73. Bonner, T., O'Brien, S. J., Nash, W. G., Rapp, U. R., Morton, C. M., and Leder, P., 1984, The human homologs of the *raf* (*mil*) oncogene are located on human chromosomes 3 and 4, *Science* **223**:71–74.
74. Yunis, J., and Ransey, N., 1978, Retinoblastoma and subband deletion of chromosome 13, *Am. J. Dis. Child.* **132**:161.

75. Knudson, A. G., Jr., 1985, Hereditary cancer, oncogenes, and antioncogenes, *Cancer Res.* **45:**1437–1443.
76. Atkin, N. B., 1985, Antioncogenes, *Lancet* **2:**1189–1190.
77. Riccardi, V., Sujansky, E., Smith, A., and Francke, U., 1978, Chromosomal imbalance in the aniridia–Wilm's tumor association: 11p interstitial deletion, *Pediatrics* **61:**604.
78. Bishop, J. M., 1985, Trends in oncogenes, *Trends Genet.* **1:**245–249.
79. Weinberg, R. A., 1985, The action of oncogenes in the cytoplasm and nucleus, *Science* **230:**770–783.
80. Ford, R., and Maizel, A. (eds.), 1985, *Mediators in Cell Growth and Differentiation,* Raven Press, New York.

2

The Structure and Function of the Normal c-*myc* Gene and Its Alteration in Malignant Cells

REBECCA A. TAUB

1. Introduction

The c-*myc* gene, first discovered as the oncogene in an acute transforming avian retrovirus,[1] has subsequently been found to have important roles in the proliferation of both tumor and normal cells. The first major analyses of the normal structure of the c-*myc* gene and its altered structure in malignant cells followed the discovery that the c-*myc* gene is involved in the chromosomal translocations observed in human Burkitt lymphomas and murine plasmacytomas.[2-5] The finding that the deregulated expression of the c-*myc* gene appears to be contributing to the abnormal proliferation of many types of tumor cells suggested that c-*myc* may have an important role in the control of normal cellular proliferation. Later studies in normal cells have supported and extended this hypothesis implicating c-*myc* as an important gene in the entry and maintenance of cells in the growth cycle.[6] However, detailed knowledge of the exact function of the c-*myc* protein has been elusive as has a precise understanding of the complex regulation of c-*myc* gene expression in both normal and tumor cells.

2. The Structure of the c-*myc* Gene in Burkitt Lymphoma and Other Malignancies

The c-*myc* gene was first described as v-*myc*, the acute transforming gene of an avian retrovirus.[1] Later its cellular homologue, c-*myc*, was discovered in normal

---

REBECCA A. TAUB • Howard Hughes Medical Institute, Department of Human Genetics, University of Pennsylvania School Of Medicine, Philadelphia, Pennsylvania 19104-6082.

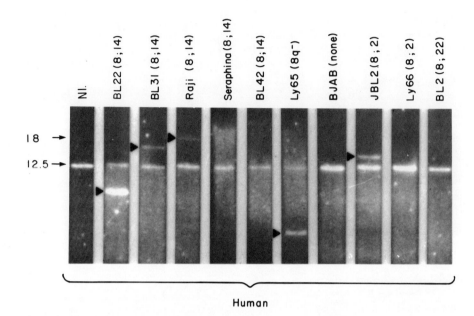

FIGURE 1. Demonstration of rearranged c-*myc*-containing fragments in Burkitt lymphoma (BL) cells. Southern blot analysis of DNA derived from N1 (normal) and BL cell lines that have been digested with restriction enzyme, *Eco*RI, and probed with a labeled human c-*myc* fragment. The normal 12.5-kb fragment is indicated by an arrow and the rearranged c-*myc* fragments are indicated by arrowheads. The designation of each Burkitt cell line and type of translocation is indicated above each lane. Bar = 1 Kb.

cells.[7] The structure of c-*myc* and the c-*myc* gene region was found to be altered in many types of malignant cells. For instance, the c-*myc* gene is amplified in a human leukemia, HL60, as well as many other types of malignant cells, in which high levels of c-*myc* are expressed.[8] In avian lymphomas and in feline leukemias, retroviruses have integrated next to the normal c-*myc* gene, resulting in its abnormal transcriptional activation.[9,10] In Burkitt lymphomas and mouse plasmacytomas, the c-*myc* gene is involved in the characteristic chromosomal translocations that occur in these cells.[2–5] Through analyses of Burkitt lymphomas and mouse plasmacytomas, the structure of the normal c-*myc* gene and its altered structure in malignant cells was characterized.

As discussed in Chapter 1, Burkitt lymphoma and certain other malignancies frequently are associated with characteristic chromosomal translocations. The finding of a characteristic crossover point in Burkitt lymphoma on chromosome 8 and the localization of the c-*myc* gene at precisely this chromosomal band allowed investigators to search for genomic rearrangements near the c-*myc* gene in these tumors that resulted directly from the translocation event.

The results of one such analysis of DNAs from several Burkitt lymphomas are shown in Fig. 1. In this analysis several of the tumor DNAs show rearrangements of the c-*myc* gene.[2] Overall, approximately 50–80% of tumors with the common

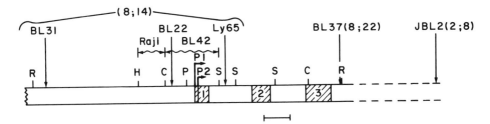

FIGURE 2. Location of translocation breakpoint with respect to the c-myc gene in seven Burkitt lymphoma cell lines. The three c-myc exons are numbered and located with respect to restriction sites.

translocation t(8;14) involving the chromosomal band bearing the immunoglobulin (Ig) heavy chain locus show a rearrangement of the c-myc gene locus, while a much smaller proportion of the tumors harboring translocations at the κ and λ light chain Ig loci [i.e., t(2;8) and t(8;22)] show rearrangements of the c-myc gene. In these cells the translocation breakpoints are farther from the c-myc gene than can be detected using genomic blot studies.

Since the chromosomal translocations in Burkitt lymphoma occur within the chromosomal bands carrying the Ig gene loci, an aberrant rearrangement found within an Ig locus could reveal the site of the chromosomal translocation breakpoint in a particular Burkitt lymphoma. Unusual rearrangements have been found in some of the t(8;14) Burkitt lymphomas that are due to translocation of the c-myc oncogene into the Ig heavy chain locus, often within the heavy chain class switching region.[2,5]

Segments of DNA from several t(8;14) cell lines containing both the c-myc gene and the Ig gene region have been isolated using molecular cloning techniques. Analysis of these segments of DNA in several laboratories showed that the translocation breakpoint is variable with respect to both the c-myc gene (see Fig. 2) and the Ig gene locus.[11-15] While the t(8;14) translocation breakpoint frequently involves the IgM switching region, it has also been found farther upstream within joining region segments or farther downstream within γ gene regions.[16,17]

Diagrams of two of these translocation breakpoints, both involving the IgM switching region, are shown in Fig. 3. The normal c-myc gene has three exons, the first of which is untranslatable into protein.[13,18] Following the translocation, the c-myc gene is oriented in a head-to-head fashion with the Ig region segments (Fig. 3A). In this case, the c-myc gene remains intact following the translocation, but frequently, the three-exon structure of the gene is interrupted by the translocation, the first, noncoding exon being removed to the reciprocal translocated chromosome (Fig. 3B). The translocation is almost exactly reciprocal, occurring within nonhomologous regions of DNA and resulting in the loss of only a small number of nucleotide base pairs at the translocation breakpoint (Fig. 3C).[13]

In Burkitt lymphomas carrying variant translocations involving the light chain Ig loci, the c-myc gene is upstream of the Ig locus and is most commonly at a greater distance from the Ig locus than can be determined from genomic blot stud-

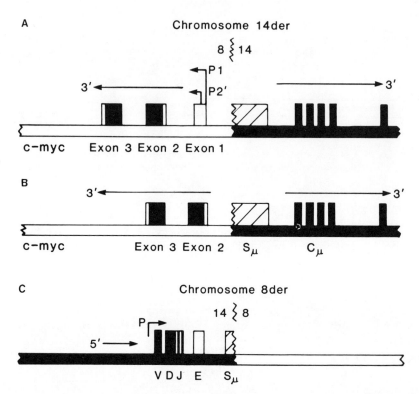

FIGURE 3. Diagrammatic representation of DNA fragments formed by translocation between chromosomes 8 and 14.[5] (A) Map of the genes as rearranged in Burkitt lymphoma cell line with an intact c-*myc* gene, with designated three-exon structure. The hatched box is the IgM switching region and downstream IgM constant region segments are indicated by solid bars. (B) Rearrangement in which the first c-*myc* exon has been moved by the translocation to the reciprocal chromosome product. (C) Diagram of a possible reciprocal product that would be expected from (A). E, location of the IgH enhancer; VDJ, variable portion of the Ig molecule.

ies. In one of the t(2;8) cell lines, the translocation breakpoint occurs within the Ig κ locus just upstream of the κ constant region.[19] This breakpoint is distant from and 3' to the c-*myc* gene. This c-*myc* gene has an altered structure, including a duplication of its untranslated first exon (Fig. 4). In this cell line, this somatic mutation may contribute to the transcriptional activation of the c-*myc* gene that is observed.

The wide variation in the location of translocation breakpoints with respect to the c-*myc* gene and Ig genes in these Burkitt lymphomas and likewise in mouse plasmacytomas[20–22] does not give us a clear picture of either the mechanism of translocation or the mechanism of activation of these c-*myc* genes. However, it is clear that in Burkitt lymphoma and mouse plasmacytomas, c-*myc* has moved into loci that are transcriptionally active. In other tumors in which activation of c-*myc* has occurred, activation of this gene apparently occurs through simple amplification of c-*myc* gene sequences resulting in a high level of c-*myc* transcripts.[8] In other tumors such as chicken lymphomas and feline leukemias, insertion of a strong

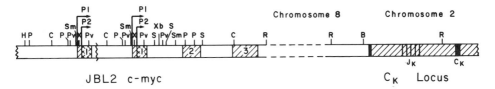

FIGURE 4. Representation of the DNA fragment formed by the translocation between chromosome 8 and chromosome 2 in the Burkitt lymphoma cell line, JBL2.[19] The c-*myc* exons are numbered demonstrating the duplication in c-*myc* exon 1. The κ constant region ($C_κ$) locus is situated 3′ to the altered c-*myc* gene at a greater distance than 20 kb.

retroviral promoter next to c-*myc* results in its deregulated expression and contributes to oncogenesis.[9,10]

3. The Function of the c-*myc* Gene Product

If translocation, amplification, or retroviral activation of the c-*myc* gene is important to the tumorigenesis of the various cells as the frequency and consistency of these c-*myc* gene alterations would imply, it should be possible to demonstrate that c-*myc* is either structurally altered or transcriptionally activated in these tumor cells. Before discussing the normal and abnormal transcriptional regulation of the c-*myc* gene, however, it is important to have some understanding of the normal function of the c-*myc* protein.

The function of c-*myc* has been examined in several ways. First, c-*myc* genes from both normal and malignant cells have been transfected into recipient cells or introduced into transgenic mice under the control of various types of transcriptional promoters. The recipient cells or animals have been examined for the development of malignancies. In another type of analysis, the role of c-*myc* has been examined in normally proliferating or differentiating cells. Finally, the properties of c-*myc* protein have been looked at directly through analysis of purified c-*myc* protein or addition of c-*myc* protein or antibody to various cells.

General information about the c-*myc* gene is that it codes for a nuclear protein with DNA-binding properties.[23] It has been highly conserved over evolution, homologous sequences being found in organisms as simple as *Drosophila*.[24] The c-*myc* protein may have a housekeeping function, in that c-*myc* mRNA is normally present in most cells that have been examined.[25] Most evidence suggests that c-*myc* has a role in normal cellular proliferation.

3.1. The Role of c-*myc* in Cellular Transformation

Since c-*myc* is involved in so many types of malignancies, investigators sought to discover the role of c-*myc* in promoting the malignant phenotype. Several types of transfections and transgenic mouse assays have been used. For instance, when transfected into rat embryo fibroblasts, c-*myc*, under the control of a strong promoter, appears to have an immortalizing function that results in the establishment of permanent cell lines from primary cells with a limited life span.[26] By itself, c-*myc* does not result in tumors, but when cotransfected with an activated *ras* gene, tumors result. Neither oncogene alone will result in tumor formation.

Several investigators have looked at the results of creating transgenic mice that contain c-*myc* genes under the control of regulatable promoters. In many cases, these mice slowly develop malignancies in tissues in which transgenic *myc* genes are highly expressed. For instance, female mice harboring c-*myc* genes under the control of a mouse mammary tumor virus (MMTV) promoter develop mammary tumors.[25] Lymphoid tumors develop in transgenic mice with c-*myc* genes driven by Ig enhancers.[27] In transgenic mice containing c-*myc* genes highly expressed constitutively, mice are developmentally normal, but develop malignancies in many tissues, although not all, in which the gene is overexpressed.[28] These studies all support the idea that c-*myc* is involved in cellular proliferation of many types of cells.

3.2. The Role of c-*myc* in the Mitogenic Response

A more controversial issue has been the role of c-*myc* in rendering quiescent cells competent to proliferate. It was first noted that resting fibroblasts and lymphocytes have almost undetectable levels of c-*myc* mRNA and protein. When these cells are stimulated with appropriate mitogens such as platelet-derived growth factor (PDGF) in the case of fibroblasts, the cells enter the G1 phase of the cell cycle. Within 2 hr of mitogen treatment, the levels of c-*myc* mRNA and protein increase dramatically. The levels of c-*myc* subsequently decline but remain elevated over the basal level while the cells continue in the growth cycle.[6] Other protooncogenes such as c-*fos* are induced by mitogen-treated quiescent cells as is discussed in more detail in another chapter. This induction of c-*myc* mRNA which appears to be at the level of increased transcription, is observed in many cell types in response to different mitogens.[29,30]

An example of this phenomenon from our own laboratory is shown in Fig. 5, in which c-*myc* mRNA is induced in quiescent rat hepatoma cells by insulin, which acts as a growth factor in these cells. Simultaneously, the level of mRNA from a control gene, β_2-microglobulin, remains unchanged, and the level of mRNA from a metabolic gene, phoshoenolpyruvate carboxykinase (PEPCK), declines in response to insulin, which is known to inhibit transcription of this gene.[31]

Experiments designed to test the importance of c-*myc* protein in making cells competent to enter the growth cycle and begin synthesizing DNA have yielded some conflicting results. Transfected c-*myc* genes under the control of an inducible promoter have been shown to stimulate DNA synthesis in BALB c/3T3 cells in the presence of platelet-poor plasma and inducing factors.[32] Microinjected c-*myc* protein can induce DNA synthesis in different cell types if platelet-poor plasma is added.[33] These studies are very conclusive in establishing the role of c-*myc* in stimulating DNA synthesis. However, c-*myc* alone is not sufficient to render cells competent to synthesize DNA. If c-*myc* mRNA is induced by agents that activate protein

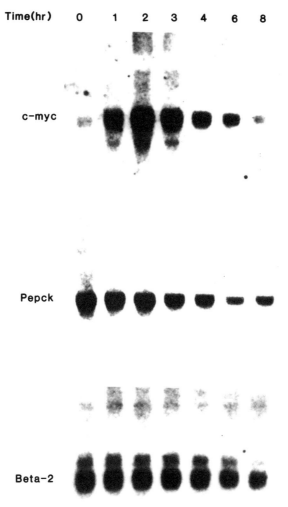

FIGURE 5. c-*myc* mRNA increases dramatically following insulin treatment of quiescent rat hepatoma (H35) cells. Insulin (1 nM) was added to quiescent H35 cells and total RNA was isolated at the indicated times. Ten micrograms of total RNA was added per lane and the Northern blots were hybridized with a c-*myc*, PEPCK, or β_2-microglobulin probe.

kinase C but do not otherwise induce DNA synthesis, no DNA synthesis results.[34] These studies indicate that c-*myc* is not sufficient to induce DNA synthesis if other unknown conditions within the cell are not appropriate.

3.3. The Role of c-*myc* in Cellular Differentiation

c-*myc* may also have a role in controlling the delicate balance between cellular proliferation and differentiation. Cell lines that can be induced to differentiate frequently (but not always) show decreases in c-*myc* mRNA as differentiation occurs.[35-37] A cell type in which this phenomenon has been examined in detail is mouse erythroleukemia cells. When these cells are induced to differentiate by various agents, c-*myc* expression is biphasic, decreasing shortly after addition of a differentiating agent, reappearing later at 12 to 24 hr and finally disappearing. Studies with transfected c-*myc* genes in these cells show that constitutive expression of c-*myc* prevents differentiation of these cells.[38] Prevention of the second burst of steady-state c-*myc* mRNA, however, also appears to block differentiation, implying that the role of c-*myc* in the differentiation process is very complex.[39] Cells may have to undergo a certain number of replicative rounds during which c-*myc* is a necessary component before differentiation can occur.

3.4. Other Studies of the Structure and Function of c-*myc* Protein

Direct studies of the structure and function of c-*myc* protein have been fairly limited. When levels of c-*myc* protein have been examined relative to levels of c-*myc* mRNA, there has been a good correlation so that most studies of c-*myc* expression have been only through examination of c-*myc* mRNA.[40] The c-*myc* protein is a phosphoprotein and appears to function in the nucleus. It was initially difficult to identify because although its predicted molecular weight is 48,000, it migrates at about 65,000 on SDS–PAGE gels due to an unusual secondary structure. Investigators agree that it binds to double-stranded DNA although the significance of this finding is not known. It appears to be a highly regulatable protein since it has a very short half-life of about 15 min.[40-43]

A recent study implicates c-*myc* directly as having a positive role in the control of DNA synthesis.[44] When antibodies to c-*myc* are added directly to nuclei isolated from many different cell types, DNA synthesis, which normally occurs, is blocked. The addition of c-*myc* protein will prevent this inhibition. The DNA polymerase activity of these nuclei appears to have been directly blocked by the addition of c-*myc* antibodies. The DNA within these nuclei remains a good template for exogenous DNA polymerase. If these results are borne out, they imply that c-*myc* controls cellular proliferation by directly controlling DNA polymerase activity.

3.5. Other *myc*-like Genes That Are Amplified in Some Malignancies

It was found that in certain malignancies, especially neuroblastomas and small cell lung carcinomas, other *myc*-like genes could be amplified with or without concomitant c-*myc* amplification.[45] These genes, designated N-*myc* and L-*myc* based on the tissue in which they were first discovered, are distinctly different genes from c-*myc* with similar genomic structures and distinct regions of homology to c-*myc* within the coding regions.[46,47] N-*myc*, which has been characterized more fully, has been shown to behave similarly to c-*myc* in that it appears to confer an immortalizing function on cells.[46] Perhaps the *myc* genes represent a class of genes that is involved in controlling certain aspects of cellular proliferation.

4. Regulation of c-*myc* Expression in Normal and Tumor Cells

Most of the early data about the regulation of c-*myc* expression came from studies comparing c-*myc* gene expression in tumor and normal cells. More recent studies have concentrated on describing c-*myc* gene expression and regulation of expression in normal cell systems. What was clearly evident from examining tumor cells in which the c-*myc* gene was activated was that the mechanism of this activation could occur by a variety of mechanisms even within a specific tumor type such as Burkitt lymphoma. In most cases the deregulation of c-*myc* is accompanied by an increased level of normal c-*myc* mRNA and protein, although in some cases, as discussed below, structural alterations in the protein have been observed as well.

4.1. Deregulation of c-*myc* in Burkitt Lymphoma and Mouse Plasmacytoma Cells

When the steady-state level of c-*myc* mRNA is measured in most Burkitt lymphoma cells, it appears higher than that observed in proliferating fibroblasts but not more than two- to fivefold elevated with respect to c-*myc* levels in EBV-immortalized lymphocytes.[14,48,49] What may be important is not the absolute level of c-*myc* mRNA but that c-*myc* mRNA is constitutively expressed in these cells and cannot be down-regulated as it is in resting or terminally differentiated cells.

In most cases this increase in the steady-state level of c-*myc* can be attributed completely to transcriptional increases. However, in certain Burkitt lymphoma and plasmacytoma cells in which the translocated c-*myc* gene is missing the first exon, the half-life of the mRNA may be increased from about 30 min to greater than 1 hr.[50] This alteration in c-*myc* half-life is not the essential change, since the most striking observation about the deregulation of expression of the translocated c-*myc* gene is the fact that in most Burkitt lymphoma and plasmacytoma cells virtually all of the transcribed c-*myc* mRNA is derived from the translocated c-*myc* gene. The normal c-*myc* gene within the same cell is transcriptionally silent.[14,49] Abnormal transcriptional activators must be affecting the translocated *myc* allele.

Another apparent alteration observed in Burkitt lymphoma and plasmacytoma cells in which the translocated c-*myc* gene retains its first exon is that there is an alteration in the utilization of the two c-*myc* promoters.[14] Normally, as shown schematically in Fig. 2, c-*myc* mRNA is transcribed from either of two promoters, encoding either 2.5- or 2.3-kb transcripts. In fibroblasts, lymphocytes, and many other normal cells, the second of these two promoters, P2, is preferentially used. In Burkitt lymphoma cells, P1 is consistently utilized at least as frequently as P2 in steady-state mRNA. Since the first exon region in both P1- and P2-derived transcripts is not translated into protein, both transcripts result in the same protein product. However, the abnormal utilization of the c-*myc* promoters by the translocated c-*myc* gene in Burkitt lymphoma cells may be a sign that the translocated c-*myc* gene is responding to unusual transcriptional control factors.

The proposed insensitivity of the translocated c-*myc* gene in Burkitt lymphoma cells to normal regulation may be due both to its location within the Ig locus and to alterations in its structure. In most t(8;14) translocations, the c-*myc* gene sits in the heavy chain locus in cells in which active transcription of this region occurs following rearrangements of the variable region segment and its accompanying transcriptional promoter. In Burkitt lymphoma cells, the c-*myc* promoter has rearranged into this locus, albeit in a head-to-head orientation with Ig sequences. It can easily be imagined that this arrangement might activate the c-*myc* gene. However, in all but a few cases,[51] the heavy chain activator or transcriptional "enhancer" is no longer present on chromosome 14 following the translocation, but has moved to the reciprocal chromosome product.[13,52] Activation of c-*myc* in these cases would have to be explained by other mechanisms. Recently it has been demonstrated in certain B cells that deletion of this enhancer did not result in reduced levels of Ig mRNA, implying that there are other important sequences near Ig genes that are transcriptional activators.[53] These sequences may be retained next to c-*myc* following the translocation event. Clearly B-cell-specific factors are important in c-*myc* transcriptional activation in Burkitt lymphoma and plasmacytoma cells. When cell fusion experiments are performed, the c-*myc* gene on the translocated chromosome remains abnormally active only when recipient cells are of B-cell lineage.[48]

In several cases, somatic mutations have been found in the putative regulatory exon of the c-*myc* gene, even though in at least one example the coding sequence of the translocated c-*myc* gene is identical to that of the normal gene.[13,14,54] In other Burkitt lymphoma cells, significant changes have occurred in the protein-coding region as well.[55,56] Presumably these mutations occurred subsequent to the translocation into the Ig heavy chain locus. They may be the result of the proximity of c-*myc* to Ig sequences, which normally appear to have a very high rate of mutation.[57] A more striking example of somatic mutations accompanying translocation are those c-*myc* genes in which the putative regulatory first c-*myc* exon has been truncated or eliminated from the protein-coding regions. These are the most commonly observed translocations in mouse plasmacytomas and are also commonly seen in Burkitt lymphomas. Another interesting example is the t(2;8) translocation, in which the c-*myc* gene has undergone duplication of its putative regulatory exon.[19] These observations indicate that the c-*myc* gene may be deregulated in these tumors as a result of both transcriptional enhancement because of proximity to the Ig locus and alterations in the structure of the c-*myc* gene.

4.2. Transcriptional Regulation of the Normal c-*myc* Gene

The c-*myc* mRNA is highly regulated at the transcriptional and posttranscriptional level. As suggested by the findings in translocated c-*myc* genes, transcriptional control regions influencing c-*myc* gene transcription are numerous and center around the first untranslated exon.[58,59] Posttranscriptional control may also influence the level of c-*myc* mRNA in the cell since the half-life of c-*myc* mRNA is very short and may be altered in different cellular environments.[60,61] As has been discussed, the c-*myc* protein also has a very short half-life so that the level of c-*myc* protein can be tightly regulated from moment to moment within the cell.

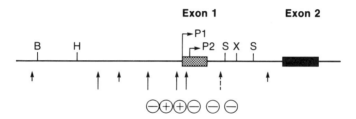

FIGURE 6. Regulatory regions around the c-*myc* gene. c-*myc* exons 1 and 2 are shown with some surrounding restriction enzyme sites. DNase I-hypersensitive sites are indicated by upward pointing arrows. Negative regulatory regions are indicated by minus signs and positive regulatory regions are indicated by plus signs.

4.2.1. Transcriptional Rate Studies

The rate of c-*myc* gene transcription has been assessed in different cells under varying conditions and the results have been conflicting. Initially it was established that c-*myc* mRNA has a very short half-life of about 20 min in different cell types, including a Burkitt lymphoma cell, a cell with amplified c-*myc* genes, and two other tumor cell lines.[60] Small changes in c-*myc* half-life could result in higher steady-state levels of c-*myc* mRNA. In mouse plasmacytoma cells, truncated c-*myc* genes appeared to produce mRNAs with longer half-lives.[50] However, appropriate controls in which both truncated c-*myc* genes and normal genes were placed in the same cell were not examined. Moreover, these truncated genes are transcriptionally active whereas the normal gene in the same cell is not,[14,48] implying that transcriptional rate is still the most important factor controlling the level of c-*myc* mRNA from the translocated gene.

Other cells in which rate of c-*myc* transcription has been examined are those that are inducible by mitogens. Initial and later studies indicated that the rate of c-*myc* transcription controlled the level of c-*myc* mRNA in the cell and that quiescent cells do not transcribe c-*myc*.[29,30] Others found that the rate of transcription as determined by transcriptional runoff assays does not change, implying that following mitogen induction, stabilization of c-*myc* mRNA accounts for the increase in steady-state c-*myc* mRNA.[61] These studies all depended on a single methodology, i.e., nuclear runoff analysis, the results of which may vary depending on how the experiment is performed. To date the best evidence suggests that for the most part, increases or decreases in c-*myc* mRNA are due to changes in transcriptional rate although stabilization of the mRNA could play a role in increasing steady-state levels of c-*myc* mRNA.

An interesting study that examined the reason for decreased c-*myc* transcripts in differentiating HL60 cells suggested that a block to elongation of c-*myc* transcripts is largely responsible for the decreased steady-state level of c-*myc* transcripts.[62] Using the runoff transcription assay, c-*myc* transcripts were found that seemed to terminate just after the first c-*myc* exon. In these cells, a DNase I-hypersensitive site (dashed arrow in Fig. 6) increased in intensity following induction of differentiation, suggesting that a regulatory protein was blocking mRNA elongation at that point. This type of transcriptional regulation of c-*myc* has not been observed in every cell that has been examined.[63] This could represent a novel form of transcriptional regulation if the results of this study can be borne out in other cell types under different conditions.

4.2.2. Transcriptional Control Regions near the Normal c-*myc* Gene

Studies of translocated and normal c-*myc* genes in Burkitt lymphoma and plasmacytoma cells suggested that transcriptional regulation of c-*myc* could be very complex. The fact that the normal c-*myc* gene is transcriptionally silent in Burkitt lymphoma cells suggested most strongly that negative control regions exist, perhaps surrounding the first c-*myc* exon, since this exon is often missing in the translocated gene. Variation in the utilization of the two c-*myc* promoters suggested that positive transcriptional factors might exist as well. Identification of the regions surrounding the c-*myc* gene that are important in transcriptional regulation has been very difficult. The likely reason for this is that there are multiple regulatory regions, both positive and negative, so that deletion of a single region might not influence the overall transcription of the c-*myc* gene.

Several laboratories performed DNase I hypersensitivity experiments in attempts to identify potential regulatory regions around the c-*myc* gene.[64,65] Figure 6 represents a composite of their findings. Investigators found two to three DNase I-hypersensitive sites upstream of the c-*myc* promoters, one site at each of the two promoters and one to two sites within c-*myc* intron 1. No definite differences in DNase I sites were found between active and inactive genes. However, the normal c-*myc* genes from a Burkitt lymphoma cell appear to have an enhanced DNase I hypersensitivity at the extreme upstream site, suggesting that this region may be a site for negative regulatory control.

The results of the DNase I studies identified regions that should be examined in transfection experiments designed to demonstrate positive and negative control regions of the c-*myc* gene. Studies of transcriptional regulation of c-*myc* have been performed using DNA transfection of putative c-*myc* regulatory sequences attached to chloramphenicol acetyltransferase (CAT) genes.[59] These genes were transfected into a Burkitt lymphoma cell with a truncated translocated c-*myc* gene. Using a competition assay, several regions were identified surrounding the c-*myc* gene that may be the sites of transcriptional regulation. As shown in Fig. 6, both positive and negative regulatory regions were identified. These regions fit well with many of the DNase I-hypersensitive sites as well as with regions that are truncated or altered in many c-*myc* translocations in Burkitt lymphomas and mouse plasmacytomas. The precise roles of these various regions remain to be determined. In addition, it has not yet been demonstrated that these same regulatory regions will be identified when transfection experiments with c-*myc* promoter region-controlled genes are carried out in normal cells.

4.2.3. Other Proteins That May Be Encoded by c-*myc* Region mRNAs

It was noted previously that a low-abundance 3.2-kb c-*myc* transcript initiating from an upstream promoter was present in some cells.[14] These observations have been extended and the upstream c-*myc* promoter (P0) has been mapped in detail.[63] It simply encodes a longer c-*myc* exon 1 of about 1200 bp. The mRNA initiating from this promoter apparently is translatable since it is present on polyribosomes. P0 may not be transcriptionally regulated in the same way as the common c-*myc* promoters in that in Burkitt lymphoma cells, P0 from the normal *myc* gene is transcriptionally active while P1 and P2 are repressed. Of particular interest is the possibility that this longer c-*myc* mRNA encodes two small upstream open reading frames one of which initiates just 5' of P1 and had been noted previously.[66]

This open reading frame terminates just before the end of c-*myc* exon 1 and could encode a 188-amino-acid protein. However, this open reading frame was not found in many of the human c-*myc* sequences that have been reported and does not exist in the mouse c-*myc* gene.[67] The existence of protein derived from this open reading frame was recently documented by careful analysis.[67] It has been found in several cell types in varying amounts and its function in these cells is completely unknown. Although this coding region may not be conserved across species, such a protein if present could potentially have a role in c-*myc* regulation or function.

5. Conclusions

The c-*myc* gene has an important role in proliferation of many tumor as well as normal cells. The importance of c-*myc* is substantiated by the fact that it contributes to tumorigenesis in many tissues in which it appears to play the role of an immortalizing gene. In most tumor cells in which c-*myc* is deregulated, the cells produce an increased and/or constitutively expressed level of normal c-*myc* protein. The c-*myc* gene is often expressed in normally growing cells and may play a significant part in controlling this growth, and in the delicate balance between cellular growth and differentiation. Although the exact function of the c-*myc* protein is unknown, the best evidence suggests that it functions within the nucleus as a positive regulator of DNA synthesis.

Regulation of expression of the c-*myc* gene is very complex, both in normal and tumor cells. The levels of c-*myc* protein and mRNA appear to be highly regulated within the cell in that they both turn over very rapidly. Both transcriptional and posttranscriptional factors may be important in controlling the level of c-*myc* mRNA although the weight of evidence favors transcriptional control as being most important. Unusual mechanisms to control the level of c-*myc* mRNA may exist in some cells in which there may be a block to elongation of c-*myc* mRNA, but these mechanisms must be further substantiated. It appears that there are multiple regulatory regions surrounding and within the first c-*myc* exon that would allow for precise control of the level of this important protein within the cell.

References

1. Bishop, J. M., 1983, Cellular oncogenes and retroviruses, *Annu. Rev. Biochemistry* **52**:301–354.
2. Taub, R., Kirsch, I., Morton, C., Lenoir, G., Swan, D., Tronick, S., Aaronson, S., and Leder, P., 1982, Translocation of the c-myc gene into the immunoglobulin heavy chain locus in human Burkitt lymphoma and mouse plasmacytoma cells, *Proc. Natl. Acad. Sci. USA* **79**:7837–7841.
3. Shen-ong, G. L. C., Keath, E. J., Piccoli, S. P., and Cole, M. D., 1982, Novel myc oncogene RNA from abortive immunoglobulin-gene recombination in mouse plasmacytomas, *Cell* **31**:443–452.
4. Dalla-Favera, R., Brogni, M., Erikson, J., Patterson, D., Gallo, R. C., and Croce, C. M., 1982, Human c-myc oncogene is located on the region of chromosome 8 that is translocated in Burkitt lymphoma cells, *Proc. Natl. Acad. Sci. USA* **79**:7824.
5. Leder, P., Battey, J., Lenoir G., Moulding, C., Murphy, W., Potter, H., Stewart, T., and Taub, R., 1983, Translocations among antibody genes in human cancer, *Science* **222**:765–771.
6. Kelly, K., Cochran, G. H., Stiles, C. D., and Leder, P., 1983, Cell-specific regulation of the c-myc gene by lymphocyte mitogens and platelet-derived growth factor, *Cell* **85**:603–610.
7. Dalla-Favera, R., Gelman, E., Martinetti, S., Franchini, G., Papas, T., Gallo, R., and Wong-Staal, F., 1982, Cloning and characteristics of different human sequences related to the onc gene (v-myc) of avian myelocytomatosis virus (MC29), *Proc. Natl. Acad. Sci. USA* **79**:6497–6501.
8. Collins, S., and Groudine, M., 1982, Amplification of endogenous myc related sequences in human myeloid leukemia cell line, *Nature* **298**:679–681.
9. Hayward, W. S., Neel, B. G., and Astrin, S. M., 1981, Activation of a cellular onc gene by promoter insertion in ALV induced lymphoid leukosis, *Nature* **290**:475–480.
10. Neil, J., Hughes, D., McFarlane, R., Wilkie, N., Onions, D., Lees, G., and Jarrett, O., 1984, Transduction and rearrangement of the myc gene by feline leukemia virus in naturally occurring T-cell leukemias, *Nature* **308**:814–820.
11. Dalla-Favera, R., Martinetti, S., Gallo, R. C., Erikson, J., and Croce, C. M., 1983, Translocation and rearrangements of the c-myc oncogene locus in human undifferentiated B-cell lymphomas, *Science* **219**:963–967.
12. Adams, J. M., Gerondakis, S., Webb, E., Corcoran, L., and Cory, S., 1983, Cellular myc oncogene is altered by chromosome translocation to an immunoglobulin locus in murine plasmacytomas and is rearranged similarly in human Burkitt lymphomas, *Proc. Natl. Acad. Sci. USA* **80**:1982–1986.
13. Battey, J., Moulding, C., Taub, R., Murphy, W., Stewart, T., Potter, H., Lenoir, G., and Leder, P., 1983, The human c-myc oncogene: Structural consequences of translocation into the IgH locus in Burkitt lymphomas, *Cell* **34**:779–787.
14. Taub, R., Moulding, C., Battey, J., Murphy, W., Vasicek, T., Lenoir, G. M., and Leder, P., 1984, Activation and somatic mutation of the translocated c-myc gene in Burkitt lymphoma cells, *Cell* **36**:339–348.
15. Hollis, G. F., Mitchell, K. F., Battey, J., Potter, H., Taub, R., Lenoir, G. M., and Leder, P., 1984, A variant translocation places the lambda immunoglobulin genes 3' to the c-myc oncogene in Burkitt lymphoma, *Nature* **307**:752–755.
16. Hamlyn, P. H., and Rabbitts, T. H., 1983, Translocation joins c-myc and immunoglobulin γ1 genes in a Burkitt lymphoma revealing a third exon in the c-myc oncogene, *Nature* **304**:135–139.
17. Neuberger, M., and Calabi, F., 1983, Reciprocal chromosome translocation between c-myc and immunoglobulin γ2b genes, *Nature* **305**:240–243.
18. Watt, R., Stanton, L. W., Marcu, K. B., Gallo, R. C., Croce, C. M., and Rovera, G., 1983, Nucleotide sequence of cloned cDNA of human c-myc oncogene, *Nature* **303**:725–728.
19. Taub, R., Kelly, K., Battey, J., Latt, S., Lenoir, G. M., Tantravahi, U., Tu, Z., and Leder, P., 1984, A novel alteration in the structure of an activated c-myc gene in a variant t(2;8) Burkitt lymphoma, *Cell* **37**:511–520.
20. Adams, J. M., Gerondakis, S., Webb, E., Mitchell, J., Bernard, O., and Cory, S., 1982, Transcriptionally active DNA region that rearranges frequently in murine lymphoid tumors, *Proc. Natl. Acad. Sci. USA* **79**:6966–6970.
21. Calame, K., Kim, S., Lalley, P., Hill, R., Davis, M., and Hood, L., 1982, Molecular cloning of trans-

locations involving chromosome 15 and the immunoglobulin Cα gene from chromosome 12 in two murine plasmacytomas, *Proc. Natl. Acad. Sci. USA* **79:**6994–6998.
22. Stanton, L. W., Watt, R., and Marcu, K. B., 1983, Translocation, breakage and truncated transcripts of c-myc oncogene in murine plasmacytomas, *Nature* **303:**401–406.
23. Donner, P., Greiser-Wilke, I., and Moelling, K., 1982, Nuclear localization and DNA binding of the transforming gene product of avian myelomatosis virus, *Nature* **296:**262–266.
24. Shilo, B., and Weinberg, R., 1981, DNA sequences homologous to vertebrate oncogenes are conserved in *Drosophila melanogaster*, *Proc. Natl. Acad. Sci. USA* **78:**6789–6792.
25. Stewart, T., Pattengale, P., and Leder, P., 1984, Spontaneous mammary adenocarcinomas in transgenic mice that carry and express MMTV/myc fusion genes, *Cell* **38:**627–637.
26. Land, H., Parada, L. F., and Weinberg, R. A., 1983, Cellular oncogenes and multistep carcinogenesis, *Science* **222:**771–778.
27. Adams, J., Harris, A., Pinkert, C., Corcoran, L., Alexander, W., Cory, S., Palmiter, R., and Brinster, R., 1985, The c-myc oncogene driven by immunoglobulin enhancers induces lymphoid malignancy in transgenic mice, *Nature* **318:**533–538.
28. Leder, A., Pattengale, P., Kuo, A., Stewart, T., and Leder, P., 1986, Consequences of widespread deregulation of the c-myc gene in transgenic mice: Multiple neoplasms and normal development, *Cell* **45:**485–495.
29. Greenberg, M., and Ziff, E., 1984, Stimulation of 3T3 cells induces transcription of the c-fos proto-oncogene, *Nature* **311:**433–438.
30. Greenberg, M., Greene, L., and Ziff, E., 1985, Nerve growth factor and epidermal growth factor induce rapid transient changes in proto-oncogene transcription in PC 12 cells, *J. Biol. Chem.* **260:**14101–14110.
31. Granner, D., Andreone, T., Sasaki, K., and Beale, E., 1983, Inhibition of transcription of the phosphoenol carboxy kinase gene by insulin, *Nature* **305:**549–551.
32. Armelin, H. A., Armelin, M. C. S., Kelly, K., Stewart, T., Leder, P., Cochran, B. H., and Stiles, C. D., 1984, Functional role for c-myc in mitogenic response to platelet derived growth factor, *Nature* **310:**655–660.
33. Kaczmarek, L., Hyland, J., Watt, R., Rosenberg, M., and Baserga, R., 1985, Microinjected c-myc as a competence factor, *Science* **228:**1313–1315.
34. Coughlin, S. R., Lee, W. M. F., Williams, P., Giels, G. M., and Williams, L. T., 1985, c-myc expression is stimulated by agents that activate protein kinase C and does not account for the mitogenic effect of PDGF, *Cell* **43:**243–251.
35. Gonda, T. J., and Metcalf, D., 1984, Expression of myb, myc and fos proto-oncogenes during the differentiation of murine myeloid leukemia, *Nature* **310:**249–251.
36. Lachman, H. M., and Skoultchi, A. I., 1984, Expression of c-myc changes during differentiation of mouse erythroleukemia cells, *Nature* **310:**592–594.
37. Endo, T., and Nadal-Ginaud, B., 1986, Transcriptional and post-transcriptional control of c-myc during myogenesis: Its mRNA remains inducible in differentiated cells and does not suppress the differentiated phenotype, *Mol. Cell. Biol.* **6:**1412–1421.
38. Prochownik, E. V., and Kukowska, J., 1986, Deregulated expression of c-myc by murine erythroleukemia cells prevents differentiation, *Nature* **322:**848–850.
39. Lachman, H. M., Cheng, G., and Skoultchi, A. I., 1986, Transfection of mouse erythroleukemia cells with myc sequences changes the rate of induced commitment to differentiate, *Proc. Natl. Acad. Sci. USA* **83:**6480–6484.
40. Persson, H., Hennighausen, L., Taub, R., DeGrado, W., and Leder, P., 1984, Antibodies to human c-myc oncogene product: Evidence of an evolutionary conserved protein induced during cell proliferation, *Science* **225:**687–693.
41. Hann, S. R., and Eisenman, R. N., 1984, Proteins encoded by the human c-myc oncogene: Differential expression in neoplastic cells, *Mol. Cell. Biol.* **4:**2486–2497.
42. Watt, R., Shatzman, A. R., and Rosenberg, M., 1985, Expression and characterization of the human c-myc DNA-binding protein, *Mol. Cell. Biol.* **5:**448–456.
43. Beimling, P., Benter, T., Sander, T., and Moelling, K., 1985, Isolation and characterization of the human cellular myc gene product, *Biochemistry* **24:**6349–6355.

44. Studzinski, G. P., Brelvi, Z. S., Feldman, S. C., and Watt, R., 1986, Participation of c-myc protein in DNA synthesis of human cells, *Science* **234:**467–470.
45. Schwab, M., Alitalo, K. H., Klempnauer, K., Varmus, H. E., Bishop, J. M., Gilbert, F., Brodeur, G., Goldstein, M., and Trent, J., 1983, Amplified DNA with limited homology to myc cellular oncogene is shared by human neuroblastoma cell lines and a neuroblastoma tumor, *Nature* **305:**245–248.
46. Kohl, N. E., Legouy, E., Depinho, R. A., Nisen, R. D., Smith, R. K., Gee, C. E., and Alt, F. W., 1986, Human M-myc is closely related in organization and nucleotide sequence to c-myc, *Nature* **319:**73–77.
47. Nau, M. M., Brooks, B. J., Battey, J., Sausville, E., Gazdar, A. F., Kirsch, I. R., McBride, O. W., Bertness, V., Hollis, G. F., and Minna, J. D., 1985, L-myc, a new myc-related gene amplified and expressed in human small cell lung cancer, *Nature* **318:**69–73.
48. Erikson, J., Ar-Rushdi, A., Drwinga, H. L., Nowell, P. C., and Croce, C. M., 1983, Transcriptional activation of the translocated c-myc oncogene in Burkitt lymphoma, *Proc. Natl. Acad. Sci. USA* **80:**820–824.
49. Ar-Rushdi, A., Nishikura, K., Erikson, J., Watt, R., Rovera, G., and Croce, C. M., Differential expression of the normal and the translocated human c-myc oncogenes in B cells, *Proc. Natl. Acad. Sci. USA* **80:**4822–4826.
50. Piechaczyk, K. M., Yang, J., Blanchard, J., Jeanteur, P., and Marcu, K. B., 1985, Posttranscriptional mechanisms are responsible for accumulation of truncated c-myc RNAs in murine plasma cell tumors, *Cell* **42:**589–597.
51. Haydoy, A. C., Gillies, S. D., Saito, H., Wood, C., Wiman, K., Hayward, W. S., and Tonegawa, S., 1984, Activation of a translocated human c-myc gene by an enhancer in the immunoglobulin heavy-chain locus, *Nature* **307:**334–340.
52. Rabbitts, T. H., Forster, A., Baer, R., and Hamlyn, P. H., 1983, Transcription enhancer identified near the human Cμ immunoglobulin heavy chain gene is unavailable to the translocated c-myc gene in a Burkitt lymphoma, *Nature* **306:**806–809.
53. Zaller, D. M., and Eckhardt, L. A., 1985, Deletion of a B-cell specific enhancer affects transfected, but not endogenous, immunoglobulin heavy-chain gene expression, *Proc. Natl. Acad. Sci. USA* **82:**5088–5092.
54. Rabbitts, T. H., Forster, A., Hamlyn, P., and Baer, R., 1984, Effect of somatic mutation within translocated c-myc genes in Burkitt lymphoma, *Nature* **309:**592–597.
55. Rabbitts, T. H., Hamlyn, P. H., and Baer, R., 1983, Altered nucleotide sequences of a translocated c-myc gene in Burkitt lymphoma, *Nature* **306:**760–765.
56. Murphy, W., Sarid, J., Taub, R., Vasicek, T., Battey, J., Lenoir, G., and Leder, P., 1986, A translocated human c-myc oncogene is altered in a conserved coding sequence, *Proc. Natl. Acad. Sci. USA* **83:**2939–2943.
57. Tonegawa, S., 1983, Somatic generation of antibody diversity, *Nature* **302:**575–581.
58. Remmers, E. I., Yang, J., and Marcu, K. B., 1986, A negative transcriptional control element located upstream of the murine c-myc gene, *EMBO J.* **5:**899–904.
59. Chung, J., Sinn, E., Reed, R. R., and Leder, P., 1986, Trans-acting elements modulate expression of the human c-myc gene in Burkitt's lymphoma cells, *Proc. Natl. Acad. Sci. USA* **83:**7918–7922.
60. Dani, C., Blanchard, J. M., Piechaczyk, K. M., El Sabouty, S., Marty, L., and Jeanteur, P., 1984, Extreme instability of the myc mRNA in normal and transformed human cells, *Proc. Natl. Acad. Sci. USA* **81:**7046–7050.
61. Blanchard, J. M., Piechaczyk, K. M., Dani, C., Charbard, J. C., Franch, A., Poyssegur, J., and Jeunteur, P., 1985, c-myc gene is transcribed at high rate in G0-arrested fibroblasts and is post-transcriptionally regulated in response to growth factors, *Nature* **317:**443–445.
62. Bentley, D. L., and Groudine, M., 1986, A block to elongation is largely responsible for decreased transcription of c-myc in differentiated HL60 cells, *Nature* **321:**702–706.
63. Bentley, D. L., and Groudine, M., 1986, Novel promoter upstream of the human c-myc gene and regulation of c-myc expression in B cell lymphomas, *Mol. Cell. Biol.* **6:**3481–3489.
64. Siebenlist, U., Hennighausen, L., Battey, J., and Leder, P., 1984, Chromatin structure and protein binding in the putative regulatory region of the c-myc gene in Burkitt lymphoma, *Cell* **37:**381–391.
65. Dyson, P. J., and Rabbitts, T. H., 1985, Chromatin structure around the c-myc gene in Burkitt

lymphomas with upstream and downstream translocation points, *Proc. Natl. Acad. Sci. USA* **82:**1984–1988.
66. Gazin, C., Dupont de Denechin, S., Hampe, A., Masson, J., Martin, P., Stehelin, D., and Galibert, F., 1984, Nucleotide sequence of the human c-myc locus: Provocative open reading home within the first exon, *EMBO J.* **3:**383–387.
67. Gazin, C., Rigolet, M., Briand, J. P., Van Regenmortef, M. H. V., and Galibert, F., 1986, Immunochemical detection of proteins related to the human c-myc exon 1, *EMBO J.* **5:**2241–2250.

3
Protooncogene Expression in Lymphoid Cells
Implications for the Regulation of Normal Cellular Growth

JOHN C. REED

1. Introduction

The transformation of mammalian cells results, at least in part, from alterations in normal cellular genes known as protooncogenes. Approximately 40 protooncogenes have been identified to data based on (1) their homology to the transforming genes of particular retroviruses (*abl, erb*-A, *erb*-B, *ets, fes/fps, fgr, fms, fos, mos, myb, myc,* L-*myc,* N-*myc, raf/mil,* Ha-*ras,* Ki-*ras, rel, ros, ski, sis, src, yes*); (2) apparent involvement in nonrandom chromosomal translocations associated with various malignancies (*bcl*-1, *bcl*-2); (3) activation by retroviral integration involving a promoter/enhancer insertion mechanism (*int*-1, *int*-2, *mlvi*-1, *mlvi*-2, *mlvi*-3, *pim*-1, *pim*-2); (4) association with products of the transforming genes of particular DNA tumor viruses such as SV40 (p53); or (5) their ability to transform a susceptible target cell line, 3T3 fibroblast cells, following DNA-mediated gene transfer (B-*lym*, T-*lym, neu,* N-*ras, tx*-1, *tx*-2, *tx*-3, *tx*-4). Through alterations in their coding sequences, regulation of their expression, or both, these genes presumably have the potential to contribute to the neoplastic process—although this remains to be formally demonstrated for many of these purported protooncogenes (see Refs. 1, 4, 10, 20, 29 for review).

Precisely how the products of activated protooncogenes (i.e., "oncogenes") render cells malignant remains to be determined. Biochemical characterizations of proteins encoded by oncogenes, however, have provided important clues regarding

---

JOHN C. REED • Department of Pathology and Laboratory Medicine, University of Pennsylvania School of Medicine, Philadelphia, Pennsylvania 19104-6082.

the mechanisms of action of these gene products (reviewed in Refs. 1, 10). For example, the products of the *fos, myc,* N-*myc, myb,* and *ski* genes reside in the nucleus and bind nucleic acids with high affinity, suggesting that they may control the expression of other genes through transcriptional or posttranscriptional mechanisms. In contrast, proteins encoded by the *abl,* erb-B, *fes/fps fgr, fms, neu, ros, src,* and *yes* oncogenes possess tyrosine kinase activity and associate with the plasma membrane, not unlike many growth factor receptors. In fact, *erb*-B, *fms, neu,* and *ros* share significant sequence homology with certain growth factor receptor genes. Besides oncogenes with homology to growth factor receptors, at least one oncogene *(sis)* is homologous at the amino acid level with a growth factor [platelet-derived growth factor (PDGF)]. Still other oncogenes encode kinases with specificity for serine and threonine residues *(raf, mos),* similar to kinases regulated by cAMP, cGMP, and Ca^{2+}. Yet another family of oncogenes, the *ras* genes (Ha-*ras,* Ki-*ras,* N-*ras*), encodes proteins that, like the G-proteins involved in the regulation of adenylate cyclase activity, bind to GTP, associate with the plasma membrane, and are thought to play a role in receptor-mediated signal transduction mechanisms. Despite these clues about oncogene-encoded proteins, a precise molecular description of the mechanisms by which any one of these gene products participates in carcinogenesis is lacking.

Additional insights into the actions of oncogene and protooncogene products have come from recent investigations of protooncogene expression in normal cells. The expression of several protooncogenes is associated with growth in normal cells, indicating that they may play a role in controlling specific events of normal cellular proliferation (reviewed in Ref. 14). For example, microinjection studies in fibroblasts have provided evidence that the c-*myc* and p53 protooncogenes can function as intracellular competence factors, rendering fibroblasts responsive to progression factors present in serum.[13,15] Other studies have suggested a role for the product of the c-*src* gene ($pp60^{c-src}$) in controlling receptor signal transduction mediated through phosphatidylinositol turnover (reviewed in Ref. 17). Though fibroblasts have been employed for most of these investigations of protooncogenes in normal cells, the more biologically relevant context for many of the protooncogenes is in hematopoietic cells, since it is in cells of hematopoietic origin that alterations in the expression of cellular genes such as c-*abl,* c-*ets,* c-*myb,* c-*myc,* and *bcl*-2 have been most clearly associated with naturally occurring neoplasms (see Ref. 20 for review).

For this reason, we have begun to investigate the regulation of the expression of protooncogenes in human peripheral blood lymphocytes (PBL). Human PBL have been employed extensively as a model system for investigating the regulation of the initial stages of cellular proliferation and entry into the cell cycle. When freshly isolated from whole blood, these cells provide a population containing $\geq$ 95% quiescent cells—the preponderance of which are T lymphocytes. When stimulated with appropriate polyclonal mitogens such as the lectin phytohemagglutinin (PHA), T cells in these cultures will enter and progress through the cell cycle in a semisynchronous fashion.[5] Using T cells as a model for normal cellular proliferation, we hope to eventually gain further insights into the normal functions of protooncogenes and their products. Additional information about the normal roles of protooncogenes should lead to an improved understanding of the altered role of these genes in neoplastic cells.

2. Several Protooncogenes Are Expressed in Mitogen-Stimulated PBL: Time Courses of mRNA Accumulation

Initially we sought to determine which of the many known protooncogenes are expressed in PBL. In addition, we wished to relate the expression of protooncogenes to other events of known importance for T-lymphocyte proliferation. Previous investigations have established that upon stimulation of resting (G0 phase) PBL with PHA, T cells enter G1 phase of the cell cycle, where they express receptors for interleukin 2 (IL2; formerly "T cell growth factor") and where some, but not all, T cells secrete this growth factor. The interaction of IL2 with its cellular receptor (IL2R) on activated T cells results in the subsequent expression of receptors for transferrin (TFR). Binding of serum transferrin to its receptor in late G1 phase of the cell cycle is then required for T cells to enter S phase where DNA synthesis occurs (reviewed in Ref. 11). Expression of genes for IL2, IL2R, and TFR, thus, represents a critical event for T-cell growth. In screening the expression of protooncogenes at the mRNA level in PHA-stimulated PBL, we therefore compared the kinetics of their expression to that of the genes for IL2, IL2R, TFR, and as an indicator of DNA synthesis, histone H3.

Figure 1 shows data from several experiments wherein PBL were cultured for various times with PHA before isolating total cellular RNA and determining the time course of expression of various genes by standard RNA-blotting techniques. Because levels of RNA increase in lymphocytes following stimulation with PHA, we compared equivalent amounts of RNA (10 μg per lane) rather than analyzing RNA from equal numbers of cells. As shown in Fig. 1 and as expected from previous investigations,[11] elevations in the levels of mRNAs for IL2 and IL2R were detectable 2–6 hr after stimulation of PBL with PHA and peaked at 6 and 14 hr, respectively. In contrast to IL2 and IL2R mRNA, increased accumulation of mRNA for TFR was not detectable until 6–14 hr after addition of PHA to cultures, and generally reached maximal levels at 14–48 hr. As also shown in Fig. 1, accumulation of mRNAs for IL2, IL2R, and TFR preceded the entry of PHA-activated PBL into S phase of the cell cycle, as determined by expression of histone H3 sequences.

With the time courses of accumulation of mRNAs for IL2, IL2R, TFR, and H3 providing temporal markers for the progression of lectin-stimulated PBL through the cell cycle, we next determined the expression of 28 protooncogenes in PHA-stimulated PBL as described in detail previously.[25] Among the protooncogenes tested, at least 12 were expressed at the mRNA level in PBL and these could be segregated roughly into four categories based on the time courses of accumulation of their mRNAs: (1) genes with detectable expression in unstimulated PBL whose mRNA levels either increased or remained approximately the same following stimulation with PHA [c-*abl*, c-*ets*, c-*fgr*, c-*raf*-1 (not shown), N-*ras*]; (2) genes expressed in unstimulated PBL whose levels of mRNA initially declined and

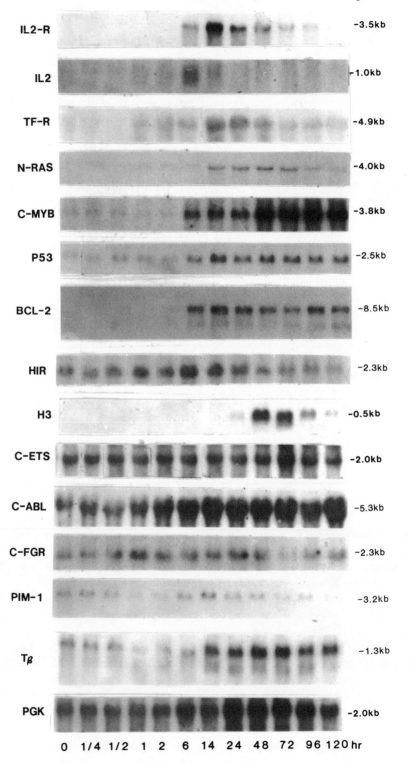

then rose following exposure of PBL to mitogenic lectin [*pim*-1, *pim*-2 (not shown)]; (3) genes whose expression was undetectable in unstimulated PBL but that underwent rapid induction of their expression following exposure to PHA (c-*fos*, c-*myc*); and (4) genes without detectable expression in unstimulated PBL that displayed increased accumulation of their mRNAs hours after stimulation by PHA (*bcl*-2, c-*myb*, p53). Comparison of the time courses of expression of those protooncogenes whose *de novo* expression is inducible by PHA (numbers 3 and 4 above) with the kinetics of the mitogen-induced expression of genes for IL2, IL2R, TRF, and H3 suggests that c-*fos* and c-*myc* expression is initiated during the G0→G1 phase transition, whereas *bcl*-2, c-*myb*, and p53 expression begins during the G1 phase.

None of the other protooncogenes was expressed at detectable levels in our RNA blot assay including B-*lym*, *erb*-A, *erb*-B, *fes*, *fms*, *fps*, *mlvi*-1, *mlvi*-2, *mos*, N-*myc*, *rel*, Ha-*ras*, Ki-*ras*, *ros*, *sis*, *ski*, *src*, T-*lym*, and *yes*. Though not scoring positively for expression at the mRNA level in PBL, each of these ^{32}P-labeled probes did hybridize to dot blots of human genomic DNA (with the exception of the v-*ski* probe), suggesting that they were at least capable of detecting specific transcripts under the hybridization conditions employed by us. It should be mentioned here that the v-*yes* probe used for these experiments, because of its extensive homology with *fgr* sequences, cross-reacted with the 2.3-kb c-*fgr* mRNA but did not hybridize to transcripts of the appropriate size for c-*yes*.[26] Also of note is the finding that despite their significant homology to human cellular oncogenes and their ability to hybridize to genomic human DNA, some viral oncogene probes (e.g., c-*raf*, c-*myb*) failed to detect transcripts, whereas the corresponding cellular probes did react positively in our RNA blot assays. Similarly, use of a human probe for *pim*-1 revealed specific mRNAs in PBL, whereas a mouse probe was nonreactive. Thus, the data presented in Fig. 1 represent a minimum estimate of the numbers of protooncogenes transcribed in PHA-stimulated human PBL.

For comparison with IL2, IL2R, TFR, H3, and protooncogenes, several other genes were assayed for expression in these cells, including human insulin receptor (HIR), phosphoglycerate kinase (PGK), and β-actin (not shown). These mRNAs were expressed at easily detectable levels in unstimulated PBL and increased approximately three- to fivefold following stimulation with PHA. Relative levels of mRNA for the β chain of the T-cell antigen receptor (T_β) were also measured. Similar to *pim*-1 and *pim*-2, T_β mRNA levels initially declined after exposure of PBL to PHA and then subsequently rose to and above prestimulation levels (Fig. 1).

←

FIGURE 1. Time course of protooncogene mRNA accumulation in PHA-stimulated PBL. PBL were cultured at 2×10^6/ml in RPMI 1640 and 10% heat-inactivated fetal calf serum ("complete medium") with 0.1% (v/v) PHA-P. At various times after stimulation, cells were recovered from culture, and total cellular RNA was isolated and analyzed by standard RNA blot assay as described previously.[25] Only protooncogenes whose expression was detectable by RNA blot analysis are shown. Expression of the c-*raf* and *pim*-2 gene was detected but is not shown here. The origins and specificities of DNA probes used for these experiments have either been described previously[25] or can be obtained upon request from the author. See text for details and abbreviations.

3. Evidence for Sequential Protooncogene Expression in PHA-Stimulated PBL

The data in Fig. 1 demonstrated a temporal sequence of inducible protooncogene expression in PBL with accumulation of mRNAs for c-*fos* and c-*myc* preceding that of mRNAs for c-*myb*, p53, and *bcl*-2. As a first step toward investigating the possibility that this temporal sequence might reflect the necessity for synthesis of c-*fos*, c-*myc*, or other gene products for accumulation of c-*myb*, p53, and *bcl*-2 mRNAs, we employed the protein synthesis inhibitor cycloheximide (CHX). Table I summarizes in semiquantitative fashion these findings derived from RNA blot analysis (see Ref. 25). As indicated in Table I, addition of CHX (15 μg/ml) to PBL cultures prior to stimulation with PHA did not impair accumulation of mature transcripts for c-*fos*, c-*myc*, IL2R, or IL2, and, instead, elevated the levels of these mRNAs in both the presence and the absence of mitogen. This finding is consistent with the hypothesis that a labile protein (or proteins) represses the transcription of these genes and/or prevents processing or stabilization of their transcripts. Levels of *bcl*-2 mRNAs were also augmented by CHX, despite the later expression of this protooncogene in PHA-stimulated PBL. In contrast to its effects on c-*fos*, c-*myc*, IL2, IL2R, and *bcl*-2 mRNAs, CHX markedly reduced the PHA-induced increase in the levels of the later-appearing mRNAs for c-*myb*, p53, and TFR. Similar results were also obtained for N-*ras* mRNA accumulation, which normally increases from low but detectable levels in resting cells to levels 10- to 15-fold above baseline within 14–48 hr after exposure of PBL to PHA (see Fig. 1); CHX completely abrogated the PHA-mediated increase in the expression of this protooncogene (Table I). That protein synthesis is necessary for the increased accumulation of c-*myb*, N-*ras*, and TFR mRNAs is consistent with the possibility that expression of c-*fos*, c-*myc*, IL2, IL2R, *bcl*-2, or any combination of these genes could be required for subsequent expression of c-*myb*, N-*ras*, and TFR genes.

Unlike the other genes whose expression was inhibited by addition of CHX to PHA-stimulated cultures (c-*myb*, N-*ras*, TFR), accumulation of p53 mRNA was augmented in unstimulated PBL by exposure to CHX. Thus, while the p53 protooncogene may be negatively regulated, similar to the earlier expressed genes such as c-*fos* and c-*myc*, protein synthesis appears necessary for the optimal expression of this gene in PHA-stimulated PBL.

Because experiments using CHX suggested that the accumulation of mRNAs for c-*myb*, p53, and TFR was dependent on protein synthesis (see Table I), we wondered whether IL2 might be one of the proteins necessary for the expression of these later genes in PHA-stimulated PBL. To investigate this possibility, cultures of PBL were stimulated with purified recombinant IL2 (gift of Cetus Corporation; Ref. 28), a suboptimal concentration of PHA (0.01% v/v), both, or neither of these reagents and RNA was isolated after various times for blot analysis. By using a suboptimal amount of PHA, T cells are induced to express IL2R and to produce IL2, but the levels of IL2 secreted are not sufficient to achieve IL2 receptor saturation. Therefore, in these suboptimally stimulated cultures of PBL, addition of exogenous IL2 produces a further increase in IL2-inducible gene expression.

TABLE I

Effects of Cycloheximide on Relative Levels of Protooncogene and Other mRNAs in PBL[a]

Culture conditions	Relative mRNA levels									
	c-fos	c-myc	c-myb	p53	bcl-2	N-ras	IL2	IL2R	TFR	PGK
Unstimulated	0	0	0	0	0	+	0	0	0	+++
CHX	+++	+−	0	++	++	+	+	++++	0	+++
PHA	++	++++	+++++	+++++	+++	+++++	++	++++	+++++	+++++
PHA + CHX	+++++	+++++	+	++	+++++	+	+++++	+++++	++	+++++

[a]PBL were incubated in medium alone (unstimulated), with 15 μg/ml cycloheximide (CHX), 0.1% PHA-P, or a combination of those reagents. CHX was always added 0.5 hr prior to PHA and inhibited protein synthesis by ≥ 95%. Total cellular RNA was analyzed for relative levels of various mRNAs by RNA blot assay[(25)] and the data expressed semiquantitatively for each probe: 0, undetectable; +, 1–24% of maximum mRNA level; ++, 25–49%; +++, 50–74%; ++++, 75–99%; +++++, maximum response (100%).

TABLE II
Effects of Purified Interleukin 2 on Levels of Protooncogene and Other mRNAs

Culture conditions[b]	Relative mRNA levels[a]								
	c-fos	c-myc	c-myb	p53	bcl-2	N-ras	IL2	IL2R	TFR	PGK
Unstimulated	0	0	0	0	0	+	0	0	0	+++
IL2	0	+++	0	0	+++	++	0	+	0	++++
PHA	+++++	++	+++	+++	++	++++	+++++	+++	+++	+++++
PHA + IL2	+++++	+++++	+++++	+++++	+++++	+++++	++++	+++++	+++++	+++

[a]Relative levels of mRNAs for various genes were determined as described for Table I.
[b]PBL were incubated in medium alone (unstimulated), with IL2 (50 units/ml), 0.1% PHA-P, or a combination of those reagents.

Table II summarizes the results. As shown, IL2 augmented the levels of mRNAs for c-*myb*, p53, N-*ras*, and TFR, thereby demonstrating that IL2 and IL2R represent at least two of the proteins required for the increased expression of these genes in PHA-stimulated PBL. In addition, IL2 also increased the levels of mRNAs for *bcl*-2, c-*myc*, and IL2R. These findings (Table II) demonstrate that the expression of the *bcl*-2, c-*myc*, and IL2R genes may be regulated through both IL2-dependent and IL2-independent mechanisms since CHX fails to inhibit the accumulation of their mRNAs (Table I). In contrast, IL2 did not up-regulate the levels of mRNAs for c-*fos*, IL2, and PGK in these cells.

Thus, the combined data in Tables I and II suggest that gene expression can be induced in PBL: (1) directly by PHA, without requirement for protein synthesis (c-*fos*, IL2, and PGK); (2) indirectly as a result of IL2 production by PHA-stimulated cells (c-*myb*, p53, N-*ras*, TFR); or (3) both by PHA and by IL2 (*bcl*-2, c-*myc*, and IL2R). That expression of some of these protooncogenes does not require new protein synthesis or IL2 (e.g., c-*fos*), whereas expression of others does (e.g., c-*myb*), illustrates conclusively that sequential expression of protooncogenes occurs in lectin-stimulated PBL.

4. Insights into Signal Transduction Mechanisms in Normal T-Lymphocytes: Mitogenic Lectin-, Phorbol Ester-, and IL2-Induced Expression of Overlapping Sets of Protooncogenes

The data presented in Tables I and II suggest that the expression of some genes, such as *bcl*-2, c-*myc*, and IL2R, may be induced in PHA-stimulated PBL both directly by PHA and indirectly as a result of IL2 production in these cultures. That PHA and IL2 stimulate expression of some of the same genes is not entirely surprising, given that signal transduction mediated by receptors for mitogenic lectins and for IL2 appears to occur, at least in part, through activation of protein kinase C.[8,18] To further compare genetic events induced by PHA and by IL2, we developed a culture system, modeled after that described by Cantrell and Smith,[2] wherein human PBL were preactivated *in vitro* and rendered responsive to both PHA and IL2. Specifically, PBL were stimulated for 3 days with PHA, then washed thoroughly and recultured in the presence of IL2-containing supernatant. After 2 additional days, these cells were washed three times and rested for 1 day in culture medium without IL2 or lectin. Such "preactivated" cells consisted of $\geq$ 95% T cells, with roughly equal proportions of $T4^+$ and $T8^+$ cells, and proliferated vigorously in response to either PHA or IL2 (J. C. Reed, unpublished results).

Figure 2 shows the proliferation kinetics of these preactivated T cells when stimulated with either PHA or IL2. Because mitogenic lectins and IL2 presumably act, in part, through activation of protein kinase C, in some experiments preactivated cells were stimulated with the phorbol ester tetradecanoate-phorbol-13-acetate (TPA), an agent that binds to and directly activates protein kinase C.[3] As shown, with appropriate adjusting of the concentrations of PHA, TPA, and IL2, these preactivated T cells undergo nearly equivalent proliferative responses to each of these agonists.

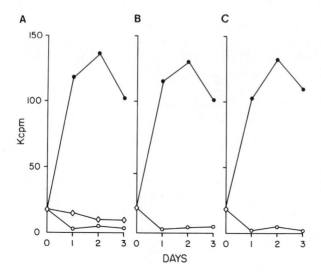

FIGURE 2. Kinetics of DNA synthesis in cultures of preactivated T lymphocytes stimulated with PHA, TPA, or IL2. Preactivated T lymphocytes, capable of responding to both lectins and IL2, were prepared as described in the text. After incubation overnight in medium to achieve quiescence, 10^6 cells per ml were stimulated with either 0.1% (v/v) PHA-P(A), 100 ng/ml TPA(B), or 100 units/ml IL2(C). Levels of DNA synthesis, determined by incorporation of [^{3}H]thymidine as described previously,[24,25] were assayed after 1, 2, or 3 days.

Using preactivated T cells under the conditions described for Fig. 2, we compared the ability of PHA, TPA, and IL2 to induce protooncogene expression. Figure 3 shows the results of these experiments. For most genes, PHA, TPA, and IL2 stimulated nearly equivalent time courses of mRNA accumulation in preactivated T cells. As before (Fig. 1), increased accumulation of mRNAs for c-*fos* and c-*myc* occurred rapidly, whereas expression of genes for c-*myb*, p53, bcl-2, IL2R, TFR, and H3 became maximal later (Fig. 3). Thus, presumably by activating protein kinase C, PHA, IL2, and TPA induce expression of many of the same genes. Despite these commonalities in signal transduction pathways used by PHA, TPA, and IL2, however, differences must also exist since: (1) IL2 failed to induce detectable c-*fos* mRNA accumulation; and (2) IL2 stimulated more rapid accumulation of c-*myb* and H3 mRNAs (Fig. 3).

A potentially complicating event in the experiments presented in Fig. 3 using preactivated T cells relates to the ability of PHA and TPA to induce IL2 production in these cultures (not shown). As a result, the patterns of gene expression observed in T cells stimulated with either PHA or TPA also reflect the actions of IL2. To avoid this complication, we made use of a cloned cytolytic murine T lymphocyte (CTB6-4A), which, like many cytolytic T cells, fails to produce IL2 (Ref. 21, and unpublished). Table III summarizes the results of experiments using CTB6 cells that were stimulated either with the mitogenic lectin concanavalin A (Con A) or with purified IL2. Both Con A and IL2 induced the accumulation of mRNAs for c-*myc*, p53, and IL2R. In contrast, expression of genes for c-*fos* was stimulated predominantly by Con A whereas expression of c-*myb*, TFR, and H3 was induced primarily by IL2. These findings thus provide an explanation for the more rapid accumulation of c-*myb* and H3 mRNAs observed in IL2-stimulated preactivated human T cells (Fig. 3) compared to cells cultured with PHA or TPA—namely, that IL2 directly increases the expression of these genes, whereas PHA and TPA indirectly stimulate their expression by inducing IL2 production. Taken together, the data in Fig. 3 and Table III provide evidence that activation of protein kinase C explains only partially the mechanisms by which mitogenic lectins and IL2 transduce their signals to the interior of activated T lymphocytes. The molecular basis for these differences in signal transduction mechanisms used by receptors for mitogenic lectins and receptors for IL2, whether quantitative or qualitative, remains to be elucidated.

5. Clues to Mechanisms of Regulation of Protooncogene Expression: Further Studies with Cycloheximide

Previous reports involving a variety of cell types have established that the protein synthesis inhibitor CHX can induce the accumulation of mRNAs for several protooncogenes.[6,14,19,25,27] Presumably, this effect of CHX reflects the presence of a short-lived protein (or proteins) that represses the accumulation of mRNAs for these genes. To further characterize the regulation of expression of various protooncogenes in T lymphocytes, we performed additional experiments with CHX in preactivated human T cells as described for Fig. 3. As expected from previous

TABLE III
Expression of Protooncogenes in Cloned T Lymphocytes (CTB6-4A) That Do Not Produce IL2[a]

		Relative mRNA levels at various times after stimulation[b]			
	Agonist	0 hr	1 hr	8 hr	24 hr
c-*fos*	Con A	0	100	0	0
	IL2		0	0	0
c-*myc*	Con A	0	22	0	0
	IL2		100	42	54
c-*myb*	Con A	0	0	0	0
	IL2		0	100	43
p53	Con A	0	29	93	78
	IL2		37	100	81
IL2R	Con A	0	7	52	31
	IL2		14	100	86
TFR	Con A	3	9	13	0
	IL2		9	100	55
H3	Con A	12	7	0	0
	IL2		0	19	100

[a] CTB6-4A cells[21] were cultured at 10^6/ml in complete medium overnight before stimulation with either 20 μg/ml Con A or 100 units/ml purified recombinant IL2 (Cetus). Total cellular RNA was isolated at 0, 1, 8, and 24 hr and relative levels of various mRNAs were determined by RNA blot analysis.[25]

[b] Data are derived from densitometric scanning of the autoradiograms and are expressed as a percentage of the maximal response wherein, for any particular mRNA, the condition yielding the greatest level of mRNA was set at 100%, that representing the lowest mRNA level at 0%, and all values between adjusted accordingly. Levels of a control mRNA, β chain of mouse T-cell antigen receptor, varied less than twofold among all conditions.

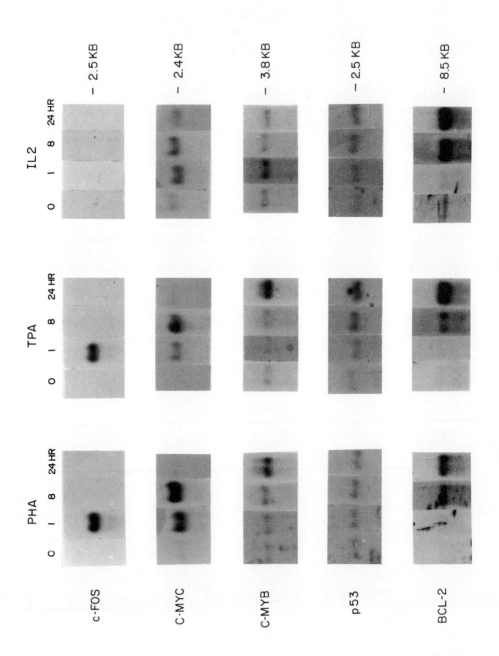

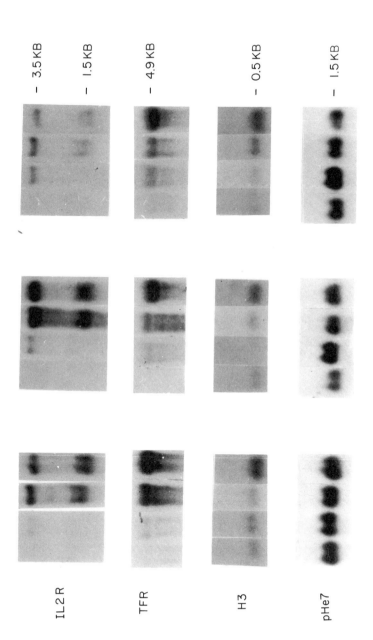

FIGURE 3. Time course of protooncogene mRNA accumulation in preactivated T lymphocytes. Preactivated T cells were cultured as described for Fig. 2 and total cellular RNA was isolated at various times after stimulation with PHA, TPA, or IL2. Relative levels of mRNAs for various genes were then assayed as described.[24,25] All RNA blots were hybridized with a control probe, pHe7, specific for a 1.5-kb mRNA present in human cells.

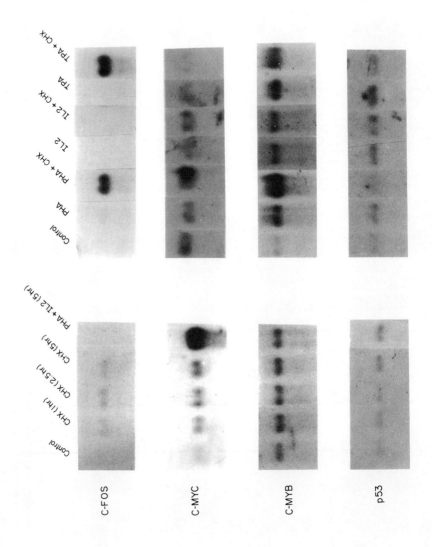

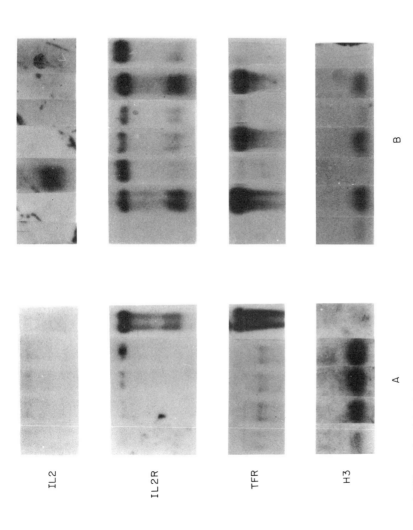

FIGURE 4. Effects of cycloheximide on mRNA levels in resting and stimulated preactivated T lymphocytes. Relative levels of various mRNAs were measured in preactivated T lymphocytes as described for Fig. 3. Cycloheximide (CHX) at 15–20 μg/ml was added to cultures for various times as shown in (A). For experiments presented in (B), CHX was added 0.5 hr prior to stimulation of cells.

experiments (Table I), CHX relieved the repression of mRNA accumulation for the c-*fos*, c-*myc*, IL2R, and p53 genes but not for the TFR gene (see Fig. 4A). CHX also appeared to augment slightly the levels of mRNA for c-*myb* (Fig. 4B) but had much more dramatic effects when added in combination with mitogens or IL2 (Fig. 4B). The influence of CHX on c-*myb* mRNA accumulation in previously activated T cells (Fig. 4) differs from its effects on freshly isolated PBL (Table I), perhaps reflecting important differences in the activation states of these cells. Similar to its effects on c-*myb* mRNA accumulation, CHX markedly augmented levels of IL2 mRNA in PHA-stimulated cells but had little effect in quiescent T cells. This protein synthesis inhibitor also elevated the levels of H3 mRNA in preactivated T cells (Fig. 4A), consistent with previous observations in fibroblast cells.[14] Thus, repression of mRNA accumulation by a labile regulatory protein(s) appears to be a component of the normal regulation of several genes in T lymphocytes (Fig. 4) and other cells.[6,14,16,19,25,27]

Investigation of the effects of CHX and mRNA levels in T cells stimulated with mitogens or IL2 (Fig. 4B) revealed that protein synthesis is required for the accumulation of TFR and H3 mRNAs—even when IL2 is used to stimulate the cells. IL2 thus directly stimulates the expression of genes for c-*myc*, c-*myb*, p53, and IL2R, without necessity for new protein synthesis, but indirectly stimulates the increased expression of genes for TFR (G1 phase) and H3 (S phase) by inducing the synthesis of new gene products. Based on the data in Fig. 4, IL2-inducible genes can be classified into protein synthesis-dependent and -independent groups. This situation is analogous to growth factor-induced gene expression in fibroblasts wherein platelet-derived growth factor (PDGF) stimulates increased accumulation of mRNAs for c-*myc* and for a gene encoding a lysosomal protein termed MEP, with the former being independent and the latter dependent on new protein synthesis.[9]

Because we measured steady-state levels of accumulated mRNAs, our data do not address questions about the relative effects of CHX on transcription, processing, and degradation in altering the levels of mRNAs for various protooncogenes in T cells. Previous investigations, however, have demonstrated that expression of genes for c-*fos*, c-*myc*, c-*myb*, and IL2 is regulated at the posttranscriptional level by CHX. Moreover, studies of IL2 gene expression in T lymphocytes have shown that CHX acts at the processing level by preventing the synthesis (presumably) of a labile protein that represses processing of IL2 primary transcripts.[16] In this regard, it is of interest that CHX altered the relative proportions of the larger (3.5 kb) and smaller (1.7 and 1.5 kb) classes of mature IL2R mRNAs in preactivated T cells (Fig. 4B). Thus, CHX may influence at the processing level the accumulation of mRNAs for other genes besides IL2, including possibly the protooncogenes studied here whose mRNA levels are augmented by CHX.

6. Summary and Speculations

The data presented here (Fig. 1) demonstrate that at least 12 protooncogenes are expressed at the mRNA level in human PBL. The biological functions of these genes in lymphocytes are presently unknown; however, considerable progress has been made toward defining the roles of protooncogenes in fibroblasts. Similarities between lymphocytes and fibroblasts with regard to protooncogene expression and cell cycle progression are striking, with, for example, PHA and PDGF, respectively, inducing a similar time course of expression of c-*fos*, c-*myc*, c-*myb*, c-*ras*, and p53 genes.[14,25] In addition, both fibroblasts and T lymphocytes appear to require three sequential growth signals to progress through cell cycle and reach S phase. Fibroblasts, for instance, require PDGF, EGF (epidermal growth factor), and IGF-I (insulinlike growth factor-I), whereas T cells sequentially require stimulation with antigen (or mitogenic lectin), IL2, and transferrin.[11,22] These similarities between the temporal sequence of events associated with the growth of T lymphocytes and fibroblasts are depicted diagrammatically in Fig. 5.

Parallels between lymphocytes and fibroblasts suggest common features in the roles of protooncogenes in regulating the growth of normal mammalian cells. Hence, investigations of protooncogenes in fibroblasts may yield clues about the functions of these genes in lymphocytes. As mentioned previously, microinjection studies in fibroblasts have demonstrated that c-*myc* and p53 can function as intracellular competence factors, abrogating the necessity for PDGF for rendering cells competent to respond to growth factors (EGF and IGF-I) present in platelet-poor plasma.[13,15] By analogy, perhaps c-*myc* and p53 play a role in inducing competence for IL2-induced growth in PHA-activated PBL. Though experiments using CHX (Table I) argue that c-*myc* and p53 are not necessary for activation of the IL2R gene, we have previously presented evidence that IL2R expression is insufficient for competence for IL2 and that other unknown events are required to render PHA-stimulated T cells responsive to IL2.[24] Thus, a role for c-*myc* and p53 for inducing competence for IL2 in activated T lymphocytes remains a tenable possibility, despite the apparent absence of a requirement for new protein synthesis for IL2R mRNA accumulation.

That c-*myc* and p53 can supplant the necessity for PDGF in fibroblasts[13–15] indicates that these protooncogenes act as intracellular mediators of PDGF-mediated growth signals. Since both mitogenic lectins and IL2 stimulate expression of the c-*myc* and p53 protooncogenes (see Table III), constitutive expression of these genes might be expected to abrogate the necessity for mitogenic lectins and/or IL2 for T-cell growth. Evidence for the latter comes from a recent report wherein infection of cloned T cells with retroviruses containing a transcriptionally active viral *myc* (v-*myc*) gene allowed continuous, IL2-independent growth of these cells.[23] It should be mentioned, however, that because significant structural differences exist between the viral and cellular *myc* genes and the proteins they encode (reviewed in Ref. 7), these observations using a v-*myc* gene may not necessarily apply for c-*myc*.

Because the protein products of the c-*fos*, c-*myc*, c-*myb*, and p53 protooncogenes associate with the nuclear matrix, it is postulated that they may directly control the expression of other genes. Since expression of these protooncogenes is induced in T cells by mitogenic lectins and/or by IL2 (see Fig. 3 and Table III), c-*fos*, c-*myc*, c-*myb*, and p53 represent excellent candidates for intracellular mediators involved in signal transduction processes that ultimately connect events at the cell

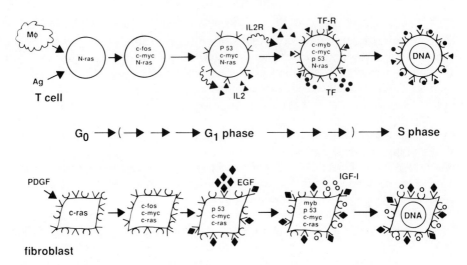

FIGURE 5. Comparison of events associated with the growth of normal T lymphocytes and fibroblasts. Diagram depicts the temporal sequence of events (not necessarily cause and effect) that occurs in T lymphocytes and in fibroblasts stimulated to enter cell cycle. IL2 receptors (IL2R) and EGF receptors are denoted by V; transferrin receptors (TF-R) and insulinlike growth factor I (IGF-I) receptors are denoted by ∪; IL2 by ▲, EGF by ♦, transferrin (TF) and IGF-I by ● and ○, respectively. See text for details.

surface with changes in the expression of groups of genes in the nucleus. In addition to protooncogenes whose products are located in the nucleus, protooncogenes encoding membrane-associated proteins also may be involved in signal transduction mechanisms in lymphocytes. For instance, the c-*fgr* gene encodes a protein with extensive amino acid homology to the product of the c-*src* gene, pp60^{src}. Previous studies in fibroblasts have suggested a role for pp60^{src} in the control of receptor signal transduction mediated through phosphatidylinositol turnover (reviewed in Ref. 17). Because both mitogenic lectins[18] and IL2[8] stimulate increased phosphatidylinositol metabolism in T lymphocytes, c-*fgr* theoretically could play a role in signal transduction events mediated through receptors for mitogenic lectins and for IL2. Another of the protooncogenes expressed in T lymphocytes, N-*ras*, encodes a GTP-binding protein that associates with the plasma membrane and that like other "G-proteins" has been postulated to participate in signal transduction mechanisms. Microinjection studies have demonstrated a requirement for *ras* protooncogene products for normal cellular growth during late G1 phase of the cell cycle,[14] thus raising the possibility that p21^{ras} may, in fact, mediate growth signals derived from the binding of transferrin to T cells or of IGF-I to fibroblasts. That p21^{ras} may noncovalently associate with receptors for transferrin has been suggested but has never been adequately addressed.[12]

Yet another possible role for protooncogene products for signal transduction regulation in normal T lymphocytes is suggested by the tyrosine kinase activities of the c-*abl* and c-*fgr* proteins. Most growth factor receptors, including receptors for PDGF, EGF, and IGF-I, possess the ability to phosphorylate particular proteins at tyrosine residues following binding of specific ligand. In contrast, receptors for antigen, IL2, and transferrin lack this capability. Given the extensive parallels between fibroblasts and T lymphocytes with regard to normal growth regulation (see Fig. 5), it is not unreasonable to speculate that c-*abl* and c-*fgr* may associate with receptors for antigen, IL2, or even, transferrin in the membranes of T cells and participate in signal transduction mechanisms mediated through these cellular receptors.

Finally, to bridge events initiated at the plasma membrane with changes in gene expression in the nucleus, cytosolic proteins are needed. At least three of the protooncogenes expressed in normal T cells encode such proteins: *bcl*-2, c-*ets*, and c-*raf*. Though little is known about the *bcl*-2 and c-*ets* gene products, the c-*raf* protooncogene encodes a kinase with specificity for serine and threonine residues (see Ref. 10). Like protein kinase C and other kinases that phosphorylate target proteins at serine and threonine, the c-*raf* gene product is an excellent candidate for a cytosolic protein that may help to bridge the membrane with the nucleus in T lymphocytes.

These speculations notwithstanding, further work is needed to determine the precise functions of the products of protooncogenes in lymphocytes. The data presented here, though largely descriptive, lay the foundation for further investigations that will ultimately reveal the molecular and cellular effects of these genes in regulating the growth of normal and neoplastic lymphocytes.

References

1. Bishop, J. M., 1985, Viral oncogenes, *Cell* **42**:23–38.
2. Cantrell, D. A., and Smith, K. A., 1984, *Science* **224**:1312–1316.
3. Castagna, M., Takai, Y., Kaibuchi, K., Sano, K., Kikkawa, U., and Nishizuka, Y., 1982, Direct activation of calcium-activated, phospholipid-dependent protein kinase by tumor promoting phorbol esters, *J. Biol. Chem.* **257**:7847–7851.
4. Cooper, G. M., and Lane, M.-A., 1984, Cellular transforming genes and oncogenesis, *Biochim. Biophys. Acta* **738**:9–20.
5. Darzykiewicz, Z., Traganos, F., Sharpless, T., and Melamed, M. R., 1976, Lymphocyte stimulation: A rapid multiparameter analysis, *Proc. Natl. Acad. Sci. USA* **73**:2881–2884.
6. Dony, C., Kessel, M., and Gruss, P., 1985, Post-transcriptional control of *myc* and p53 expression during differentiation of the embryonal carcinoma cell line 9, *Nature* **317**:636–639.
7. Duesberg, P. H., 1983, Retroviral transforming genes in normal cells? *Nature* **304**:219–226.
8. Farrar, W. L., Evan, S. W., Rascetti, F. W., Bonvinni, E., Young, H. A., and Sparks, B. M., 1986, Biochemical mechanisms of interleukin-2 regulation of lymphocyte growth, in: *Leukocytes and Host Defense* (J. J. Oppenheim and D. M. Jacobs, eds.), Liss, New York, pp. 75–82.
9. Frick, K. K., Doherty, P. J., Gottesman, M. M., and Scher, C. D., 1985, Regulation of the transcript for a lysosomal protein: Evidence for a gene program modified by platelet-derived growth factor, *Mol. Cell. Biol.* **5**:2582–2589.
10. Gordon, H., 1985, Oncogenes, *Mayo Clin. Proc.* **60**:697–713.

11. Greene, W. C., and Robb, R. J., 1985, Receptors for T cell growth factor: Structure, function, and expression in normal and neoplastic cells, *Contemp. Top. Mol. Immunol.* **10:**1–34.
12. Harford, J., 1984, An artefact explains the apparent association of the transferrin receptor with a *ras* gene product, *Nature* **311:**673–675.
13. Kaczmarek, L., Hyland, J. K., Watt, R., Rosenberg, M., and Baserga, R., 1983, Microinjected c-*myc* as a competence factor, *Science* **228:**1313–1315.
14. Kaczmarek, L., 1986, Proto-oncogene expression during the cell cycle, *Lab. Invest.* **54:**365–376.
15. Kaczmarek, L., Oren, M., and Baserga, R., 1986, Co-operation between the p53 protein tumor antigen and platelet-poor plasma in the induction of cellular DNA synthesis, *Exp. Cell Res.* **162:**268–272.
16. Kaempfer, R., and Efrat, S., 1986, Regulation of human interleukin 2 gene expression, *Leukocytes and Host Defense* (J. J. Oppenheim and D. M. Jacobs, eds.), Liss, New York, pp. 57–68.
17. Macara, I. G., 1985, Oncogenes, ions, and phospholipids, *Am. J. Physiol.* **248:**C3–C11.
18. Maino, V. C., Hayman, M. J., and Crumpton, M. J., 1975, Relationship between enhanced turnover of phosphatidylinositol and lymphocyte activation by mitogens, *Biochem. J.* **146:**247–252.
19. Müller, R., Bravo, R., Burckhardt, J., and Curran, T., 1984, Induction of c-*fos* gene and protein by growth factors precedes activation of c-*myc*, *Nature* **312:**716–720.
20. Nowell, P. C., Emanuel, B. S., and Croce, C. M., 1986, Chromosome and genetic changes in leukemia, in: *Leukocytes and Host Defense* (J. J. Oppenheim and D. M. Jacobs, eds.), Liss, New York, pp. 83–86.
21. Pan, S., and Knowles, C. B., 1983, Monoclonal antibody to SV40 T antigen blocks lysis of cloned cytolytic T cell line specific for SV40 TASA, *Virology* **125:**1–7.
22. Pledger, W. J., Stiles, C. D., Antoniades, H. N., and Scher, C. D., 1978, An ordered sequence of events is required before BALB/c-3T3 cells become committed to DNA synthesis, *Proc. Natl. Acad. Sci. USA* **75:**2839–2843.
23. Rapp, U. R., Cleveland, J. L., Brightman, K., Scott, A., and Ihle, J. N., 1985, Abrogation of IL3 and IL2 dependence by recombinant murine retroviruses expressing v-*myc* oncogenes, *Nature* **317:**434–436.
24. Reed, J. C., Abidi, A. H., Alpers, J. D., Hoover, R. G., Robb, R. J., and Nowell, P. C., 1986, Effect of cyclosporin A and dexamethasone on interleukin 2 receptor gene expression, *J. Immunol.* **137:**150–154.
25. Reed, J. C., Alpers, J. D., Nowell, P. C., Hoover, R. G., 1986, Sequential expression of proto-oncogenes during lectin-stimulated mitogenesis of normal human lymphocytes, *Proc. Natl. Acad. Sci. USA* **83:**3982–3986.
26. Semba, K., Yamanashi, Y., Nishizawa, M., Sukegawa, J., Yoshida, M., Sasaki, M., Yamamoto, T., and Toyoshima, K., 1985, Location of the c-*yes* gene on the human chromosome and its expression in various tissues, *Science* **227:**1038–1040.
27. Thompson, C. B., Challoner, P. B., Neiman, P. S., and Groudine, M., 1986, Expression of the c-*myb* proto-oncogene during cellular proliferation, *Nature* **319:**374–380.
28. Wang, A., Lu, S. D., and Mark, D. F., 1984, Site-specific mutagenesis of the human interleukin-2 gene: Structure–function analysis of the cysteine residues, *Science* **224:**1431–1433.
29. Willecke, K., and Schafer, R., 1984, Human oncogenes, *Human Genetics* **66:**132–142.

4
Oncogene Products as Receptors

ERICA M. S. SIBINGA, GAIL R. MASSEY, AND MARK I. GREENE

1. Introduction

By studying the changes in cancerous cells that result in their ability to resist normal growth control mechanisms, it may be possible to elucidate certain processes relevant to carcinogenesis. An important realization involved the recognition of the role of changes in DNA in the induction of cancer. In karyotypic studies of patients with chronic granulocytic leukemia, Nowell and Hungerford described chromosomal aberrations associated with a type of neoplasia.[1] The connection between altered DNA and cancer was demonstrated more conclusively 20 years later, with the use of DNA transfection. Shih *et al.* applied high-molecular-weight DNA from carcinogen-induced tumors to monolayers of NIH 3T3 cells, which took up the DNA. Cells that developed malignant phenotypic features were isolated. Extracting the DNA from the primary transfectant cells and applying it to NIH 3T3 monolayers allowed for a second selection of the transformed phenotype.[2] The DNA of the secondary transfection retained the ability to transform the NIH 3T3 cells and therefore also retained the genetic information necessary for the transformation. Because the secondary transfectants contain only a small amount of DNA from the original tumorigenic donor, this DNA is thought to be critical for the development of malignancy (oncogenesis).

Using transfection, more than 20 oncogenes have been identified. Both viral oncogenes (for review, see Ref. 3) and those resulting from chemical mutagenesis have been isolated. Although it seemed initially that viral and nonviral oncogenes might represent distinct oncogenetic mechanisms, they now often appear to be homologous.[4] The characterization of many of these oncogenes has included localizing the cellular region to which the protein product migrates. The gene products

---

ERICA M. S. SIBINGA AND GAIL R. MASSEY • Department of Pathology and Laboratory Medicine, University of Pennsylvania School of Medicine, Philadelphia, Pennsylvania 19104-6082. MARK I. GREENE • Division of Immunology, Department of Pathology and Laboratory Medicine, University of Pennsylvania School of Medicine, Philadelphia, Pennsylvania 19104-6082.

may migrate to any region of the cell; they may remain in the nucleus, be found in the cytoplasm or at the plasma membrane, or be secreted. It is also inferred that the cellular localization of the protein product is important in transformation. This review will consider oncogenes that code for proteins residing at the surface of the cell.

The oncogenes that are known to encode cell-surface proteins are *erb*-B, *fms*, and *neu*. They offer a unique opportunity to study both malignant and normal cell-surface signaling. Cells transformed by these oncogenes seem to have obtained a growth advantage by mimicking mechanisms of normal cellular growth. However, in these transformed cells, the normal system of regulation is evaded, and the growth advantage is achieved by the cells' growth autonomy. The relationship of these oncogenes to receptor molecules is useful in understanding their mechanisms; therefore, we will present a brief review of the properties of growth factor receptors, followed by a consideration of the oncogenes that encode cell-surface molecules.

2. Features of Cell-Surface Receptors

One of the earliest references to the concept of receptors was made by Langley at the turn of the century to explain the effects of curare on skeletal muscle contractions.[5] Since that time there has been a tremendous accumulation of knowledge about receptors. This section is a summary of some of the common properties of growth factor or growth factor-related receptors, some of which are related to oncogene-encoded receptors. The receptors that will be mentioned include: colony-stimulating factor-1 receptor (CSF-1R), epidermal growth factor receptor (EGFR), insulin receptor, insulinlike growth factor-1 receptor (IGF-1), low-density lipoprotein receptor (LDLR), and platelet-derived growth factor receptor (PDGFR). HER2, which has the properties of a growth factor receptor but no known ligand, will also be considered.

The structural properties of these receptors show a familial relationship. The first similarity is the presence of a signal sequence of the nascent membrane protein that is recognized by the signal recognition particle that brings it to the endoplasmic reticulum translocation system.[6] The recognized consensus feature of signal sequences remains unknown. Because the sequences lack primary structural homology but are generally hydrophobic in nature,[7] it is probably secondary or tertiary structure that is important for recognition.[8] The presumptive signal sequences (which are eventually cleaved) are 21 amino acids in length for HER2[9], 32 amino acids for PDGFR,[10] and 27 amino acids for human insulin receptor.[11]

The amino-terminal portion of receptor molecules is usually extracellular and is characterized by cysteine-rich regions. A comparison of the extracellular ligand-binding domains of HER2 (632 amino acids) and EGFR (621 amino acids) yields about 40% homology including two cysteine-rich subdomains of 26 and 21 cysteine residues.[9] The insulin receptor α subunit also contains one homologous cysteine-rich domain.[11] Human LDLR has eight cysteine-rich repeat units in its extracellular portions.[7] PDGFR and CSF-1R contain 10 cysteine residues homologous to each other,[10,12] but they do not show the same type of clustering as above. The role of the cysteine clusters is thought to be one of forming an essential structural backbone[9] and stabilizing but not forming the ligand-binding site.[7]

The extracellular regions of these receptors have varying numbers of sites for possible N-linked glycosylation (Asn-X-Thr/Ser) including 11, 8, 12, and 15 for PDGFR, HER2, EGFR, and insulin receptor α chain, respectively.[9–11]

The hydrophobic membrane anchor of these proteins consists of slightly more than 20 amino acids. This is followed by a sequence of predominantly basic amino acids in HER2,[9] EGFR,[13] insulin receptor,[11] and LDLR.[7] In this region of EGFR is a Thr residue that may be involved in receptor modulation by protein kinase C.[14] The HER2 molecule has a homologously positioned Thr.[9]

The cytoplasmic region of many growth factor receptors contains a tyrosine kinase domain. Some of these receptors have been shown to have ligand-stimulated autophosphorylating activity specific for Tyr, including the EGF, PDGF, insulin, IGF-1, and CSF-1 receptors. Tyrosine kinase activity is not inherent to all growth factor receptors, however, as it has not been detected in the interleukin-2, NGF, or IGF-2 receptors.[15,16] The growth factor receptors with tyrosine kinase activity have a conserved Lys that is thought to be involved in the nucleotide binding site and a conserved Tyr that is autophosphorylated.[10] Structurally, the protein tyrosine kinases fall into two subgroups. The first subgroup has uninterrupted kinase domains while the second has kinase domains with long inserted sequences. Members of the first subgroup include the insulin receptor, EGFR, and HER2. PDGFR and CSF-1R comprise the second subgroup.[10]

Some cell-surface receptors such as the LDL, EGF, and insulin receptors concentrate in coated pits and recycle. One of the better studied systems is the LDLR.[17] A brief summary of its recycling pattern follows: LDLRs localize to clathrin-coated pits both in the presence and absence of ligand[6,18] (it is not clear whether other recycling receptors, such as EGFR, require the binding of ligand in order to recycle[19]); the coated pits internalize, become coated vesicles, lose their clathrin coats, and fuse to become endosomes. The acid pH of the endosomes is thought to cause the dissociation of the ligand from the receptor, thereby allowing the receptor to recycle to the cell surface without sustaining proteolytic damage from lysosomal enzymes. Indeed, LDLR can make 150 complete trips and still remain intact.[17] The mechanism of EGFR recycling may not be as efficient as that of LDLR because its half-life decreases tenfold in the presence of ligand.[20] Ligand also apparently accelerates the degradation of the insulin receptor, which moves to coated pits only after ligand binding and then is degraded in the lysosomes along with the insulin.[21] Although not all receptors utilize coated pits to move their ligands into the cell, it is a mechanism thought to be characteristic of growth factor receptors.

Although some of the events following ligand binding to growth factors are now known, the exact mechanisms by which they elicit their various effects remain to be elucidated.

3. Oncogenes That Encode Receptors

3.1. erb-B

The erb-B oncogene product bears close similarity to EGFR[22] and illustrates the potential for neoplasia in aberrant receptor molecules. The erb-B gene encodes the transforming protein of avian erythroblastosis virus (AEV), consisting of 604 amino acids.[23] Comparison of the v-erb-B protein with peptides from human EGFR shows a large degree of homology. That is, with six peptides from EGFR containing 83 amino acids in all, 74 are identical to v-erb-B. EGFR is thought to contain about 1250 amino acids and has three domains: an EGF-binding domain external to the plasma membrane, a transmembrane domain, and a cytoplasmic domain, characterized by kinase activity and sites for autophosphorylation. Since there were eight EGFR-derived peptides that could not be matched with v-erb-B and the predicted number of amino acids is greatly different (about half as many in v-erb-B as in EGFR), the v-erb-B product represents a truncated EGFR. v-erb-B encodes the cytoplasmic and transmembrane domains of EGFR, but lacks the greater part of the EGF-binding region by which normal growth is regulated.[22] The v-erb-B protein has been found to induce Tyr phosphorylation *in vivo* and *in vitro* and contains sequences homologous to the kinase domain of EGFR and corresponding regions of retroviral kinases[24] suggesting that the unregulated kinase activity may be responsible for transformation. Thus, the structure of the v-erb-B product—consisting of transmembrane and cytoplasmic, but not ligand-binding, domains—allows for an unregulated mitogenic signal. (For further review of erb-B, see the chapter by Yamamoto and Toyoshima in this volume.)

3.2. fms

The oncogene v-fms, associated with the McDonough strain of feline sarcoma virus (SM-FeSV), also encodes a protein exhibiting receptor characteristics. The v-fms protein product is a 140,000-dalton transmembrane glycoprotein[25–27] with the amino-terminal domain oriented outside the cell and the carboxy-terminal region in the cytoplasm.[27,28] Using antibodies to v-fms glycoproteins and subsequent indirect immunofluorescence, it has been shown that the v-fms products are expressed on the cell surface and are associated with clathrin, clathrin-coated pits, and endocytotic vesicles,[29] characteristics typical of receptor molecules.[17] Common to growth-related receptors, v-fms displays *in vitro* phosphorylation of the glycoprotein at Tyr.[26,30] Using a nontransforming mutation of v-fms, which retains the tyrosine protein kinase activity but is not detected at the cell surface, it was shown that cell-surface expression is necessary for v-fms transformation, indicating the importance of the plasma membrane in v-fms transformation.[26]

Nucleotide sequencing of the v-fms gene shows further similarities to growth factor receptors. The protein is predicted to be 160,000 daltons, with 14 potential sites for glycosylation in the putative extracellular domain and a hydrophobic region that correlates with the transmembrane orientation of the molecule. Additionally, the sequence of the cytoplasmic domain shows a close relationship to tyrosine protein kinases known in several other retroviral oncogenes and growth-related receptors.[28]

The cellular counterpart of the transforming *fms* gene seems to be the receptor for the murine colony-stimulating factor, CSF-1.[16] The c-*fms* product is a 170,000-dalton glycoprotein that is associated with tyrosine kinase activity *in vitro* and found at high levels in mature cat macrophages. Rabbit antisera to the v-*fms* product precipitates a glycoprotein that has been shown to represent the CSF-1R, indicating that the c-*fms* gene product is closely related, if not identical, to the receptor for CSF-1.[31] The relationship between CSF-1R and SM-FeSV remains perplexing, however, since v-*fms* transforms fibroblasts most efficiently, and to an extent epithelial cell lines,[31] neither of which requires CSF-1 for growth. Thus, although v-*fms* encodes an altered CSF-1R, the mechanism of the subsequent transformation remains unclear. (See chapter by Sherr.)

3.3. *neu*

As opposed to the two previously mentioned virus-induced oncogenes, v-*fms* and v-*erb*-B, the *neu* oncogene results from a chemically induced mutagenesis. Fetal BDIX rats were exposed to ethylnitrosourea *in utero* on day 15 following conception. Neuroblastoma cell lines were derived from the ensuing tumors and DNA from these lines was used in two cycles of transfection.[32] A phosphoprotein of 185,000 daltons (p185) was found to be specifically associated with the *neu* transforming sequence [33] and monoclonal antibodies (7.16.4) reactive with this cell-surface protein were developed.[34] Treatment of NIH 3T3 cells transformed with *neu* with these antibodies causes a marked decrease in levels of cell-surface p185, an increase in p185 degradation, a decrease in overall cellular levels of p185, and a reversion of anchorage-independent colonization in soft agar, an important characteristic for distinguishing the transformed phenotype. Thus, the expression of p185 is required to maintain the *neu*-induced transformation.[35]

Homology between *erb*-B and *neu* was found using Southern blot hybridization[36] and has been extremely useful in the characterization of *neu*. Although related, *erb*-B and *neu* are distinct genes, mapping to different human chromosomes in *in situ* hybridization studies.[37] The similarity to *erb*-B, which encodes the receptor for EGF, encouraged the idea that *neu* codes for a development-related receptor.

Nucleotide sequence analysis strongly supports the possibility that *neu* encodes a receptorlike molecule. The predicted product is a 1260-amino-acid transmembrane protein with a structure similar to EGFR.[38] The homology between EGFR and *neu* products is 50% overall, with greater than 80% amino acid identity in the tyrosine kinase domain. The sequence shows a hydrophobic signal sequence at the amino terminus, as is associated with membrane-bound molecules, followed by a 640-amino-acid extracellular domain, a hydrophobic transmembrane region, and a cytoplasmic domain. The extracellular domain has two cysteine-rich areas, which are highly conserved in relation to both the insulin receptor and EGFR. The 25-

amino-acid transmembrane region precedes the 250-residue cytoplasmic region. The cytoplasmic domain shows high homology with EGFR (82%) and with proteins encoded by the tyrosine kinase gene family, and *neu* Tyr residues are in the same positions as EGFR phosphotyrosine sites. Additionally, there is a Thr site that may serve as a site for negative regulation of receptor expression.[38] The nucleotide analysis of the *neu* gene gives a clear view of the receptorlike structure of the *neu* protein product.

The mutation that causes *neu*-induced transformation does not cause major structural alterations[39] or drastic changes in DNA sequence.[40] Analysis of restriction enzyme patterns[40] and partial proteolytic mapping[39] show that the alteration(s) involved in *neu* transformation is subtle at the DNA and structural levels. It is clear, however, that transformation is due to mutation(s) in the protein-encoding region and not a regulatory domain from studies of overexpression of *neu* and p185 without related transformation.[40] Comparison of the transforming *neu* sequence with the cellular human *neu* gene shows that the oncogene results from a point mutation in the transmembrane region.[9,38]

Isolated by homology with *erb*-B, the human cellular *neu* gene has been loctated and sequenced. *In situ* hybridization identified the chromosomal location of this gene related to *erb*-B as q21 of chromosome 17, coincident with the *neu* locus. It is therefore thought to be *neu*.[9] Sequence analysis shows a 21-amino-acid signal sequence, an extracellular ligand-binding domain, a hydrophobic transmembrane region, and a carboxy-terminal cytoplasmic domain. The amino-terminal ligand-binding region includes two cysteine-rich subdomains characteristic of receptor molecules. The hydrophobic region is followed by a Thr residue that may play an important part in receptor regulation mediated by protein kinase C. The cytoplasmic domain, which shows the most homology to *erb*-B, encodes the ATP-binding and tyrosine kinase regions and is followed by a hydrophobic region, as is seen in EGFR.[9]

The data strongly suggest that the nontransforming *neu* gene (the protooncogene) encodes a receptor for a factor involved in growth regulation. In attempting to identify the ligand for p185, Stern *et al.* tested a panel of receptors for reactivity with 7.16.4 antibody and examined the stability of p185 in the presence of various ligands. 7.16.4 was not significantly reactive with receptors for PDGF, TGF-β, and insulin; no substantial decrease was found in p185 stability (as would be expected with ligand-receptor binding) with addition of EGF, TGF-α, TGF-β, PDGF, insulin, IGF-1, IGF-2, FGF, and NGF.[39] Thus, the ligand for the cell-surface protein encoded by *neu* is as yet unidentified. (See chapter by Drebin *et al.*)

4. Conclusion

The isolation and subsequent characterization of oncogenes have linked their mechanisms of transformation with various aspects of growth factor activation. It seems that some oncogenes obtain their growth advantage by conferring growth factor independence on the transformed cells. This is possible at a number of points in the normal mitogenic pathway—the growth factor, its transmembrane receptor, and intracellular signaling mechanisms—and oncogenes affecting each level are known (for reviews, see Refs. 41 and 42).

We have focused on those oncogenes relating to growth factor receptors and can clearly identify characteristics shared among them and with known growth factor receptors: cysteine-rich extracellular region (except for v-*erb*-B, which lacks this region altogether) presumably involved in ligand binding, the transmembrane portion essential for transduction of the mitogenic signal, and the intracellular tyrosine kinase domain typical of growth factor receptors. In addition, the data relating to the oncogenes *erb*-B and *fms* are convincing that these are closely associated with EGFR and CSF-1R, respectively. Although it is very likely that *neu* encodes a receptor also, the ligand that it binds remains elusive. The *neu* protooncogene may encode a receptor for an as yet unidentified growth factor, a hormone of developmental importance, or even a mitogenic mechanism that is so far unclear. Characterization of oncogenic evasion of growth regulation and an understanding of the developmental conditions under which these ligands function may help in the clarification of both carcinogenic and normal receptor-mediated mitogenesis.

References

1. Nowell, P. C., and Hungerford, D. A., 1960, Chromosome studies on normal and leukemic human leukocytes, *J. Natl. Cancer Inst.* **25**:85–110.
2. Shih, C., Shilo, B., Goldfarb, M. P., Dannenberg, A., and Weinberg, R. A., 1979, Passage of phenotypes of chemically transformed cells in a transfection of DNA and chromatin, *Proc. Natl. Acad. Sci. USA* **76**:5714–5718.
3. Bishop, J. M., 1985, Viral oncogenes, *Cell* **42**:23–38.
4. Weinberg, R. A., 1982, Fewer and fewer oncogenes, *Cell* **40**:3–4.
5. Langley, J. N., 1905, On the reaction of cells and membrane endings to certain poisons, *J. Physiol. (London)* **33**:374–413.
6. Anderson, D. J., Walter, P., and Blobel, G., 1982, Signal recognition protein is required for the integration of acetylcholine receptor and subunit, a transmembrane glycoprotein, into the endoplasmic reticulum membrane, *J. Cell Biol.* **93**:501–506.
7. Yamamoto, T., Davis, C. G., Brown, M. S., Schneider, W. J., Casey, M. L., Goldstein, J. L., and Russell, D. W., 1984, The human LDL receptor: A cysteine-rich protein with multiple Alu sequences in its mRNA, *Cell* **39**:27–38.
8. Walter, P., Gilmore, R., and Blobel, G., 1984, Protein translocation across the endoplasmic reticulum, *Cell* **38**:5–8.
9. Coussens, L., Yeng-Feng, T. L., Liao, Y., Chen, E., Gray, A., McGrath, J., Seeburg, P. H., Liberman, T. A. Schlessinger, J., Francke, U., Levinson, A., and Ullrich, A., 1985, Tyrosine kinase receptor with extensive homology to EGF receptor shares chromosomal location with *neu* oncogene, *Science* **230**:1132–1139.
10. Yarden, Y., Escobedo, J. A., Kuang, W.-J., Yang-Feng, T. L., Daniel, T. O., Tremble, P. M., Chen, E. Y., Ando, M. E., Harkins, R. N., Francke, U., Fried, V. A., Ullrich, A., and Williams, L. T., 1986, Structure of the receptor for platelet-derived growth factor helps define a family of closely related growth factor receptors, *Nature* **323**:226–232.
11. Ullrich, A., Bell, J. R., Chen, E. Y., Herrera, R., Petruzzelli, L. M., Dull, T. J., Gray, A., Coussens, L., Liao, Y.-C., Tsubokawa, M., Mason, A., Seeburg, P. H., Grunfeld, C., Rosen, D. M., and Ramachandran, J., 1985, Human insulin receptor and its relationship to the tyrosine kinase family of oncogenes, *Nature* **313**:756–761.

12. Coussens, L., Van Beveren, C., Smith, D., Chen, E., Mitchell, R. L., Isacki, C. M., Verma, I. M., and Ullrich, A., 1986, Structural alteration of viral homologue of receptor proto-oncogene *fms* at carboxyl terminus, *Nature* **320:**277-280.
13. Ullrich, A., Coussens, L., Hayflick, J. S., Dull, T. J., Gray, A., Tam, A. W., Lee, J., Yarden, T., Liberman, T. A., Schlessinger, J., Downward, J., Mayes, E. L. V., Whittle, N., Waterfield, M. D., and Seeburg, P. H., 1984, Human epidermal growth factor receptor cDNA sequence and aberrant expression of the amplified gene in A431 epidermoid carcinoma cells, *Nature* **309:**418-425.
14. Hunter, T., Ling, N., and Cooper, J. A., 1984, Protein kinase C phosphorylation of the EGF receptor at a threonine residue close to the cytoplasmic face of the plasma membrane, *Nature* **311:**480-483.
15. Hunter, T., and Cooper, J. A., 1985, Protein-tyrosine kinases, *Annu. Rev. Biochem.* **54:**897-930.
16. Sherr, C. J., Rettenmier, C. W., Sacca, R., Roussel, M. F., Look, A. T., and Stanley, E. R., 1985, The *c-fms* proto-oncogene product is related to the receptor for the mononuclear phagocyte growth factor, CSF-1, *Cell* **41:**665-676.
17. Brown, M. S., Anderson, R. G. W., and Goldstein, J. L., 1983, Recycling receptors: The round trip itinerary of migrant membrane proteins, *Cell* **32:**663-667.
18. Pearse, B. M. F., 1975, Coated vesicles from pig brain: purification and biochemical characterization, *J. Mol. Biol.* **97:**93-98.
19. Pastan, I. H., and Willingham, M. C., 1981, Journey to the center of the cell: Role of the receptosome, *Science* **214:**504-509.
20. Stoscheck, C. M., and Carpenter, G., 1984, Down regulation of epidermal growth factor receptors: Direct demonstration of receptor degradation in human fibroblasts, *J. Cell Biol.* **98:**1048-1053.
21. Pastan, I. H., and Willingham, M. C., 1981, Receptor-mediated endocytosis of hormones in cultured cells, *Annu. Rev. Physiol.* **43:**239-250.
22. Downward, J., Yarden, Y., Mayes, E., Scrace, G., Totty, N., Stockwell, P., Ullrich, A., Schlessinger, J., and Waterfield, M. D., 1984, Close similarity of epidermal growth factor receptor and v-*erb*-B oncogene protein sequences, *Nature* **307:**521-527.
23. Yamamoto, T., Nishida, T., Miyajima, N., Kawai, S., Ooi, T., and Toyoshima, K., 1983, The *erbB* gene of avian erythroblastosis virus is a member of the *src* gene family *Cell* **35:**71-78.
24. Gilmore, T., DeClue, J. E., and Martin, G. S., 1985, Protein phosphorylation at tyrosine is induced by the v-*erbB* gene product *in vivo* and *in vitro*, *Cell* **40:**609-618.
25. Anderson, S. J., Gonda, M. A., Rettenmier, C. W., and Sherr, C. J., 1984, Subcellular localization of glycoproteins encoded by the viral oncogene v-*fms*, *J. Virol.* **51:**730-741.
26. Roussel, M. F., Rettenmier, C. W., Look, A. T., and Sherr, C. J., 1984, Cell surface expression of v-*fms*-coded glycoproteins is required for transformation, *Mol. Cell. Biol.* **4:**1999-2009.
27. Rettenmier, C. W., Roussel, M. F., Quinn, C. O., Kitchingman, G. R., Look, A. T., and Sherr, C. J., 1985, Transmembrane orientation of glycoproteins encoded by the v-*fms* oncogene, *Cell* **40:**971-981.
28. Hampe, A., Gobet, M., Sherr, C. J., and Galibert, F., 1984, Nucleotide sequence of the feline retroviral oncogene v-*fms* shows unexpected homology with oncogenes encoding tyrosine-specific protein kinases, *Proc. Natl. Acad. Sci. USA* **81:**85-89.
29. Marger, R., Najita, L., Nichols, E. J., Hakomori, S., and Rohrschneider, L., 1984, Cell surface expression of the McDonough strain of feline sarcoma virus *fms* gene product (gp140^{fms}), *Cell* **39:**327-337.
30. Barbacid, M., and Lauver, A. V., 1981, Gene products of McDonough feline sarcoma virus have an *in vitro*-associated protein kinase that phosphorylates tyrosine residues: Lack of detection of this enzymatic activity *in vivo*, *J. Virol.* **40:**812-821.
31. Rettenmier, C. W., Chen, J. H., Roussel, M. F., and Sherr, C. J., 1985, The product of the c-*fms* proto-oncogene: A glycoprotein with associated tyrosine kinase activity, *Science* **228:**320-322.
32. Shih, C., Padhy, L. C., Murray, M., and Weinberg, R. A., 1981, Transforming genes of carcinomas and neuroblastomas introduced into mouse fibroblasts, *Nature* **290:**261-264.
33. Padhy, L. C., Shih, C., Cowing, D., Finkelstein, R., and Weinberg, R. A., 1982, Identification of a phosphoprotein specifically induced by the transforming DNA of rat neuroblastomas, *Cell* **28:**865-871.

34. Drebin, J. A., Stern, D. F., Link, V. C., Weinberg, R. A., and Greene, M. I., 1984, Monoclonal antibodies identify a cell-surface antigen associated with an activated cellular oncogene, *Nature* **312**:545–548.
35. Drebin, J. A., Link, V. C., Stern, D. F., Weinberg, R. A., and Greene, M. I., 1985, Down-modulation of an oncogene protein product and reversion of the transformed phenotype by monoclonal antibodies, *Cell* **41**:695–706.
36. Schechter, A. L., Stern, D. F., Vaidyanathan, L., Decker, S. J., Drebin, J. A., Greene, M. I., and Weinberg, R. A., 1984, The *neu* oncogene: An *erb-B*-related gene encoding a 185,000-M_r tumor antigen, *Nature* **312**:513–516.
37. Schechter, A. L. Hung, M. -C., Vaidyanathan, L., Weinberg, R. A., Yang-Feng, T. L., Francke, U., Ullrich, A., and Coussens, L., 1985, The *neu* gene: An *erbB*-homologous gene distinct from and unlinked to the gene encoding the EGF receptor, *Science* **229**:976–978.
38. Bargmann, C. I., Hung, M., and Weinberg, R. A., 1986, The *neu* oncogene encodes an epidermal growth factor receptor-related protein, *Nature* **319**:226–230.
39. Stern, D. F., Heffernan, P. A., and Weinberg, R. A., 1986, p185, a product of the *neu* proto-oncogene, is a receptor-like protein associated with tyrosine kinase activity, *Mol. Cell. Biol.* **6**:1729–1740.
40. Hung, M., Schechter, A. L., Chevray, P. M., Stern, D. F., and Weinberg, R. A., 1986, Molecular cloning of the *neu* gene: Absence of gross structural alteration in oncogenic alleles, *Proc. Natl. Acad. Sci. USA* **83**:261–264.
41. Heldin, C., and Westermark, B., 1984, Growth factors: Mechanism of action and relation to oncogenes, *Cell* **37**:9–20.
42. Sporn, M. B., and Roberts, A. B., 1985, Autocrine growth factors and cancer, *Nature* **313**:745–747.

5
Monoclonal Antibodies Reactive with the *neu* Oncogene Product Inhibit the Neoplastic Properties of *neu*-Transformed Cells

JEFFREY A. DREBIN, VICTORIA C. LINK, AND MARK I. GREENE

1. Introduction

The past decade has seen major advances in our understanding of the molecular events involved in malignant transformation. Studies of RNA tumor viruses initially identified specific genes that were able to confer neoplastic properties on cells in tissue culture and that were responsible for tumor formation *in vivo*.[1] These genes were termed oncogenes. Subsequently, it was shown that genes closely related to retroviral oncogenes (protooncogenes) exist in the genomes of all eukaryotic cells. These protooncogenes have been highly conserved in evolution and are likely to play important roles in normal growth and development.[1,2] It is now clear that retroviruses have acquired their oncogenes by retroviral transduction of cellular protooncogenes. It is thought that elevated expression and/or structural alterations of protooncogenes within the retroviral genome are responsible for the abilities of some retroviruses to cause neoplastic transformation.

Recent studies have implicated structural and functional abnormalities involving cellular protooncogenes in the etiology of spontaneous neoplasia as well. Approximately 20% of human tumors contain activated oncogenes capable of transforming NIH 3T3 cells in DNA transfection assays.[3,4] Many of the cellular

---

JEFFREY A. DREBIN AND VICTORIA C. LINK • Department of Pathology, Harvard Medical School, Boston, Massachusetts 02115. MARK I. GREENE • Division of Immunology, Department of Pathology and Laboratory Medicine, University of Pennsylvania School of Medicine, Philadelphia, Pennsylvania 19104–6082.

oncogenes identified in transfection assays have been members of the *ras* oncogene family.[3-5] Point mutations at certain critical sites have been shown to be responsible for activating the malignant potential of the *ras* genes.[5] Mutated *ras* oncogenes have only been found in neoplastic cells; they have not been identified in normal tissues from individuals whose tumors contain the mutant genes.[6,7]

Other types of genetic anomalies involving protooncogenes have been described in tumor cells that may lack oncogenes active in transfection assays. For example, chromosomal rearrangements involving the *myc* and *abl* protooncogenes, respectively, have been described in a majority of the Burkitt's lymphomas and chronic myelogenous leukemias that have been examined.[8,9] Protooncogene alterations by yet another mechanism, gene amplification, has been identified in a substantial number of solid tumors.[10,11] Collectively, such data suggest that genetic lesions involving cellular protooncogenes may play an important role in the process of neoplastic transformation.

Reflecting these genetic alterations, the protein products of cellular oncogenes in tumor cells appear to differ qualitatively or quantitatively from those of the corresponding protooncogenes in nontransformed cells.[12] Such oncogene-encoded proteins may prove useful targets against which to direct antitumor chemotherapy or immunotherapy. In an effort to develop a model system for such therapy we have investigated the protein product of an oncogene initially identified by transfection of DNA from chemically induced rat neuroblastomas.[13] Here we will describe the results of studies that have shown: that this oncogene (termed *neu*) encodes a cell-surface protein of 185,000 molecular weight that can be identified by immunofluorescence and immunoprecipitation using monoclonal antibodies; that monoclonal antibodies reactive with cell-surface domains of the p185 molecule cause the rapid internalization and degradation of p185 and correspondingly cause *neu*-transformed cells to revert to a nontransformed phenotype; and that anti-p185 monoclonal antibodies exert immunological antitumor effects *in vitro* and *in vivo*.

2. The *neu* Oncogene: An Oncogene Encoding A 185,000-Dalton Tumor Antigen

Studies from the laboratory of Robert A. Weinberg at MIT originally identified an activated oncogene present in four independently derived ethlynitrosourea-induced rat neuroblastomas.[13] This oncogene was shown by Schecter *et al.* to be related to, but distinct from, the *erb*B oncogene; it was termed *neu* because of its identification in chemically induced neuroblastomas.[14] A homologous gene, isolated from the human genome by two independent laboratories, has been termed HER-2 and c-*erb*B-2.[15,16] Both the rat *neu* gene and its human homologue encode 185,000-dalton glycoproteins that possess intrinsic tyrosine kinase activity.[14,17]

Subsequent studies by Bargmann *et al.* have shown that the activated rat *neu* oncogene identified in DNA transfection assays has undergone a single point mutation that results in a change of one amino acid of the p185 molecule from valine to glutamic acid.[18] This subtle alteration appears to be responsible for the transforming activity of this gene in transfection assays. Whether this mutation alters the tyrosine kinase activity of the p185 molecule or induces some other change in the function of the p185 molecule is currently the subject of active study.

3. Development of Monoclonal Antibodies Reactive with the *neu* Oncogene Product

In order to facilitate structural and functional studies of the p185 molecule, we have generated a panel of monoclonal antibodies that specifically bind p185. Preliminary immunofluorescence studies utilizing polyclonal antiserum reactive with the p185 molecule suggested that a portion of the molecule might be exposed on the surface of intact *neu*-transformed cells.[19] We therefore generated hybridomas by fusing NS1 myeloma cells and spleen cells from mice immunized with

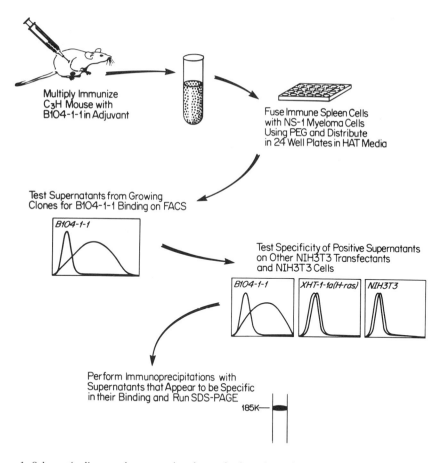

FIGURE 1. Schematic diagram demonstrating the method used to select hybridomas secreting anti-p185 monoclonal antibodies reactive with domains of p185 expressed on the surface of *neu*-transformed cells (cell line B104-1-1).

TABLE I
Anti-p185 Monoclonal Antibodies

Antibody	(Isotype[a])	Immunofluorescent staining of cell lines[a]			Immunopreciptation of p185[a]
		B104-1-1[b]	XHT-1-1a[c]	NIH 3T3	
7.5.5	(IgG2b)	+	−	−	+
7.9.5	(IgG1)	+	−	−	+
7.16.4	(IgG2a)	+	−	−	+
7.16.5	(IgM)	+	−	−	+
7.21.2	(IgG1)	+	−	−	+

[a]Determined as in Ref. 20.
[b]NIH 3T3 cells transformed by transfection with an activated rat *neu* oncogene.
[c]NIH 3T3 cells transformed by transfection with a v-H-*ras* oncogene.

the *neu*-transformed NIH 3T3 cell line B104-1-1 (Fig. 1). These hybridomas were screened for secretion of antibodies reactive with B104-1-1 cell-surface antigens as determined by quantitative immunofluroescence tests using a fluorescence-activated cell sorter (FACS). Hybridomas secreting antibodies that showed reactivity with *neu*-transformed NIH 3T3 cells, and that did not react with normal NIH 3T3 cells or the H-*ras*-transformed NIH 3T3 cell line XHT-1-1a were then tested for the ability to immunoprecipitate the *neu* oncogene-encoded p185 molecule from metabolically labeled B104-1-1 cell lysates. We have screened approximately 10,000 hybridomas and have identified 5 that bind cells expressing p185 at the cell surface and immunoprecipitate p185 from metabolically labeled cell lysates.[20] The names and isotypes of these antibodies are shown in Table I.

The anti-p185 antibodies have been used in immunofluorescence and immunohistochemistry tests to identify cells expressing p185. As shown in Table II, these antibodies bind to a number of rat neuroblastomas containing activated *neu* oncogenes and also bind NIH 3T3 cells transformed by DNA extracted from these neuroblastomas. The transfectants display variably increased levels of p185 expression, as has been seen for the products of other transfected genes. There is no binding to normal NIH 3T3 cells or to NIH 3T3 cells transformed by a variety of oncogenes unrelated to *neu*. Certain of the anti-p185 antibodies have also been used in immunohistochemical studies to identify *neu*-transformed cells in tumors induced by the implantation of *neu*-transformed cells into nude mice (data now shown). We are currently using the anti-p185 monoclonal antibodies to determine whether p185 expression correlates with the invasive properties of neuroblastomas or any other tumor types.

4. Down-Modulation of Cell-Surface p185 Induced by Monoclonal Anti-p185 Antibodies

Binding of antibodies to cell-surface macromolecules may induce the rapid reorganization of these structures, processes known as patching and capping.[21]

TABLE II
Immunofluorescent Staining of Cell Lines Using Anti-p185 Monoclonal Antibody 7.16.4

Cell line	Description	Immunofluroescent staining[a]
NIH 3T3	Murine fibroblast	−
B104	Rat neuroblastoma[b]	+
B104-2	NIH 3T3 cells transformed with B104 DNA	+
B104-1-1	NIH 3T3 cells transformed with B104 DNA	++++
B104-1-2	NIH 3T3 cells transformed with B104 DNA	+
B50	Rat neuroblastoma[b]	+
B50-1	NIH 3T3 cells transformed with B50 DNA	+
B50-6a-2	NIH 3T3 cells transformed with B50 DNA	++++
B50-6a-3	NIH 3T3 cells transformed with B50 DNA	++
B82	Rab neuroblastoma[b]	+
B82-1	NIH 3T3 cells transformed with B82 DNA	+
B82-3	NIH 3T3 cells transformed with B82 DNA	+++
B82-3-2	NIH 3T3 cells transformed with B82 DNA	+
XHT-1-1a	NIH 3T3 cells transformed with a v-H-*ras* oncogene	−
RSV 3T3	NIH 3T3 cells transformed by Rous sarcoma virus	−
MSV 3T3	NIH 3T3 cells transformed by Moloney sarcoma virus	−
PyD2	Murine fibroblasts transformed by polyoma virus	−
SVD2	Murine fibroblasts transformed by SV40 virus	−

[a]Determined on the FACS as determined in Ref. 20.
[b]Cell lines containing *neu* oncogenes active in transfection assays.[13]

This, in turn, may be followed by internalization or shedding of the macromole-cule–antibody complex, resulting in its disappearance from the cell surface.[22] The process of antibody-induced loss of cell-surface antigens has been referred to as down-modulation.[23] This process appears to require cross-linking (or at least dimerization) of cell-surface structures by antibody, because multivalent antibody but not monovalent F(ab) fragments are able to induce down-modulation of cell-surface antigens.[21,22]

In order to investigate whether anti-p185 antibodies could induce down-modulation of cell-surface p185, we exposed the *neu*-transformed NIH 3T3 cell line B104-1-1 to intact anti-p185 antibody 7.16.4 molecules, or to monovalent 7.16.4 F(ab) fragments, and examined cell-surface p185 expression by quantitative immunofluorescence using a flow cytometer. As shown in Table III, incubation of B104-1-1 cells with antibody 7.16.4 caused the rapid down-modulation of cell-surface p185 expression. This effect persisted as long as cells were exposed to antibody, and was completely reversible following removal of antibody from the culture medium. In contrast, monovalent F(ab) fragments of antibody 7.16.4 had no effect on p185 expression. Furthermore, immunoprecipitable p185 was decreased in parallel with the loss of cell-surface p185,[24] suggesting that total p185 levels in cells was decreased as a result of antibody-induced down-modulation. Subsequent studies have shown that exposure to anti-p185 antibodies results in the rapid internalization of cell-surface antibody–p185 complexes, presumably reflecting entry

TABLE III
Down-Modulation of Cell-Surface p185 by Monoclonal Antibody 7.16.4[a]

	Treatment	Cell-surface p185
Experiment 1	None	100%
	7.16.4—15 min	55%
	7.16.4—30 min	50%
	7.16.4—60 min	35%
	7.16.4—4 hr	30%
	7.16.4—20 hr	30%
	7.16.4—4 hr, then washed free of antibody and incubated an additional 16 hr	90%
Experiment 2	None	100%
	7.16.4 (whole antibody)—20 hr	45%
	7.16.4 F(ab) fragments—20 hr	110%

[a] For experimental details see Ref. 24.

into lysosomes, in a temperature-dependent manner (data now shown). Thus, it appears that anti-p185 monoclonal antibodies can induce the rapid internalization and degradation of cell-surface p185, resulting in lower steady-state levels within *neu*-transformed cells.

5. Reversion of the Transformed Phenotype by Monoclonal Anti-p185 Antibodies

Tumor cells display a number of *in vitro* characteristics that distinguish them from nontumor cells.[25] These characteristics are collectively known as the transformed phenotype. Several studies have shown that the ability of cells to grow in the absence of anchorage (when suspended in agar, for example) is the aspect of the transformed phenotype most closely correlated with the ability to form tumors *in vivo*.[25] Introduction of an activated *neu* oncogene into NIH 3T3 cells by the process of DNA transfection results in p185 expression by the transfected cells and their acquisition of the transformed phenotype.[13,24] Since exposure of *neu*-transformed cells to anti-p185 monoclonal antibody results in the rapid down-modulation of cell-surface p185 and lowering of total cellular p185 levels, we examined whether exposure to anti-p185 antibodies might have an effect on the transformed phenotype of *neu*-transformed cells.

As shown in Table IV, exposure of the *neu*-transformed NIH 3T3 cell lines B104-1-1 and B104-1-2 and of the rat neuroblastoma cell line B104 (in which the *neu* oncogene was originally activated by chemical carcinogenesis) to monoclonal anti-p185 antibody 7.16.4 results in dramatic reduction of their ability to form anchorage-independent colonies when suspended in soft agar. In contrast, exposure to antibody 7.16.4 has no effect on the anchorage-independent growth of cells

TABLE IV
Inhibition of the Anchorage-Independent Growth of neu-Transformed Cells by Monoclonal Antibody 7.16.4[a]

Cell line[b]	Anchorage-independent colonies		Percent inhibition
	Control	+7.16.4	
B104-1-1	95	2	98%
B104-1-2	26	6	77%
B104	38	<1	98%
RSV 3T3	74	75	0%
XHT-1-1a	102	90	10%

[a]For experimental details see Ref. 24.
[b]For a description of cell lines see Table II.

transformed by oncogenes unrelated to *neu* (Table IV). Antibody 7.16.4 exposure of B104-1-1 cells attached to tissue culture dishes has no effect on adherent cell growth, demonstrating that the antibody is not simply toxic to the cells (data not shown). Rather, it appears that antibody 7.16.4 is able to selectively cause the reversion of *neu*-transformed cells to a nontransformed phenotype, as determined by anchorage-independent growth.

Further evidence of the specificity of the effect of anti-p185 antibodies on the anchorage-independent growth of *neu*-transformed cells is shown in Table V. A

TABLE V
Anti-p185 Antibodies Inhibit the Anchorage-Independent Growth of neu-Transformed Cells[a]

	Antibody/specificity	Anchorage-independent colonies	Percent inhibition
Experiment 1	None	25	—
	7.5.5/anti-p185	6	76%
	7.9.5/anti-p185	7	72%
	7.16.4/anti-p185	1	96%
	7.16.5/anti-p185	1	96%
	7.21.2/anti-p185	12	52%
	9BG5/antireovirus	21	16%
	87.92.6/antireovirus receptor	29	0%
Experiment 2	None	43	—
	7.16.4 (whole antibody)	1	98%
	7.16.4 F(ab) fragments	44	0%
	7.16.4 F(ab) + anti-immunoglobulin	13	70%
	Anti-immunoglobulin alone	41	5%

[a]For experimental details see Ref. 24.

number of control antibodies have no effect on the anchorage-independent growth of *neu*-transformed cells, while all of the anti-p185 antibodies that we have examined inhibit the anchorage-independent growth of *neu*-transformed cells. Furthermore, monovalent F(ab) fragments of antibody 7.16.4, which bind p185 on the surface of *neu*-transformed cells, are unable to inhibit the anchorage-independent growth of *neu*-transformed cells unless they are cross-linked by anti-immunoglobulin antibodies (Table V). Because the requirements for the inhibition of anchorage-independent growth by anti-p185 antibodies parallel the requirements for down modulation of p185 by these same antibodies, we suggest that the inhibition of anchorage-independent growth reflects the reduction of cellular p185 levels resulting from antibody-mediated p185 down-modulation.

6. Immunologic Antitumor Effects of Anti-p185 Antibodies

The experiments described above have identified a direct cytostatic effect on the anchorage-independent growth of *neu*-transformed cells mediated by anti-p185 monoclonal antibodies. In order to characterize additional mechanisms by which these antibodies might mediate antitumor effects, we have examined the abilities of anti-p185 monoclonal antibodies to target *neu*-transformed cells for complement-dependent or antibody-dependent cellular lysis *in vitro*. As shown in Table VI, several antibodies were able to cause significant killing of *neu*-transformed cells in the presence of rabbit complement. In addition, one antibody, the IgG2a antibody 7.16.4, was able to cause a modest level of tumor cell killing in the presence of nonimmune spleen cells, which serve as effectors of antibody-dependent cellular cytotoxicity (ADCC). The lysis of *neu*-transformed cells by both mechanisms was specific, in that there was no lysis of the *ras*-transformed cell line XHT-1-1a in experiments conducted in parallel. Thus, in addition to their cytostatic effects on

TABLE VI
In Vitro Cytotoxic Effects of Anti-p185 Monoclonal Antibodies[a]

Antibody	Complement-dependent lysis of cell lines[b]:		Antibody-dependent cellular lysis of cell lines[c]:	
	B104-1-1	XHT-1-1a	B104-1-1	XHT-1-1a
7.5.5	+++	−	−	−
7.9.5	−	−	−	−
7.16.4	++++	−	+	−
7.16.5	++	−	−	−
7.21.2	−	−	−	−

[a]Cytolysis in $^{51}CrO_4$ release microcytotoxicity assays containing 10^4 tumor targets/well was quantitated by the following scale: −, < 10% specific lysis; +, 10–20% specific lysis; ++, 20–40% specific lysis; +++, 40–60% specific lysis; ++++, > 60% specific lysis.
[b]One-hour cytotoxicity in the presence of 5 µg/ml antibody and 1:20 rabbit complement.
[c]Eighteen-hour cytotoxicity in the presence of 5 µg/ml antibody and 100:1 spleen cells:tumor targets.

the anchorage-independent growth of *neu*-transformed cells, certain anti-p185 monoclonal antibodies were able to target *neu*-transformed cells for lysis by immunologic effectors.

7. Effects of Anti-p185 Antibodies on the Tumorigenic Growth of *neu*-Transformed Cells.

Because it appeared that anti-p185 monoclonal antibodies could mediate *in vitro* antitumor effects by multiple mechanisms, we examined their effects on the *in vivo* tumorigenic growth of *neu*-transformed cells. For these experiments we implanted *neu*-transformed NIH 3T3 cells (cell line B104-1-1) subcutaneously into groups of nude mice and then injected the mice intravenously with either normal saline or ascites fluid containing monoclonal antibodies. As shown in Fig. 2, treatment with antibodies 7.5.5, 7.9.5, or 7.16.4 was able to significantly inhibit the growth of B104-1-1 cell tumors ($p < 0.05$ at all days measured.) Antibodies 7.16.5 and 7.21.2 had a more modest effect on tumor growth (data not shown). Subsequent studies have shown that injection of a control ascites fluid (containing anti-reovirus monoclonal antibodies) has no effect on the growth of B104-1-1 tumors, and that treatment with anti-p185 monoclonal antibodies has no effect on the tumorigenic growth of the *ras*-transformed cell line XHT-1-1a.[26] Thus, anti-p185 monoclonal antibody treatment is able to specifically inhibit the tumorigenic growth of *neu*-transformed cells.

8. Conclusions

Abnormalities involving cellular oncogenes and their products have been described in a number of tumors.[1-13,16,30] These oncogene products represent very attractive targets for antitumor therapy because they appear to play an important role in the neoplastic state. We have shown that monoclonal antibodies reactive with the *neu* oncogene-encoded p185 molecule can be used to identify cells expressing high levels of p185 by immunofluorescent and immunohistochemical techniques. These antibodies are able to induce the down-modulation of p185 from the surfaces of *neu*-transformed cells, and cause these cells to revert to a nontransformed phenotype *in vitro*. Furthermore, some anti-p185 monoclonal antibodies are able to mediate significant levels of cytotoxicity against *neu*-transformed cells in the presence of complement or nonimmune spleen cells. Thus, these antibodies exert *in vitro* antitumor effects by multiple mechanisms.

Perhaps most significantly, certain anti-p185 antibodies are able to inhibit the tumorigenic growth of *neu*-transformed cells *in vivo*. While our studies have not entirely clarified which antitumor mechanisms account for this inhibition of tumor growth, it appears that neither the ability to lyse cells in the presence of complement nor the ability to target ADCC is necessary, since antibody 7.9.5 (complement lysis$^-$, ADCC$^-$) exerts an *in vivo* antitumor effect quite similar to antibodies 7.5.5 (complement lysis$^+$, ADCC$^-$) and 7.16.4 (complement lysis$^+$, ADCC$^+$). It seems

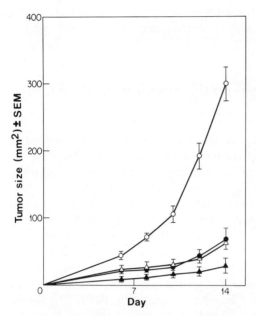

FIGURE 2. Inhibition of the tumorigenic growth of *neu*-transformed NIH 3T3 cells by anti-p185 monoclonal antibodies. BALB/c nude mice were injected subcutaneously in the middorsum with 1×10^6 B104-1-1 tumor cells on day 0. Groups of five mice received 0.5-ml intravenous injections of either saline (O), ascites fluid containing antibody 7.5.5 (Δ), ascites fluid containing antibody 7.9.5 (●), or ascites fluid containing antibody 7.16.4 (▲), on days 0 and 7. Growing tumors were measured with calipers, and tumor size was calculated as the product of tumor length and width. Statistical significance of differences in tumor size between treatment groups was determined using Student's *t* test.

likely that the principal mechanism accounting for the inhibition of tumor growth is a direct effect on the transformed phenotype resulting from p185 down-modulation, along with a lesser contribution form immunological cytotoxic effects.

A number of oncogenes, including *erb*B, *fms*, and *neu*, encode protein expressed on the cell surface.[20,27,28] Among human tumors, elevated expression of the *erb*B oncogene has been described in a number of squamous cell carcinomas, and elevated expression of the *neu* oncogene has been demonstrated in some adenocarcinomas.[29] Monoclonal antibodies reactive with the products of these human oncogenes may exert antitumor effects in cancer patients similar to those seen in experimental animals treated with anti-p185 monoclonal antibodies in our studies. The administration of monoclonal antibodies reactive with cell-surface domains of oncogene-encoded proteins represents a novel and potentially efficacious approach to the therapy of neoplasia.

References

1. Bishop, J. M., and Varmus, H. E., 1982, Function and origin of retroviral transforming genes, in: *RNA Tumor Viruses*. (R. Weiss, N. Teich, H. Varmus, and J. Coffin, eds.), Cold Spring Harbor Laboratory, Cold Spring Harbor, N.Y., pp. 999–1109.
2. Bishop, J. M., 1983, Cellular oncogenes and retroviruses, *Annu. Rev. Biochem.* **52**:301–354.
3. Weinberg, R. A., 1985, The action of oncogenes in the cytoplasm and the nucleus, *Science* **230**:770–775.
4. Cooper, G. M., 1982, Cellular transforming genes, *Science* **217**:801–806.
5. Land, H., Parada, L. F., and Weinberg, R. A., 1983, Cellular oncogenes and multistep carcinogenesis, *Science* **222**:771–778.

6. Santos, E., Martin-Zanca, D., Reddy, E. P., Pierotti, M. A., Porta, G. D., and Barbacid, M., 1984, Malignant activation of a K-*ras* oncogene in lung carcinoma but not in normal tissue of the same patient, *Science* **223**:661–664.
7. Feig, L. A., Bast, R. C., Knapp, R. C., and Cooper, G. M., 1984, Somatic activation of ras^k gene in a human ovarian carcinoma, *Science* **223**:698–700.
8. Taub, R., Kirsch, I., Morton, C., Lenoir, G., Swan, D., Tronick, S., Aaronson, S., and Leder, P., 1982, Translocation of the c-*myc* gene into the immunoglobulin heavy chain locus in human Burkitt lymphoma and murine plasmacytoma cells, *Proc. Natl. Acad. Sci. USA* **79**:7837–7841.
9. de Klein, A., van Kessel, A. G., Grosveld, G., Bartram, C. R., Hagemeijer, A., Bootsma, D., Spurr, N. K., Heisterkamp, N., Groffen, J., and Stephenson, J. R., 1982, A cellular oncogene is translocated to the Philadelphia chromosome in the chronic myelogenous leukemia, *Nature* **300**:765–767.
10. Xu, Y., Richert, N. Ito, S., Merlino, G. T., and Pastan, I., 1984, Characterization of epidermal growth factor receptor gene expression in malignant and normal human cell lines, *Proc. Natl. Acad. Sci. USA* **81**:7308–7312.
11. Hendler, F. J., and Ozanne, B. W., 1984, Human squamous cell lung cancers express increased epidermal growth factor receptors, *J. Clin. Invest.* **74**:647–651.
12. Hunter, T. J., 1984, Oncogenes and proto-oncogenes: How do they differ? *J. Natl. Cancer Inst.* **73**:773–786.
13. Shih, C., Padhy, L. C., Murray, M., and Weinberg, R. A., 1981, Transforming genes of carcinomas and neuroblastomas introduced into mouse fibroblasts, *Nature* **290**:261–264.
14. Schecter, A. L., Stern, D. F., Vaidyanathan, L., Decker, S. J., Drebin, J. A., Greene, M. I., and Weinberg, R. A., 1984, The *neu* oncogene: An *erbB* related gene encoding a 185,000-M_r tumor antigen, *Nature* **312**:513–516.
15. Coussens, L., Yang-Feng, T. L., Liao, Y., Chen, E., Gray, A., McGrath, J., Seeburg, P. H., Libermann, T. A., Schlessinger, J., Franke, U., Levinson, A., and Ullrich, A., 1985, Tyrosine kinase receptor with extensive homology to EGF receptor shares chromosomal location with *neu* oncogene, *Science* **230**:1132–1139.
16. Yamamoto, T., Ikawa, S., Akiyama, T., Semba, K., Nomura, N., Miyajima, N., Saito, T., and Toyoshima, K., 1986, Similarity of protein encoded by the human c-*erbB*-2 gene to epidermal growth factor receptor, *Nature* **319**:230–234.
17. Akiyama, T., Sudo, C., Ogawara, H., Toyoshima, K., and Yamamoto, T., 1986, The product of the human c-*erbB*-2 gene: A 185-kilodalton glycoprotein with tyrosine kinase activity, *Science* **232**:1644–1646.
18. Bargmann, C., Hung, M. C., and Weinberg, R. A., 1986, Multiple independent activations of the *neu* oncogene by a point mutation altering the transmembrane domain of p185, *Cell* **45**:649–657.
19. Drebin, J. A., Link, V. C., Stern, D. F., Weinberg, R. A., and Greene, M. I., 1984, Immune responses against transforming gene associated antigens, in: *Regulation of the Immune System* (E. Sercarz, H. Cantor, and L. Chess, eds.), Liss, New York, pp. 919–928.
20. Drebin, J. A., Stern, D. F., Link, V. C., Weinberg, R. A., and Greene, M. I., 1984, Monoclonal antibodies identify a cell surface antigen associated with an activated cellular oncogene, *Nature* **312**:545–548.
21. Schreiner, G. F., and Unanue, E. R., 1977, Capping and the lymphocyte: Models for membrane reorganization, *J. Immunol.* **119**:1549–1551.
22. Baumann, H., and Doyle, D., 1980, Metabolic fate of cell surface glycoproteins during immunoglobulin-induced internalization, *Cell* **21**:897–907.
23. Boyse, E. A., Stockert, E., and Old, L. J., 1967, Modification of the antigenic structure of the cell membrane by thymus-leukemia (TL) antibody, *Proc. Natl. Acad. Sci. USA* **58**:954–959.
24. Drebin, J. A., Link, V. C., Stern, D. F., Weinberg, R. A., and Green, M. I., 1985, Down-modulation of an oncogene protein product and reversion of the transformed phenotype by monoclonal antibodies, *Cell* **41**:695–706.
25. Pollack, R., Chen, S., Powers, S., and Verderame, M., 1984, Transformation mechanisms at the cellular level, in: *Advances in Viral Oncology*, Vol. 3 (G. Klein, ed.), Raven Press, New York, pp. 3–28.
26. Drebin, J. A., Link, V. C., Weinberg, R. A., and Greene, M. I., 1987, Inhibition of tumor growth by a monoclonal antibody reactive with an oncogene-encoded tumor antigen, *Proc. Natl. Acad. Sci. USA* (in press).

27. Downward, J., Yarden, Y., Mayes, E., Scrace, G., Totty, N., Stockwell, P., Ullrich, A., Schlessinger, J., and Waterfield, M. D., 1984, Close similarity of epidermal growth factor receptor and v-*erbB* oncogene protein sequences, *Nature* **307:**521–527.
28. Roussel, M. F., Rettenmier, C. W., Look, A. T., and Sherr, C. J., 1984, Cell surface expression of v-*fms*-coded glycoproteins is required for transformation, *Mol. Cell. Biol.* **4:**1999–2009.
29. Yokata, J., Yamamoto, T., Toyoshima, K., Terada, M., Sugimura, T., Battifora, H., and Cline, M. J., 1986 Amplification of c-*erbB*-2 oncogene in human adenocarcinomas in vivo, *Lancet* **1:**765–767.

6
Relationship of the c-*fms* Protooncogene Product to the CSF-1 Receptor

CHARLES J. SHERR

The genomes of RNA tumor viruses contain viral oncogene sequences derived by recombination from protooncogenes present in all normal cells.[1] As the biochemical functions of protooncogene products are elucidated, we are beginning to formulate a better understanding of the processes that govern cell proliferation and are gaining parallel insights into how aberrant stimuli for growth predispose to malignancy. The products of at least two classes of retroviral oncogenes, including those encoding tyrosine-specific kinases (e.g., v-*src*, v-*abl*, v-*fes*) and guanine nucleotide-binding proteins (the v-*ras* genes), exert their transforming functions at the plasma membrane. These products are thought to act by emulating the functions of cell surface proteins that transduce extracellular hormonal signals. The fact that two members of the tyrosine kinase gene family (v-*erb*B and v-*fms*) encode aberrant forms of cell surface receptors for polypeptide growth factors[2,3] has underscored the possibility that critical alterations in receptor function might directly contribute to neoplasia.

1. The v-*fms* Oncogene Product

The McDonough strain of feline sarcoma virus (SM-FeSV) containing the v-*fms* oncogene[4,5] was isolated from a multicentric fibrosarcoma of a domestic cat.[6] The virus both transforms fibroblast cell lines in culture and induces fibrosarcomas when reinoculated into animals.[7] Recombination between feline leukemia virus

---

CHARLES J. SHERR • Department of Tumor Cell Biology, St. Jude Children's Research Hospital, Memphis, Tennessee 38105.

(FeLV) and c-*fms* protooncogene sequences in cat cellular DNA resulted in the insertion of v-*fms* sequences into the open reading frame of the FeLV "group-specific antigen" *(gag)* gene. The fused *gag* and v-*fms* sequences are expressed as a 180,000-dalton polyprotein[8-10] whose amino-terminal 459 amino acids are specified by *gag* sequences and whose carboxy-terminal 975 amino acids are encoded by v-*fms*.[11] The polyprotein is synthesized on membrane-bound polyribosomes, so that elongating polypeptides are translocated into the cisternae of the rough endoplasmic reticulum (ER). Translocation of nascent chains is arrested before chain termination by the translation of a hydrophobic membrane-spanning segment, located in the middle of the v-*fms*-coded portion of the molecule, that immobilizes the polypeptide in the ER membrane. The synthesis of the carboxy-terminal 406 amino acids is then completed in the cytoplasm. The full-length polypeptide is an integral transmembrane protein, oriented with its amino-terminal portion in the lumen of the ER and its carboxy-terminal domain in the cytoplasm.[12]

Proteolysis of the polyprotein during its translation liberates the amino-terminal *gag*-coded fragment[8,9] into the ER cisternae.[12] The 120,000-dalton v-*fms*-coded portion of the polyprotein (gp120v-*fms*) remains membrane-associated and is rapidly glycosylated in its amino-terminal domain.[13-15] All of the carbohydrate chains are asparagine-linked,[14] and of 11 canonical sites of carbohydrate addition predicted by the nucleotide sequence of cloned SM-FeSV DNA,[11] most appear to be glycosylated.[15] The protein is transported through the ER–Golgi complex to the plasma membrane and undergoes concomitant modification of its N-linked oligosaccharides during transit. Remodeling of carbohydrate chains from their mannose-rich form to more complex chains containing terminal fucose and sialic acid residues increases the apparent molecular mass of the glycoprotein so that the mature cell surface form is 140,000 daltons (gp140v-*fms*).[14,15] On the cell surface, gp140v-*fms* segregates into clathrin-coated pits, is internalized into endosomes, and degraded.[16] Because it is transported to the cell surface in membranous vesicles, the plasma membrane form of the glycoprotein is oriented with its glycosylated amino-terminal portion (ca. 450 amino acids) outside the cell and its carboxy-terminal domain (ca. 400 amino acids) in the cytoplasm.[12] Hence, antisera raised by inoculating live SM-FeSV-transformed rat cells into syngeneic animals contain antibodies directed exclusively to epitopes in the extracellular amino-terminal domain of gp140v-*fms*.[12,17] By fusing splenocytes from animals immunized by this procedure with rat myeloma cells, monoclonal antibodies directed to extracellular epitopes have been obtained[14] and have been used to identify and isolate SM-FeSV-transformed cells by flow cytometric techniques.[12,17]

The carboxy-terminal portion of v-*fms*-coded molecules was predicted to be related to similar regions of prototypic tyrosine kinases,[11] consistent with experimental evidence demonstrating enzymatic activity.[18] Immune complexes containing the v-*fms*-coded glycoproteins, when incubated in the presence of [γ-^{32}P]-ATP and a suitable divalent cation, exhibit an associated kinase activity that phosphorylates the glycoproteins on tyrosine. When incubated with admixed substrates like casein, the immune complexes catalyze the phosphotransfer of phosphate from ATP to heterologous proteins, although "autophosphorylation" of the v-*fms* products predominates. Similarly, the polypeptide chains of immunoglobulin coprecipitated in the complexes are not detectably phosphorylated. In this regard, the v-*fms* products differ from more promiscuous oncogene-coded tyrosine kinases of the *src* gene family that efficiently phosphorylate both immunoglobulin heavy chains and exogenous substrates such as casein, angiotensin, or denatured enolase.[19] In agreement with the results of *in vitro* enzyme assays, SM-FeSV-transformed cells metabolically labeled with [^{32}P]phosphoric acid exhibit no apparent increase in their total level of protein phosphotyrosine as compared to nontransformed control cells. Moreover, when labeled *in vivo*, the v-*fms* gene products are phosphorylated on serine residues, and tyrosine phosphorylation is barely detected.[18] These results suggest that the v-*fms* gene products are only transiently phosphorylated on tyrosine *in vivo*, and imply that tyrosine phosphatases may play an important role in regulating the extent of tyrosine phosphorylation of these molecules.

A series of SM-FeSV mutants lacking tyrosine kinase activity was found to be inactive in transformation. More interestingly, a mutant glycoprotein that retained wild-type kinase activity but was blocked in its transport to the plasma membrane was nontransforming.[17] Similarly, drugs that inhibit the maturation of *N*-linked oligosaccharide chains and thereby inhibit transport of the v-*fms*-coded glycoproteins to the cell surface reverse the transformed phenotype.[20] Thus, tyrosine kinase activity per se is insufficient for transformation, and the relevant physiologic targets of the enzyme are probably at the plasma membrane.

2. The c-*fms* Product and Its Ligand

The biochemical and topological properties of the v-*fms* gene product were reminiscent of those of the receptors for epidermal growth factor (EGF), platelet-derived growth factor (PDGF), insulin, and the insulinlike growth factors (IGFs). In particular, each of these receptors is a plasma membrane glycoprotein with an associated tyrosine-specific protein kinase activity, and each is autophosphorylated on tyrosine in response to ligand stimulation.[21–25] We reasoned that an understanding of the differentiated phenotype of cells expressing the c-*fms* protooncogene might offer some clue as to the nature of this putative receptor. When tissues from adult cats were screened by Northern blotting for the presence of c-*fms* mRNA, transcripts of about 4 kb in length were detected at high levels in spleen, and at very much lower levels in bone marrow, liver, and brain.[26] Using the immune complex kinase reaction performed with monoclonal antibodies to the v-*fms* gene products, we screened lysates of cat splenocytes for c-*fms* gene products and identified two radiolabeled glycoproteins, gp130c-*fms* and gp170c-*fms*, that were each phosphorylated on tyrosine. These two products were found to be analogous to the v-*fms* gene products, gp120v-*fms* and gp140v-*fms*, with gp130c-*fms* representing the immature precursor of the cell surface form, gp170c-*fms*.[26] The expression of the c-*fms* protooncogene in spleen cells was particularly intriguing since the tissue consists primarily of lymphoid and erythroid elements. However, when splenocytes were fractionated on Percoll gradients, we were surprised to find that all of the kinase activity associated with the c-*fms* gene products was localized to a light-density cell population corresponding to granulocytes, macrophages, and blast cells of unknown origin. Since these primarily represented inflammatory cells, cats were inoculated with intraperitoneal irritants, and cells in peritoneal exudates were harvested by lavage 4 days later and examined for c-*fms* expression. When the cells were sorted by flow cytometry based on fluorescence with monoclonal antibodies to extracellular v-*fms*-coded epitopes, the fluorescence-positive cells were identified morphologically as macrophages.[3]

These results pointed to the possibility that the c-*fms* gene encoded a receptor for a macrophage growth factor, the best candidate being the receptor for the macrophage colony-stimulating factor (CSF-1). Since the CSF-1 receptor expressed in murine macrophages had been partially characterized,[27] we attempted to precipitate mouse c-*fms* products with monoclonal antibodies reactive to v-*fms*-coded epitopes. These reagents proved to be species-restricted in their reactivity, and did not precipitate c-*fms* gene products from species other than the cat. However, antisera raised to a recombinant v-*fms*-coded polypeptide expressed in bacteria precipitated two murine c-*fms*-coded glycoproteins analogous to those identified in cat splenocytes. By several independent criteria, these molecules proved to represent differentially glycosylated forms of the mouse CSF-1 receptor. Thus, we concluded that the c-*fms* gene product was related, or possibly identical, to the CSF-1 receptor.[3]

Antisera to the recombinant v-*fms*-coded product also precipitated analogous glycoproteins from human peripheral blood monocytes, but not from granulocytes or erythrocytes. Because Müller and co-workers had previously reported that c-*fms* transcripts could be detected at high levels in placental tissues and in malignant trophoblasts,[28,29] we tested two human choriocarcinoma cell lines for c-*fms* gene products. Polypeptides biochemically and immunologically indistinguishable from those detected in peripheral blood monocytes were precipitated from lysates of human choriocarcinoma cell lines, and indeed these cells were found to specifically bind human urinary CSF-1. Thus, while CSF-1 has been defined through its action on hematopoietic cells, it may also function outside the context of hematopoiesis.

3. Transformation by the v-*fms* Gene

Because the v-*fms* and c-*fms* products were closely related in many of their biochemical properties, we tested the ability of the v-*fms* gene product to bind CSF-1. A potential difficulty in these experiments is that murine CSF-1 is species-restricted in its action and does not stimulate the proliferation or survival of feline machrophages. However, cells transformed by SM-FeSV displayed specific binding sites for murine CSF-1.[30] The affinity of binding of the murine growth factor to v-*fms* transformants was indistinguishable from that seen with feline macrophages, but was considerably lower than that detected with homologous murine macrophages. By contrast, no binding was observed using uninfected parental cell lines or cells transformed by another FeSV strain containing the oncogene v-*fes*. Chemical cross-linking of CSF-1 to its receptor on SM-FeSV-transformed cells established that CSF-1 bound specifically to gp140v-*fms*. In addition, a monoclonal antibody to a v-*fms*-coded epitope was found to interfere with CSF-1 binding. These results showed that the v-*fms* gene product, unlike the glycoprotein encoded by the v-*erb*B oncogene,[31] retains a competent ligand-binding domain so that cells infected with SM-FeSV can specifically interact with the growth factor.

To test the possibility that SM-FeSV transformation involved the transduction of a competent v-*fms*-coded receptor into cells that synthesize the ligand, we tested fibroblast cell lines susceptible to SM-FeSV transformation for CSF-1 production. Using either radioreceptor or radioimmune assays, CSF-1 production was detected by each of three mouse, rat, and mink cell lines studied.[30] This is of particular interest since SM-FeSV produces fibrosarcomas in cats.[7] Although these findings suggested that the mechanism of SM-FeSV transformation could be autocrine, the transformed cells were found to grow in the absence of exogenous CSF-1, and antibodies to the v-*fms* gene product that inhibit CSF-1 binding, or antibodies to CSF-1 itself, did not affect the transformed phenotype.[30] Because the v-*fms* gene product was derived from a cat cellular protooncogene, an effect of feline CSF-1 on the transformed phenotype cannot be excluded. Nor can we preclude the possibility of an interaction between the hormone and the oncogene-coded receptor within membranous vesicles during intracellular transport.

Given the ambiguities posed by introduction of the v-*fms* gene into CSF-1-producing fibroblasts, we infected a CSF-1-dependent mouse macrophage cell line, BAC1.2F5, with SM-FeSV. The latter cells are strictly dependent on exogenous CSF-1 for their proliferation and survival in culture and express normal CSF-1 receptors.[32] Infected cells that expressed high levels of the v-*fms* gene product at their cell surface grew in a factor-independent manner and were tumorigenic in nude mice, giving risk to histiocytic sarcomas. The transformed cells did not synthesize CSF-1 mRNA nor produce factors that supported the growth of parental BAC1.2F5 cells. Moreover, transformation by SM-FeSV affected neither the synthesis nor turnover of normal CSF-1 receptors coexpressed by the same cells.[51] Thus, v-*fms*-induced transformation by a nonautocrine mechanism, presumably due to signals mediated by its constitutive tyrosine kinase that bypassed the proximal receptor pathway.

In assays using membrane preparation, the c-*fms* gene product was active as a substrate for its associated tyrosine kinase activity, and its phosphorylation on tyrosine was enhanced in the presence of CSF-1.[3] In contrast, the v-*fms*-coded molecules appeared to act constitutively as kinases, and their phosphorylation on tyrosine was not enhanced by the ligand.[30] The simplest interpretation of these data is that certain modifications in the kinase domain of the v-*fms*-coded glycoproteins activate the enzyme and enable it to function in the absence of the growth factor. Nucleotide sequencing of c-*fms* cDNA clones predicted that the carboxylterminal end of the c-*fms* product was longer than its v-*fms* counterpart, and that its terminal 40 amino acids were replaced by 11 unrelated residues.[52] The unique c-*fms*-coded carboxylterminus includes a single tyrosine residue that was deleted from the v-*fms*-coded molecule. Recent experiments using retroviral constructs containing the c-*fms* gene have indicated that conversion of c-*fms* to a fully active oncogene (i.e., one as transforming as v-*fms*) requires two genetic alterations: (1) an "activating mutation" in the body of the gene that renders the receptor kinase activity independent of CSF-1, and (2) deletion of the unique carboxylterminal tyrosine residue which appears to function as a negative regulatory site of phosphorylation.[53] The latter experiments also indicated that expression of the human c-*fms* gene in mouse NIH-3T3 cells enabled them to form colonies in semisolid medium in response to human recombinant CSF-1. Thus, c-*fms* was formally demonstrated to encode a CSF-1 receptor, since transduction of the gene into fibroblasts was sufficient to confer a CSF-1-dependent mitogenic response.

4. The CSF-1 Gene

CSF-1 is a typical growth factor and is necessary for the proliferation, differentiation, and survival of cells of the mononuclear phagocyte lineage.[33] The murine hormone is composed of two 14,000-dalton polypeptide chains that are covalently linked through disulfide bonds; both chains are presumed to be identical in sequence, but may be variably glycosylated.[34,35] Based on its structure, CSF-1 appears to be analogous to dimeric growth factors like PDGF rather than single-chain polypeptides like EGF, suggesting that structural motifs in the ligand-binding domain of the CSF-1 and PDGF receptors may prove to be similar.

A 1.6-kb cDNA encoding biologically active human CSF-1 was recently obtained and expressed in mammalian cells.[36] The cDNA hybridizes with a simple pattern of restriction fragments in the human genome which appear to be derived from a unique chromosomal locus.[55] However, the cDNA clone anneals with a spectrum of polyadenylated RNAs in CSF-1-producing cells, ranging in size up to 4.5 kb. The existence of related genes cannot therefore be rigorously excluded. Nor is it clear that a single molecular species of the growth factor is produced. If larger forms of the hormone are synthesized and transported to the plasma membrane, some species of CSF-1 that remain cell-associated could act to stimulate responding cells by cell–cell interactions, rather than by paracrine mechanisms.

5. The c-*fms* Gene

The human c-*fms* gene has been molecularly cloned and is approximately 35 kb in length.[37,38] Because of its complexity, its precise exon–intron organization remains unknown, and only small portions of the genomic clones have been sequenced. The gene maps to the long arm of chromosome 5 at bands 5q33–34.[39,40] Deletion of this region has been associated with several hematopoietic disorders including refractory anemia, myelodysplasias, and therapy-associated myelogenous leukemia.[41,42] Recent analysis indicates that the gene coding for the granuloctye–macrophage colony-stimulating factor maps close to c-*fms*, more proximal to the centromere.[40,43] Deletions of both genes have been detected in the 5q$^-$ refractory anemia syndrome, either by Southern blot analysis of somatic cell hybrids containing the 5q$^-$ chromosome[44] or by *in situ* hybridization.[40] Thus, one or both genes may contribute to defects in hematopoiesis associated with these syndromes.

The site of recombination between FeLV and c-*fms* has been deduced from an analysis of the human c-*fms* gene. The nucleotide sequence of a 2.6-kb *Eco*RI fragment that hybridizes to the extreme 5' end of the v-*fms* gene was recently determined and was found to contain a single short open reading frame homologous to v-*fms*. The first ATG codon in the open reading frame is 102 base pairs downstream of the apparent site of recombination between FeLV and c-*fms*, and corresponds to a consensus initiator codon. Directly downstream of the ATG codon is a region coding for a hydrophobic peptide that represents the "signal peptide" for the c-*fms*-coded polypeptide.[52] These considerations suggested that recombination may have occurred in a sequence encoding the 5' untranslated region of c-*fms* RNA. The SM-FeSV-coded polyprotein is therefore predicted to contain amino acids at the *gag–fms* junction that are not ordinarily translated from c-*fms* mRNA. Since the polyprotein is cleaved at the *gag–fms* junction during translation, possibly by a "signal peptidase," the amino-termini of the viral and cellular oncogene products are identical.[51]

6. Future Perspectives

The oncogene hypothesis predicts that a restricted number of normal cellular genes, when altered or inappropriately expressed, contribute directly to cell transformation and tumorigenesis. The underlying implication is that protooncogenes play pivotal roles in regulating normal cell growth, and in some way describe a mitogenic pathway. In the case of the c-*fms* gene, it is now clear that this protooncogene encodes a receptor whose role is to regulate the growth and differentiation of cells in response to a typical polypeptide growth factor. An unanticipated conclusion, perhaps, is that the normal c-*fms* gene product appears to be highly restricted in its expression, at least in adult animals, to one of the hematopoietic lineages. This contrasts quite sharply with the expression of other protooncogenes—c-*ras* genes, for example—that are expressed in many tissues and are found to undergo mutational activation in a high proportion of tumors of different histologic types.[1,45] At issue is whether *in situ* alterations that convert c-*fms* to an active oncogene can occur only in hematopoietic cells or in other cells as well.

The simplest model is that modifications in c-*fms*-coding sequences that alter the kinase activity or the ligand-binding properties of the receptor could affect the differentiation of mononuclear phagocytes. Genetic alterations that "up-regulate" kinase activity (similar to those associated with the v-*fms* product) could potentially contribute proliferative signals in the absence of ligand, thereby leading to myeloid leukemia. Alternatively, mutations or rearrangements that lower the affinity of growth factor binding or diminish kinase activity in response to ligand stimulation could inhibit the proliferation and differentiation of mononuclear phagocytes and their committed bone marrow progenitors. In the latter case, the affected cells would not have a proliferative advantage, and should be replaced by their normal counterparts. However, stem cell defects in the c-*fms* gene might lead to significant physiologic consequences. This may be of particular importance, since monocytes and macrophages participate in a variety of host defense functions including antigen processing and presentation, phagocytosis, and secretion of cytokines that regulate other hematopoietic cells.

An alternative model is that critical alterations in regulatory control region(s) of the c-*fms* gene could result in inappropriate gene expression. If this occurred in mesenchymal cells that produce CSF-1, an autocrine loop would be created. An analogous situation would involve the untimely expression of the CSF-1 gene in macrophages. The consequences of these models are directly testable. Since both the CSF-1 and c-*fms* cDNA clones are now available, it is possible to experimentally insert these genes into different cell types and study their expression in various contexts. Analogous experiments in which the granulocyte-macrophage colony stimulating factor (GM-CSF) gene was transferred into factor-dependent myeloid cells demonstrated that the recipient cells became both factor-independent and tumorigenic.[46] Similarly, transformation of hematopoietic cells with v-*abl*, a prototypic member of the tyrosine kinase gene family, released them from factor-dependence through nonautocrine mechanisms.[47-49] Model systems of this type involving proliferating transformed macrophages[51] might prove important in testing the therapeutic efficacy of toxin immunoconjugates or of drugs targeted to the ligand-binding domain of the receptor.

Our experiments, like those from many other laboratories, focus attention on the roles of tyrosine kinases in providing membrane signals for growth. Although numerous cellular proteins are phosphorylated on tyrosine in response to transforming genes of this family,[19] the quest for physiologically relevant substrates has not proven particularly fruitful. It will be important to demonstrate that phosphorylation of certain molecules not only accompanies the transformed phenotype but is essential for transformation, and such studies are limited by the conspicuous absence of appropriate genetic systems. Of particular interest might be conditional cellular mutants that are refractory to the action of transforming genes of the tyrosine kinase gene family. At least theoretically, an analysis of such mutants and their complementation groups would pinpoint relevant targets for these enzymes, although the derivation of such cells is probably not a trivial undertaking.

The lack of cellular mutants for transformation places a premium on intuition and luck. It seems quite clear that signals for cell proliferation originating at the cell surface must institute an array of secondary events that ultimately lead to cell division. At least some rapid secondary responses have been identified, and include such phenomena as increased hexose uptake, a rapid rise in intracellular pH, and an increase in the turnover of phosphatidylinositol (PI). The cleavage of one of the intermediates of the PI cycle, PI-4,5-diphosphate, by phospholipase C generates two "second messengers" including diacylglycerol, the activator of protein kinase C, and inositol triphosphate, a mobilizer of intracellular calcium. PI turnover is increased in v-*fms* transformants, apparently due to the activation of a guanine nucleotide-dependent, membrane-associated phospholipase C that specifically hydrolyzes PI-4,5-diphosphate in the absence of added calcium.[50] The v-*fms*-coded kinase may be somehow coupled to this enzyme, possibly through a G protein, and it is attractive to speculate that tyrosine-phosphorylation provides the mechanism of activation. The lack of definitive data leaves more than adequate room for imagination, serendipity, and the occasional good experiment.

ACKNOWLEDGMENTS. The author wishes to thank both his colleagues and collaborators who have contributed to these studies and whose work is cited herein. The studies were supported in part by NIH Grant CA 38187, and by ALSAC of St. Jude Children's Research Hospital.

References

1. Bishop, J. M., 1983, Cellular oncogenes and retroviruses, *Annu. Rev. Biochem.* **52**:301–354.
2. Downward, J., Yarden, Y., Mayes, E., Scrace, G., Totty, N., Stockwell, P., Ullrich, A., Schlessinger, J., and Waterfield, M. D., 1984, Close similarity of epidermal growth factor receptor and v-*erb*B oncogene protein sequences, *Nature* **307**:521–527.
3. Sherr, C. J., Rettenmier, C. W., Sacca, R., Roussel, M. F., Look, A. T., and Stanley, E. R., 1985, The c-*fms* proto-oncogene product is related to the receptor for the mononuclear phagocyte growth factor, CSF-1, *Cell* **41**:665–676.
4. Frankel, A. E., Gilbert, J. H. Porzig, K. J., Scolnick, E. M., and Aaronson, S. A., 1979, Nature and distribution of feline sarcoma virus nucleotide sequences, *J. Virol.* **30**:821–827.
5. Donner, L., Fedele, L. A., Garon, C. F., Anderson, S., and Sherr, C. J., 1982, McDonough feline sarcoma virus: Characterization of the cloned provirus and its feline oncogene (v-*fms*), *J. Virol.* **41**:489–500.
6. McDonough, S. K., Larsen, S., Brodey, R. S., Stock, N. D., and Hardy, W. D., Jr., 1971, A transmissible feline fibrosarcoma of viral origin, *Cancer Res.* **31**:953–956.
7. Hardy, W. D., Jr., 1981, The feline sarcoma viruses, *J. Am. Hosp. Assoc.* **17**:981–996.
8. Barbacid, M., Lauver, A. V., and Devare, S. G., 1980, Biochemical and immunological characterization of polyproteins coded for by the McDonough, Gardner–Arnstein, and Snyder–Theilen strains of feline sarcoma virus, *J. Virol.* **33**:196–207.
9. Ruscetti, S. K., Turek, L. P., and Sherr, C. J., 1980, Three independent isolates of feline sarcoma virus code for three distinct *gag*-X polyproteins, *J. Virol.* **35**:259–264.
10. Van de Ven, W. J. M., Reynolds, F. H., Jr., Nalewaik, R. P., and Stephenson, J. R., 1980, Characterization of a 170,000-dalton polyprotein encoded by the McDonough strain of feline sarcoma virus, *J. Virol.* **35**:165–175.
11. Hampe, A., Gobet, M., Sherr, C. J., and Galibert, F., 1984, The nucleotide sequence of the feline retroviral oncogene v-*fms* shows unexpected homology with oncogenes encoding tyrosine-specific protein kinases, *Proc. Natl. Acad. Sci. USA* **81**:85–89.
12. Rettenmier, C. W., Roussel, M. F., Quinn, C. O., Kitchingman, G. R., Look, A. T., and Sherr, C. J., 1985, Transmembrane orientation of glycoproteins encoded by the v-*fms* oncogene, *Cell* **40**:971–981.
13. Sherr, C. J., Donner, L., Fedele, L. A., Turek, L. P., Even, J., and Ruscetti, S. K., 1980, Molecular structure and products of feline sarcoma and leukemia viruses; relationship to FOCMA expression, in: *Feline Leukemia Viruses* (W. D. Hardy, Jr., M. Essex, and A. J. McClelland, eds.), Elsevier/North-Holland, Amsterdam, pp. 293–307.
14. Anderson, S. J., Furth, M., Wolff, L., Ruscetti, S. K., and Sherr, C. J., 1982, Monoclonal antibodies to the transformation-specific glycoprotein encoded by the feline retroviral oncogene v-*fms*, *J. Virol.* **44**:696–702.
15. Anderson, S. J., Gonda, M. A., Rettenmier, C. W., and Sherr, C. J., 1984, Subcellular localization of glycoproteins encoded by the viral oncogene v-*fms*, *J. Virol.* **51**:730–741.
16. Manger, R., Najita, L., Nichols, E. J., Hakomori, S. -I., and Rohrschneider, L., 1984, Cell surface expression of the McDonough strain of feline sarcoma virus *fms* gene product (gp140*fms*), *Cell* **39**:327–337.
17. Roussel, M. F., Rettenmier, C. W., Look, A. T., and Sherr, C. J., 1984, Cell surface expression of v-*fms*-coded glycoproteins is required for transformation, *Mol. Cell. Biol.* **4**:1999–2009.
18. Barbacid, M., and Lauver, A. V., 1981, Gene products of McDonough feline sarcoma virus have an *in vitro*-associated protein kinase that phosphorylates tyrosine residues: Lack of detection of this enzymatic activity *in vivo*, *J. Virol.* **40**:812–821.

19. Hunter, T., and Cooper, J. A., 1985, Protein tyrosine kinases, *Annu. Rev. Biochem.* **54:**897–931.
20. Nichols, E. J., Manger, R., Hakomori, S., Herscovics, A., and Rohrschneider, R. L., 1985, Transformation by the v-*fms* oncogene product: Role of glycosylational processes and cell surface expression, *Mol. Cell. Biol.* **5:**3467–3475.
21. Ushiro, H., and Cohen, S., 1980, Identification of phosphotyrosine as a product of epidermal growth factor-activated protein kinase in A431 cell membranes, *J. Biol. Chem.* **255:**8363–8365.
22. Kasuga, M., Zick, Y., Blithe, D. L., Crettaz, M., and Kahn, C. R., 1982, Insulin stimulates tyrosine phosphorylation of the insulin receptor in a cell-free system, *Nature* **298:**667–669.
23. Nishimura, J., Huang, J. S., and Deuel, T. F., 1982, Platelet-derived growth factor stimulates tyrosine-specific protein kinase activity in Swiss mouse 3T3 cell membrane, *Proc. Natl. Acad. Sci. USA* **79:**4303–4307.
24. Ek, B., Westermark, B., Wasteson, A., and Heldin, C. -H., 1982, Stimulation of tyrosine-specific phosphorylation by platelet-derived growth factor, *Nature* **295:**419–420.
25. Jacobs, S., Kull, F. C., Jr., Earp, H. S., Svoboda, M. E., van Wyk, J. J., and Cuatrecasas, P., 1983, Somatomedin C stimulates the phosphorylation of the beta subunit of its own receptor, *J. Biol. Chem.* **253:**9581–9584.
26. Rettenmier, C. W., Chen, J. H., Roussel, M. F., and Sherr, C. J., 1985, The product of the c-*fms* proto-oncogene: A glycoprotein with associated tyrosine kinase activity, *Science* **228:**320–322.
27. Morgan, C. J., and Stanley, E. R., 1984, Chemical crosslinking of the mononuclear phagocyte specific growth factor CSF-1 to its receptor at the cell surface, *Biochem. Biophys. Res. Commun.* **119:**35–41.
28. Müller, R., Tremblay, J. M., Adamson, E. D., and Verma, I. M., 1983, Tissue and cell type specific expression of two human c-*onc* genes, *Nature* **304:**454–456.
29. Müller, R., Slamon, D. J., Adamson, E. D., Tremblay, J. M., Muller, D., Cline, M. J., and Verma, I. M., 1983, Transcription of c-*onc* genes c-*ras*^{ki} and c-*fms* during mouse development, *Mol. Cell. Biol.* **3:**1062–1069.
30. Sacca, R., Stanley, E. R., Sherr, C. J., and Rettenmier, C. W., 1986, Specific binding of the mononuclear phagocyte colony stimulating factor, CSF-1, to the product of the v-*fms* oncogene, *Proc. Natl. Acad. Sci. USA* **83:**3331–3335.
31. Yamamoto, T., Nishida, T., Miyajima, N., Kawai, S., Ooi, T., and Toyoshima, K., 1982, The *erb*B gene of avian erythroblastosis virus is a member of the *src* gene family, *Cell* **35:**71–78.
32. Schwartzbaum, S., Halpern, R., and Diamond, B., 1984, The generation of macrophage-like cell lines by transfection with SV40 origin defective DNA, *J. Immunol.* **132:**1158–1162.
33. Stanley, E. R., Guilbert, L. J., Tushinski, R. J., and Bartelmez, S. H., 1984, Growth factors regulating mononuclear phagocyte production, in: *Mononuclear Phagocyte Biology* (A. Volkman, ed.), Dekker, New York, pp. 373–387.
34. Stanley, E. R., and Heard, P. M., 1977, Factors regulating macrophage production and growth: Purification and some properties of the colony stimulating factor from medium conditioned by mouse L cells, *J. Biol. Chem.* **252:**4305–4312.
35. Das, S. K., and Stanley, E. R., 1982, Structure–function studies of a colony stimulating factor (CSF-1), *J. Biol. Chem.* **257:**13679–13684.
36. Kawasaki, E. S., Ladner, M. B., Wang, A. M., Van Arsdell, J., Warren, M. K., Coyne, M. Y., Schweickart, V. L., Lee, M. T., Wilson, K. J., Boosman, A., Stanley, E. R., Ralph, P., and Mark, D. F., 1985, Molecular cloning of a complementary DNA encoding human macrophage-specific colony stimulating factor (CSF-1), *Science* **230:**291–296.
37. Roussel, M. F., Sherr, C. J., Barker, P. E., and Ruddle, F. H., 1983, Molecular cloning of the c-*fms* locus and its assignment to human chromosome 5, *J. Virol.* **48:**770–773.
38. Heisterkamp, N., Groffen, J., and Stephenson, J. R., 1983, Isolation of v-*fms* and its human cellular homolog, *Virology* **126:**248–258.
39. Groffen, J., Heisterkamp, N., Spurr, N., Dana, S., Wasmuth, J. J., and Stephenson, J. R., 1983, Chromosomal localization of the human c-*fms* oncogene, *Nucleic Acids Res.* **11:**6331–6339.
40. LeBeau, M. M., Westbrook, C. A., Diaz, M. O., Larson, R. A., Rowley, J. D., Gasson, J. C., Golde, D. W., and Sherr, C. J., 1986, Evidence for the involvement of GM-CSF and c-*fms* in the deletion (5q) in myeloid disorders, *Science* **231:**984–987.

41. Wisniewski, L. P., and Hirschhorn, K., 1983, Acquired partial deletions of the long arm of chromosome 5 in hematologic disorders, *Am. J. Hematol.* **15**:295–310.
42. Van den Berghe, H., Vermaelen, K., Mecucci, C., Barbieri, D., and Tricot, G., 1985, The 5q⁻ anomaly, *Cancer Genet. Cytogenet.* **17**:189–255.
43. Huebner, K., Isobe, M., Croce, C. M., Golde, D. W., Kaufman, S. E., and Gasson, J. C., 1985, The human gene encoding GM-CSF is at 5q21–q32, the chromosome region deleted in the 5q⁻ anomaly, *Science* **230**:1282–1285.
44. Nienhuis, A. W., Bunn, H. F., Turner, P. H., Gopal, T. V., Nash, W. G., O'Brien, S. J., and Sherr, C. J., 1985, Expression of the human c-*fms* proto-oncogene in hematopoietic cells and its deletion in the 5q⁻ syndrome, *Cell* **42**:421–428.
45. Ellis, R. W., Lowy, D. R., and Scolnick, E. M., 1982, The viral and cellular p21(ras) gene family, in: *Advances in Viral Oncology*, Vol. 1 (G. Klein, ed.), Raven Press, New York, pp. 107–126.
46. Lang, R. A., Metcalf, D., Gough, N. M., Dunn, A. R., and Gonda, T. J., 1985, Expression of a hemopoietic growth factor cDNA in a factor-dependent cell line results in autonomous growth and tumorigenicity, *Cell* **43**:531–542.
47. Cook, W. D., Metcalf, D., Nicola, N. A., Burgess, A. W., and Walker, F., 1985, Malignant transformation of a growth factor-dependent myeloid cell line by Abelson virus without evidence of an autocrine mechanism, *Cell* **41**:677–683.
48. Oliff, A., Agranovsky, O., McKinney, M. D., Murty, V. V. V. S., and Bauchwitz, R., 1985, Friend murine leukemia virus-immortalized myeloid cells are converted into tumorigenic cell lines by Abelson leukemia virus, *Proc. Natl. Acad. Sci. USA* **82**:3306–3310.
49. Pierce, J. H., Di Fiore, P. P., Aaronson, S. A., Potter, M., Pumphrey, J., Scott, A., and Ihle, J. N., 1985, Neoplastic transformation of mast cells by Abelson MuLV: Abrogation of IL-3 dependence by a nonautocrine mechanism, *Cell* **41**:685–693.
50. Jackowski, S., Rettenmier, C. W., Sherr, C. J., and Rock, C. O., 1986, A guanine nucleotide-dependent phosphatidylinositol-4,5-diphosphate-specific phospholipase C in cells transformed by the v-*fms* and v-*fes* oncogenes, *J. Biol. Chem.* **261**:4978–4985.
51. Wheeler, E. F., Rettenmier, C. W., Look, A. T., and Sherr, C. J., 1986, The v-*fms* oncogene induces factor independence and tumorigenicity in a CSF-1 dependent macrophage cell line, *Nature (Lond.)* **324**:377–380.
52. Coussens, L., Van Beveren, C., Smith, D., Chen, E., Mitchell, R. L., Isacke, C., Verma, I. M., and Ullrich, A., 1986, Structural alteration of viral homologue of receptor proto-oncogene *fms* at carboxyl terminus, *Nature (Lond.)* **320**:277–281.
53. Roussel, M. F., Dull, T. J., Rettenmier, C. W., Ralph, P., Ullrich, A., and Sherr, C. J., 1987, Transforming potential of the c-*fms* proto-oncogene (CSF-1 receptor), *Nature* **325**:549–552.
54. Wheeler, E. F., Roussel, M. F., Hampe, A., Walker, M. H., Fried, V. A., Look, A. T., Rettenmier, C. W., and Sherr, C. J., 1986, The aminoterminal domain of the v-*fms* oncogene product includes a functional signal peptide that directs synthesis of a transforming glycoprotein in the absence of feline leukemia virus *gag* sequences, *J. Virol.* **59**:224–233.
55. Pattenati, M. J., Le Beau, M. M., Lemons, R. S., Shima, E. A., Kawasaki, E. S., Larson, R. A., Sherr, C. J., Diaz, M. O., and Rowley, J. D., 1987, Assignment of CSF-1 to 5q 33.1: evidence for clustering of genes regulating hematopoiesis and for their involvement in the deletion of the long arm of chromosome 5 in myeloid disorders, *Proc. Natl. Acad. Sci USA* **84**:2970–2974.

7
Two *erbB*-Related Protooncogenes Encoding Growth Factor Receptors

Tadashi Yamamoto and Kumao Toyoshima

1. Introduction

Oncogenes were first found in the genome of RNA tumor viruses and termed viral oncogenes. To date, as many as 20 viral oncogenes have been identified.[1] All these genes originated from genes carried by host cells, which were tentatively named cellular oncogenes or protooncogenes.[2] The remarkable conservation of protooncogenes over vast distances of evolutionary time[3] had led to the idea that protooncogenes play key roles in vital functions in organisms. This became apparent with the discovery that the v-*sis* gene product of simian sarcoma virus showed extensive identity with platelet-derived growth factor (PDGF)[4,5] and that the v-*erbB* gene of avian erythroblastosis virus was derived from the epidermal growth factor (EGF) receptor gene.[6-8] These findings, followed by reports of close relationships between growth-controlling proteins and oncogene products, shed light on oncogene expression and the outgrowth of tumor cells. Certain oncogenes are expressed in differentiated cells,[9-11] and their protein products may be important in cellular functions that are not directly related to cell growth.

This chapter reviews studies on *erbB*-related protooncogenes in an attempt at an understanding of one of the mechanisms of neoplastic transformation. Since little is known about the biochemical pathway of neoplastic transformation or of growth control by *erbB*-related genes, the findings reported represent only an initial step toward the final goal. We first describe the role of the v-*erbB* gene of avian

---

Tadashi Yamamoto and Kumao Toyoshima • Institute of Medical Science, University of Tokyo, Tokyo 108, Japan.

erythroblastosis virus (AEV) in cell transformation and then describe studies on cellular homologues of the v-*erbB* gene, the EGF receptor gene, and the c-*erbB*-2 gene. Finally, we refer to the oncogene/protooncogene products of the tyrosine kinase family, which are not members of the family of receptors of polypeptide growth factors.

2. The *erbB* Gene of Avian Erythroblastosis Virus: Its Transforming Ability

AEV is a representative acute leukemia virus isolated from chickens and is replication defective. A unique character of acute leukemia viruses is their specific ability to transform hematopoietic cells of different lineages.[12] The transformation of hematopoietic cells by AEV is suggested to begin with the infection of burst-forming unit cells and to become evident only when the infected cells have matured to erythroid colony-forming unit cells,[13] resulting in the induction of erythroblastosis *in vivo*. In addition, AEV occasionally induces sarcomas *in vivo* and transforms fibroblasts *in vitro*. Three independently isolated strains of AEV, strains ES4, R, and H, have been extensively characterized both virologically and molecularly (Fig. 1). Of these, AEV-ES4 and AEV-R are indistinguishable by restriction-map analysis of their cloned DNAs.[14,15] Two lines of evidence provided by molecular analysis of the three AEV strains apparently showed that the *erbB* gene is responsible for the induction of both erythroblastosis and sarcomas: (1) Both types of tumor are induced by AEV-ES4 (R), which carries both the *erbA* gene and the *erbB* gene, and by AEV-H, which carries only the *erbB* gene.[16] (2) Examination of the

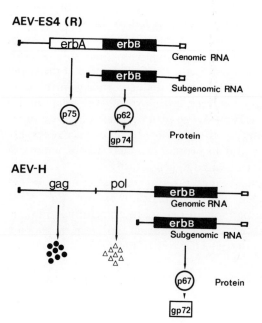

FIGURE 1. Genomic organization of AEV-ES4(R) and AEV-H. Both gp74 and gp72 of AEV-ES4(R) and AEV-H, respectively, seem to be further modified by carbohydrates.

transforming capacity of deletion mutants of the *erbA* gene and *erbB* gene in AEV-ES4, which were constructed by *in vitro* manipulation of molecularly cloned AEV-ES4 DNA, showed that *erbA⁻B⁺* mutants transformed both erythroblasts and fibroblasts in tissue culture, whereas *erbA⁺B⁻* mutants were unable to transform these cells.[17] However, leukemic cells transformed by *erbA⁻B⁺* mutants or by AEV-H frequently differentiated into mature cells, suggesting that the *erbA* gene may function in inhibiting differentiation of transformed erythroblasts. Interestingly, some oncogenes of the protein tyrosine kinase family, such as v-*src* and v-*fps*, are also reported to be capable of transforming erythroblasts *in vitro*,[18,19] although less efficiently than v-*erbB*. Again the transformed cells could differentiate, probably due to lack of v-*erbA*.

Another line of evidence for the erythroblast-transforming ability of the *erbB* gene emerged from studies on activation of the protooncogene by LTR of the avian leukosis virus (ALV).[20,21] Southern blot hybridization analysis of genomic DNA prepared from chickens with erythroblastosis due to ALV infection revealed that ALV-LTR induces transcription of a cellular counterpart of v-*erbB*.

3. The *erbB* Gene Is a Member of the *src* Gene Family

The v-*erbB* gene product of AEV-ES4 was first reported to be a 45,000-dalton protein,[22–25] but subsequently found to be a 62,000-dalton protein that becomes phosphorylated and glycosylated to yield a series of glycoproteins, $gp66^{erbB}$, $gp68^{erbB}$, $gp74^{erbB}$, and $gp82^{erbB}$.[26–29] Both $gp66^{erbB}$ and $gp68^{erbB}$ are present on the rough endoplasmic reticulum, while $gp74^{erbB}$, derived from $gp68^{erbB}$ and representing a very small fraction of the *erbB* protein ($\geq 5\%$), is located in the plasma membrane. Studies using thermosensitive mutants of AEV-ES4, which do not express $gp74^{erbB}$ at the nonpermissive temperature, suggested that expression of $gp74^{erbB}$ at the cell surface is required for cell transformation.[30]

Nucleotide sequencing showed that the *erbB* gene of AEV-H consists of 1812 nucleotides, encoding a protein with a calculated molecular mass of 67,638 daltons.[6] Consistent with this, the primary product of the *erbB* gene in AEV-H-infected chick embryo fibroblasts is a 67,000-dalton protein ($p67^{erbB}$), which is then modified to a glycoprotein of 72,000 daltons ($gp72^{erbB}$).[15] The *erbB* gene of AEV-H carries a longer c-*erbB*-derived sequence at the 3' end than that of AEV-ES4. Consequently, its *erbB* protein is larger than that of the *erbB* gene of AEV-ES4. There are also a series of glycosylated *erbB* proteins besides $gp72^{erbB}$ in cells infected with AEV-H, but these have not as yet been well characterized.

From its predicted amino acid sequence, $p67^{erbB}$ consists of three domains: an extracellular domain, a kinase domain, and an erythroblastosis domain (Figs. 2 and 3). The extracellular domain and the kinase domain are separated by a stretch of 23 hydrophobic amino acid residues, which are believed to serve for membrane anchoring. There are three possible sites of *N*-glycosidic carbohydrate linkage in the extracellular domain of the protein molecule. A sequence of 285 amino acids in the intracellular domain of $p67^{erbB}$ is significantly homologous (38% identity of amino acids) with the sequence of the carboxy-half of $pp60^{src}$, a transforming pro-

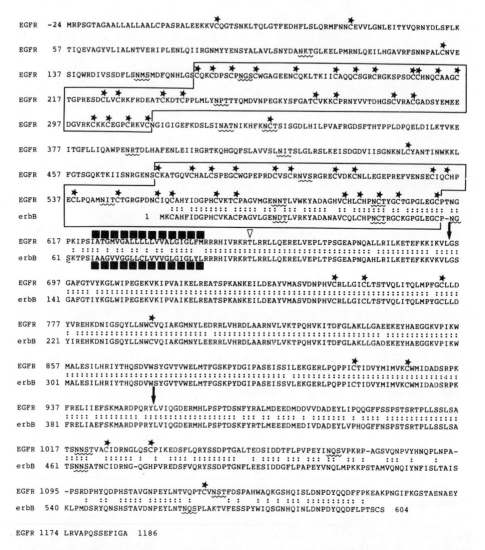

FIGURE 2. Amino acid sequences of the *erbB*-related proteins. Amino acid sequence of EGF receptor is compared with those of the *erbB* protein (A) and the c-*erbB*-2 protein (B). Identities in sequences are marked by two dots between the two lines. The predicted transmembrane regions are shown by solid squares. Stars show cysteine residues and the sequences rich in cysteine residues are boxed. Wavy lines indicate possible N-linked glycosylation sites. The major sites of threonine phosphorylation of the EGF receptor by TPA are conserved in c-*erbB*-2 and are shown by open triangles.

tein of Rous sarcoma virus, in which protein tyrosine kinase activity is encoded. The *erbB* protein itself has recently been demonstrated to exhibit protein tyrosine kinase activity.[31,32] Thus, by analogy to pp60^{src}, the transforming ability of the *erbB* protein is suggested to reside in the kinase domain.

A mutant of AEV-H, td-130, has a deletion of 169 nucleotides in the *erbB* gene

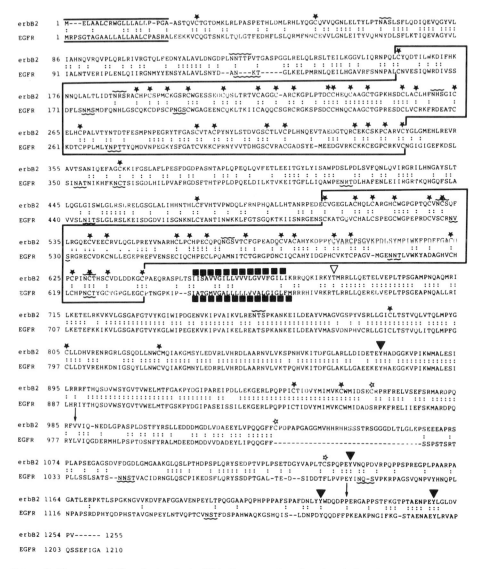

FIGURE 2. (CONTINUED) Closed triangles in (B) indicate tyrosine phosphorylation sites of the EGF receptor and pp60[src]. The kinase domain is flanked by two vertical arrows in (A). Two vertical arrows in (B) indicate the amino acid residues corresponding to the extreme carboxy-termini of pp60[src] and the v-erbB protein of AEV-H. Two horizontal bars at the amino-termini (B) show signal peptides characteristic of transmembrane protein.

and produces a truncated *erbB* protein of 42,000 daltons that does not have the carboxy one-third of the wild-type *erbB* protein.[6] This mutant does not induce erythroblastosis, but causes sarcomas in chickens, suggesting that this domain (termed the erythroblastosis domain or E domain) may be important in transformation of erythroid cells but not of fibroblasts. This idea is supported by the iso-

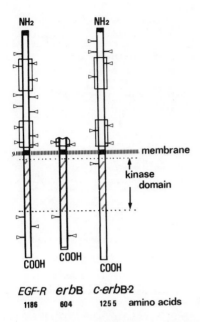

FIGURE 3. Schematic illustration of the EGF receptor, *erbB* protein, and c-*erbB*-2 protein. Cysteine-rich regions are boxed. Solid squares at the amino-termini and at the membrane show the signal peptides and transmembrane sequences, respectively. Triangles represent possible glycosylation sites.

lation of a series of AEVs that carry deletions in the erythroblastosis domain and are unable to induce erythroblastosis.[33] These viruses retain the ability to induce angiosarcomas or fibrosarcomas in chicken.

Since the *erbB* protein is a transmembrane protein and is presumed to exert its protein tyrosine kinase activity at the surface membrane, it has been thought to be related to a receptor for a ligand that induces cell growth or cell differentiation.

4. The Cellular Homologue of *erbB* Encodes the EGF Receptor

EGF, a 53-amino-acid mitogen, is one of the most thoroughly studied growth factors. By binding to its own receptor on the cell surface, this peptide triggers a cascade of intracellular events that eventually leads to initiation of the mitotic cycle.[34] The EGF receptor is a 170,000-dalton transmembrane glycoprotein associated with protein tyrosine kinase activity. The amino acid sequences of tryptic peptides generated from the EGF receptor are highly homologous to those of the *erbB* protein.[7] Direct comparison of the amino acid sequence deduced from the sequence of a cDNA clone of the EGF receptor[8] with that of the *erbB* protein[6] provided strong evidence that the gene encoding the EGF receptor is the protooncogene (c-*erbB*-1) of *erbB*. The *erbB* protein was shown to be a truncated form of the EGF receptor consisting of a short stretch of the extracellular sequence, the transmembrane sequence, and the tyrosine kinase domain but lacking most of the extracellular domain of the receptor (Figs. 2 and 3). The *erbB* protein also lacks 32 amino acid residues at the carboxy-terminus of the receptor, which are replaced by 4 amino acids translated from the *env* gene sequence of AEV-H in a wrong frame. Truncation at the carboxy-terminus does not seem to be necessary for the transforming ability (at least for erythroid cells), since the ALV-LTR-activated c-*erbB*-1 sequence, which is capable of transforming erythroid cells, codes for a truncated EGF receptor, which is similar to the v-*erbB* protein and lacks most of the extracellular domain but has no truncation at the carboxy-terminus.[35] These data suggest that transformation of cells by AEV may result from deregulated expression of a truncated EGF receptor molecule that expresses protein tyrosine kinase activity without EGF binding.

5. Another *erbB*-Related Cellular Protooncogene Distinct from the EGF Receptor Gene

The human squamous carcinoma cell line A431 expresses elevated levels of EGF receptor[36] as a consequence of amplification of the EGF receptor gene[8,37,38] (see below), as shown by Southern hybridization analysis with either *erbB* DNA or cloned EGF receptor cDNA as a probe. In this analysis, the *erbB* DNA probe, but not the EGF receptor cDNA probe, was found to hybridize under nonstringent conditions, with an *erbB*-related cellular sequence that is not amplified in A431 cells.[39] We assumed that this sequence represents a novel *erbB*-related cellular gene termed c-*erbB*-2 that is distinct from the EGF receptor gene.

To examine this possibility, we searched a human genomic library for *erbB*-related sequences using *erbB* probe under nonstringent condition and isolated six independent clones. Restriction mapping showed that all these clones except λ107 represented one gene, c-*erbB*-1. Thus, we tentatively concluded that λ107 represented the c-*erbB*-2 gene. By hybridizing both sorted chromosomes and metaphase spreads with DNA probe prepared from λ107, we mapped the c-*erbB*-2 locus to human chromosome 17 at q21.[40] This also indicates that the c-*erbB*-2 gene is distinct from the EGF receptor gene, which has been mapped to human chromosome 7.

Recently, the *neu* oncogene, which is active in a series of rat neuroblastomas, was found to be an *erbB*-related gene.[41] The human counterpart of the *neu* gene was also mapped to the same locus as the c-*erbB*-2 gene.[42] Direct comparison of the nucleotide sequences and the deduced amino acid sequences of the human c-*erbB*-2 and rat *neu* revealed marked similarity of the two genes, which suggests that they are the same gene.[43,44] Genes similar to c-*erbB*-2 have been identified in the human genome by other investigators and variously named *HER*-2 and v-*erbB*-related genes.[45,46] *HER*-2 has also been mapped to the same locus (17q21) as c-*erbB*-2. In addition, direct comparison of the nucleotide sequences of these genes revealed that they are the same gene. Thus, for the sake of simplicity, we shall continue to use the designation c-*erbB*-2.

From the analysis of genomic clones, the exon–intron structure of the c-*erbB*-2 gene was found to be identical to that of the EGF receptor gene, at least at the kinase domain[39] (M. Suzuki, personal communication). This structure, however, was different from those of the chicken and hyman c-*src* genes, which are conserved among the *src*-related genes such as c-*yes*, c-*fgr*, and *syn* as far as examined (Refs. 47, 48, and our unpublished data). In addition, recent studies showed that the splicing points of cellular counterparts of other members of the oncogenes of the tyrosine kinase family, mouse c-*abl*, human c-*raf*, and chicken c-*fps*, were different from those of the human c-*erbB*-2 gene.[49–51] An *erbB*-related gene (DER) was also identified in the genome of *Drosophila melanogaster*.[52] The DER gene contains two introns that interrupt nucleotide sequence coding for the extracellular domain. Accordingly, the c-*erbB*-2 gene and EGF receptor gene seem to be formed in the most recent split within the protooncogenes of the tyrosine kinase family.

6. Transcripts of *erbB*-Related Genes

c-*erbB*-1 mRNA was first detected in chicken embryo fibroblasts by RNA blot hybridization.[53] Two species of transcripts, 9- and 12-kb mRNAs, hybridized with the *erbB* DNA probe under conditions that did not allow cross hybridization with related transcripts. The 12-kb mRNA does not seem to be a precursor of the 9-kb mRNA, since both species are present in the cytoplasm as well as the nucleus. Two species of c-*erbB* transcripts (5.6 and 10 kb) were also detected in human RNAs.[8,37,38] Analysis of EGF receptor cDNA clones suggested that translation of the 5.6-kb mRNA resulted in production of the EGF receptor. The 5.6-kb mRNA consists of a short 5' noncoding stretch (ca. 300 nucleotides), a coding sequence of 3.6 kb, and a 3' noncoding sequence of about 1.7 kb. Although the 10-kb mRNA has not been analyzed in detail, it could be generated from the same gene as that from which 5.6-kb mRNA is synthesized by a different splicing mechanism or by carrying a longer noncoding sequence. Minor mRNA species of 13, 8.9, 6.1, and 1.8 kb were also detectable with the EGF receptor cDNA probe, giving weak hybridization signals, but their nature is unknown.[8] These RNAs might be synthesized from a yet unidentified *erbB*-related gene(s).

The transcript of another *erbB*-related gene, c-*erbB*-2, is a single species of 4.6-kb mRNA[39] similar to the 5.6-kb EGF receptor mRNA in that the coding sequence is flanked by short and long stretches of noncoding sequences at the 5' and 3' ends, respectively. The 3' noncoding sequence of the c-*erbB*-2 mRNA is a little shorter than that of the EGF receptor mRNA.

The expressions of the two *erbB*-related genes are rather cell-type specific. Neither the 10- and 5.6-kb EGF receptor mRNA nor the 4.6-kb c-*erbB*-2 mRNA was detectable by hybridization analysis of RNAs from the leukemic cell lines K562 (chronic myelogenous leukemia cells) and MT2 (adult T lymphatic leukemia cells) or a B-cell line (IM-9) established by Epstein–Barr virus infection. The two genes are transcribed at similar levels in epidermal cells, embryo fibroblasts, and placenta.

7. c-*erbB*-2 Protein

7.1. Structural Aspect

A 4480-bp-long nucleotide sequence of c-*erbB*-2 was obtained from MKN-7 cDNA clones. This sequence contains an open reading frame of 1255 amino acid codons, starting with an initiation methionine codon (ATG), which is flanked by nucleotides that match Kozak's criteria for a translation initiation site. The primary translation product of the c-*erbB*-2 gene was calculated to have a relative molecular weight of 137,895.[44]

Inspection of the predicted amino acid sequence showed that the c-*erbB*-2 protein has several of the structural features of receptors (Fig. 3). In addition, the sequence of the c-*erbB*-2 protein was found to be extremely similar to that of the EGF receptor. A sequence of 22 amino acid residues (654–675) is strongly hydrophobic and would serve as a membrane-anchoring domain. This sequence is followed immediately by basic amino acids (Lys-Arg-Arg), which would help in the correct allocation of the protein at the cell surface. The first 21 amino acid residues are also hydrophobic, suggesting that they represent a signal sequence for transfer of the protein across the endoplasmic reticulum membrane. In addition, eight possible sites of N-linked glycosylation were identified in the amino-terminal moiety of the c-*erbB*-2 protein. These data together suggest that the c-*erbB*-2 protein is a receptor for an unknown growth factor. The putative extracellular domain shows 44% homology in amino acid sequence with the ligand-binding domain of the EGF receptor. A striking similarity is the presence of two cysteine-rich regions, in which the spatial distribution of cysteine residues is virtually identical to that in the EGF receptor. Therefore, we imagine that the extracellular domains of the c-*erbB*-2 protein and EGF receptor form similar configurations and bind to structurally related ligands. However, neither EGF, nor ligands such as tumor growth factor (TGF)-β, TGF-γ, fibroblast growth factor, nerve growth factor, insulin, or PDGF bound to the c-*erbB*-2 protein.[54,55]

7.2. A Tyrosine Kinase

A sequence of 260 amino acids of the c-*erbB*-2 protein (residues 727–986) is highly homologous with the kinase domain of the EGF receptor (80% homology) and with that of retroviral oncogene products of the *src* family (25–40% homology). Thus, the c-*erbB*-2 protein seems to have tyrosine kinase activity. This possibility was examined using antibodies raised against a synthetic peptide corresponding to 14 amino acid residues at the carboxy-terminus deduced from the c-*erbB*-2 nucleotide sequence.[55] These antibodies immunoprecipitated a 185,000-dalton glycoprotein from MKN-7 human adenocarcinoma cells, in which the c-*erbB*-2 gene is amplified and overexpressed (see Section 9). Incubation of the immunoprecipitates with [γ-^{32}P]-ATP resulted in phosphorylation of the c-*erbB*-2 protein on tyrosine residues, indicating that this protein is associated with protein tyrosine kinase activity. As shown in Fig. 2, the tyrosine residues at positions 1139, 1222, and 1248 of the c-*erbB*-2 protein are at positions equivalent to those of the EGF receptor autophosphorylated by its own kinase activity.[56] Addition of EGF to the kinase reaction mixture did not enhance the phosphorylation of the 185,000-dalton protein, consistent with the observation that [^{125}I]-EGF did not bind to the c-*erbB*-2 protein at the surface of the MKN-7 cells.

8. Phosphorylation of the *erbB*-Related Proteins by C-kinase

12-O-Tetradecanoyl-phorbol-13-acetate (TPA) is reported to prevent increase of tyrosine phosphorylation induced in A431 cells by binding of EGF to its receptor.[57] Since the major receptor for TPA has been identified as the serine/threonine-specific protein kinase, C-kinase, phosphorylation of the EGF receptor by C-kinase is suggested to result in reduction of EGF receptor kinase activity. A major phosphorylation site for C-kinase in the EGF receptor was determined to reside at amino acid 645, a threonine residue in a very basic sequence of 9 residues from the cytoplasmic face of the plasma membrane.[58] Thus, the C-kinase seems to modulate signaling between the external EGF-binding domain and internal kinase domain of the EGF receptor. A threonine residue in the corresponding position of the *erbB* protein, Thr-98, is reasonably expected to be phosphorylated via C-kinase in TPA-treated cells. Since TPA inhibited growth of AEV-transformed cells,[59] phosphorylation of Thr-98 of the *erbB* protein may cause negative regulation of the outgrowth of transformed cells. It should be noted that TPA also inhibits growth of A431 cells.[60] Corresponding to Thr-654 of the EGF receptor, the c-*erbB*-2 gene product possesses a threonine residue at position 686, that is also surrounded by basic amino acid residues. We found that the c-*erbB*-2 protein of MKN-7 cells was phosphorylated at serine/threonine and that the level of this phosphorylation was elevated by TPA treatment (T. Akiyama, T. Saito, H. Ogawara, K. Toyoshima, and T. Yamamoto unpublished data). Therefore, it is likely that Thr-686 of the c-*erbB*-2 protein is phosphorylated by the action of activated C-kinase. Interestingly, treatment of MKN-7 cells with EGF also stimulated phosphorylation of the c-*erbB*-2 protein on serine/threonine residues. Tryptic phosphopeptide analysis of the c-*erbB*-2 protein from TPA- and EGF-treated cells revealed that both treatments induced phosphorylation of two unique peptides that were not observed in untreated cells. This was not due to the direct binding of EGF to the c-*erbB*-2 protein (discussed above). In A431 cells, EGF, by binding to the EGF receptor, induces enhanced Ca^{2+} influx and phosphatidylinositol turnover which would result in activation of C-kinase.[61,62] These data suggest that EGF stimulation of c-*erbB*-2 phosphorylation may occur via C-kinase, which would lead to inhibition of c-*erbB*-2 kinase activity (Fig. 4). However, we detected no effect of EGF on the immune-complex kinase activity of the c-*erbB*-2 protein. Possibly immunoprecipitation causes alteration of the kinase activity.

9. Aberrant Expression of *erbB*-Related Proteins in Cancerous Cells

There is accumulating evidence that overexpression of protooncogenes is associated with human tumors, which presumably plays an important part in the neoplastic process. Since transformation of chicken cells by AEV is apparently due to increased expression of a truncated form of the EGF receptor, qualitatively and/or quantitatively abnormal expression of the EGF receptor gene in human cells may be involved in some stage of tumorigenesis. Therefore, we examined human tumors for aberrant expression of the EGF receptor gene as well as the c-*erbB*-2 gene. We tested a total of 118 fresh samples of human malignant tumors of 25 different types for amplification of the EGF receptor and c-*erbB*-2 genes.[63,64] Hybridization analysis of high-molecular-weight DNAs from these tumors after their digestion with restriction endonucleases showed that neither the EGF receptor gene nor the c-*erbB*-2 gene was amplified or rearranged grossly in 14 sarcomas, 20 leukemias, or 4 malignant lymphomas examined. However, amplification of the

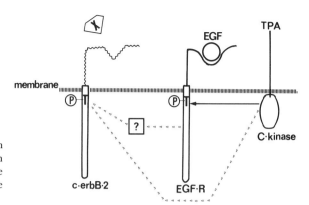

FIGURE 4. Possible phosphorylation mechanism of the c-*erbB*-2 protein in EGF- or TPA-treated cells. We assume that the factor "?" might be C-kinase.

EGF receptor gene was seen in 2 of 14 squamous cell carcinomas and in 1 of 66 adenocarcinomas. This is consistent with previous observations of other investigators that EGF receptors are expressed at high levels on the cell surface of some squamous cell carcinomas.[65] In contrast, the c-*erbB*-2 gene was amplified in 6 of 66 adenocarcinomas, but not in 14 squamous cell carcinomas. The 6 samples that showed amplification of the c-*erbB*-2 gene were from adenocarcinomas of the salivary gland (1 case), the stomach (2 cases), the breast (2 cases), and the kidney (1 case). Because amplification of the c-*erbB*-2 gene was restricted to these carcinomas, we assume that this gene may encode a growth factor receptor whose expression is associated with cell growth of glandular epithelium. An example of the analysis is shown in Table I.

We also examined cell lines established from malignant tumors for the amplification of the *erbB*-related genes. In addition to a squamous cell carcinoma cell line, A431, the EGF receptor gene was found to be amplified and overexpressed in 10 of 12 squamous cell carcinoma cell lines to various degrees (3- to 60-fold that in normal keratinocytes). The higher incidence of amplification of the EGF receptor gene in cell lines established from squamous cell carcinomas than in primary tumors suggests that malignant squamous cells with an amplified EGF receptor gene adapt more readily than those without an amplified EGF receptor gene to growth in tissue culture leading to their establishment as cell lines. On the contrary, amplification of the c-*erbB*-2 gene was observed in cell lines established from adenocarcinomas as frequently as in primary tumors; the c-*erbB*-2 gene was amplified in one gastric cancer cell line (MKN-7) among ten cell lines of adenocarcinomas tested.

The above observations suggest that aberrant expression of the EGF receptor gene tends to be associated with induction or progression of malignancy in squamous epithelium. Moreover, altered expression of the c-*erbB*-2 gene has thus far been observed only in adenocarcinomas of glandular epithelium.

10. Implications

The *erbB* gene of AEV was shown to be responsible for the induction of both erythroblastosis and sarcomas. Now that the EGF receptor gene is known to be the

TABLE I
Amplification of Cellular Oncogenes in Human Cancers[a]

Tumor type	Number of samples	Amplification		
		c-erbB-2	c-erbB-1/EGFR	c-myc
Adenocarcinoma				
Lung	3	—	—	1
Stomach	9	2	—	—
Cecum	1	—	—	—
Colon	29(7)	—	—	1
Rectum	2(1)	—	—	1 (1)
Kidney	4	1	—	—
Breast	10(2)	2 (1)	—	2
Ovary	5(4)	—	—	1 (1)
Squamous cell carcinoma				
Skin, head, and neck	7(2)	—	1	2 (1)
Esophagus	1	—	—	—
Sarcoma				
Osteogenic	2	—	—	1
Chondrosarcoma	1	—	—	—
Pleomorphic	1	—	—	—
Liposarcoma	3	—	—	—
Rhabdomyoscarcoma	1	—	—	1
Fibroxanthosarcoma	1	—	—	—
Leukemia				
Acute myelocytic	3	—	—	—
Acute lymphocytic	1	—	—	—
Chronic myelocytic	8	—	—	—
Chronic lymphocytic	6	—	—	—
Malignant lymphoma	3	—	—	—
Total	101(16)	5 (1)	1	10 (3)

[a]From Ref. 64.
[b]Numbers in parentheses are number of metastatic tumors.

protooncogene of the *erbB* gene, i.e., that the *erbB* protein is a truncated version of the EGF receptor, an important issue is being addressed. Why does the *erbB* gene induce these diseases? In human tumors, altered expression of the EGF receptor gene is often associated with squamous cell carcinomas and amplification of the second gene related to *erbB*, the c-*erbB*-2 gene, has so far been observed only in adenocarcinomas. None of these cellular genes are related to sarcomas and erythroleukemia by virtue of their unusual expression, except the EGF receptor gene, which was reported to be amplified and/or rearranged in glial tumors. In addition, we could not detect transcripts of these genes in hematopoietic cells. Important information should be obtained by examining expression of the *erbB*-related cellular gene at each stage of differentiation of hematopoietic cells of erythroid lineage. Although we have not discussed the function of v-*erbA*, it has become apparent that cellular genes that encode receptors for a series of steroid hormones,[66,67] characterized as DNA-binding proteins, are related to the v-*erbA* gene. Using antibodies that recognize *gag* sequence, a major fraction of the *gag* fused v-*erbA* protein, p75$^{gag-erbA}$, has been found in the cytoplasm of AEV-transformed cells.[68] However, a small fraction of this protein may be localized in the nucleus due to the nature of the sequence characteristic of the DNA-binding protein. Collaboration of two oncogene products, the nuclear protein and the cytoplasmic protein, is effective for cellular transformation in many cases.[69] Thus, the cooperative function of the v-*erbB* protein located at the cell surface and the v-*erbA* protein possibly located at the nucleus might be required for full transformation of erythroblasts.

As suggested by a study of ALV-LTR-induced erythroblastosis, truncation of the EGF receptor at its carboxy-terminus is not required for the induction of erythroleukemia in chickens. However, this truncation may be required to potentiate the effect of *erbB* protein in transforming fibroblasts. This possibility is supported by the findings that alteration of c-*src* at its carboxy-terminus seems to be one mechanism of converting c-*src* lacking transforming ability to viral *src* possessing the ability to transform fibroblasts.[70–75] Similar alterations have been observed in oncogene products of the tyrosine kinase family, which are closely related to *src*.

Although the precise role of altered expression of protooncogenes in carcinogenesis is still uncertain, the transforming abilities of activated protooncogenes have been tested directly by gene transfer. Among the protooncogenes related to the retroviral oncogenes, members of the *ras* family have been repeatedly identified as the active oncogenes in human tumors. Initially, only protooncogenes of the *ras* family were thought to be involved in the process of neoplastic diseases. However, extensive analyses of transforming genes from far more tumors have indicated that protooncogenes of the tyrosine kinase family are also active in tumor cells. The best example is rat *neu*, detected as an active oncogene in a series of rat neuro/glioblastomas, although we do not know whether the human version of *neu*, the c-*erbB*-2 gene, is similarly activated in some human tumors. Based on the results of many studies of *ras* activation, we believe that structural alteration, point mutation, or amplification of the c-*erbB*-2 gene plays a part in the genesis of human tumors. The possible involvement of protooncogenes of the tyrosine kinase family was also demonstrated by the identifications of the *met* oncogene in a human cell line transformed *in vitro* by treatment with *N*-methyl-*N'*-nitronitrosoguanidine[76] and of the *trk* oncogene in a carcinoma of the ascending colon.[77] Nucleotide sequence analysis showed considerable sequence homology of these genes with the tyrosine kinase family of oncogenes. In any case, the activated oncogenes of this family (*neu*, *met*, and *trk*) encode proteins that are characteristic of receptors for polypeptide growth factor. However, non-receptor-type protooncogenes of this family, such as *src*, *yes*, *syn*, *lyn*,[78] and *fgr*, which are structurally similar to each other, have not been shown to be active in human tumors. In addition, amplification of these genes in tumors seems to be rare, because we could not detect any amplification of the *yes* gene in analyses of more than 100 samples of various human tumors. As the c-*src* gene has been shown to be expressed at high levels in differentiated cells, such as neurons, these protooncogenes may not be involved in growth control of cells. This may account for the rare incidence of activation or alteration of this type of protooncogene in tumors. It is important to study the target proteins of tyrosine phosphorylation induced by tyrosine kinases of the receptor type and nonreceptor type.

Ligands binding to receptors of polypeptide growth factor have been shown to activate their own protein tyrosine kinase activity. In addition, in some cases agonist–receptor interaction has been reported to activate turnover of phosphatidylinositol, which in turn activates C-kinase and changes the intracellular level of calcium. As a consequence, many pleiotropic changes are induced and cells begin to proliferate. Possibly, lipid kinases are activated by phosphorylation of tyrosine residues by the receptor-kinase, resulting in induction of phosphatidylinositol turnover.

A number of antitumor monoclonal antibodies have been developed for treatment of malignancies. Successful application of this method for a wide variety of tumors is anticipated. As described above, amplification of the c-erbB-2 gene is frequent in adenocarcinomas found in several human organs. Our preliminary data showed that this amplification is associated with overexpression of the c-erbB-2 protein at the cell surface. In NIH 3T3 cells transformed by the neu oncogene, a rat counterpart of the c-erbB-2 gene, down-regulation of gp185^{neu} by the monoclonal antibodies against this protein was shown to cause reversion of the transformed phenotype.[79] Thus, continued expression of neu-encoded gp185 at the cell surface seems to be essential for malignancy. Assuming that overexpression of c-erbB-2 proteins at the surface of adenocarcinomas is important for tumorigenesis and that these proteins function in maintaining the malignant phenotype, immunotherapy of these adenocarcinomas using monoclonal antibodies against c-erbB-2 protein should be an effective novel approach. This treatment should also be useful in therapy of malignancies in which the growth factor receptor is overexpressed.

References

1. Bishop, J. M., 1983, Cellular oncogenes and retroviruses, *Annu. Rev. Biochem.* **52**:301–354.
2. Land, H., Parada, L. F., and Weinberg, R. A., 1983, Cellular oncogenes and multistep carcinogenesis, *Science* **222**:771–778.
3. Bishop, J. M., and Varmus, H. E., 1985, Functions and origins of retroviral transforming genes, in: *RNA Tumor Viruses-2* (R. Weiss, N. Teich, H. Varmus, and J. Coffin, eds.), Cold Spring Harbor Laboratory, Cold Spring Harbor, N.Y., pp. 294–356.
4. Waterfield, M. D., Scrace, T., Whittle, N., Stroobant, P., Johnsson, A., Wasteson, A., Westermark, B., Heldin, C.-H., Huang, J. S., and Deuel, T. F., 1983, Platelet-derived growth factor is structurally related to the putative transforming protein p28^{sis} of simian sarcoma virus, *Nature* **304**:35–39.
5. Doolittle, R. F., Hunkapiller, M. W., Hood, L. E., Devare, S. G., Robbins, K. C., Aaronson, S. A., and Antoniades, H. N., 1983, Simian sarcoma virus oncogene, v-sis, is derived from the gene (or genes) encoding a platelet-derived growth factor, *Science* **221**:275–277.
6. Yamamoto, T., Nishida, N., Kawai, S., Ooi, T., and Toyoshima, K., 1983, The erbB gene of avian erythroblastosis virus is a member of the src gene family, *Cell* **35**:71–78.
7. Downward, J., Yarden, Y., Mayes, E., Scrace, G., Totty, N., Stockwell, P., Ullrich, A., Schlessinger, J., and Waterfield, M. D., 1984, Close similarity of epidermal growth factor receptor and v-erbB oncogene protein sequences, *Nature* **307**:521–527.
8. Ullrich, A., Coussens, L., Hayflick, J. S., Dull, T. J., Gray, A., Tam, A. W., Lee, J., Yarden, Y., Libermann, T. A., Schlessinger, J., Downward, J., Mayes, E. L. V., Whittle, N., Waterfield, M. D., and Seeburg, P. H., 1984, Human epidermal growth factor receptor cDNA sequence and aberrant expression of the amplified gene in A431 epidermal carcinoma cells, *Nature* **309**:418–425.

9. Brugghe, J. S., Cotton, A. E., Barrett, J. N., Nonner, D., and Keane, R. W., 1985, Neurones express high levels of a structurally modified, activated form of pp60^{c-src}, *Nature* **316:**554–557.
10. Simon, M. A., Drees, B., Kornberg, T., and Bishop, J. M., 1985, The nucleotide sequence and the tissue-specific expression of Drosophila c-*src*, *Cell* **42:**831–840.
11. Golden, A., Nemeth, S. P., and Brugghe, J. S., 1986, Blood platelets express high levels of the pp60^{c-src}-specific tyrosine kinase activity, *Proc. Natl. Acad. Sci. USA* **83:**852–856.
12. Graf, T., and Beug, H., 1978, Avian leukemia viruses: Interactions with their target cells *in vitro* and *in vivo*, *Biochim. Biophys. Acta* **516:**269–299.
13. Samarut, J., and Gazzolo, L., 1982, Target cells infected by avian erythroblastosis virus differentiate and become transformed, *Cell* **28:**921–929.
14. Vennstrom, B., Fanshier, L., Moscovici, C., and Bishop, J. M., 1980, Molecular cloning of avian erythroblastosis virus genome and recovery of oncogenic virus by transfection of chicken cells, *J. Virol.* **36:**575–585.
15. Nishida, T., Sakamoto, S., Yamamoto, T., Hayman, M., Kawai, S., and Toyoshima, K., 1984, Comparison of genome structure among three different strains of avian erythroblastosis virus, *Gann* **75:**325–333.
16. Yamamoto, T., Hihara, H., Nishida, T., Kawai, S., and Toyoshima, K., 1983, A new avian erythro blastosis virus, AEV-H, carries *erbB* gene responsible for the induction of both erythroblastosis and sarcomas, *Cell* **34:**225–232.
17. Frykberg, L., Palmieri, S., Beug, H., Graf, T., Hayman, M. J., and Vennstrom, B., 1983, Transforming capacities of avian erythroblastosis virus mutants deleted in the *erbA* or *erbB* oncogenes, *Cell* **32:**227–283.
18. Pierce, J. H., Aaronson, S. A., and Anderson, S., 1984, Hematopoietic cell transformation by a murine recombinant retrovirus containing the *src* gene of Rous sarcoma virus, *Proc. Natl. Acad. Sci. USA* **81:**2374–2378.
19. Kahn, P., Adkins, B., Beug, H., and Graf, T., 1984, *src*- and *fps*-containing avian sarcoma viruses transform chicken erythroid cells, *Proc. Natl. Acad. Sci. USA* **81:**7122–7126.
20. Fung, Y. K. T., Lewis, W. G., Kung, H.-J., and Crittenden, L. B., 1983, Activation of the cellular oncogene c-*erbB* by LTR insertion: Molecular basis for induction of erythroblastosis by avian leukosis virus, *Cell* **33:**357–368.
21. Raines, M. A., Lewis, W. G., Crittenden, L. B., and Kung, H.-J., 1985, c-*erbB* activation in avian leukosis virus-induced erythroblastosis: Clustered integration sites and the arrangement of provirus in the c-*erbB* alleles, *Proc. Natl. Acad. Sci. USA* **82:**2287–2291.
22. Anderson, S. M., Hayward, W. S., Neel, B. G., and Hanafusa, H., 1980, Avian erythroblastosis virus produces two mRNA's, *J. Virol.* **36:**676–683.
23. Lai, M. M. C., Neil, J. C., and Vogt, P. K., 1980, Cell free translation of avian erythroblastosis virus RNA yields two specific and distinct proteins with molecular weights of 75,000 and 40,000, *Virology* **100:**475–483.
24. Pawson, T., and Martin, G. S., 1980, Cell-free translation of avian erythroblastosis virus RNA, *J. Virol.* **34:**280–284.
25. Yoshida, M., and Toyoshima, K., 1980, *In vitro* translation of avian erythroblastosis virus RNA: Identification of two major polypeptides, *Virology* **100:**484–487.
26. Privalsky, M. L., Sealy, L., Bishop, J. M., McGrath, J. P., and Levinson, A. D., 1983, The product of the avian erythroblastosis virus *erbB* locus is a glycoprotein, *Cell* **32:**1257–1267.
27. Hayman, M. J., Ramsay, G. M., Savin, K., Kitchener, G., Graf, T., and Beug, H., 1983, Identification and characterization of the avian erythroblastosis virus *erb-B* gene product as membrane glycoprotein, *Cell* **32:**579–588.
28. Hayman, M. J., and Beug, H., 1984, Identification of a form of the avian erythroblastosis virus *erb-B* gene product at the cell surface, *Nature* **309:**460–462.
29. Decker, S. J., 1985, Phosphorylation of the *erbB* gene product from an avian erythroblastosis virus-transformed chick fibroblast cell line, *J. Biol. Chem.* **260:**2003–2006.
30. Beug, H., and Hayman, M. J., 1984, Temperature-sensitive mutants of avian erythroblastosis virus: Surface expression of the *erbB* product correlates with transformation, *Cell* **36:**963–972.
31. Gilmore, T., Declue, J. E., and Martin, G. S., 1985, Protein phosphorylation at tyrosine is induced by the v-*erbB* gene product *in vivo* and *in vitro*, *Cell* **40:**609–618.

32. Kris, R. M., Lax, I., Gullick, W., Waterfield, M. D., Ullrich, A., Fridkin, M., and Schlessinger, J., 1985, Antibodies against a synthetic peptide as a probe for the kinase activity of the avian EGF receptor and v-*erbB* protein, *Cell* **40**:619–625.
33. Tracy, S. E., Woda, B. A., and Robinson, H. L., 1985, Induction of angiosarcoma by a c-*erbB* transducing virus, *J. Virol.* **54**:304–310.
34. Janes, R., and Bradshaw, R. A., 1984, Polypeptide growth factors, *Annu. Rev. Biochem.* **53**:259–292.
35. Nilsen, T. W., Maroney, P. A., Goodwin, R. G., Rottman, F. M., Crittenden, L. B., Raines, M. A., and Kung, H.-J., 1985, c-*erbB* activation in ALV-induced erythroblastosis: Novel RNA processing and promoter insertion result in expression of an amino-truncated EGF receptor, *Cell* **41**:719–726.
36. Fablicant, R. N., DeLarco, J. E., and Todaro, G. J., 1977, Nerve growth factor receptor on human melanoma cells in culture, *Proc. Natl. Acad. Sci. USA* **74**:565–569.
37. Merlino, G. T., Ishii, S., Xu, Y.-H., Clark, A. J. L., Semba, K., Toyoshima, K., Yamamoto, T., and Pastan, I., 1984, Amplification and enhanced expression of the epidermal growth factor receptor gene in A431 human carcinoma cells, *Science* **224**:417–419.
38. Lin, C. R., Chen, W. S., Kruiger, W., Stolarsky, L. S., Weber, W., Evans, R. M., Verma, I. M., Gill, G. N., and Rosenfeld, M. G., 1984, Expression cloning of human EGF receptor complementary DNA: Gene amplification and three related messenger RNA products in A431 cells, *Science* **224**:843–848.
39. Semba, K., Kamata, N., Toyoshima, K., and Yamamoto, T., 1985, A v-*erbB*-related protooncogene, c-*erbB*-2, is distinct from the c-erbB-1/epidermal growth factor-receptor gene and is amplified in a human salivary gland adenocarcinoma, *Proc. Natl. Acad. Sci. USA* **82**:6497–6501.
40. Fukushige, S., Matsubara, K., Yoshida, M., Sasaki, M., Suzuki, T., Semba, K., Toyoshima, K., and Yamamoto, T., 1986, Localization of a novel v-*erbB*-related gene, c-*erbB*-2, on human chromosome 17 and its amplification in a gastric cancer cell line, *Mol. Cell. Biol.* **6**:955–958.
41. Schechter, A. L., Stern, D. F., Vaidyanathan, L., Decker, S. J., Drebin, J. A., Greene, M. I., and Weinberg, R. A., 1984, The *neu* oncogene: An *erbB*-related gene encoding a 185,000-M_r tumor antigen, *Nature* **312**:513–516.
42. Schechter, A. L., Hung, M.-C., Vaidyanathan, L., Weinberg, R. A., Yang-Feng, T. L., Francke, U., Ullrich, A., and Coussens, L., 1985, The *neu* gene: An *erbB*-homologous gene distinct from and unlinked to the gene encoding the EGF receptor, *Science* **229**:976–978.
43. Bargmann, C. I., Hung, M.-C., and Weinberg, R. A., 1986, The *neu* oncogene encodes an epidermal growth factor receptor-related protein, *Nature* **319**:226–230.
44. Yamamoto, T., Ikawa, S., Akiyama, T., Semba, K., Nomura, N., Miyajima, N., Saito, T., and Toyoshima, K., 1986, Similarity of protein encoded by the human c-*erbB*-2 gene to epidermal growth factor receptor, *Nature* **319**:230–234.
45. Coussens, L., Yang-Feng, T. L., Liao, Y.-C., Chen, E., Gray, A., McGrath, J., Seeburg, P. H., Libermann, T. A., Schlessinger, J., Francke, U., Levison, A., and Ullrich, A., 1985, Tyrosine kinase receptor with extensive homology to EGF receptor shares chromosomal location with *neu* oncogene, *Science* **230**:1132–1139.
46. King, C. R., Kraus, M. H., and Anderson, S. A., 1985, Amplification of a novel v-*erbB*-related gene in a human mammary carcinoma, *Science* **229**:974–976.
47. Nishizawa, M., Semba, K., Yoshida, M. C., Yamamoto, T., Sasaki, M., and Toyoshima, K., 1986, Structure, expression, and chromosomal location of the human c-*fgr* gene, *Mol. Cell. Biol.* **6**:511–517.
48. Semba, K., Nishizawa, M., Miyajima, N., Yoshida, M. C., Sukegawa, J., Yamanashi, Y., Sasaki, M., Yamamoto, T., and Toyoshima, K., 1986, *yes*-Related protooncogene, *syn*, belongs to the protein-tyrosine kinase family, *Proc. Natl. Acad. Sci. USA* **83**:5459–5463.
49. Wang, J. Y. J., Ledley, F., Goff, S., Lee, R., Groner, Y., and Baltimore, D., 1984, The mouse c-*abl* locus: Molecular cloning and characterization, *Cell* **36**:349–356.
50. Bonner, T. I., Kerby, S. B., Sutrave, P., Gunnell, M. A., Mark, G., and Rapp, U. R., 1985, Structure and biological activity of human homologs of the *raf/mil* oncogene, *Mol. Cell. Biol.* **5**:1400–1407.
51. Huang, C.-C., Hammond, C., and Bishop, J. M., 1985, Nucleotide sequence and topography of chicken c-*fps:* Genesis of a retroviral oncogene encoding a tyrosine-specific protein kinase, *J. Mol. Biol.* **181**:175–186.

52. Livneh, E., Glazer, L., Segal, D., Schlessinger, J., and Shilo, B.-Z., 1985, The Drosophila EGF receptor gene homolog: Conservation of both hormone binding and kinase domain, *Cell* **40**:599–607.
53. Vennstrom, B., and Bishop, J. M., 1982, Isolation and characterization of chicken DNA homologous to the two putative oncogenes of avian erythroblastosis virus, *Cell* **28**:135–143.
54. Stern, D. F., Hefferman, P. A., and Weinberg, R. A., 1986, p185, a product of the *neu* protooncogene, is a receptor like protein associated with tyrosine kinase activity, *Mol. Cell. Biol.* **6**:1729–1740.
55. Akiyama, T., Sudo, C., Ogawara, H., Toyoshima, K., and Yamamoto, T., 1986, The product of the human c-*erbB*-2 gene: A 185,000 dalton glycoprotein with tyrosine kinase activity, *Science* **232**:1644–1646.
56. Downward, J., Parker, P., and Waterfield, M. D., 1985, Autophosphorylation sites on the epidermal growth factor receptor, *Nature* **311**:483–485.
57. Cochet, C., Gill, G. N., Meisengelder, J., Cooper, J. A., and Hunter, T., 1984, C kinase phosphorylates the epidermal growth factor receptor and reduces its epidermal growth factor stimulated tyrosine protein kinase activity, *J. Biol. Chem.* **259**:2553–2558.
58. Hunter, T., Ling, N., and Cooper, J. A., 1984, Protein kinase C phosphorylation of the EGF receptor at a threonine residue close to a cytoplasmic face of the plasma membrane, *Nature* **311**:480–483.
59. Decker, S. J., 1985, Phosphorylation of the *erbB* gene product from an avian erythroblastosis virus-transformed chick fibroblast cell line, *J. Biol. Chem.* **260**:2003–2006.
60. Decker, S., 1984, Effects of epidermal growth factor and 12-*O*-tetradecanoylphorbol-13-acetate on metabolism of the epidermal growth factor receptor in normal human fibroblasts, *Mol. Cell. Biol.* **4**:1718–1723.
61. Sauyer, S. T., and Cohen, S., 1981, Enhancement of calcium uptake and phosphatidylinositol turnover by epidermal growth factor in A431 cells, *Biochemistry* **20**:6280–6286.
62. Moolenaar, W. H., Aberts, R. J., Tertoolen, L. G. J., and de Laat, S. W., 1986, The epidermal growth factor-induced calcium signal in A431 cells, *J. Biol. Chem.* **261**:279–284.
63. Yamamoto, T., Kamata, N., Kawano, H., Shimizu, S., Kuroki, T., Toyoshima, K., Rikimaru, K., Nomura, N., Ishizaki, R., Pastan, I., Gamou, S., and Shimizu, N., 1986, High incidence of amplification of the epidermal growth factor receptor gene in human squamous carcinoma cell lines, *Cancer Res.* **46**:414–416.
64. Yokota, J., Yamamoto, T., Toyoshima, K., Terada, M., Sugimura, T., Battifora, H., and Cline, M. J., 1986, Amplification of c-*erbB*-2 oncogene in human adenocarcinomas *in vivo*, *Lancet* **2**:765–767.
65. Cowley, O., Smith, J. A., Gusherson, B., Hendler, F., and Ozanne, B., 1984, The amount of EGF receptors is elevated on squamous cell carcinomas, in: *Cancer Cells*, Vol. I (A. T. Levine, G. F. Vande Woude, W. C. Topp, and J. D. Watson, eds.), Cold Spring Harbor Laboratory, Cold Spring Harbor, N.Y., pp. 5–10.
66. Weinberger, C., Hollenberg, S. M., Rosenfeld, M. G., and Evans, R. M., 1985, Domain structure of human glucocorticoid receptor and its relationship to the v-*erbA* oncogene product, *Nature* **318**:670–672.
67. Green, S., Walter, P., Kumar, P., Krust, A., Borment, J.-M., Argos, P., and Chambon, P., 1986, Human aestrogen receptor cDNA: Sequence, expression and homology to v-*erbA*, *Nature* **320**:134–139.
68. Bunte, T., Greiser-Wilke, I., and Moelling, K., 1982, Association of *gag-myc* proteins from avian myelocytomatosis virus wild-type and mutants with chromatin, *EMBO J.* **1**:919–928.
69. Weinberg, R. A., 1986, The action of oncogene in the cytoplasm and nucleus, *Science* **230**:770–776.
70. Iba, H., Cross, F. R., Garber, E. A., and Hanafusa, H., 1985, Low level of cellular protein phosphorylation by nontransforming overproduced p60^{c-src}, *Mol. Cell. Biol.* **5**:1058–1066.
71. Iba, H., Takeya, T., Cross, F. R., Hanafusa, T., and Hanafusa, H., 1984, Rous sarcoma virus variants which carry the cellular *src* gene instead of the viral *src* gene cannot transform chicken embryo fibroblasts, *Proc. Natl. Acad. Sci. USA* **81**:4424–4429.
72. Parker, R. C., Varmus, H. E., and Bishop, J. M., 1984, Expression of v-*src* and chicken c-*src* in rat cells demonstrates qualitative differences between pp60^{c-src}, *Cell* **37**:131–139.

73. Takeya, T., and Hanafusa, H., 1982, DNA sequence of the viral and cellular *src* gene of chickens. II. Comparison of the *src* genes of two strains of avian sarcoma virus and of the cellular homolog, *J. Virol.* **44**:12–18.
74. Ikawa, S., Hagino-Yamanashi, K., Kawai, S., Yamamoto, T., and Toyoshima, K., 1986, Activation of the cellular *src* gene by transducing retrovirus, *Mol. Cell. Biol.* **6**:2420–2428.
75. Cooper, J. A., Gould, K. L., Cartwright, C. A., and Hunter, T., 1986, Tyr-527 is phosphorylated in pp60^{c-src}: Implications for regulation, *Science* **231**:1431–1434.
76. Dean, M., Park, M., Le Beau, M. M., Robin, T. S., Diaz, M. D., Rouley, J. D., Blair, D. G., and Vande Woude, G. F., 1985, The human *met* oncogene is related to the tyrosine kinase oncogenes, *Nature* **318**:385–388.
77. Martin-Zanca, D., Hughes, S. H., and Barbacid, M., 1986, A human oncogene formed by the fusion of truncated tropomyosin and protein tyrosine kinase sequence, *Nature* **319**:743–748.
78. Yamanashi, Y., Fukushige, S., Semba, K., Sukegawa, J., Miyajima, N., Matsubara, K., Yamamoto, T., and Toyoshima, K., 1987, The *yes*-related cellular gene *lyn* encodes a possible tyrosine kinase similar to p56^{ck}, *Mol. Cell Biol.* **7**:237–243.
79. Drebin, J. A., Link, V. C., Stern, D. F., Weinberg, R. A., and Greene, M. I., 1985, Down-modulation of an oncogene protein product and reversion of the transformed phenotype by monoclonal antibodies, *Cell* **41**:695–706.

8

The Receptor for Epidermal Growth Factor

WENDELYN H. INMAN AND GRAHAM CARPENTER

1. Introduction

Epidermal growth factor (EGF) is a well-characterized growth factor that was initially isolated from mouse submaxillary glands.[1-3] The low-molecular-weight growth factor (M_r 6000) can be isolated as a component of a high-molecular-weight complex with an associated arginine esterase activity.[4,5] EGF has been found in all mammals and, in humans, is a component of nearly all body fluids.[6,7] A variety of biological responses are mediated by EGF *in vivo* and *in vitro*, in a diverse number of cells including epidermal and epithelial cells, which are particularly responsive in the intact animal.[7] It is important to note that almost 90% of all malignancies are derived from epithelial cells.

Specific receptors in the cell membrane of target cells interact with EGF to initiate and maintain a complex series of biochemical and morphological events leading to cell growth and multiplication *in vivo*.[8-10] The receptors for EGF have been detected in a number of different cell types.[11-14] Embryonic tissues have also been found that express the EGF receptor, implying that the EGF receptor has a functional role in embryogenesis.[11,15]

The growth factor's, perhaps direct, involvement in neoplastic transformation is implicated by the discovery of an EGF-like molecule, termed transforming growth factor, type α, or TGF-α.[16] TGF-α is characterized by its capacity to bind to the EGF receptor and to trigger an EGF-like set of responses in the cell, including activation of the EGF receptor-associated tyrosine kinase.[17,18] Interestingly, TGF-α (or EGF) in the presence of another growth factor, TGF-β,[19,20] can reversibly transform normal indicator cells in culture.[21,22] In this instance, both TGF-β

WENDELYN H. INMAN AND GRAHAM CARPENTER • Departments of Medicine and Biochemistry, Vanderbilt University School of Medicine, Nashville, Tennessee 37232.

and TGF-α (or EGF) are required for the maintenance of the transformed phenotype *in vitro*. Other evidence for the involvement of components of the mechanisms of action of EGF in the transformation of cells is the high degree of structural homology between the product of the viral oncogene v-*erbB* and the cytoplasmic portion of the EGF receptor.[23] It is very likely that the gene for the EGF receptor represents the protooncogene for v-*erbB*.[24]

EGF is a single polypeptide which consists of 53 amino acid residues containing three intramolecular disulfide bonds that are essential for biological activity.[25] It is interesting to note that the greatest degree of sequence conservation among EGF-related growth factors, TGF-α and vaccinia growth factor (VGF), is the retention and location of these three disulfide linkages.[26,27] Mature TGF-α, which was initially identified by its ability to compete with [^{125}I]-EGF,[28,29] mature EGF, and mature VGF are all derived from large precursor molecules, and have relatively little sequence homology (approximately 22%) as mature growth factors, even though they are ligands for the same receptor molecule.[30,31]

2. EGF Receptors

2.1. Isolation

The EGF receptor has been purified, from A-431 cells, to near homogeneity by the use of EGF affinity chromatography.[32] A-431 cells have been used almost exclusively for structural studies of the EGF receptor as this cell line overexpresses the EGF receptor by a factor of 20- to 50-fold.[59] There is no reason to suspect that the information gained from the study of the EGF receptors in A-431 cells is particularly unusual; however, certain details such as the structure of oligosaccharide side chains can be expected to differ in other cell types. In the initial studies, the purified receptor from A-431 cells had an apparent molecular weight of 150,000; however, subsequent purification methods that eliminated calcium from the assay buffers indicated that the actual size of the receptor was 170,000.[33] Most cell homogenates contain a calcium-activated protease that cleaves the native receptor molecule to the lower-molecular-weight species.[34-36] Interestingly, isolation of the EGF receptor resulted in the copurification of a growth factor-sensitive tyrosine kinase activity.[37] Previous studies had shown that addition of EGF to membranes prepared from A-431 cells resulted in the activation of a protein tyrosine kinase.[66] The purification results suggested that the ligand-binding site and the kinase activity might be physically coupled. This was consistent with studies showing that antibodies to the 170,000-dalton form of the receptor were able to coprecipitate the kinase activity.[28] Also, treatment of the EGF receptor in membrane vesicles with a radiolabeled ATP analogue which forms a covalent bond to proteins at their ATP-binding site, demonstrated the presence of such a binding site on the 170,000-dalton receptor.[38,39] Importantly, following mild heating or *N*-ethylmaleimide exposure the tyrosine kinase was inactivated and subsequent incubation with the labeled ATP analogue failed to label the EGF receptor. These studies led to the unusual conclusion, at that time, that the EGF receptor molecule was a single polypeptide containing both a ligand-binding site and growth factor-sensitive tyrosine kinase activity.

2.2. Characterization

The sequence analysis of tryptic peptides derived from the EGF receptor[20] led to cDNA cloning of the receptor and to the deducted amino acid sequence of the entire receptor molecule[40] (Fig. 1). These studies have provided, in addition to the complete amino acid sequence, an indication of the relationship between ligand-binding and tyrosine kinase domains within the EGF receptor. The receptor precursor contains a signal peptide at the N-terminus of 24 residues. Based on amino acid sequence analysis, the mature receptor, excluding oligosaccharide chains, has a molecular weight of 131,600 and is composed of a single polypeptide chain of 1186 amino acid residues that is divided into two domains by one hydrophobic membrane-spanning sequence.

2.2.1. Extracellular Domain

The extracellular domain of the EGF receptor contains 622 amino acid residues, as determined from sequence data. The region must fold in such a manner that high-affinity ligand binding is possible. The extracellular domain is characterized by two structural features: a high cysteine content (approximately 10%) dispersed in two clusters and a relatively large number of canonical sequences for N-linked glycosylation. The extracellular cysteine residues are most likely in the form of disulfides, but the actual content of disulfides in the receptor has not been quantitated. Due to the cysteine content in this region of the receptor, it is likely that thermodynamically stable regions exist in the extracellular domain. It is not known if these regions participate in the actual ligand-binding site.

The ligand-binding domain of the EGF receptor is also characterized by a relatively high level of carbohydrate. Direct[41] and indirect[42,43] studies do not detect the presence of O-linked carbohydrate. Therefore, the carbohydrate which modifies the EGF receptor (approximately 30,000 daltons) is present only as N-linked oligosaccharide chains. It is estimated that 10 to 11 of the 12 potential N-linked oligosaccharide sites are actually occupied by oligosaccharide chains.[42,44] Con A lectin chromatography and endoglycosidase H sensitivity of the mature receptor indicate that the N-linked oligosaccharides consist mostly of complex-type chains and approximately three high-mannose chains.[42-44]

2.2.2. Cytoplasmic Domain

The cytoplasmic portion of the EGF receptor has been intensively studied since the tyrosine kinase activity that is encoded in this region of the molecule is considered to be the primary effector system in the transmembrane signaling process. A similar tyrosine kinase activity is present in several other growth factor receptors and in several oncogene products.[21] There is a large region (250 amino

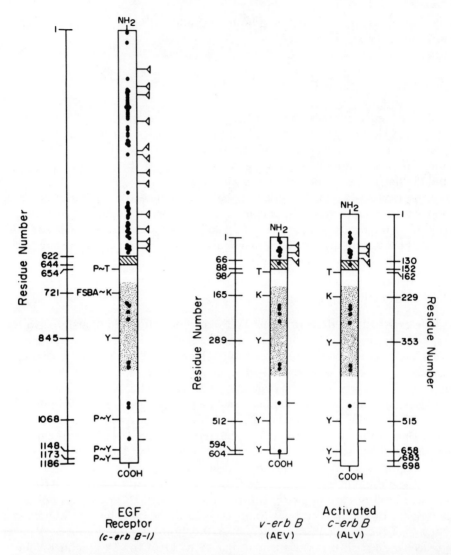

FIGURE 1. Comparison of the structural features of the EGF receptor and related oncogene products. The following symbols apply: ———◁, sites of N-linked glycosylation; ———, canonical sequences for N-linked glycosylation; ●, cysteine residues; Y, tyrosine residues; P∼Y, phosphotyrosine residues; T, threonine residues; P∼T, phosphothreonine residues; K, lysine residues; FSBA∼K, lysine covalently labeled with p-fluorosulfonylbenzoyl adenosine; crosshatched area, membrane-spanning region; stippled area, sequences similar to src kinase; AEV, avian erythroblastosis virus; ALV, avian leukosis virus. Data from Refs. 40, 45, 53, 88–90.

acid residues) in the cytoplasmic domain that shares sequence homology with other members of the tyrosine kinase family (the src kinase of the Rous sarcoma virus being considered the prototype). This region, indicated by stippling in Fig. 1, is considered to constitute the catalytic domain (tyrosine kinases) of the receptor. It contains the binding site for ATP, at lysine residue 721 as determined by labeling with ATP analogues which covalently bind to the receptor.[45] Twenty-five residues (residues 695–700) proximal to Lys-721 are a group of residues (Gly-X-Gly-X-X-Gly), which are highly conserved in nucleotide-binding proteins.[46] Studies in which the lysine residue is replaced in the v-*fps* tyrosine kinase with either a glycine or an arginine residue show that the molecule is inactive as a tyrosine kinase and as a transforming gene product.[47] When the corresponding lysine (Lys-295) was substituted with methionine in v-*src*, similar results were obtained.[48]

The carboxy-terminal region of the EGF receptor contains sequences which are not located in other tyrosine kinases. The C-terminal portion of the receptor is thought to be important because it contains the major sites of autophosphorylation. These sites have been determined by sequencing phosphotyrosine-containing tryptic peptides of the EGF receptor from A-431 cells that had been labeled with ^{32}P in the presence of EGF.[49] The autophosphorylation sites correspond to tyrosine residues 1173, 1148, and 1068. Apparently, the residue at site 1173 is the major site of EGF-induced autophosphorylation of the receptor. Another site of tyrosine phosphorylation has been labeled but its position has not been determined. The autophosphorylation sites that are produced *in vitro* correspond to the sites of autophosphorylation *in vivo*.

Tryptic digestion of the EGF receptor results in the recovery of a 42,000-dalton fragment that is capable of growth factor-independent tyrosine kinase activity.[50] Protease digestion generates similar-sized catalytically active fragments from viral oncogene tyrosine kinases.[51,52]

Immediately after the membrane-spanning domain of the EGF receptor, at the cytoplasmic interface, is a 13-residue sequence which is highly enriched in basic amino acids. This sequence likely functions as a "stop transfer" sequence following the insertion of the amino-terminal portion of the EGF receptor polypeptide chain into the lumen of the endoplasmic reticulum. This sequence, which is a common feature in other growth factor receptors, is probably important for the correct translation and membrane insertion of the receptor. However, it has not been determined whether or not this sequence is important for the functioning of mature receptors. The potential functional role of this sequence in the mature receptor is of interest because it contains the site of phosphorylation by protein kinase C (Thr-654), which is in the midst of these basic residues.[53,54]

2.3. Cell Physiology

Human fibroblasts contain between 40,000 and 100,000 specific EGF receptors on the cell surface.[57,58] The human epidermoid carcinoma cell line, A-431, has an exaggerated number of EGF-binding sites, nearly 3 million, per cell.[59–61] It is not clear what role the expression of such a large number of EGF receptors plays in metabolism in A-431 cells. However, overexpression of the EGF receptor has now been reported for a substantial number of carcinoma cell lines and tissues. The initiation of cellular responses by EGF is a direct result of the growth factor's binding to specific membrane receptors.

The interaction of EGF with specific, saturable, receptors has been almost exclusively examined by using ^{125}I-labeled EGF. By this method specific receptors for [^{125}I]-EGF have been detected in a variety of cell types.[4,55] With the exception of hematopoietic cells, nearly all cell types have been reported to display some level of [^{125}I]-EGF binding capacity.[5] In studies characterizing EGF receptor interactions in human fibroblasts, it was noted that the cell-bound growth factor became less susceptible to protease action and less accessible to antibodies against EGF as the cells were incubated at 37°C.[56] Also, it was shown that most of the inaccessible growth factor was endocytosed at 37°C and was rapidly degraded to mono [^{125}I]iodotyrosine. Because this labeled amino acid is not retained by the cell, growth factor binding is characterized by a decrease in cell-associated radioactivity with time, which eventually reaches a level close to 20% of the maximal amount bound. Fibroblasts incubated at 37°C in the presence of inhibitors of lysosomal function, however, had no apparent loss of radioactivity or degradation of the growth factor. At 4°C all the radiolabeled EGF remained associated with the cell surface, was not endocytosed, and there was no appreciable loss of radioactivity, indicating no degradation of the growth factor.

The fate of the EGF–receptor complex during cellular processing of EGF has been studied extensively. Preexposure of cells to unlabeled EGF resulted in a dramatic decrease in the number of surface receptors for labeled EGF, suggesting receptor "down-regulation."[56] Following the binding of EGF to its receptor, the endocytosis of the entire complex prevents any further interaction between growth factor outside the cell and the receptor. After the growth factor binds to the receptor, the receptors aggregate and become localized into coated pit regions on the cell surface.[62] Substances, such as dansylcadaverine, which can inhibit receptor clustering, also prevent internalization of the EGF–receptor complex by cells.[63] Internalization of aggregated EGF–receptor complexes may be required for the biological activities of the growth factor, but this point is not clear.

The rates of turnover and metabolic processing of the EGF receptor have been investigated in the presence and absence of EGF. This was done to determine whether the internalized EGF receptor was recycled back to the cell surface or whether the receptor, like the growth factor, was subject to intracellular degradation. By metabolically labeling the receptor in intact cells with [^{35}S]methionine and performing a chase with unlabeled methionine, the half-life of the receptor in the cell was determined.[64] The levels of prelabeled protein present at various times during the "chase" period reflects the physical turnover of the receptor. In human fibroblasts, the half-life of the receptor is 10 hr when EGF is not added during the chase period. If, however, EGF is present during the chase, the half-life of the receptor is shortened considerably to 1 hr. The same is true for other cell types. For example, the half-life of the EGF receptor in A-431 cells drops from 20 hr in the absence of EGF to 7 hr when EGF is added.[65] The EGF-induced acceleration of receptor turnover is ligand specific, and can be inhibited by reagents which inhibit lysosomal function. Attempts to detect prelysosomal processing of the receptor have not been successful. However, two laboratories report that the endosome-derived EGF receptor is a more effective tyrosine kinase toward exogenous substrates than is the plasma membrane-derived receptor.[67,68]

2.4. Mechanisms of Action

Efforts are under way to determine the mechanism of EGF regulation and its modulation of cellular responses. The binding of the growth factor to its receptor causes a rapid phosphorylation of several cellular proteins, including the receptor itself.[66,69–73] In addition, the receptor stimulates the phosphorylation of several synthetic and exogenous proteins *in vitro*.[74,75] A major substrate for the EGF receptor is a 36,000-dalton protein that has amino acid sequence homology with a family of proteins, which possess phospholipase A_2 inhibitory activity, termed lipocortins.[76] Phosphorylation of similar molecules by EGF receptors *in vivo* is speculated to be important in the mediation of the enzymatic activities of such regulatory proteins. Another 81,000-dalton cellular protein phosphorylated by the EGF receptor has been identified as ezrin, a cellular structural protein.[77] It is reasonable to guess that the phosphorylation of structural components by the EGF receptor may be important in the mediation of a number of cellular events effected by EGF, such as membrane ruffling, transport activities, and cell division, among others.

Normally, the phosphorylation of tyrosine residues is a rare event. It is generally believed that tyrosine phosphorylation is infrequent enough to constitute the major signaling event directed by the EGF receptor. Other growth factor receptors such as platelet-derived growth factor, insulin, and insulinlike growth factor-I also have detectable levels of tyrosine kinase activity.[78–85] The fact that other growth factor receptors and a large number of oncogene products have tyrosine kinase activity strengthens the argument that tyrosine kinases have a principal role in growth factor-regulated activities.

It has not been determined whether the mitogenic effects of EGF are dependent on the phosphorylation of specific substrates by the receptor as well as the autophosphorylation of the receptor itself. Phosphorylation of the EGF receptor by protein kinase C, specifically on Thr-654, results in a direct reduction of the EGF receptor's affinity for its ligand and diminishes the autophosphorylation activity of the receptor.[24] The mechanism(s) by which these modulations occurs is not clear, particularly in view of the synergistic activities of C-kinase activators (TPA) and EGF.[87,86]

How the EGF receptor communicates with other cellular processes to regulate and control cell functions is not clear. However, it is clear that a considerable amount of effort is still required to unravel the particular details of EGF-mediated events in the intact cell.

ACKNOWLEDGMENTS. The authors thank Drs. R. Gates and L. King for reading the manuscript and Mrs. S. Heaver for typing. W.H.I. was supported by NIH Training Grant 5T32 AM07491. G.C. was supported by NCI Grant CA24071.

References

1. Cohen, S., 1962, Isolation of a mouse submaxillary gland protein accelerating incisor eruption and eyelid opening in the newborn animal, *J. Biol. Chem.* **237**:1555–1562.

2. Cohen, S., and Elliott, G. A., 1963, The stimulation of epidermal keratinization by a protein isolated from the submaxillary glands of the mouse, *J. Invest. Dermatol.* **40:**1–5.
3. Taylor, J. M., Mitchell, W. M., and Cohen, S., 1974, Characterization of the high molecular weight form of epidermal growth factor, *J. Biol. Chem.* **249:**3198–3203.
4. Cohen, S., 1965, The stimulation of epidermal proliferation by a specific protein, *Dev. Biol.* **12:**394–407.
5. Carpenter, G., and Cohen, S., 1979, Epidermal growth factor, *Annu. Rev. Biochem.* **48:**193–216.
6. Carpenter, G., 1985, Epidermal growth factor: Biology and receptor metabolism, *J. Cell Sci. Suppl.* **3:**1–9.
7. Carpenter, G., Goodman, L., and Shaver, L., 1986, The physiology of epidermal growth factor, in: *Oncogenes and Growth Control* (P. Kahn and T. Graf, eds.), Springer-Verlag, Berlin, pp. 65–69.
8. Carpenter, G., Lembach, K. J., Morrison, M. M., and Cohen, S., 1975, Characterization of the binding of ^{125}I-labeled epidermal growth factor to human fibroblasts, *J. Biol. Chem.* **250:**4297–4304.
9. Das, M., Miyakawa, T., Fox, C. F., Pruss, R. M., Aharonov, A., and Herschman, H. R., 1977, Specific radiolabeling of a cell surface receptor for epidermal growth factor, *Proc. Natl. Acad. Sci. USA* **74:**2790–2794.
10. Carpenter, G., and Cohen, S., 1978, Biological and molecular studies of the mitogenic effects of human epidermal growth factor, in: *Molecular Control of Proliferation and Differentiation* (J. Papaconstantinou and W. J. Rutter, eds.), Academic Press, New York, pp. 13–31.
11. Gospodarowicz, D., 1981, Epidermal and nerve growth factors in mammalian development, *Annu. Rev. Physiol.* **43:**251–263.
12. O'Keefe, E., Hollenberg, M. D., and Cuatrecasas, P., 1974, Epidermal growth factor: Characteristics of specific binding in membranes from liver, placenta, and other target tissue, *Arch. Biochem. Biophys.* **164:**518–526.
13. Carpenter, G., 1980, Epidermal growth factor, *Hand. Exp. Pharmacol.* **57:**89–132.
14. Das, M., 1982, Epidermal growth factor: Mechanism of action, *Int. Rev. Cytol.* **78:**233–256.
15. Nexo, E., Hollenberg, M. D., Figueroa, A., and Pratt, R. M., 1980, Detection of epidermal growth factor–urogastrone and its receptor during fetal mouse development, *Proc. Natl. Acad. Sci. USA* **77:**2782–2785.
16. Marquardt, H., Hunkapillar, M. W., Hood, L. E., Twardzik, D. R., DeLarco, J. E., Stephenson, J. R., and Todaro, G. J., 1983, Transforming growth factors produced by retro virus transformed rodent fibroblasts and human melanoma cells: Amino acid sequence homology with epidermal growth factor, *Proc. Natl. Acad. Sci. USA* **80:**4684–4688.
17. Roberts, A. B., Frolik, C. A., Anzano, M. A., and Sporn, M. B., 1983, Transforming growth factors from neoplastic and non neoplastic tissues, *Fed. Proc. Red. Am. Soc. Exp. Biol.* **42:**103–108.
18. Rhodes, J. A., Tam, J. P., Finke, U., Saunders, M., Bernanke, J., Silen, W., and Murphy, R. A., 1986, Transforming growth factor inhibits secretion of gastric acid, *Proc. Natl. Acad, Sci. USA* **83:**3844–3846.
19. Childs, C. B., Proper, J. A., Tucker, R. F., and Moses, H. L., 1982, Serum contains a platelet-derived transforming growth factor, *Proc. Natl. Acad. Sci. USA* **79:**5312–5316.
20. Roberts, A. B., Anzano, M. A., Lamb, L. C., Smith, J. M., and Sporn, M. B., 1981, New class of transforming growth factors potentiated by epidermal growth factor: Isolation from non-neoplastic tissues, *Proc. Natl. Acad. Sci. USA* **78:**5339–5343.
21. Anzano, M. A., Roberts, A. B., Lamb, L. C., Smith, J. M., and Sporn, M. B., 1982, Purification by reversed-phase high performance liquid chromatography of an epidermal growth factor-dependent transforming growth factor, *Anal. Biochem.* **125:**217–224.
22. Anzano, M. A., Roberts, A. B., Meyers, C. A., Komoliya, A., Lamb, L. C., Smith, J. M., and Sporn, M. B., 1982, Synergistic interaction of two classes of transforming growth factors from murine sarcoma cells, *Cancer Res.* **42:**4776–4778.
23. Downward, J., Yarden, I., Mayes, E., Scrace, G., Totty, N., Stockwell, P., Ullrich, A., Schlessinger, J., and Waterfield, M. D., 1984, Close similarity of epidermal growth factor receptor and v-erb B oncogene protein sequence, *Nature* **307:**521–527.
24. Hunter, T., and Cooper, J. H., 1985, Protein-tyrosine kinases, *Annu. Rev. Biochem.* **54:**897–930.
25. Savage, C. R., Jr., Hash, J. H., and Cohen, S., 1973, Epidermal growth factor: Location of disulfide bonds, *J. Biol. Chem.* **248:**7669–7672.

26. Strobant, P., Rice, A. P., Gullick, W. J., Cheng, D. J., Kerr, I. M., and Waterfield, M. D., 1985, Purification and characterization of vaccinia virus growth factor, *Cell* **42**:383–393.
27. Brown, J. P., Twardzik, D. R., Marquardt, H., and Todaro, G. J., 1985, Vaccinia virus encodes a polypeptide homologous to epidermal growth factor and transforming growth factor, *Nature* **313**:491–492.
28. DeLarco, J. E., and Todaro, G. J., 1978, Growth factors from murine sarcoma virus-transformed cells, *Proc. Natl. Acad. Sci. USA* **75**:4001–4005.
29. Todaro, G. J., Fryling, C., and DeLarco, J. E., 1980, Transforming growth factors produced by certain human tumor cells: Polypeptides that interact with epidermal growth factor receptor, *Proc. Natl. Acad. Sci. USA* **77**:5258–5262.
30. Carpenter, G., and Zendegui, J. G., 1986, Epidermal growth factor, its receptor, and related proteins, *Exp. Cell Res.* **164**:1–10.
31. Hirata, Y., and Orth, D. N., 1979, Conversion of high molecular weight human epidermal growth factor (hEGF)/urogastrone (UG) to small molecular weight hEGF/UG by mouse EGF-associated arginine esterase, *J. Clin. Endocrinol. Metab.* **49**:481–483.
32. Cohen, S., Carpenter, G., and King, L. E., Jr., 1980, Epidermal Growth factor-receptor protein kinase interactions, *J. Biol. Chem.* **255**:4834–4842.
33. Cohen, S., Ushiro, H., Stoscheck, C., and Chinkers, M., 1982, A native 170,000 epidermal growth factor receptor kinase complex from plasma membrane vesicles, *J. Biol. Chem.* **257**:1523–1531.
34. Cassel, D., and Glaser, L., 1982, Proteolytic cleavage of epidermal growth factor receptor, *J. Biol. Chem.* **257**:9845–9848.
35. Gates, R. E., and King, L. E., Jr., 1982, Calcium facilitates endogenous proteolysis of the EGF receptor-kinase, *Mol. Cell. Endocrinol.* **27**:263–276.
36. Gates, R. E., and King, L., 1979, Proteolysis of the epidermal growth factor receptor by endogenous calcium activated neutral protease from rat liver, *Biochem. Biophys. Res. Commun.* **113**:255–261.
37. Carpenter, G., King, L., and Cohen, S., 1979, Rapid enhancement of protein phosphorylation in A-431 cell membrane preparations by epidermal growth factor, *J. Biol. Chem.* **254**:4884–4891.
38. Buhrow, S. A., Cohen, S., and Staros, J. V., 1982, Affinity labeling of the protein kinases associated with the EGF receptor in membrane residues, *J. Biol. Chem.* **257**:4019–4022.
39. Buhrow, S. A., Cohen, S., Garbers, D. L., and Staros, J. V., 1983, Characterization of the interaction of 5'-p-fluorosulfonylbenzoyl adenosine with epidermal growth factor/protein kinase in A-431 cell membranes, *J. Biol. Chem.* **258**:7824–7827.
40. Ullrich, A., Coussens, L., Hayflick, J. S., Dull, T. J., Gray, A., Tarn, A. W., Lee, J., Yarden, Y., Libermann, T. A., Schlessinger, J., Downward, J., Mayes, E. L. V., Whittle, N., Waterfield, M. D., and Seeburg, P. H., 1984, Human epidermal growth factor receptor cDNA sequence and aberrant expression on the amplified gene in A-431 epidermoid carcinoma cells, *Nature* **309**:418–425.
41. Soderquist, A. M., and Carpenter, G., 1984, Glycosylation of the epidermal growth factor receptor in A-431 cell, *J. Biol. Chem.* **259**:12586–12594.
42. Cummings, R. D., Soderquist, A. M., and Carpenter, G., 1985, The oligosaccharide moieties of the epidermal growth factor receptor in A-431 cells: Presence of both high mannose N-linked chains and complex that contain terminal N-acetyl galactosamine residues, *J. Biol. Chem.* **260**:11944–11952.
43. Childs, R. A., Gregoriou, M., Scudder, P., Thorpe, S. J., Rees, A. R., and Feizi, T., 1984, Blood group-active carbohydrate chains on the receptor for epidermal growth factor of A-431 cells, *EMBO J.* **3**:2227–2233.
44. Mayes, E. L. V., and Waterfield, M. D., 1984, Biosynthesis of the epidermal growth factor receptor in A-431 cells, *EMBO J.* **3**:531–537.
45. Russo, M. W., Lukas, T. J., Cohen, S., and Staros, J. V., 1985, Identification of residues in the nucleotide binding site of the epidermal growth factor receptor/kinase, *J. Biol. Chem.* **269**:5205–5208.
46. Wierenga, R., and Hol, W., 1983, Predicted nucleotide binding properties of p21 protein and its cancer-associated variant, *Nature* **302**:842–844.
47. Wienmaster, G., Zoller, M. J., and Pawson, T., 1986, A lysine in the ATP binding site of p130gag-fps is essential for protein-tyrosine kinase activity, *EMBO J.* **5**:69–76.
48. Snyder, M. A., Bishop, M., McGrath, J. P., and Levison, A. D., 1985, A mutation at the ATP binding

site of pp60 v-src abolishes kinase activity, transformation, and tumorigenicity, *Mol. Cell. Biol.* **5:**1772–1779.
49. Downward, J., Parker, P., and Waterfield, M. D., 1984, Autophosphorylation sites on the epidermal growth factor receptor, *Nature* **311:**483–485.
50. Basu, M., Biswas, R., and Das, M., 1984, 42,000-molecular weight EGF receptor has protein kinase activity, *Nature* **311:**477–480
51. Levison, A. D., Courtneidge, S. A., and Bishop, J. M., 1981, Structural and functional domains of the Rous sarcoma virus transforming protein (pp609src), *Proc. Natl. Acad. Sci. USA* **78:**1624–1628.
52. Brugge, J. S., and Darrow, D., 1984, Analysis of the catalytic domain of phosphotransferase activity of two avian sarcoma virus transforming proteins, *J. Biol. Chem.* **259:**4550–4557.
53. Hunter, T., Ling, N., and Cooper, J. A., 1984, Protein kinase C phosphorylation of the EGF receptor at a threonine residue close to the cytoplasmic face of the plasma membrane, *Nature* **311:**480–483.
54. Davis, R. J., and Czech, M. P., 1985, Tumor promoting phorbol diesters cause the phosphorylation of epidermal growth factor receptors in normal human fibroblasts at threonine-654, *Proc. Natl. Acad. Sci. USA* **82:**1974–1978.
55. Comens, P. G., Simmer, R. L., and Baker, J. B., 1982, Direct linkage of ^{125}I-EGF to cell surface receptors, *J. Biol. Chem* **257:**42–45.
56. Carpenter, G., and Cohen, S., 1976, ^{125}I-labeled human epidermal growth factor: Binding and internalization of epidermal growth factor in human carcinoma cells A-431, *J. Cell Biol.* **71:**159–171.
57. Hollenberg, M. D., and Cuatrecasas, P., 1975, Insulin and epidermal growth factor: Human fibroblast receptors related to deoxyribonucleic acid synthesis and amino acid uptake, *J. Biol. Chem.* **250:**3845–3853.
58. Carpenter, G., Lembach, K. J., Morrison, M. M., and Cohen, S., 1975, Characterization of the binding of ^{125}I-labeled epidermal growth factor to human fibroblasts, *J. Biol. Chem.* **250:**4297–4304
59. Haigler, H., Ash, J. F., Singer, S. J., and Cohen, S., 1978, Visualization by fluorescence of the binding and internalization of epidermal growth factor in human carcinoma cells A-431, *Proc. Natl. Acad. Sci. USA* **75:**3317–3321.
60. Fabricant, R. N., DeLarco, J. E., and Todaro, G. J., 1977, Nerve growth factor receptors on human melanoma cells in culture, *Proc. Natl. Acad. Sci. USA* **75:**565–569.
61. Stoscheck, C. M., and Carpenter, G., 1983, Biology of the A-431 cell: A useful organism for hormone research, *J. Cell. Biochem.* **23:**191–202.
62. Gorden, P., Carpentier, J., Cohen, S., and Orci, L., 1978, Epidermal growth factor: Morphological demonstration of binding, internalization, and lysosomal association in human fibroblasts, *Proc. Natl. Acad. Sci. USA* **75:**5025–5029.
63. Haigler, H. T., Maxfield, F. R., Willingham, M. C., and Pastan, I., 1980, Dansylcadaverine inhibits internalization of ^{125}I-epidermal growth factor in Balb 3T3 cells, *J. Biol. Chem.* **255:**1239–1241.
64. Stoscheck, C. M., and Carpenter, G., 1984, Down regulation of epidermal growth factor receptor: Direct demonstration of receptor degradation in human fibroblasts, *J. Cell Biol.* **98:**1048–1053.
65. Stoscheck, C. M., and Carpenter, G., 1984, Characterization of the metabolic turnover of epidermal growth factor receptor protein in A-431 cells, *J. Cell. Physiol.* **120:**296–302.
66. Carpenter, G., King, L., and Cohen, S., 1978, Epidermal growth factor stimulates phosphorylation in membrane preparations *in vitro, Nature* **276:**409–410.
67. Cohen, S., and Fava, R. A., 1985, Internalization of functional epidermal growth factor: Receptor/kinase complexes in A-431 cells, *J. Biol. Chem.* **260:**12351–12358.
68. Kay, D. G., Wei, H. L., Uchihashi, M., Khan, M. M., Posner, B. I., and Bergeron, J. J. M., 1986, Epidermal growth factor receptor kinase translocation and activation *in vivo, J. Biol. Chem.* **261:**8473–8480.
69. Sawyer, S. T., and Cohen, S., 1985, Epidermal growth factor stimulates the phosphorylation of the calcium-dependent 35,000-dalton substrate in intact A-431 cells, *J. Biol. Chem.* **260:**8233–8236.
70. King, L. E., Jr., Carpenter, G., and Cohen, S., 1980, Characterization by electrophoresis of epidermal growth factor stimulated phosphorylation using A-431 membranes, *Biochemistry* **19:**1524–1528.
71. Cooper, J. A., Bowen-Pope, D., Raines, E., Ross, R., and Hunter, T., 1982, Similar effects of plate-

let-derived growth factor and epidermal growth factor on the phosphorylation of tyrosine in cellular proteins, *Cell* **31**:263–275.
72. Valentine-Brown, K. A., Northrup, J. K., and Hollenberg, M. D., 1986, Epidermal growth factor (urogastrone)-mediated phosphorylation of a 35-kDa substrate in human placenta membranes: Relationship to the β subunit of the guanine nucleotide regulatory complex, *Proc. Natl. Acad. Sci. USA* **83**:236–240.
73. Gates, R. E., and King, L. E., Jr., 1982, The EGF receptor-kinase has multiple phosphorylation sites, *Biochem. Biophys. Res. Commun.* **105**:57–66.
74. Cassel, D., Pike, L. J., Grant, G. A., Krebs, E. G., and Glaser, L., 1983, Interaction of epidermal growth factor dependent protein kinase with endogenous membrane proteins and soluble peptide substrate, *J. Biol. Chem.* **258**:2945–2950.
75. Zendegui, J. G., and Carpenter, G., 1984, Substrates of the epidermal growth factor receptor-kinase, *Cell Biol. Int. Rep.* **8**:619–633.
76. Pepinsky, R. B., and Sinclair, L. K., 1986, Epidermal growth factor dependent phosphorylation of lipocortin, *Nature* **321**:81–84.
77. Gould, K. L., Cooper J. A., Bretscher, A., and Hunter, T., 1986, The protein tyrosine kinase substrate, p81, is homologous to a chicken microvillar core protein, *J. Cell Biol.* **102**:660–669.
78. Ek, B., Westermark, B., Wasteson, A., and Heldin, C. H., 1982, Stimulation of tyrosine-specific phosphorylation by platelet-derived growth factor, *Nature* **295**:419–420.
79. Nishimura, J., Huang, J. S., and Deuel, T. F., 1982, Platelet-derived growth factor stimulates tyrosine-specific protein kinase activity in Swiss mouse 3T3 cell membranes, *Proc. Natl. Acad. Sci. USA* **79**:4303–4307.
80. Kasuga, M., Zick, Y., Blith, D. L., Karlsson, J. A., Karing, H. U., and Kahn, C. R., 1982, Insulin stimulation of phosphorylation of the β subunit of the insulin receptor: Formation of both phosphoserine and phosphotyrosine, *J. Biol. Chem.* **257**:9891–9894.
81. Petruzzelli, L. M., Ganguly, S., Smith, C. J., Cobb, M. H., Rubin, C. S., and Rosen, O. M., 1982, Insulin activates a tyrosine-specific protein kinase in extracts of 3T3-L1 adipocytes and human placenta, *Proc. Natl. Acad. Sci. USA* **79**:6792–6796.
82. Aruch, J., Nemenoff, R. A., Blackshear, P. J., Pierce, M. W., and Osathanondh, 1982, Insulin-stimulated tyrosine phosphorylation of the insulin receptor in detergent extracts of human placenta membranes, *J. Biol. Chem.* **257**:15162–15166.
83. Shia, M. A., and Pilch, P.F., 1983, The β subunit of the insulin receptor is an insulin-activated protein kinase, *Biochemistry* **22**:717–721.
84. Jacobs, S., Keel, F. C., Earp, H. S., Svoboda, M. E., Van Wyk, J. J., and Cuatrecasas, P., 1983, Somatomedin-C stimulates the phosphorylation of the beta-subunit of its own receptor, *J. Biol. Chem.* **258**:9581–9584.
85. Rubin, J. B., Shea, M. A., and Pilch, P. F., 1983, Stimulation of tyrosine-specific phosphorylation *in vitro* by insulin-like growth factor I, *Nature* **305**:438–440.
86. Matrisian, L., Bowden G. T., and Magun, B. E., 1981, Mechanism of synergistic induction of DNA synthesis by epidermal growth factor and tumor promoters, *J. Cell. Physiol.* **108**:417–425.
87. Dicker, P., and Rozengurt, E., 1980, Phorbol esters and vasopressin stimulate DNA synthesis by a common mechanism, *Nature* **287**:607–612.
88. Yamamoto, T., Nishida, T., Miyajima, N., Kawai, S., Ooi, T., and Toyoshima, K., 1983, The erb B gene of avian erythroblastosis virus is a member of the src gene family, *Cell* **35**:71–78.
89. Nilsen, T. W., Maroney, P. A., Goodwin, R. G., Rottman, F. M., Cruttender, L. B., Raines, M. A., and Kung, H., 1985, c-erb B activation in ALV-induced erythroblastosis: Novel RNA processing and promoter insertion result in expression of an amino-truncated EGF receptor, *Cell* **47**:719–726.
90. Downward, J., Parker, D., and Waterfield, M. D., 1984, Autophosphorylation sites on the epidermal growth factor receptor, *Nature* **311**:483–485.

9

The Role of the *abl* Gene in Transformation

NAOMI ROSENBERG

1. Introduction

The *onc* gene *abl* is a member of the tyrosine kinase family of oncogenes and was first isolated in a murine retrovirus, Abelson leukemia virus (A-MuLV).[1] This virus has been of interest as one of a number of transforming viruses that encode tyrosine protein kinases. In addition, because of its ability to transform both fibroblasts and hematopoietic cells, A-MuLV has been a useful tool to probe growth and differentiation of hematopoietic cells and to examine questions of virus–target cell specificity. Since the isolation of A-MuLV, a second virus, HZ-FeSVII, containing a portion of the *abl* gene has been isolated from a spontaneous feline sarcoma.[2] In addition, analyses of c-*onc* genes revealed that the chromosomal translocation associated with human chronic myelogenous leukemia (CML)[3] alters the location and expression of the c-*abl* protooncogene.[4–6] Although in each of these settings, expression of the tyrosine kinase activity associated with the *abl* gene product[7–10] appears to play a key role in oncogenesis, the exact mechanism by which this is accomplished remains elusive. Because a general discussion of the origin and biology of A-MuLV has been presented recently,[11–13] this review will focus on recent developments concerning the transformation potential of v-*abl* genes and the expression of both the normal c-*abl* gene and the altered *bcr/abl* gene found in CML.

NAOMI ROSENBERG • Departments of Pathology and Molecular Biology and Microbiology, Tufts University School of Medicine, Boston, Massachusetts 02111.

2. A-MuLV

2.1. A-MuLV Genome Structure

A-MuLV arose by recombination of the replication-competent virus, Moloney murine leukemia virus (M-MuLV), and the protooncogene c-*abl*. This event involved the joining of base 1330 in the p30 portion of the M-MuLV *gag* gene to sequences in the fourth exon of c-*abl* at the 5' end and the joining of base 7614 in the M-MuLV *env* gene to sequences in the last exon of c-*abl* at the 3' end.[14–16] A four-base homology exists at the 5' recombination site while no homology exists at the 3' recombination site. Thus, while homologous recombination may have played a role in generating the left joint, it is likely that a second event, at the RNA level, was required for the generation of A-MuLV.

When the genome structure of A-MuLV became clear, it was apparent that two strains of A-MuLV existed. One has a provirus of 6.3 kb and the second, a provirus of 5.5 kb.[17,18] Sequence analysis revealed that the latter contains a 789-base in-frame deletion in v-*abl*.[17,19] Although early passages of A-MuLV stocks have been lost, the fact that the information deleted in the smaller virus is present in c-*abl* mRNA and that the sequence at the M-MuLV–*abl* junctions is identical in both isolates[15,20] argues that this virus is a deletion mutant of the larger isolate.

2.2. Expression of the Viral Genome

The virus synthesizes a single genome length transcript that is translated to form a fusion protein containing the p15 and p12 proteins and 21 amino acids of the p30 protein encoded by the M-MuLV-derived *gag* gene fused to v-*abl* sequences.[21–23] The molecular mass of the protein encoded by the 6.3-kb strain is 160,000 daltons and the strain is called A-MuLV-p160 while that encoded by the smaller isolate is 120,000 daltons.[17] In each case, translation terminates at the same position in v-*abl*. The remaining 856 bases of v-*abl* appear not to encode protein. An open reading frame of 489 bases exists in this region, but no protein product has been found. Because multiple strains of A-MuLV exist and most encode proteins of different sizes, the term Abelson protein will be used for the product of the viral genome.

2.3. Structure of the Abelson Protein

In transformed cells, the Abelson protein is present in both a soluble and a membrane-bound form in approximately equal proportions.[24,25] A fraction of the protein is myristylated on glycine, the second amino acid of p15,[26,27] and it is believed that this modification is important in membrane association of the protein. However, direct evidence that membrane association is required for transformation as is the case with the avian transforming virus Rous sarcoma virus (RSV)[28] has not been presented. In addition to myristylation, the p160, but not the p120, protein exists in both a glycosylated and a nonglycosylated form.[29] The glycosylation signals normally present in the M-MuLV-derived *gag* gene direct this modification and the presence of a point mutation in the *gag* leader region in A-MuLV-p120 prevents expression of a glycosylated form of p120.[23,30] Glycosylated p160 is not phosphorylated on tyrosine *in vivo* or *in vitro* and an A-MuLV-p120 recombinant strain containing *gag* gene sequences derived from A-MuLV-p160 does not differ from this latter virus in its ability to transform cells.[30] Thus, it appears that the glycosylation is not important in the biology of the virus.

Immunofluorescence and adsorption studies using an anti-Abelson cell tumor regressor serum[24] indicated that a portion of the *abl*-encoded determinants are expressed on the surface of transformed cells.[24,25] However, attempts to detect these determinants with anti-*abl* monoclonal antibodies have been unsuccessful.[31] In contrast, monoclonal antibodies selected for binding to A-MuLV-transformed cells react with sequences present in the *gag*-derived p15 protein,[31] indicating that a small portion of this part of the Abelson protein is expressed on the surface of most transformed cells. Because the sequence of the Abelson protein does not resemble that characteristic of a membrane protein,[23] it is likely that p15 and not v-*abl* determinants were detected in the studies using tumor regressor sera. It is possible that myristylation plays a key role in directing the protein to the membrane. Glycosylation does not play a role in directing the molecule to the surface because both p120 and p160 are expressed on the surface of cells.[31] Based on analogies with pp60*src*[32,33] it is likely that membrane anchorage is important for the function of Abelson protein. However, additional experiments are needed to determine the functional significance of the surface expression of p15 determinants.

2.4. Function of the Abelson Protein

The protein encoded by A-MuLV is a tyrosine protein kinase with the ability to phosphorylate itself in immune complex reactions *in vitro*.[7] Alignment of the protein sequence of p160 with that of pp60*src*, the transforming protein of RSV, shows that 43% of the residues in the carboxy-terminal portion of pp60*src* are conserved in the amino-terminal portion of v-*abl*.[19,23,33] Expression of a fragment containing the conserved region in *E. coli* induces phosphorylation of bacterial proteins on tyrosine residues[34,35] and the Abelson protein fragment purified from bacteria functions as a tyrosine protein kinase *in vitro*.[36]

In transformed cells, the Abelson protein is phosphorylated on tyrosine residues[37,38] and multiple cellular proteins have increased levels of p-Tyr.[37,39,40] Indeed, many of these same proteins are phosphorylated in cells transformed by other viruses that encode tyrosine protein kinases.[39] However, the cellular substrate(s) of the Abelson protein remains to be identified.

2.5. Abelson Protein Structure and Transformation

As noted earlier, A-MuLV can transform some fibroblastoid cell lines and several types of hematopoietic cells. The isolation of protein kinase-negative, transformation-defective strains of A-MuLV demonstrated the key role of the kinase activity in transformation.[8,9] Examination of a series of viruses encoding Abelson proteins deleted at the carboxy-terminal end of the molecule separated the requirements for efficient transformation of NIH 3T3 and lymphoid cells.[41] Based on these and other studies, the protein has been divided into four regions[29]: region I, the *gag*-derived portion; region II, the region required for protein kinase activity; region III, the portion of A-MuLV-p160 that is deleted in A-MuLV-p120; and region IV, the carboxy-terminal portion. The role of these regions in transformation of various cell types has been the focus of a number of recent investigations.

2.5.1. The Role of Region II in Transformation

The isolation of a protein kinase-negative virus that contained a large deletion in region II[8,18] and the failure of cloned A-MuLV plasmids digested within this region to transform cells[42] pointed to the key role of this portion of the protein in kinase activity and transformation. These studies have been extended to identify the minimal v-*abl* region required for transformation of NIH 3T3 cells using a series of A-MuLV strains engineered by recombinant DNA techniques. A virus containing all of the *gag* and v-*abl* sequences up to the *Pst*I site at 2550 (approximately 1.2 kb of v-*abl*) retains the ability to transform NIH 3T3, while a virus truncated at the *Sac*I site 159 bases upstream does not.[43] Further delineation of the transforming potential was accomplished using *Eco*RI or *Bam*HI linkers to insert four in-frame amino acids into the Abelson protein. Insertions after amino acid 4 and 15 of v-*abl* do not affect transformation while insertions after amino acid 136 either reduce or abolish transforming function as do insertions at other points farther into region II.[43] In all these cases, the ability to transform NIH 3T3 cells correlates closely with the ability of v-*abl* to function as a protein kinase in *E. coli*,[35,43] solidifying a central role for the protein kinase in transformation.

Although the virus truncated at the *Pst*I site mentioned above transforms NIH 3T3 and the transformed cells grow in agar,[43] the cells in primary foci induced by this virus are usually more fusiform than cells transformed by wild-type A-MuLV strains (Prywes, Rosenberg, and Baltimore, unpublished data). This virus also transforms lymphoid cells, but at an extremely low frequency after an extended latent period. Because the titer of these stocks is low, it was impossible to measure transformation frequencies in these experiments.[44] Whether these results indicated that the virus is impaired in its ability to transform cells in a subtle way requires further study.

2.5.2. The Role of Region I in Fibroblast Transformation

Analysis of an A-MuLV strain that encodes a protein with the first 15 amino acids of the p15 *gag* protein fused to sequences encoded by v-*abl* information from the *Hinc*II site 15 bases 3′ of the start of v-*abl* to the *Pst*I site at 2550, demonstrated that most *gag*-encoded sequences were not required for transformation of NIH 3T3.[42] Like the strain containing the normal *gag*-encoded information, NIH 3T3 cells infected with this virus also fail to assume a fully transformed morphology (Prywes, Rosenberg, and Baltimore, unpublished data). Nonetheless, the ability of this virus to transform NIH 3T3 cells correlates with the ability of the HincII–PstI fragment to function as a protein kinase in E. coli and indicates that most of gag is not required for transformation.

2.5.3. The Lethal Effect of Region IV in Fibroblasts

While region II is sufficient for transformation of NIH 3T3 cells, transfection experiments indicate that sequences in region IV are important in determining the outcome of the virus–cell interaction. Cotransfection of NIH 3T3 cells with wild-type virus and a selectable marker gene in the absence of helper virus DNA results in production of fewer drug-resistant foci than anticipated while deletion of sequences in region IV abolishes this effect.[45] Consistent with this, Ziegler and co-workers[29] showed that most BALB/c 3T3 cells infected with wild-type A-MuLV underwent a transient transformation and died. Stable transformants could only be derived when viruses with deletions in region IV were used.[29,46] This lethal effect of the virus is thought to stem from either overproduction of the transforming protein in the case of transfected cells or an acute sensitivity of some cell lines to the effects of the protein in the case of the infected BALB/c 3T3 cells. Whether other cell types are sensitive to lethal effects of the virus remains to be investigated.

2.5.4. The Role of Region IV in Lymphoid Transformation

Although highly deleted viruses are capable of transforming NIH 3T3 cells, many of these are compromised in their ability to transform lymphoid cells *in vitro* and to induce Abelson disease *in vivo* indicating that the requirements for these transforming functions are more stringent. Sequences in region IV, the carboxy-terminal portion of the Abelson protein, appear to mediate this phenomenon. Mutants derived from A-MuLV-p120 using biological approaches and from A-MuLV-p160 using recombinant DNA approaches have been studied. Although in each instance a strong case for the role of region IV in lymphoid cell transformation is evident, the biological properties of the two series of viruses are distinct.

2.5.4a. Region IV in the Context of Region III. A series of viruses isolated from A-MuLV-p160 has allowed examination of the role of region IV in the context of part or all of region III. In one set, viruses lacking all of region IV and a portion of region III (truncated at the NarI or XhoI site in region III) were found to transform lymphoid cells with a tenfold decreased efficiency.[44,47] These viruses were also compromised in their ability to induce Abelson disease *in vivo*. As noted earlier, a virus truncated at the PstI site at 2550 retained the ability to transform lymphoid cells *in vitro*, indicating that the minimal transforming region required for both fibroblasts and lymphoid cells was the same. However, the inability to obtain a high-titer stock of this strain precluded quantitative studies and an assessment of its ability to induce Abelson disease.[44]

A third virus, A-MuLV-p130, generated by *Bal*31 mutagenesis at the *Sal*I site in region IV[30] also fails to transform large numbers of lymphoid cells *in vitro* and *in vivo* (Rosenberg and Witte, unpublished data). This virus retains a lethal function in 3T3 cell lines, indicating that sequences required for these two functions are distinct[30] (N. Rosenberg and O. N. Witte, unpublished data). Precise mapping of the deletion in this strain has not been done, but the p130 protein appears to lack about the last 100 amino acids of p160. More precise mapping of the sequences required for lethality versus lymphoid transformation would be useful. However, the necessity of recovering viruses with small deletions engineered in region IV by NIH 3T3 cell transfection and the lethality of these viruses in transfection systems have complicated these experiments.

2.5.4b. Region IV in the Absence of Region III. A series of spontaneous mutants that encode proteins deleted in the carboxy-terminal portion of region IV have also been isolated from the A-MuLV-p120 strain, the wild-type strain in which region III is deleted and region IV sequences are fused, in frame, to those of region II. While most of these viruses behave like those isolated from A-MuLV-p160 in that they are compromised in their ability to transform lymphoid cells but not NIH 3T3[17,30] (N. Rosenberg, unpublished data), study of one of these has allowed separation of as yet unidentified functions required for *in vitro* versus *in vivo* transformation of lymphoid cells. This strain, called A-MuLV-p90A, encodes a 90,000-dalton transforming protein and has a 20-base deletion from base 2874 to base 2894 in region IV. This deletion causes a shift in reading frame and results in a truncated protein with 35 unique amino acids at the carboxy-terminus (R. Huebner and N. Rosenberg, in preparation).

A-MuLV-p90A, like the region IV mutants derived from A-MuLV-p160, transforms about tenfold fewer lymphoid cells *in vitro* and induces Abelson disease at a lower frequency than A-MuLV-p120.[41] However, unlike most of the A-MuLV-p160-derived region IV mutants and the closely related A-MuLV-p120-derived p90B mutant,[19] tumors arising in animals injected with A-MuLV-p90A almost always contain variant viruses that encode either larger or smaller Abelson proteins than p90.[48] These variant viruses are stable on passage both *in vitro* and *in vivo*. Variants that encode larger proteins are similar to A-MuLV-p120 in transforming potential *in vivo* and *in vitro* and appear to arise by small mutations that shift the reading frame of the mutant back to that of the wild-type virus. In contrast, the variants that encode transforming proteins smaller than p90 transform a low frequency of lymphoid cells *in vitro*, but they are highly tumorigenic *in vivo*. These viruses appear to arise by mutations that induce termination before the A-MuLV-p90A deletion. In contrast, the A-MuLV-p90B strain, a virus with a single base deletion at 2946 which causes a shift to the third reading frame resulting in a truncated protein with 12 unique amino acids at the carboxy-terminus,[19] does not generate smaller variants *in vivo* (R. Huebner and N. Rosenberg, in preparation). This argues that the sequences at the extreme carboxy-terminus of the Abelson protein play an important role in the ability of infected cells to form tumors in mice. Further studies are needed to determine the mechanism(s) controlling this phenomenon, but the observation that recombinant viruses carrying the A-MuLV-p90A deletion generate variants *in vivo* (R. Huebner and N. Rosenberg, in preparation) argues that only this region of the genome is required.

2.5.5. The Role of Region I in Lymphoid Transformation

In addition to sequences in region IV of the genome, sequences in region I, the *gag*-derived portion of the Abelson protein, are also important in transformation of lymphoid cells, at least in the context of viruses deleted in region IV and a part of region III.[44,47] Viruses lacking all but coding information for the first 15 amino acids of *gag* and truncated at the *Xho*I or *Nar*I site in region III fail to transform lymphoid cells *in vivo* or *in vitro*. The sequences required for lymphoid transformation have been mapped to a region of p15 between amino acids 38 and 115. Because the Abelson protein encoded by this virus is not stable in a chemically transformed lymphoid cell line, it is likely that the inability of the *gag*-deleted viruses to transform lymphoid cells is due to rapid turnover of the transforming protein in this cell type.[47]

2.5.6. The Role of Region III in Lymphoid Transformation

Historically, A-MuLV-p160 and A-MuLV-p120 have been considered wild-type strains of virus. Comparison of the ability of these two strains to transform fibroblast cell lines has, as expected from the results obtained with viruses containing only region II sequences, failed to reveal a difference between the two viruses. In addition, comparisons of their ability to induce tumors in BALB/c mice or to transform BALB/c bone marrow cells *in vitro* have suggested there is little difference between the two strains. However, the observation that A-MuLV-p160, but not A-MuLV-p120, transforms large numbers of lymphoid cells from DBA/2 mice (L. Schiff-Maker and N. Rosenberg, unpublished data) suggests that sequences in region III may be important in transformation *in vitro* in some instances. Consistent with this, tumor cells from DBA/2 mice infected with A-MuLV-p120 are difficult to grow *in vitro*.[49] Further comparisons of these two strains are needed to determine the way in which region III sequences impact on transformation *in vitro*.

2.5.7. Summary

The studies with A-MuLV mutants can be summarized as follows: (1) the portion of the Abelson protein that is minimally required for protein kinase activity is necessary and sufficient to transform NIH 3T3 and bone marrow cells; (2) the carboxy-terminal portion of the Abelson protein is responsible for a lethal effect in fibroblasts but this region is required for high-efficiency transformation of lymphoid cells *in vitro;* (3) while the carboxy-terminus of region IV is usually required to induce a high frequency of Abelson disease, a subset of mutants encoding proteins that lack these sequences is highly oncogenic *in vivo;* (4) a portion of the *gag*-derived sequences are required for lymphoid transformation, but not for fibroblast transformation; and (5) sequences in region III may be important for lymphoid transformation *in vitro*. Although the picture that emerges is satisfying because a great deal is known concerning which sequences are important in which biological systems, the situation is less satisfying when viewed from a mechanistic standpoint. Although it has been proposed that carboxy-terminal sequences are important in controlling the level of kinase activity of the Abelson protein,[29,41] a clear demonstration of this remains to be presented. The carboxy-terminal portion of the protein is rapidly degraded in *E. coli*[35] and purification of intact molecules from mammalian cells has not been reported. In addition, the mechanism by which changes in kinase activity could affect lymphoid cell transformation is still unclear. Such a change might affect interactions with substrate molecules. Finally, the difficulties of transfection systems have necessitated examining the role of some regions of the genome in the context of other deletions. Whether this has amplified the effects of the mutations under study remains to be determined.

3. A-MuLV–Hematopoietic Cell Interaction

The ability of A-MuLV to transform lymphoid cells *in vitro*[50-52] has allowed investigators to manipulate target cell populations and to isolate and characterize clonally derived transformed cells. Using this system, questions concerning both the type(s) of cells with which the virus interacts as well as the effect of transformation on the differentiation program of the target cells have been addressed. The first area has been approached by direct analysis of the target cells using either cells from different tissues, cells from animals deficient in certain lymphoid precursors, or cell populations selected on the basis of expression of certain differentiation antigens. Two strategies have been used to approach the second question: analysis of transformed cells using monoclonal antibodies that react with antigens characteristic of cells at particular stages of differentiation; and analysis of Ig gene structure.

3.1. An Overview of Mouse Hematopoiesis

To fully appreciate the complexities of A-MuLV–hematopoietic cell interaction, a working knowledge of hematopoiesis is required. An extensive review of this area is beyond the scope of this chapter. However, a brief discussion of the origin of hematopoietic cells and especially B cells will help to orient the uninitiated. To supplement the following discussion, the reader is referred to other recent reviews centering on this topic.[52-54] All hematopoietic cells arise from a common progenitor called the pluripotent stem cell. This cell is believed to give rise to two committed stem cells, the myeloid stem cell which differentiates to form granulocytes, macrophages, erythrocytes, and platelets and the lymphoid stem cell which gives rise to T and B lymphocytes.

The earliest cell identified in the B-lymphocyte lineage is the pro-B cell.[55,56] This cell expresses the B-lineage differentiation marker, B220, and is thought to have Ig genes in an embryonic or unrearranged configuration. This cell probably arises from an earlier, Thy-1$^+$ cell because cells expressing this determinant at a low level in the absence of B220 can give rise to pre-B and B lymphocytes when grown in Whitlock–Witte cultures.[57] The pro-B cell undergoes heavy chain gene rearrangement forming a complete V gene, by first joining a D segment to a J_H segment and subsequently appending a V_H segment to the DJ_H unit, and differentiates to give rise to the pre-B lymphocyte.[54,56] This cell expresses immunoglobulin in the form of cytoplasmic μ protein.[58-60] A subsequent differentiation step, characterized by the rearrangement of light chain genes, gives rise to virgin B cells that express complete Ig molecules in the form of IgM on their surface. These cells can interact with antigen and differentiate further to form either memory cells or the effector cells of the humoral immune response, the Ig-secreting plasma cell.[61] These cells may continue to express IgM or one of the other six heavy chain constant region genes in the context of the variable region gene originally expressed with the μ constant region and the light chain originally expressed in the IgM molecule.[62–64] Differentiation stages characterized by particular Ig gene structural configurations are associated with specific differentiation markers and cellular functions.

3.2. Differentiation Markers Expressed by A-MuLV-Transformed Lymphoid Cells

The frequency of nucleated bone marrow cells susceptible to A-MuLV-induced transformation is about 1 in 10^3 [51], making direct analysis of target cells difficult. This fact, coupled with the ease of growing large numbers of transformants, has made analysis of the phenotypic characteristics of the cell lines an attractive area of investigation. Analysis of transformants has helped to set the stage for direct examination of the A-MuLV target cells in complex populations of normal cells. These studies have shown that most lymphoid transformants derived from either adult bone marrow or fetal liver using the agar assay[51] are similar to pre-B lymphocytes. For example, most of the isolates express the B-lineage markers lyb-2[65] (L. Ramakrishnan and N. Rosenberg, unpublished data), ly-5 (B220),[65,66] ly-17 (Fc receptor),[65,67] and other differentiation markers characterized by the monoclonal antibodies 19B5,[68] J11d,[66] AA4,[66] GF1,[66] and 6C3 and BP-1.[69,70] Generally, the cells lack Ia,[66] a marker present on differentiated B cells, ThB[65] (L. Ramakrishnan and N. Rosenberg, unpublished data), a marker present on some pre-B cells and more differentiated B cells and T cells, Mac-1,[65] a marker present on macrophages, and Thy-1, lyt-2, and other T-lymphocyte markers.[65,66] At least some isolates appear to express ly-1, a marker characteristic of certain B- and T-cell subpopulations[71,72] and whether this is a general feature of most of the transformants remains to be examined.

Depending on the determinant, the percentage of positive cells in different clones can vary and some markers appear to be lost with increasing time in culture. For example, when analyzed early after isolation, about 60% of primary transformed clones contain cells that synthesize cytoplasmic μ but no light chain protein[73,74] while only about 10% of these still express μ after 4 to 6 months of culture[74] (N. Rosenberg, unpublished data). In addition, most primary transformants contain 20–30% Thy-1$^+$ cells when isolated from agar but only rare isolates (< 10%) retain Thy-1 expression after several months of culture[75] (L. Ramakrishnan and N. Rosenberg, unpublished data). Whether the loss of some differentiation antigens and the heterogeneity in expression noted in some clones relates to the differentiation state of these cells remains an open question.

Expression of the determinant recognized by the 6C3 monoclonal antibody deserves special mention. This antibody recognizes a 163,000-dalton surface glycoprotein present on some normal T and B lymphocytes.[69] Most established A-

MuLV-transformed lymphoid cells also express this determinant. However, lymphoid cells derived by growing A-MuLV-infected bone marrow cells on bone marrow-derived stromal monolayers in the presence of a low concentration (5%) of fetal calf serum fail to express the 6C3 antigen.[69,76] The feeder-dependent cells are nontumorigenic but become tumorigenic as they become independent of the feeder cells, a change that is not accompanied by any detectable changes in Abelson protein expression or cellular phosphotyrosine.[77] However, the 6C3 antigen does appear coincident with feeder cell independence.[69] Whether its expression in this setting reflects a change in differentiation or whether the molecule plays an active role in the acquisition of tumorigenic potential remains an open question.

3.3. Ig Gene Structure in A-MuLV-Transformed Lymphoid Cells

Examination of Ig gene structure in a large series of cell lines derived from adult bone marrow has revealed that all have completed at least DJ_H joining on both alleles and that at least 50% of cell lines derived from adult bone marrow have undergone $V_H DJ_H$ joining on at least one allele while only rare isolates have rearranged their light chain genes[78] (L. Ramakrishnan and N. Rosenberg, in preparation). None of the cell lines have rearranged the α and β T-cell receptor genes.[79] This pattern of gene structure is consistent with the phenotypic marker characterization and supports a relationship of the cells to pre-B cells. Whether cells that have both heavy chain genes in a DJ_H configuration are analogous to a normal cell type remains to be determined. Many clones of this type transcribe a C_μ mRNA using a promoter 5' of the D segment and two have been shown to produce a truncated μ protein.[80] The fact that subclones of these cell lines will complete rearrangement of the heavy chain locus[81] has aroused speculation that this type of cell may be of functional significance *in vivo*.

Although most lymphoid cell lines derived from adult bone marrow are phenotypically similar to pre-B cells, some are similar to more differentiated B cells. About 5% of the isolates derived using the classical agar transformation system have rearranged light chain genes and some of these express light chain protein, either constitutively or after stimulation with the B-cell mitogen LPS.[60,82,83] Subclones of these or other cell lines have been isolated that have undergone heavy chain class switching and express other H chain proteins, especially γ 2b.[83-87] Whether the isolates that give rise to cells with more differentiated phenotypes have undergone an altered virus–cell interaction or whether the cell initially infected with the virus was at a later point in differentiation is not known. All of the more differentiated clones continue to express Abelson protein, demonstrating that a more differentiated phenotype can exist in the presence of the transforming protein. Often the more differentiated cells are recognized as minor populations in typical transformed clonal cell lines and selected by subcloning. It is possible that many isolates retain the ability to differentiate further especially early in their passage but that the continual selection for cell growth selects against these cells.

In contrast to the rare, more differentiated adult bone marrow-derived transformants, cell lines derived from fetal liver appear to be analogous to cells between the normal pro-B and pre-B cell stages. While these transformants cannot be distinguished from others on the basis of phenotypic markers,[66] examination of the heavy chain gene structure in these cell lines indicates that they are at an earlier stage of differentiation. Most of these cells are in the process of secondary D to J_H joining and some of them will append V_H genes to the DJ_H unit during culture *in vitro*.[78,88] Cell lines with similar Ig gene structure have been isolated from the bone marrow of adult animals undergoing repopulation of the B-cell compartment (L. Ramakrishnan and N. Rosenberg, in preparation). The observation that these cells transcribe germline V_H genes at a high level (L. Ramakrishnan and N. Rosenberg, in preparation), a developmental characteristic of normal fetal lymphoid cells,[89] further supports the contention that these cells are less differentiated.

3.4. Analysis of A-MuLV Target Cells

The ability to isolate transformants that appear less differentiated from some tissues suggests that the target cells available to the virus in those tissues are less mature and that cells at several stages of differentiation are susceptible to A-MuLV transformation. A complete definition of the properties of the lymphoid cells(s) susceptible to A-MuLV in the standard *in vitro* transformation system has yet to be presented. However, it is clear that the first lymphoid cells susceptible to *in vitro* transformation appear in the fetal liver at about day 13 of gestation[74]—around the same time as the appearance of the first pre-B lymphocytes.[90,91] Precursors capable of differentiating into B lymphocytes are present prior to this time[92,93] but these cells are not able to function as target cells, suggesting that cells must reach a particular point in the B-cell pathway before they are susceptible to the virus. The absence of A-MuLV-susceptible cells in the *nu/xid* mouse,[94] an animal that lacks both B and T cells but contains pro-B cells,[94,95] also supports this notion.

A second approach to analysis of the types of cells susceptible to A-MuLV involves the use of monoclonal antibodies to fractionate complex normal cell populations. This technology was first used to demonstrate that the determinant recognized by the rat anti-mouse brain monoclonal antibody 19B5 was expressed on A-MuLV target cells present in mouse bone marrow.[68] A more elegant application, involving separation using the fluorescence-activated cell sorter, has recently shown that A-MuLV target cells can be divided into two populations, one that expresses B220 and a second that expresses both B220 and a low amount of Thy-1[75] with the latter population being highly enriched for susceptible cells. Whether the transformed cells isolated from the two populations differ with respect to Ig gene structure is still being examined. The transformants from the Thy-1$^+$ population may contain cells at an earlier point in differentiation since some normal Thy-1$^+$ cell populations contain precursor cells that are able to differentiate into normal pre-B cells *in vitro*.[57] If this hypothesis is correct, transformants from the Thy-1$^+$ population may have their heavy chain locus in a DJ_H rearranging configuration. Alternatively, because Thy-1 expression is unstable in these and other A-MuLV transformants[75] (Ramakrishnan and Rosenberg, unpublished data), it is possible that these transformants are similar to the classically described cells and that detection of Thy-1 escaped the notice of earlier investigators because the determinant is lost after a short time *in vitro*.

The changes observed in the phenotype of A-MuLV-transformed cells early in their culture history raise the central question of how closely the phenotype of the transformant reflects that of the cell when it was infected and whether individual infected cells differentiate following infection to resemble cells at different points early in the B-cell pathway. A direct analysis of this question using clones of normal pre-B and B cells isolated from Whitlock–Witte cultures as target cells has shown that both of these cell populations can serve as target cells.[96] Comparison of the Ig gene structure and Ig production in the clones before and after transformation showed that while some remained the same, a significant percentage had undergone further differentiation as judged by light chain gene rearrangements.[96,97] Thus, in at least some instances, differentiation can occur during the transformation process.

3.5. A-MuLV Interaction with Erythroid Cells

Although infection of hematopoietic cells *in vitro* is classically associated with transformation of pre-B cells, infection of other target cell populations or of bone marrow cells under different conditions allows growth of other cell types. For example, when fetal liver, or placenta from day 9 to day 13 gestation mice is infected and plated in the pre-B agar transformation assay, colonies of erythroid cells arise.[74] The cells in these colonies differentiate and die within 2 to 3 weeks postinfection. Analysis of the role of A-MuLV in colony induction has shown that highly oncogenic strains of virus induce more colonies and that all of the cells in each of the colonies express Abelson protein. Normally, erythroid colony induction is dependent on the presence of a hormone such as erythropoietin. Analysis of the growth requirements of A-MuLV-induced erythroid colonies suggests that A-MuLV supports the growth and differentiation of these cells by replacing the need for a normal hormonal signal.[98] Thus, in this instance infection with a virus that normally transforms cells leads to normal growth and differentiation. The factors that control the response of the cell to the infection are still unknown. However, it seems reasonable to assume that the virus interacts with similar cellular substrates in the case of lymphoid and erythroid cells but that impact on the cells' differentiation program is quite different.

3.6. A-MuLV Infection and IL-3

The notion that in part, A-MuLV infection alters the growth requirements of cells is substantiated not only by the experiments using lymphoid cells and feeder layers [76,77] and the erythroid cell experiments[98] but also by a series of experiments demonstrating that A-MuLV infection can substitute for a requirement of the lymphokine IL-3. IL-3-dependent mast cell lines and tumor cells isolated from Friend virus-infected mice become IL-3 independent by a nonautocrine mechanism after A-MuLV infection.[99–101] As yet it is not clear how the virus relieves the need for IL-3. Because IL-3 also acts on normal erythroid precursors,[102] it is possible that the effect of the virus on these cells is mediated via the same pathway. Whether this effect plays a role in pre-B-cell transformation remains to be determined. IL-3 appears to stimulate normal pre-B cells or stromal cells often present in cultures used to grow these cells.[56]

3.7. Tumors Induced by A-MuLV

Although most mice injected with A-MuLV develop tumors composed of pre-B cells similar in phenotype to those isolated from *in vitro* agar transformation systems, the number of exceptions to this rule has grown in recent years. Prominent among these is the observation by Cook,[103,104] subsequently confirmed by others,[105,106] that intrathymic inoculation of C57BL mice with A-MuLV leads to a high frequency of thymic tumors. Adult C57BL mice are unique in their susceptibility to this type of tumor, perhaps because they are refractory to A-MuLV-induced pre B-cell tumors.[107] The tumor cells express T-cell markers such as Thy-1, L3T4, and lyt-2 and lack B220.[103-106] In addition, cells in these tumors have Ig genes in an embryonic configuration while the gene for the T-cell receptor is rearranged.[106] This series of experiments, added to those showing that A-MuLV is associated with plasmacytoma acceleration,[108] mastocytomas,[109] and transformed macrophages,[110] strongly supports that idea that the microenvironment in which infection occurs plays a strong role in determining the outcome of the infection.

The role of A-MuLV in some of these systems is hard to assess. While the thymomas[103-106] express A-MuLV, tumors isolated by others have lost the viral genome. Several of the tumor cell lines isolated from pristane-primed, A-MuLV-infected mice do not contain the viral genome.[111] An *in vitro* system using LPS and A-MuLV in which mature B-cell lines are stimulated to grow indefinitely[112] may represent a similar case. While the role of the virus in the *in vitro* system remains unclear, in at least one case in the pristane system, insertion of the M-MuLV helper virus present in the A-MuLV stocks near the c-*myb* protooncogene may be involved in the tumorigenesis.[113] Several tumor cell lines isolated from C57BL also appear to lose the A-MuLV genome with passage *in vivo*.[114] All of these experiments point to a need for clarification of the role of A-MuLV and environmental factors in establishing and maintaining the transformed state.

4. HZ-FeSVII

In the course of screening for new oncogenes in spontaneous feline sarcomas, Besmer and co-workers[2] showed that one of their isolates contained the *onc* gene *abl*. This isolate, HZ-FeSVII, contains information from the feline c-*abl* gene fused to feline leukemia virus-derived *gag* and *pol* sequences at the 5′ and 3′ ends, respectively (Besmer, personal communication). The amount of c-*abl* information present in HZ-FeSVII represents a region roughly corresponding to region II of A-MuLV. However, sequences 5′ of the v-*abl* information contained in A-MuLV are present in HZ-FeSVII. As expected, this virus encodes a tyrosine protein kinase. HZ-

FeSVII induces sarcomas in kittens and transforms fibroblast cells *in vitro*.[2] Infection of murine bone marrow cells induces transformation of lymphoid cells that appear similar to those induced by A-MuLV (N. Rosenberg and P. Besmer, unpublished observations) but the frequency of transformation is significantly lower than that obtained with wild-type A-MuLV strains. This result is consistent with the amount of c-*abl* present in HZ-FeSVII. Whether the ability of this virus to cause sarcomas is an intrinsic property of the virus or relates to virus–cell interaction in the cat requires further study.

5. C-ABL

5.1. The c-*abl* Gene

The c-*abl* gene is the normal homologue of v-*abl*. All vertebrate species examined[14] and *Drosophila*[115,116] contain a c-*abl* gene. In both human and murine cells, the gene is a single locus made up of 11 exons that span over 30 kb.[15,16,117] In the mouse, c-*abl* is located on chromosome 2[118]; in humans the gene is on chromosome 9.[119] The normal gene contains sequences at both 5′ and 3′ ends that are not present in v-*abl*, but it appears that only those at the 5′ end contribute to the gene product of the locus.[15,16] Comparison of the regions that have been sequenced shows that the human and murine c-*abl* genes are highly related and that very few changes distinguish c-*abl* and v-*abl*.[15,16,117] The c-*abl* gene in *Drosophila* shares homology with the kinase region of mammalian c-*abl* and v-*abl* but diverges significantly in the 3′ portion of the gene.[116]

5.2. Expression of c-*abl* in Normal Cells

c-*abl* is expressed in a variety of murine cells and at least three mRNA species have been identified. Two of these, the 6.5- and 5.3-kb species, are found in most tissues examined, while a third of 4.2 kb is prominent in testis.[20,120] The differences in size of these mRNAs are accounted for, at least in part, by the presence of four alternative exon I sequences.[16] Each of these (designated types I–IV) has been identified by examining *abl*-containing cDNA clones isolated from the murine lymphoma line 70/Z. Mapping using ribonuclease S1 indicates that all four of the exons are expressed by splicing to a common site at the beginning of exon 2. Type II/IV exons give rise to the 6.5-kb mRNA while the 5.3-kb species appears to involve use of the type I exon. The origin of the smaller 4.2-kb RNA found in testis[120] is not clear, but presumably it does not contain type II/IV exons.[16]

5.3. The Product of the c-*abl* Gene

A protein that shares homology with v-*abl*-encoded proteins can be isolated from normal murine and human cells.[121–123] The molecular mass of the protein isolated from human cells is 145,000 daltons and that of the murine protein is 150,000 daltons and like the v-*abl* proteins, these molecules function as tyrosine protein kinases.[124] The specific activity of the c-*abl* proteins appears to be similar to that of v-*abl* proteins, but the amount of the molecule per cell is about 50–100 times less than the amount of v-*abl* in a typical transformed cell.[122,123] The function of the c-*abl* protein is unknown, but the fact that it is expressed in many cell types suggests that it plays a central role in cell growth, presumably via phosphorylation of other cellular proteins. The impact of alternative use of the 5' exons on the subcellular localization of the molecule and on the protein kinase activity is a subject of active investigation. The sequence of exon 1, type IV predicts an N-terminal Met-Gly-Gln sequence[16] identical to that at the *gag*-derived amino-terminus of the Abelson protein and other *onc* gene products that are myristylated.[26–28] Thus, this form of the c-*abl*-encoded protein may be similarly modified and associated with cell membranes.

5.4. Expression of c-*abl* in CML

Interest in the function of c-*abl* was heightened by the observation that this gene is involved in the translocation that characterizes human chronic myelogenous leukemia (CML). This reciprocal translocation, involving chromosomes 9 and 22, invariably moves the segment of chromosome 9 containing the c-*abl* gene to a specific region of chromosome 22 originally called the breakpoint cluster region or *bcr*.[4,125] Cells carrying the translocation synthesize, in addition to normal c-*abl* transcripts, an abnormally large 8-kb c-*abl* transcript.[5,126] Cloning of cDNA from the tumor cells has revealed that sequences derived from the *bcr* region are fused to sequences from exon II of c-*abl* at the same position that the variable exon I sequences are normally spliced onto the body of c-*abl* mRNA.[6,16,127]

A *bcr* cDNA has been isolated from a normal human fibroblast cDNA library and used to analyze the structure of the *bcr* region.[6] These studies have revealed the existence of a gene, now termed *bcr* of about 45 kb in length, containing at least 13 exons. Analysis of the sequence of the cDNA does not reveal homology to any known proteins and the normal product of the gene has not been identified. The hybrid *bcr*/*abl* transcript contains the first two and sometimes the third exon of *bcr*.

5.5. The Product of *bcr*/c-*abl*

A protein of 210,000 daltons has been identified in tumor cells from CML patients carrying the 9;22 translocation using antibodies reactive with Abelson protein.[123] This molecule shares tryptic peptides with the v-*abl* and c-*abl* proteins and presumably represents the fusion of *bcr*- and c-*abl*-encoded determinants. The molecule has tyrosine protein kinase activity and is present at levels similar to those of Abelson protein in virus-transformed cells. The presence of the *bcr*/c-*abl* transcript and its product in almost all tumor cells from patients carrying the translocation, suggests that expression of this protein plays a critical role in the pathogenesis of the disease. The mechanism by which this protein participates in the disease process is a subject of intense investigation. It is possible that the *bcr*-derived sequences alter its interaction with cellular substrates. p210 is phosphorylated on tyrosine residues *in vivo* while p145 and p150, the products of the normal c-*abl* gene, are not,[123,124] suggesting that the N-terminal sequences of the protein may play an important role in the activity of the molecule *in vivo*.

6. Future Directions

Despite the great strides made in understanding the biology of A-MuLV and in dissecting the structure of the c-*abl* gene, many unanswered questions remain. First, it is necessary to understand the function of c-*abl* in normal cellular growth. The presence of multiple forms of exon I suggests that expression of the gene is subject to complex control mechanisms. It is important to determine the function of the various forms of the c-*abl* protein. Not only will this information further our understanding of growth and differentiation, but it will serve as a basis to understand the mechanism of transformation by v-*abl* and *bcr/abl*. Investigating the role that altered structure, localization, and expression in appropriate cell types plays in *abl*-associated neoplasia should contribute to our understanding of this process. A key area of investigation revolves around identification of the substrate molecules of all forms of the *abl* proteins. The notion that these molecules transduce signals from a growth factor receptor is extremely attractive. The fact that A-MuLV can replace the need for IL-3 in several experimental systems lends support to this idea. Finally, the roles of v-*abl* and *bcr/abl* in initiation and maintenance of the transformed state are poorly understood. It is clear that, in certain situations, A-MuLV can stimulate growth in the absence of malignant state. Also, some tumor cells appear to lose A-MuLV, indicating that the virus may play a more limited role in neoplastic transformation than previously thought. Identification of second events, acting in concert with *abl*, will complete our understanding of the neoplastic process. With the technology and information currently available, one can anticipate that the next few years will bring major advances in these areas.

ACKNOWLEDGMENTS. The assistance of Dr. Lalita Ramakrishnan in preparing the manuscript is gratefully acknowledged. The author is supported by grants from the National Cancer Institute and the National Institute of General Medical Sciences.

References

1. Abelson, H. T., and Rabstein, L. S., 1979, Lymphosarcoma: Virus-induced thymic-independent disease in mice, *Cancer Res.* **30:**2213–2222.
2. Besmer, P., Hardy, W. D., Jr., Zuckerman, E. E., Bergold, P., Lederman, L., and Snyder, H. W., Jr., 1983, The Hardy–Zuckerman 2-FeSV, a new feline retrovirus with oncogene homology to Abelson-MuLV, *Nature* **303:**825–828.

3. Nowell, P. C., and Hungerford, D. A., 1960, Chromosome studies on normal and leukemic human leukocytes, *J. Natl. Cancer Inst.* **25**:85–109.
 4. Heisterkamp, N., Stephenson, J. R., Groffen, J., Hansen, P. F., de Klein, A., Bartram, C. R., and Grosveld, G., 1983, Localization of the c-*abl* oncogene adjacent to a translocation breakpoint in chronic myelocytic leukemia, *Nature* **306**:239–246.
 5. Gale, R. P., and Canaani, E., 1984, A 8-kilobase *abl* RNA transcript in chronic myelogenous leukemia, *Proc. Natl. Acad. Sci. USA* **81**:5648–5662.
 6. Heisterkamp, N., Stam, K., Groffen, J., de Klein, A., and Grosveld, G., 1985, Structural organization of the *bcr* gene and its role in the Ph' translocation, *Nature* **315**:758–761.
 7. Witte, O. N., Dasgupta, A., and Baltimore, D., 1980, Abelson murine leukemia virus protein is phosphorylated in vitro to form phosphotyrosine, *Nature* **283**:826–831.
 8. Witte, O. N., Goff, S., Rosenberg, N., and Baltimore, D., 1980, A transformation-defective mutant of Abelson murine leukemia virus lacks protein kinase activity, *Proc. Natl. Acad. Sci. USA* **77**:4993–4997.
 9. Reynolds, F. H., Van de Ven, W. J. M., and Stephenson, J. R., 1980, Abelson murine leukemia virus transformation-defective mutants with impaired p120-associated protein kinase activity, *J. Virol.* **36**:374–386.
10. Konopka, J. B., Watanabe, S. M., and Witte, O. N., 1984, An alteration of the human c-*abl* protein in K562 leukemia cells unmasks associated tyrosine kinase activity, *Cell* **37**:1035–1042.
11. Rosenberg, N., 1982, Abelson leukemia virus. *Curr. Top. Microbiol. Immunol.* **101**:163–194.
12. Konopka, J. B., and Witte, O. N., 1985, Activation of the *abl* oncogene in murine and human leukemias. *Biochim. Biophys. Acta* **823**:1–17.
13. Risser, R., 1982, The pathogenesis of Abelson virus lymphomas of the mouse, *Biochim. Biophys. Acta* **651**:213–233.
14. Goff, S. P., Gilboa, E., Witte, O. N., and Baltimore, D., 1980, Structure of Abelson murine leukemia virus genome and the homologous cellular gene: Studies with cloned viral DNA, *Cell* **22**:777–785.
15. Wang, J. Y. J., Ledley, F., Goff, S. P., Lee, R., Groner, Y., and Baltimore, D., 1984, The mouse c-*abl* locus: Molecular cloning and characterization, *Cell* **36**:349–356.
16. Ben-Neriah, Y., Bernards, A., Paskind, M., Daley, G. Q., and Baltimore, D., 1986, Alternative 5' exons in c-*abl* mRNA, *Cell* **44**:577–586.
17. Rosenberg, N., and Witte, O. N., 1980, Abelson murine leukemia virus mutants with alterations in the A-MuLV-specific p120 molecule, *J. Virol.* **33**:340–348.
18. Goff, S. P., Witte, O. N., Gilboa, E., Rosenberg, N., and Baltimore, D., 1981, Genome structure of Abelson murine leukemia virus variants: Provirus in fibroblast and lymphoid cells, *J. Virol.* **38**:460–468.
19. Lee, R., Paskind, M., Wang, J. Y. J., and Baltimore, D., 1985, Abelson (P160) murine leukemia virus (Ab-MLV) *abl* gene, in : *RNA Tumor Viruses* (R. Weiss, N. Teich, H. Varmus, and J. Coffin, eds.), Cold Springs Harbor Laboratory, Cold Spring Harbor, N.Y., pp. 861–868.
20. Wang, J. Y. J., and Baltimore, D., 1983, Cellular RNA homologous to the Abelson murine leukemia virus transforming gene: Expression and relationship to the viral sequence, *Mol. Cell. Biol.* **3**:773–779.
21. Witte, O. N., Rosenberg, N., Paskind, M., Shields, A., and Baltimore, D., 1978, Identification of an Abelson murine leukemia virus-encoded protein present in transformed fibroblasts and lymphoid cells, *Proc. Natl. Acad. Sci. USA* **75**:2488–2492.
22. Reynolds, F. H., Jr., Sacks, T. S., Deogagkar, D. N., and Stephenson, J. R., 1978, Cells nonproductively infected by Abelson murine leukemia virus express a high molecular weight polyprotein containing structural and nonstructural components, *Proc. Natl. Acad. Sci. USA* **75**:3974–3978.
23. Reddy, E. P., Smith, M. J., and Srinivasan, A., 1983, Abelson murine leukemia virus genome: Structural similarity of its transforming gene product to other *onc* products with tyrosine-specific kinase activity, *Proc. Natl. Acad. Sci. USA* **80**:3623–3627.
24. Witte, O. N., Rosenberg, N., and Baltimore, D., 1979, Preparation of syngeneic tumor regressor serum reactive with the unique determinants of the Abelson MuLV encoded P120 protein at the cell surface, *J. Virol.* **31**:776–784.
25. Boss, M. A., Dreyfuss, G., and Baltimore, D., 1981, Localization of the Abelson murine leukemia

virus protein in a detergent-insoluble subcellular matrix: Architecture of the protein, *J. Virol.* **40:**472–481.
26. Sefton, B. M., Trowbridge, I. S., Cooper, J. A., and Scolnick, E. M., 1982, The transforming proteins of Rous sarcoma virus, Harvey sarcoma virus and Abelson virus contain tightly bound lipid, *Cell* **31:**465–474.
27. Schultz, A., and Oroszlan, S., 1984, Myristylation of *gag–onc* fusion proteins in mammalian transforming retroviruses, *Virology* **133:**431–437.
28. Cross, F. R., Garber, E. A., Pellman, D., and Hanafusa, H., 1984, A short sequence in the p60*src* N terminus is required for p60*src* myristylation and membrane association and for cell transformation, *Mol. Cell. Biol.* **4:**1834–1842.
29. Ziegler, S. F., Whitlock, C. A., Goff, S. P., Gifford, A., and Witte, O. N., 1981, Lethal effect of the Abelson murine leukemia virus transforming gene product, *Cell* **27:**477–486.
30. Watanabe, S. M., Rosenberg, N., and Witte, O. N., 1984, A membrane-associated, carbohydrate-modified form of the v-*abl* protein that cannot be phosphorylated in vivo or in vitro, *J. Virol.* **51:**620–627.
31. Schiff-Maker, L., and Rosenberg, N., 1986, *Gag*-derived but not *abl*-derived determinants are exposed on the surface of Abelson virus-transformed cells, *Virology* **154:**286–301.
32. Pellman, D., Garber, E. A., Cross, F. R., and Hanafusa, H., 1985, Fine structure mapping of a critical NH_2-terminal region of p60*src*, *Proc. Natl. Acad. Sci. USA* **82:**1623–1627.
33. Bishop, J. M., 1983, Cellular oncogenes and retroviruses, *Annu. Rev. Biochem.* **52:**301–354.
34. Wang, J. Y. J., Queen, C., and Baltimore, D., 1982, Expression of an Abelson murine leukemia virus-encoded protein in *Escherichia coli* causes extensive phosphorylation of tyrosine residues, *J. Biol. Chem.* **257:**13181–13184.
35. Wang, J. Y. J., and Baltimore, D., 1985, Localization of tyrosine kinase-coding region in v-*abl* oncogene by the expression of v-*abl*-encoded proteins in bacteria, *J. Biol. Chem.* **260:**64–71.
36. Foulkes, J. G., Chow, M., Gorka, C., Frackelton, R., Jr., and Baltimore, D., 1985, Purification and characterization of a protein-tyrosine kinase encoded by the Abelson murine leukemia virus, *J. Biol. Chem.* **260:**8070–8077.
37. Sefton, B. M., Hunter T., and Raschke, W. C., 1981, Evidence that the Abelson virus protein functions in vivo as a protein kinase that phosphorylates tyrosine, *Proc. Natl. Acad. Sci. USA* **78:**1552–1556.
38. Witte, O. N., Ponticelli, A., Gifford, A., Baltimore, D., Rosenberg, N., and Elder, J., 1981, Phosphorylation of the Abelson murine leukemia virus transforming protein, *J. Virol.* **39:**870–878.
39. Cooper, J. A., and Hunter, T., 1981, Four different classes of retroviruses induce phosphorylation of tyrosine present in similar cellular proteins, *Mol. Cell. Biol.* **1:**394–407.
40. Cooper, J. A., and Hunter, T., 1983, Regulation of cell growth and transformation by tyrosine-specific protein kinases: The search for important cellular substrate proteins, *Curr. Top. Microbiol. Immunol.* **107:**125–161.
41. Rosenberg, N., Clark, D. R., and Witte, O. N., 1980, Abelson murine leukemia mutants deficient in kinase activity and lymphoid cell transformation, *J. Virol.* **36:**766–774.
42. Srinivasan, A., Dunn, C. Y., Yuasa, Y., Devare, S. G., Premkumar, E., Reddy, E. P., and Aaronson, S. A., 1982, Abelson murine leukemia virus: Structural requirements for transforming gene function, *Proc. Natl. Acad. Sci. USA* **79:**5508–5512.
43. Prywes, R., Foulkes, J. G., and Baltimore, D., 1985, The minimum transforming region of v-*abl* is the segment encoding tyrosine kinase, *J. Virol.* **53:**114–122.
44. Prywes, R., Hoag, J., Rosenberg, N., and Baltimore, D., 1985, Protein stabilization explains the *gag* requirement for transformation of lymphoid cells by Abelson murine leukemia virus, *J. Virol.* **53:**123–132.
45. Goff, S. P., Tabin, C. J., Wang, J., Weinberg, R., and Baltimore, D., 1982, Transfection of fibroblasts by cloned Abelson murine leukemia virus DNA and recovery of transmissible virus by recombination with helper virus, *J. Virol.* **41:**271–285.
46. Watanabe, S. M., and Witte, O. N., 1983, Site-directed deletions of Abelson murine leukemia virus define 3' sequences essential for transformation and lethality, *J. Virol.* **45:**1028–1036.
47. Prywes, R., Foulkes, J. G., Rosenberg, N., and Baltimore, D., 1983, Sequences of the A-MuLV protein needed for fibroblast and lymphoid cell transformation, *Cell* **34:**569–579.

48. Murtagh, K., Skladany, G., Hoag, J., and Rosenberg, N., 1986, Abelson virus variants with increased oncogenic potential, *J. Virol.* **60:**599–606.
49. Earl, C. D., and Scher, C. D., 1980, Detection of lymphoid leukemia colony-forming cells in Abelson virus-infected mice: Differences in inbred strains, *J. Cell. Physiol.* **104:**153–162.
50. Rosenberg, N., Baltimore, D., and Scher, C. D., 1975, In vitro transformation of lymphoid cells by Abelson murine leukemia virus, *Proc. Natl. Acad. Sci. USA* **72:**1932–1936.
51. Rosenberg, N., and Baltimore, D., A quantitative assay for transformation of bone marrow cells by Abelson murine leukemia virus, *J. Exp. Med.* **143:**1453–1463.
52. Till, J. E., and McCulloch, E. A., 1980, Hematopoietic cell differentiation, *Biochim. Biophys. Acta* **605:**431–459.
53. Whitlock, C. A., Denis, K., Robertson, D., and Witte, O., 1985, In vitro analysis of murine B cell development, *Annu. Rev. Immunol.* **3:**213–236.
54. Alt, F. W., Blackwell, T. K., DePinho, R. A., Reth, M. G., and Yancopoulos, G. D., 1986, Regulation of genome rearrangement events during lymphocyte differentiation, *Immunol. Rev.* **89:**5–30.
55. Palacios, R., Henson, G., Steinmetz, M., and McKearn, J. P., 1984, Interleukin-3 supports growth of mouse pre-B cell clones in vitro, *Nature* **309:**126–134.
56. Palacios, R., and Steinmetz, M., 1985, IL-3 dependent mouse clones that express B-220 surface antigen, contain Ig genes in germline configuration and generate B lymphocytes in vivo, *Cell* **41:**727–734.
57. Muller-Seiburg, C. E., Whitlock, C. A., and Weissman, I. L., 1986, Isolation of two early B lymphocyte progenitors from mouse marrow: A committed pre-pre-B cell and a clonogenic Thy-1^{lo} hematopoietic stem cell, *Cell* **44:**653–662.
58. Burrows, P., LeJune, M., and Kearney, J. F., 1979, Evidence that murine pre-B cells synthesize μ heavy chains but no light chains, *Nature* **280:**838–841.
59. Levitt, D., and Cooper, M. D., 1980, Mouse pre-B cells synthesize and secrete μ heavy chains but not light chains, *Cell* **19:**617–626.
60. Alt, F. W., Rosenberg, N., Lewis, S., Thomas, E., and Baltimore, D., 1981, Organization and reorganization of immunoglobulin genes in A-MuLV-transformed cells: Rearrangement of heavy but not light chain genes, *Cell* **27:**381–390.
61. Melchers, F., and Andersson, J., 1984, B cell activation: Three steps and their variations, *Cell* **37:**715–720.
62. Cory, S., and Adams, J. M., 1980, Deletions are associated with somatic rearrangements of immunoglobulin heavy chain genes, *Cell* **19:**37–51.
63. Cory, S., Jackson, J., and Adams, J. M., 1980, Deletions in the constant region locus can account for switches in immunoglobulin heavy chain expression, *Nature* **285:**450–456.
64. Rabbitts, T. H., Forster, A., Dunnick, W., and Bentley, D. L., 1980, The role of gene deletion in the immunoglobulin heavy chain switch, *Nature* **283:**351–356.
65. Holmes, K. L., Pierce, J. H., Davidson, W. F., and Morse, H. C., III, 1986, Murine hematopoietic cells with pre-B or pre-B/myeloid characteristics are generated by in vitro transformation with retroviruses containing *fes, ras, abl* and *src* oncogenes, *J. Exp. Med.* **164:**443–457.
66. McKearn, J. P., and Rosenberg, N., 1985, Mapping cell surface antigens on mouse pre-B cell lines, *Eur. J. Immunol.* **15:**295–298.
67. Pratt, D. M., Strominger, J., Parkman, R., Kaplan, D., Schwaber, J., Rosenberg, N., and Scher, C. D., 1977, Abelson virus-transformed lymphocytes: Null cells that modulate H-2, *Cell* **12:**683–690.
68. Shinefeld, L., Sato, V., and Rosenberg, N., 1980, Monoclonal rat anti-mouse brain antibody detects Abelson murine leukemia virus target cells in mouse bone marrow, *Cell* **20:**11–18.
69. Pillemer, E., Whitlock, C., and Weissman, I. L., 1984, Transformation-associated proteins in murine B cell lymphomas that are distinct from Abelson virus gene products, *Proc. Natl. Acad. Sci. USA* **81:**4434–4438.
70. Cooper, M. D., Mulvaney, D., Coutinho, A., and Cazenave, P. A., 1986, A novel cell surface molecule on early B-lineage cells, *Nature* **321:**616–618.
71. Manohar, V., Brown, E., Leiserson, W. M., and Chused, T. M., 1982, Expression of Lyt-1 by a subset of B lymphocytes, *J. Immunol.* **129:**532–538.

72. Hayakawa, K., Hardy, R., Parks, D. R., and Herzenberg, L. A., 1983, The Ly-1 B cell subpopulation in normal, immunodefective and auto-immune mice, *J. Exp. Med.* **157**:202–218.
73. Siden, E. J., Baltimore, D., Clark, D., and Rosenberg, N., 1979, Immunoglobulin synthesis by lymphoid cells transformed in vitro by Abelson murine leukemia virus, *Cell* **16**:389–396.
74. Waneck, G. L., and Rosenberg, N., 1981, Abelson leukemia virus induces lymphoid and erythroid colonies in infected fetal cell cultures, *Cell* **26**:79–89.
75. Tidmarsh, G. F., Heimfeld, S., Whitlock, C. A., Weissman, I. L., and Muller-Sieburg, C. A., 1987, Bone marrow target cells for Abelson leukemia virus in vitro transformation co-express B-200 and Thy-1 antigens (submitted for publication).
76. Whitlock, C. A., and Witte, O. N., 1981, Abelson virus-infected cells can exhibit restricted in vitro growth and low oncogenic potential, *J. Virol.* **40**:577–584.
77. Whitlock, C. A., Ziegler, S. F., and Witte, O. N., 1983, Progression of the transformed phenotype in clonal lines of Abelson virus-infected lymphocytes, *Mol. Cell. Biol.* **3**:596–604.
78. Alt, F. W., Yancopoulos, G. D., Blackwell, T. K., Wood, C., Thomas, E., Boss, M., Coffman, R., Rosenberg, N., Tonegawa, S., and Baltimore, D., 1984, Ordered rearrangement of immunoglobulin heavy chain variable region segments, *EMBO J.* **3**:1209–1219.
79. Yancopoulos, G., D., Blackwell, T. K., Suh, H., Hood, L., and Alt, F. W., 1986, Introduced T cell receptor variable region gene segments in pre-B cells: Evidence that B and T cells use a common recombinase, *Cell* **44**:251–259.
80. Reth, M. G., and Alt, F. W., 1984, Novel immunoglobulin heavy chains are produced from DJ_H gene segment rearrangement in lymphoid cells, *Nature* **312**:418–420.
81. Reth, M. G., Amirati, P., Jackson, S., and Alt, F. W., 1985, Regulated progression of a cultured pre-B cell line to the B cell stage, *Nature* **317**:353–355.
82. Clark, D. R., and Rosenberg, N., 1980, Immunoglobulin synthesis in LPS-stimulated lymphoid cells transformed by Abelson murine leukemia virus, in: *Erythropoiesis and Differentiation in Friend Leukemia Cells* (G. B. Rossi, ed.) Elsevier/North-Holland, Amsterdam, pp. 517–536.
83. Alt, F. W., Rosenberg, N., Casanova, R. J., Thomas, E. J., and Baltimore, D., 1982, Immunoglobulin heavy chain expression and class switching in a murine leukemia cell line, *Nature* **296**:325–331.
84. Lewis, S., Rosenberg, N., Alt, F., and Baltimore, D., 1982, Continuing gene rearrangement in an Abelson murine leukemia virus transformed cell line, *Cell* **30**:807–816.
85. Burrows, P. D., Beck-Engeser, G. B., and Wabl, M. R., 1983, Immunoglobulin heavy-chain class switching in a pre-B cell line is accompanied by DNA rearrangement, *Nature* **306**:243–246.
86. Akira, S., Sugiyama, H., Yoshida, N., Kitutani, H., Yamamura, Y., and Kishimoto, T., 1983, Isotype switching in murine pre-B cell lines, *Cell* **34**:545–556.
87. Sugiyama, H., Maeda, T., Akira, S., and Kishimoto, T., 1986, Class-switching from μ to γ_3 or γ_{2b} production at pre-B cell stage, *J. Immunol.* **136**:3092–3097.
88. Desiderio, S. V., Yancopoulos, G. D., Paskind, M., Thomas, E., Boss, M. A., Landau, N., Alt, F. W., and Baltimore, D., 1984, Insertion of N regions into heavy chain genes is correlated with expression of terminal deoxytransferase in B cells, *Nature* **311**:752–755.
89. Yancopoulos, G. D., and Alt, F. W., 1985, Developmentally controlled and tissue-specific expression of unrearranged V_H gene segments, *Cell* **40**:271–281.
90. Raff, M. C., Megson, M., Owen, J. T. T., and Cooper, M. D., 1976, Early production of intracellular IgM by B-lymphocyte precursors in mouse, *Nature* **259**:224–226.
91. Andrew, T. A., and Owen, J. T. T., 1978, Studies on the earliest sites of B cell differentiation in the mouse embryo, *Dev. Comp. Immunol.* **2**:339–346.
92. Melchers, F., 1977, B lymphocyte development in fetal liver. II. Frequencies of precursor B cells during gestation, *Eur. J. Immunol.* **7**:482–486.
93. Melchers, F., 1979, Murine embryonic B lymphocyte development in the placenta, *Nature* **277**:219–221.
94. Karagogeos, D., Rosenberg, N., and Wortis, H., 1986, Early arrest of B cell development in nude X-linked immunodeficient mice, *Eur. J. Immunol.* **16**:1125–1130.
95. Wortis, H. H., Burkly, L., Hughes, D., Roschelle, S., and Waneck, G., 1982, Lack of mature B cells in nude mice with X-linked immune deficiency, *J. Exp. Med.* **155**:903–913.
96. Whitlock, C. A., Ziegler, S. F., Treiman, L. J., Stafford, J. I., and Witte, O. N., 1983, Differentia-

tion of cloned populations of immature B cells after transformation with Abelson murine leukemia virus, *Cell* **32**:903–911.

97. Ziegler, S. F., Treiman, L. J., and Witte, O. N., 1984, Gene diversity among the clonal progeny of pre-B lymphocytes, *Proc. Natl. Acad. Sci. USA* **81**:1529–1533.
98. Waneck, G. L., Keyes, L., and Rosenberg, N., 1986, Abelson virus drives the differentiation of Harvey virus-infected erythroid cells, *Cell* **44**:337–344.
99. Oliff, A., Agranovsky, O., McKinney, M. D., Murty, V. V. V. S., and Bauchwitz, R., 1985, Friend murine leukemia virus-immortalized myeloid cells are converted into tumorigenic cells by Abelson leukemia virus, *Proc. Natl. Acad. Sci. USA* **82**:3306–3310.
100. Pierce, J. H., DiFiore, P. P., Aaronson, S. A., Potter, M., Pumphrey, J., Scott, A., and Ihle, J. N., 1985, Neoplastic transformation of mast cells by Abelson-MuLV: Abrogation of IL-3 dependence by a nonautocrine mechanism, *Cell* **41**:685–693.
101. Cook, W. D., Metcalf, D., Nicola, N. A., Burgess, A. W., and Walker, F., 1985, Malignant transformation of a growth factor-dependent myeloid cell line by Abelson virus without evidence of an autocrine mechanism, *Cell* **41**:677–683.
102. Iscove, N. N., Roitsch, C. A., Williams, N., and Guilbert, L. J., 1982, Molecules stimulating early red cell, granulocyte, macrophage and megakaryocyte precursors in culture: Similarity in size, hydrophobicity and charge, *J. Cell. Physiol. Suppl.* **1**:65–78
103. Cook, W. D., 1982, Rapid thymomas induced by Abelson murine leukemia virus, *Proc. Natl. Acad. Sci. USA* **79**:2917–2921.
104. Cook W. D., 1985, Thymocyte subsets transformed by Abelson murine leukemia virus, *Mol. Cell. Biol.* **5**:390–397.
105. Risser, R., Kaehler, D., and Lamph, W., 1985, Different genes control the susceptibility of mice to Moloney or Abelson murine leukemia virus, *J. Virol.* **55**:547–553.
106. Scott, M. L., Davis, M. M., and Feinberg, M. B., 1986, Transformation of T lymphoid cells by Abelson murine leukemia virus, *J. Virol.* **59**:434–443.
107. Risser, R., Potter, M., and Rowe, W. P., 1978, Abelson virus-induced lymphoma genesis in mice, *J. Exp. Med.* **148**:714–726.
108. Potter, M., Sklar, M. D., and Rowe, W. P., 1973, Rapid viral induction of plasmacytomas in pristane-primed Balb/c mice, *Science* **182**:592–594.
109. Mendoza, G. R., and Metzer, H., 1976, Disparity of IgE binding between normal and tumor mouse mast cells, *J. Immunol.* **117**:1573–1578.
110. Raschke, W. C., Baird, S., Ralph, P., and Nakoinz, I., 1978, Functional macrophage cell lines transformed by Abelson murine leukemia virus, *Cell* **15**:261–267.
111. Mushinski, J. F., Potter, M., Bauer, S. R., and Reddy, E. P., 1983, DNA rearrangement and altered RNA expression of the c-*myb* oncogene in mouse plasmcytoid lymphosarcomas, *Science* **220**:795–799.
112. Serunian, L. A., and Rosenberg, N., 1986, Abelson virus potentiates long term growth of mature B lymphocytes, *Mol. Cell. Biol.* **6**:183–194.
113. Shen-Ong, G. L. C., Potter, M., Mushinski, J. F., Lavu, S., and Reddy, E. P., 1984, Activation of the c-*myb* locus by viral insertional mutagenesis in plasmacytoid lymphosarcomas, *Science* **226**:1077–1080.
114. Grunwald, D., Dale, B., Dudley, J., Lamph, W., Sugden, B., Ozanne, B., and Risser, R., 1982, Loss of viral gene expression and retention of tumorigenicity by Abelson lymphoma cells, *J. Virol.* **42**:92–103.
115. Hoffman-Falk, H., Einat, P., Shilo, B.-Z., and Hoffman, F. M., 1983, *Drosophila melanogaster* DNA clones homologous to vertebrate oncogenes: Evidence for a common ancestor to the *src* and *abl* cellular genes, *Cell* **32**:589–598.
116. Hoffman, F. M., Fresco, L. D., Hoffman-Falk, H., and Shilo, B.-Z., 1983, Nucleotide sequences of the Drosophila *src* and *abl* homologs: Conservation and variability in the *src* family oncogenes, *Cell* **35**:393–401.
117. Heisterkamp, N., Groffen, J., and Stephenson, J. R., 1983, The human v-*abl* cellular homologue, *J. Appll. Genet.* **2**:57–68.
118. Goff, S. P., D'Eustachio, P., Ruddle, F. H., and Baltimore, D., 1982, Chromosomal assignment of the endogenous proto-oncogene c-*abl*, *Science* **218**:1317–1319.

119. Heisterkamp, N., Groffen, J., Stephenson, J. R., Spurr, N. K., Goodfellow, P. N., Solomon, E., Carritt, B., and Bodmer, W. F., 1982, Chromosomal localization of human cellular homologues of two viral oncogenes, *Nature* **299**:747–749.
120. Muller, R., Salmon, D. J., Tremblay, J. M., Cline, M. J., and Verma, I. M., 1982, Differential expression of cellular oncogenes during pre- and post-natal development of the mouse, *Nature* **299**:640–643.
121. Witte, O. N., Rosenberg, N., and Baltimore, D., 1979, Identification of a normal cellular protein cross-reactive to the major Abelson leukemia virus gene product, *Nature* **281**:396–398.
122. Ponticelli, A. S., Whitlock, C. A., Rosenberg, N., and Witte, O. N., 1982, In vivo tyrosine phosphorylations of the Abelson virus transforming protein are absent in its normal cellular homolog, *Cell* **29**:953–960.
123. Konopka, J. B., Watanabe, S. M., and Witte, O. N., 1984, An alteration of the human c-*abl* protein in K562 leukemia cells unmasks associated tyrosine kinase activity, *Cell* **37**:1035–1042.
124. Konopka, J. B., and Witte, O. N., 1985, Detection of c-*abl* tyrosine kinase activity in vitro permits direct comparison of normal and altered *abl* gene products, *Mol. Cell. Biol.* **5**:3116–3123.
125. Groffen, J., Stephenson, J. R., Heisterkamp, N., deKlein, A., Bartram, C. R., and Grosveld, G., 1984, Philadelphia chromosomal breakpoints are clustered within a limited region, *bcr*, on chromosome 22, *Cell* **36**:93–99.
126. Collins, S. J., Kubonishi, I., Miyoshi, I., and Groudine, M. T., 1984, Altered transcription of the c-*abl* oncogene in K-562 and other chronic myelogenous leukemia cells, *Science* **225**:72–74.
127. Shtivelman, E., Lifshitz, B., Gale, R. P., and Canaani, E., 1985, Fused transcript of *abl* and *bcr* genes in chronic myelogenous leukemia, *Nature* **315**:550–554.

10

Comparison of the Structural and Functional Properties of the Viral and Cellular *src* Gene Products

PAUL COUSSENS AND JOAN S. BRUGGE

1. Introduction

The oncogenic transformation of cells in culture and the induction of tumors in animals by some retroviruses are due to the expression of transformation-specific viral genes (v-*onc* genes; reviewed in Ref. 1). These genes were acquired from normal cells via recombination between the retroviral genome and genes in host cell DNA (c-*onc* or protooncogenes; reviewed in Ref. 2). The v-*onc* genes are homologous to their cellular progenitor, but often contain mutations that are responsible for activating the oncogenic potential of the c-*onc* genes. Cellular protooncogenes have been highly conserved throughout evolution, suggesting that their protein products perform essential functions in normal cells. This review will focus on investigations of the viral and cellular *src* genes and their protein products, which have provided a model system for the analysis of oncogenic transformation. This will include (1) a comparison of the structural and functional differences between the nononcogenic cellular *src* gene product and the oncogenic viral *src* protein of Rous sarcoma virus (RSV), (2) a discussion of the expression of the cellular *src* gene product in various normal cells, and (3) an analysis of the regulation of $pp60^{c-src}$ functional activity in different cell types.

PAUL COUSSENS AND JOAN S. BRUGGE • Department of Microbiology, State University of New York, Stony Brook, New York 11794-8621. *Present address of P.C.:* Department of Microbiology and Public Health, Michigan State University, East Lansing, Michigan 48824.

2. Transformation by the v-*src* Protein

RSV induces the rapid production of fibrosarcomas in chickens and the transformation of a variety of cell types in culture. Multiple phenotypic alterations in cellular physiology accompany transformation by RSV, including changes in cell morphology, an increased rate of glycolysis and transport of simple sugars, an increased production of extracellular proteases, and loss of anchorage-dependence of growth (reviewed by Hanafusa[3]). All of these phenotypic changes are induced by the expression of the transforming gene, v-*src*, which is encoded by RSV. The protein product of the *src* gene is a 60,000-dalton phosphoprotein ($pp60^{v-src}$)[4,5] that possesses a protein phosphotransferase activity specific for tyrosine residues.[6,7] Many other structurally distinct retroviral transforming proteins also possess tyrosine-specific protein kinase activity, and share varying, but significant, homology with the amino acid sequence of $pp60^{v-src}$. The most highly conserved region is that surrounding the putative catalytic domain, representing the carboxy-half of $pp60^{v-src}$. Thus, $pp60^{v-src}$ is a member of a family of tyrosine-specific protein kinases encoded by oncogenic retroviruses.[2,8]

It is likely that the phenotypic alterations associated with RSV-induced transformation result from $pp60^{v-src}$-mediated phosphorylation of cellular proteins on tyrosine. RSV-transformed cells characteristically display elevated levels of total cellular phosphotyrosine and the phosphorylation of at least 30 different proteins on tyrosine has been shown to increase following expression of $pp60^{v-src}$ (reviewed in Refs. 8, 9). Studies on various RSV mutants that are temperature sensitive for transformation have shown that the presence of a kinase-active $pp60^{v-src}$ protein is essential for the maintenance of these high phosphotyrosine levels and the expression of the transformed phenotype.[10] Furthermore, the analysis of mutant revertant viruses that display partially transformed phenotypes suggests that interaction of $pp60^{v-src}$ with multiple substrates is responsible for induction of the various alterations in cell physiology which accompany RSV transformation.[11,12] Unfortunately, there is no direct biochemical evidence that links any of the phosphorylated substrates in $pp60^{v-src}$-transformed cells to the expression of any of these phenotypic changes. Indeed, several lines of evidence suggest that many of the phosphorylation events that occur in v-*src*-transformed cells are of no physiological relevance and merely reflect the activity of an unregulated tyrosine kinase.

3. Structural Comparison of the Cellular and Viral *src* Gene Products

The c-*src* gene product is structurally similar to the v-*src* protein.[13,14] Amino acid sequence data, derived from the DNA sequence of the v-*src* and c-*src* genes, have shown that the v-*src* gene contains approximately 10 single amino acid substitutions scattered throughout the molecule (the exact number and location depend on the strain of virus) and that the last 19 amino acids of $pp60^{c-src}$ have been replaced by a different sequence of 12 amino acids in $pp60^{v-src}$.[15] Both viral and cellular *src* proteins are phosphorylated on serine and tyrosine residues.

Although the site of serine phosphorylation appears to be identical for the cellular and viral forms of pp60^{src}, the *in vivo* tyrosine phosphorylation site of the c-*src* protein is residue 527,[16] whereas the v-*src* gene product is phosphorylated on Tyr-416.[17] Another posttranslational modification, which is shared by both forms of the *src* gene products, is the addition of the fatty acid, myristate,[18,19] to the amino-terminal glycine residue. This modification appears to facilitate the association of the *src* gene product with the plasma membrane since mutants containing amino acid substitutions at the glycine acceptor of myristate are not capable of fatty acylation and do not associate with the membrane.[20]

4. Functional Comparison of the Viral and Cellular *src* Gene Products

Investigations of other protooncogenes and their viral counterparts have indicated that certain protooncogenes are oncogenic when expressed at high levels relative to their normal expression.[21–23] Alternatively, other oncogenes require mutagenic alterations to activate their oncogenic potential.[24] In fibroblasts transformed by RSV, pp60^{v-src} is expressed at levels 15- to 30-fold higher than the c-*src* gene product.[25] Since the v-*src* gene product contains multiple amino acid differences compared to the c-*src* gene product, it was not clear whether one can account for the oncogenicity of the v-*src* gene product merely on the basis of its overexpression relative to pp60^{c-src}, or whether the mutational differences in the v-*src* protein were necessary to alter the functional activity of the *src* gene. In order to distinguish between these possibilities, several laboratories constructed recombinant DNA vectors that allowed the expression of the c-*src* gene product at levels comparable to those of the pp60^{v-src} protein in RSV-transformed cells. Iba *et al.*[26] constructed a variant of RSV in which the c-*src* gene was substituted for the v-*src* gene. Parker *et al.*[27] and Shalloway *et al.*[28] generated plasmids that carry the c-*src* gene under the transcriptional control of foreign promoters that allow high-level expression after transfection of mammalian cells. All three approaches yielded similar results, indicating that the expression of the c-*src* gene product at levels comparable to those found in RSV-transformed cells does not result in tumor formation in animals or cellular transformation in culture. In order to determine which mutational difference between pp60^{c-src} and pp60^{v-src} was responsible for the differences in oncogenicity of these gene products, several hybrid genes were constructed which contained different portions of the c-and v-*src* genes.[26,28] While the substitution of the cluster of unique sequences at the 3′ end of the v-*src* gene was sufficient to activate the oncogenicity of the c-*src* gene product, single or double amino acid substitutions were also able to activate the c-*src* gene[26,28,29] (Takeya, Kato, Granderi, Levy, Iba, and Hanafusa, unpublished results). This conclusion was substantiated by the analysis of transforming variants which were generated during passage of the c-*src*-containing RSV.[30] These variants were found to possess single, unique amino acid substitutions at amino acid residues 378 and 441.[30] It is of interest that the mutagenic alterations, which cause the c-*src* gene to be oncogenic, are contained within both amino- and carboxy-halves of the *src* gene. Considering this large target size, it is perhaps surprising that no human tumors have been shown to contain an activated c-*src* protooncogene product.

5. Comparison of the Tyrosine Kinase Activity of the c-*src* and v-*src* Gene Products

Despite the high degree of amino acid conservation in the putative catalytic domains of pp60^{c-src} and pp60^{v-src}, the specific activity of pp60^{c-src} is only 1–10% that of pp60^{v-src}.[31,32] This low specificity activity of pp60^{c-src} observed in immune complex kinase assays *in vitro* is also reflected in the phosphorylation of protein on tyrosine *in vivo*. Cells infected with the c-*src*-RSV variant contained levels of total cellular phosphotyrosine (0.1–0.13%) only slightly higher than those found in uninfected cells (0.07%), despite the 15- to 20-fold higher levels of the c-*src* gene product.[31] The phosphorylation of the 34,000-dalton substrate on tyrosine was also 30-fold lower than the levels detected in wild-type RSV-infected cells which express equivalent levels of the v-*src* protein.[31] Coussens and co-workers detected slightly higher levels of phosphotyrosine in mouse cells that expressed the c-*src* gene product at levels 30-fold higher than those found in normal cells; however, these levels were significantly lower than in cells that expressed 15-fold lower levels of the v-*src* protein.[32] Thus, it is clear that the kinase activity of the c-*src* gene product is severely restricted compared to its viral homologue, and the analyses of c-*src* variant viruses, which possess transforming activity, have shown a strong correlation between activation of pp60^{src} kinase activity and oncogenicity.

Another distinction between the nononcogenic and oncogenic variants of the c-*src* protein was a shift in the site of tyrosine phosphorylation *in vivo* from Tyr-527 to Tyr 416, the site of the phosphorylation of pp60^{v-src}.[31] This suggests the possibility that tyrosine phosphorylation at Tyr-527 may regulate the kinase activity of pp60^{c-src}. Cartwright *et al.*[33] have shown recently that Tyr-527 is not phosphorylated *in vivo* on the activated form of pp60^{c-src} which is associated in a complex with the middle T antigen of polyoma virus (see Section 6). As in other activated forms of pp60^{c-src}, Tyr-416 was the acceptor of phosphate *in vivo*. Courtneidge and co-workers have also reported that the c-*src* protein extracted from cells under conditions which do not preserve phosphorylation on Tyr-527 possesses two- to three-fold higher levels of kinase activity.[34] In addition, Cooper and King have shown that dephosphorylation of Tyr-527 using potato acid phosphatase causes a 6- to 10-fold activation of pp60^{c-src} kinase activity.[35] These results suggest that phosphorylation of Tyr-527 might repress the activity of pp60^{c-src}.

The effect of Tyr-416 phosphorylation on pp60^{c-src} activity has not been clarified. While all the activated forms of pp60^{c-src} expressed in fibroblasts are phosphorylated on Tyr-416 (rather than on Tyr-527),[30,31,32] it is not clear whether this phosphorylation is essential for the activation, or if the phosphorylation of Tyr-416 is merely a consequence of alterations in the c-*src* catalytic domain. Mutant v-*src* gene products, which lack Tyr-416, possess undiminished kinase activity.[36,37] This suggests that phosphorylation at this site is not essential for high levels of pp60^{v-src} kinase activity; however, interpretation of these results is complicated by the existence of multiple mutagenic alterations in the v-*src* gene, which might contribute to the activation of the RSV-encoded *src* gene product.[15] A better evaluation of the influence of Tyr-416 phosphorylation on pp60^{c-src} kinase activity could be derived from mutagenic alteration of this site in the activated variants of pp60^{c-src} which possess only a single mutation in the c-*src* gene.[30]

Both c-*src* and v-*src* proteins are also phosphorylated on Ser-17.[38] Current evidence suggests that this site is phosphorylated by cellular cAMP-dependent protein kinases[39]; however, this has not been rigidly evaluated. Elimination of Ser-17 within pp60^{v-src} does not have any effect on the kinase or transforming activities of this protein.[37] Gottesman and co-workers have reported an increase in pp60^{v-src} kinase activity after treatment of RSV-transformed cells with dibutyryl-cAMP under conditions which lead to increased phosphorylation at this site.[40] As with Tyr-416, the effects of Ser-17 phosphorylation–dephosphorylation on pp60^{src} activity might be better examined using unaltered or activated pp60^{c-src} molecules. The c-*src* and v-*src* gene products have also been shown to be phosphorylated on a novel serine site (Ser-12) after treatment of cells with phorbol esters that activate the cellular kinase, protein kinase C.[41,42] This site is also phosphorylated *in vitro* by purified protein kinase C. There is no evidence that this phosphorylation influences the protein kinase activity of pp60^{src}.

Although the results cited above provide clues that pp60^{c-src} may be regulated *in vivo* by specific phosphorylation events, direct evidence for this is lacking. It is clear that further analysis of pp60^{c-src} variants containing amino acid substitutions for Tyr residues 527 and 416 would help to clarify this issue. Analyses of the phosphorylation of various forms of c-*src* expressed in different cell types would also help to evaluate the mechanism of regulation more precisely.

6. Activation of the pp60^{c-src} Tyrosine Kinase Activity by Polyoma Virus Middle Tumor Antigen

One of the most interesting aspects of pp60^{c-src} regulation involves its interaction with the polyoma virus (Py) transforming protein, middle T antigen (MTAg). It was shown recently that MTAg associates in a complex with pp60^{c-src} in Py-infected and -transformed cells.[43,44] In collaboration with J. Bolen and M. Israel, we have demonstrated that this interaction causes a 30- to 100-fold activation of the tyrosine kinase activity of the pp60^{c-src}.[44] Only a small percentage of intracellular pp60^{c-src} molecules are bound to MTAg in Py-transformed cells, and MTAg binding is necessary for the activation of pp60^{c-src} kinase activity.[44] These results suggested the possibility that certain aspects of MTAg-induced transformation might be due to the MTAg-mediated activation of pp60^{c-src} protein kinase activity. Two questions raised by these findings are: what is the molecular basis for the activation of the pp60^{c-src} kinase activity, and what are the consequences of the activation of pp60^{c-src} kinase activity *in vivo*?

In addressing the first question, we and others have attempted to determine if the pp60^{c-src} protein, which is associated with MTAg, can be distinguished from the unbound form of the c-*src* protein. These studies have identified differences in both the intracellular sites of phosphorylation *in vivo* and the sites phosphorylated in *in vitro* kinase assays. Cartwright *et al.*[(33)] have found that pp60^{c-src} molecules which are bound to MTAg are phosphorylated primarily on Tyr-416, whereas the pp60^{c-src} molecules that are not associated with MTAg are phosphorylated on Tyr-527.[(33)] As mentioned above, it has also been suggested that the phosphorylation on Tyr-527 may suppress the kinase activity of pp60^{c-src}.[(34)] However, the reported 2- to 3-fold activation in pp60^{c-src} kinase activity,[(34)] which resulted from removal of phosphate from Tyr-527, would not appear to account for the 30- to 100-fold activation of pp60^{c-src} associated with MTAg binding. Therefore, it is unlikely that the absence of phosphate on Tyr-527 is solely responsible for the MTAg-induced activation of pp60^{c-src} kinase activity.

We have also found a difference in the site of phosphorylation *in vitro* of pp60^{c-src} protein molecules bound to MTAg.[(45)] The pp60^{c-src} molecules that are immunoprecipitated with antibody directed against the MTAg are phosphorylated on a novel tyrosine residue within the amino-terminal 16,000 daltons of the protein, while the unbound pp60^{c-src} protein molecules are phosphorylated exclusively on Tyr-416. It is not clear if this tyrosine phosphorylation site within the amino-terminal region of pp60^{c-src} is phosphorylated *in vivo*. Phosphorylation at this site is not detectable by standard methods of *in vivo* labeling with ^{32}P-labeled orthophosphate. However, we have been able to detect pp60^{c-src} molecules that are phosphorylated on tyrosine within the amino-terminal 16,000 daltons of the protein by labeling Py-transformed cells with ^{32}P-orthophosphate in the presence of sodium orthovanadate, a potent inhibitor of tyrosine phosphate-specific phosphatases.[(46)] This result raises the possibility that tyrosine phosphorylation at this site might turn over very rapidly *in vivo* and is, therefore, undetectable under standard labeling conditions. Using this method, a similar form of the v-*src* gene product has been detected in RSV-transformed cells.[(47,48)] These results must be viewed with caution, however, since it is possible that orthovanadate induces phosphorylation at sites which are not physiologically relevant.

If the activation of pp60^{c-src} protein kinase activity, which is detected in *in vitro* kinase assays, mimics the intracellular activity of pp60^{c-src}, one would expect to see increased tyrosine phosphorylation of cellular proteins in Py-transformed cells. However, using the methods employed to identify substrates in RSV-transformed cells, no detectable increase in either the content of total cell phosphotyrosine or the phosphorylation of specific substrates of pp60^{v-src} was observed. Two considerations led us to employ sodium orthovanadate to stabilize tyrosine phosphorylation in Py-transformed cells: (1) The dosage of activated pp60^{src} in RSV-transformed cells is at least 150- to 200-fold higher than in Py-transformed cells, since RSV-transformed cells contain 15- to 20-fold more pp60^{v-src} than pp60^{c-src}, and since MTAg activates only one-tenth of the pp60^{c-src} molecules in fibroblasts. (2) In the absence of a phosphatase inhibitor, such as vanadate, tyrosine phosphorylation events turn over very rapidly.

We have examined tyrosine phosphorylation in MTAg-transformed and normal rat F1-11 cells after incubation with vanadate.[(49)] Three methods of analysis were employed: total cell phosphoaminoacid analysis, analysis of tyrosine phosphorylation of the 34,000-dalton substrate phosphorylated in RSV-transformed cells,[(50,51)] and analysis of individual species of phosphotyrosine-containing proteins using antibodies that recognize phosphotyrosine in an immunoblot assay.[52] These studies revealed that vanadate-treated, MTAg-transformed rat F1-11 cells displayed a 16-fold elevation in total cellular phosphotyrosine compared to untreated MTAg-transformed cells. Normal F1-11 cells treated with vanadate showed only a 2-fold increase in cellular phosphotyrosine. The level of 34k phosphorylation on tyrosine in vanadate-treated MTAg-F1-11 cells was higher than that detected in RSV-transformed F1-11 cells and undetectable in untreated MTAg-F1-11 cells or vanadate-treated normal F1-11 cells. Finally, the profile of phosphotyrosine-containing proteins in the vanadate-treated MTAg-F1-11 cells was similar to that from RSV-F1-11 cells, suggesting that the substrates phosphorylated in vanadate-treated MTAg-transformed cells are similar to those phosphorylated in untreated RSV-transformed cells. The fact that normal F1-11 cells showed only slight increases (at the most 2- to 3-fold) in all of the assays described above suggests that this vanadate-induced effect was specifically due to the expression of the MTAg. We have found that vanadate treatment of Py-infected mouse embryo fibroblasts also leads to the detection of high levels of tyrosine phosphorylation of the 34k substrate. This supports the evidence that the increase in detection of phosphorylation on tyrosine in the cells described above is a consequence of MTAg expression. The increased level of phosphotyrosine detected in MTAg-transformed cells does not appear to be due to activation of the $pp60^{c-src}$: MTAg complex kinase activity since we detected no increase in the activity of the $pp60^{c-src}$: MTAg complex in vanadate-treated cells.

All of the above lines of evidence suggest that Py transformation might involve the phosphorylation of substrates similar to those which are phosphorylated in RSV-transformed cells. The fact that these substrates are not detectable in the absence of vanadate could reflect differences in the dosage of activated $pp60^{src}$ in Py- versus RSV-transformed cells. It is clear that the level of $pp60^{v-src}$ necessary to induce oncogenic transformation is much lower than the levels detectable in RSV-transformed cells. Jakobovits et al. have shown that levels of $pp60^{v-src}$ expression, which do not result in detectable tyrosine phosphorylation of 34k, are sufficient to allow growth of the cells in soft agar.[53] These cells might be comparable to Py-transformed cells that display no detectable tyrosine phosphorylation in the absence of an inhibitorlike vanadate that would allow the accumulation of tyrosine phosphorylation events to a detectable level.

It is not clear whether activation of $pp60^{c-src}$ kinase activity by MTAg represents only a quantitative increase in activity, and not an alteration in substrate specificity, or possible other changes in the expression of this enzymatic activity. Until the normal physiological substrates of $pp60^{c-src}$ and the substrates that mediate the events involved in transformation are identified, it is very difficult to evaluate this question adequately.

7. Cell-Type-Specific Expression of the c-src Gene Product

All of the analyses of the c-src protein described above were performed on the c-src protein extracted from either primary embryo fibroblasts or stable lines of rodent fibroblastic cells. These cells represent the typical host cells for RSV infection or transfection by v-*src* expression vectors and are easily established in culture. The levels of the pp60^{c-src} protein were found to be very low in fibroblasts, representing approximately 0.002% of total cellular proteins.[25] To assess the expression of the c-*src* gene product in other cell types, several groups have examined the levels of pp60^{c-src} protein and pp60^{c-src}-specific kinase activity in developing embryos,[54–56] primary cultures of cells from embryos,[57] or established cell lines displaying the phenotype of defined cell lineages. The pattern of c-*src* expression was found to vary both quantitatively and qualitatively in different cell types.

In our examination of chicken embryos, all neural tissues (including brain, retina, and spinal ganglion) were found to contain eight- to ten-fold higher levels of the c-*src* protein than detected in most embryonic tissues and in embryonic fibroblasts.[55] Other investigators have found similarly high levels of pp60^{c-src} in chicken and human embryonic tissues,[58–60] and *src*-related RNA has been shown to be expressed at high levels in neural tissues from *Drosophila*.[62]

Examination of primary cultures of pure neurons and astrocytes indicated that both cell types expressed 15- to 20-fold higher levels of pp60^{c-src} than fibroblasts; however, the pp60^{c-src} protein expressed in the neuronal cultures displayed a retarded mobility on SDS–polyacrylamide gels. Peptide analysis of this neuron-specific variant indicated that the structural difference responsible for the altered mobility was present within the amino-terminal 16,000 daltons of the molecule.[57] The alteration does not appear to result from a posttranslational modification since the c-*src* gene product translated *in vitro* by rabbit reticulocyte lysates programmed with RNA from brain tissues also displayed a retarded electrophoretic mobility compared to the translation product from limb RNA.[61] These results suggest a neuron-specific alteration in the processing of c-*src* mRNA since chicken cells contain only one genomic copy of the c-*src* gene.

Recently, Martinez, Mathey-Prevot, and Baltimore have isolated and characterized a c-*src*-related cDNA clone derived from mouse brain mRNA molecules. The nucleotide sequence of this cDNA is closely related to the coding sequences of the mouse and human c-*src* genes except that it contains an additional 18 nucleotides located at the junction between the sequences which encode exons 3 and 4 of the genomic copy of chicken c-*src* gene. The predicted amino acid sequence of the protein encoded by this cDNA contains few amino acid differences compared to the human and chicken pp60^{c-src} proteins, with the exception of the additional 6 amino acids between amino acids 117 and 118. It is likely that this insertion is responsible for the altered mobility of the neuronal c-*src* gene product since the pp60^{c-src} protein expressed in cells carrying this cDNA displays an altered mobility on SDS-polyacrylamide gels compared to the mobility of mouse fibroblastic pp60^{c-src} (Martinez, Mathey-Prevot, and Baltimore, personal communication). Thus, neurons appear to express high levels of a structurally distinct form of the c-*src* gene product. This finding raises the question of whether c-*src* is involved in triggering neuronal differentiation, or if the c-*src* protein has a specific function in mature neurons.

Recently, we have identified another type of cell that contains high levels of the *src* protein. Peripheral blood platelets contain approximately 20-fold higher levels of the pp60^{c-src} than those found in RSV-transformed cells or 400-fold more than fibroblasts.[63] The platelet form of the *src* gene product does not display the retarded electrophoretic mobility of the neuronal form of pp60^{c-src}. Analysis of the phosphorylation profile of isolated platelet plasma membranes following *in vitro* kinase reactions revealed that tyrosine phosphorylation represented 80 to 90% of the kinase activity, suggesting that pp60^{c-src} might provide a function in events that take place in platelet membranes. We are currently investigating whether platelet activation is accompanied by changes in the phosphorylation of any platelet proteins on tyrosine. Interestingly, most stable lines of megakaryocytes also possess high levels of the c-*src* protein, approximately 15- to 20-fold higher than fibroblasts (A. Golden, D. Morgan, and J. S. Brugge, unpublished results).

The identification of high levels of the pp60^{c-src} in postmitotic, terminally differentiated cells, like neurons and platelets, was unexpected. Expression of the v-*src* gene product in immature stem cells has been shown to stimulate cell proliferation and to interfere with the differentiation of these cells in culture. This behavior of the v-*src* protein led to the speculation that the c-*src* gene product might be involved in the regulation of normal cell growth and proliferation. If high levels of expression of pp60^{c-src} in neurons and platelets are an indication that this protein provides specific functions in these mature cells, it is likely that the c-*src* gene product is not involved in growth control, unless it signals the terminal differentiation of some cells. It is likely that tyrosine kinases are involved in many different aspects of cellular signal transduction. While it is clear that several tyrosine kinases, such as the growth factor receptors, regulate growth control, other tyrosine kinases might regulate distinct cellular events that are triggered by signals from the extracellular environment. The c-*src* gene does not possess an extracellular domain, but might communicate with a protein that contains an extracellular ligand-binding site.

Given this scenario, how does the v-*src* gene product cause cellular transformation? Perhaps the stimulation of kinase activity associated with the oncogenic activation of the *src* gene allows the v-*src* gene product to phosphorylate cellular substrates of other tyrosine kinases. This could lead to multiple pleiotropic alterations in the cell that would include a constitutive stimulation of cell proliferation as a result of the phosphorylation of growth hormone receptor substrates.

References

1. Bishop, J. M., 1985, Viral oncogenes, *Cell* **42**:23–38.
2. Bishop, J. M., 1983, Cellular oncogenes and retroviruses, *Annu. Rev. Biochem.* **52**:301–354.
3. Hanafusa, H., 1977, Cell transformation by RNA tumor viruses, in: *Comprehensive Virology*, Vol. 10 (H. Fraenkel-Conrat and R. P. Wagner, eds.), Plenum Press, New York, pp. 401–483.
4. Brugge, J. S., and Erikson, R. L., 1977, Identification of a transformation-specific antigen induced by avian sarcoma virus, *Nature* **269**:346–348.
5. Purchio, A. F., Erikson, E., Brugge, J. S., and Erikson, R. L., 1978, Identification of a polypeptide encoded by the avian sarcoma virus *src* gene, *Proc. Natl. Acad. Sci. USA* **75**:1567–1571.
6. Collett, M. S., and Erikson, R. L., 1978, Protein kinase activity associated with the avian sarcoma virus *src* gene product, *Proc. Natl. Acad. Sci. USA* **75**:2021–2024.
7. Levinson, A. D., Opperman, H., Varmus, H. E., and Bishop, J. M., 1980, Evidence that the trans-

forming gene of avian sarcoma virus encodes a protein kinase associated with a phosphoprotein, *Cell* **15**:561–572.
8. Sefton, B. M., and Hunter, T., 1984, Tyrosine protein kinases, in: *Advances in Cyclic Nucleotide and Protein Phosphorylation Research,* Vol. 18 (P. Greengard and G. A. Robinson, eds.), Raven Press, New York, pp. 195–226.
9. Sefton, B. M., and Hunter, T., 1984, Tyrosine protein kinases, in: *Advances in Cyclic Nucleotide and Protein Phosphorylation Research,* Vol 18 (P. Greengard and G. A. Robinson, eds.), Raven Press, New York, pp. 195–226.
10. Sefton, B. M., Hunter, T., Beemon, K., and Eckhart, W., 1980, Evidence that the phosphorylation of tyrosine is essential for cellular transformation by Rous sarcoma virus, *Cell* **20**:807–816.
11. Anderson, D. D., Beckmann, R. P., Harms, E. H., Nakamura, K., and Weber, M. J., 1980, Biological properties of "partial" transformation mutants of Rous sarcoma virus and characterization of their pp60^{src} kinase, *J. Virol.* **37**:445–458.
12. Nakamura, K. D., and Weber, M. J., 1982, Phosphorylation of a 36,000 M_r cellular protein in cells infected with partial transformation mutants of Rous sarcoma virus, *Mol. Cell. Biol.* **2**:147–153.
13. Collett, M. S., Erikson, E., Purchio, A. F., Brugge, J. S., and Erikson, R. L., 1979, A normal cell protein similar in structure and function to the avian sarcoma virus *src* gene product, *Proc. Natl. Acad. Sci. USA* **76**:3159–3163.
14. Sefton, B. M., Hunter, T., and Beemon, K., 1980, Relationship of polypeptide products of the transforming gene of Rous sarcoma virus and the homologous gene of vertebrates, *Proc. Natl. Acad. Sci. USA* **77**:2059–2063.
15. Takeya, T., and Hanafusa, H., 1983, Structure and sequence of the cellular gene homologous to the *src* gene of Rous sarcoma virus and the mechanism for the generation of a viral transforming gene, *Cell* **32**:881–890.
16. Cooper, J. A., Gould, K. L., Cartwright, C. A., and Hunter, T., 1986, Tyr^{527} is phosphorylated in pp60^{c-src}:Implications for regulation, *Science* **231**:1431–1434.
17. Smart, J. E., Opperman, H., Czernilofsky, A. P., Purchio, A. F., Erikson, R. L., and Bishop, J. M., 1981, Characterization of sites for tyrosine phosphorylation in the transforming protein of Rous sarcoma virus (pp60^{v-src}) and its normal cellular homologue (pp60^{c-src}), *Proc. Natl. Acad. Sci. USA* **78**:6013–6017.
18. Sefton, B. M., Trowbridge, I. S., Cooper, J. A., and Scolnick, E. M., 1982, The transforming proteins of Rous sarcoma virus, Harvey sarcoma virus, and Abelson virus contain tightly bound lipid, *Cell* **31**:465–474.
19. Buss, J. E., Kamps, M. P., and Sefton, B. M., 1984, Myristic acid is attached to the transforming protein of Rous sarcoma virus during or immediately after synthesis and is present in both the soluble and membrane-bound forms of the protein, *Mol. Cell. Biol.* **4**:2697–2704.
20. Kamps, M. P., Buss, J. E., and Sefton, B. M., 1985, Mutation of the NH$_2$-terminal glycine of pp60^{src} prevents both fatty acylation and morphological transformation, *Proc. Natl. Acad. Sci. USA* **82**:4625–4628.
21. Blair, D. G., Oskarsson, M., Wood, T. G., McClements, W. L., Fischinger, P. J., and Van de Woude, G. F., 1981, Activation of the transforming potential of a normal cell sequence: A model for oncogenesis, *Science* **212**:941–943.
22. DeFeo, D., Gonda, M. A., Yong, H. A., Chang, E. H., Long, D. R., Scolnick, E. M., and Ellis, R. W., 1981, Analysis of two divergent rat genomic clones homologous to the transforming gene of Harvey murine sarcoma virus, *Proc. Natl. Acad. Sci. USA* **78**:3328–3332.
23. Miller, A. D., Curran, T., and Verma, I. M., 1984, c-*fos* protein can induce transformation: A novel mechanism of activation of a cellular oncogene, *Cell* **36**:51–60.
24. Reddy, E. P., Reynolds, R. K., Santos, E., and Barbacid, M., 1982, A point mutation is responsible for the acquisition of transforming properties by the T24 bladder carcinoma oncogene, *Nature* **300**:143–149.
25. Collett, M. S., Brugge, J. S., and Erikson, R. L., 1978, Characterization of a normal avian cell protein related to the avian sarcoma virus-transforming gene product, *Cell* **15**:1363–1369.
26. Iba, H., Takeya, T., Cross, F. R., Hanafusa, T., and Hanafusa, H., 1984, Rous sarcoma virus variants that carry the cellular *src* gene instead of the viral *src* gene cannot transform chicken embryo fibroblasts, *Proc. Natl. Acad. Sci. USA* **81**:4424–4428.

27. Parker, R. C., Varmus, H. E., and Bishop, J. M., 1984, Expression of v-*src* and chicken c-*src* in rat cells demonstrates qualitative differences between pp60^{v-src} and pp60^{c-src}, *Cell* **37**:131–139.
28. Shalloway, D., Coussens, P. M., and Yaciuk, P., 1984, c-*src* and *src* homolog overexpression in mouse cells, in: *Cancer Cell*, Vol. 2 (G. F. Van de Woude, A. J. Levine, W. C. Topp, and J. D. Watson, eds.), Cold Spring Harbor Laboratory, Cold Spring Harbor, N.Y., pp. 9-17.
29. Kato, J., Takeya, T., Grandori, C., Iba, H., Levy, J., and Hanafusa, H., 1986, Amino acid substitutions sufficient to convert the nontransforming p60^{c-src} protein to a transforming protein, *Mol. Cell. Biol.* **6**:4155–4160.
30. Levy, J. B., Iba, H., and Hanafusa, H., 1986, Activation of the transforming potential of pp60^{c-src} by a single amino acid change, *Proc. Natl. Acad. Sci. USA* **83**:4228–4232.
31. Iba, H., Cross, F. R., Garber, E. A., and Hanafusa, H., 1985, Low level of cellular protein phosphorylation by nontransforming, overproduced pp60^{c-src}, *Mol. Cell. Biol.* **5**:1058–1066.
32. Coussens, P. M., Cooper, J. A., Hunter, T., and Shalloway, D., 1985, Restriction of the *in vitro* and *in vivo* tyrosine protein kinase activities of pp60^{c-src} relative to pp60^{v-src}, *Mol. Cell. Biol.* **5**:2753–2763.
33. Cartwright, C. A., Kaplan, P. L., Cooper, J. A., Hunter, T., and Eckhart, W., 1986, Altered sites of tyrosine phosphorylation in pp60^{c-src} associated with polyoma middle tumor antigen, *Mol. Cell. Biol.* **6**:1562–1570.
34. Courtneidge, S. A., 1985, Activation of the pp60^{c-src} kinase by middle T antigen binding or by dephosphorylation, *EMBO J.* **4**:1471–1477.
35. Cooper, J., and King, C., 1986, Dephosphorylation or antibody binding to the carboxy terminus stimulates pp60^{c-src}, *Mol. Cell. Biol.* **6**:4467–4477.
36. Snyder, M. A., Bishop, J. M., Colby, W. W., and Levinson, A. D., 1983, Phosphorylation of tyrosine-416 is not required for the transforming properties and kinase activity of pp60^{v-src}, *Cell* **32**:891–901.
37. Cross, F. R., and Hanafusa, H., 1983, Local mutagenesis of Rous sarcoma virus: The major sites of tyrosine and serine phosphorylation of pp60^{v-src} are dispensable for transformation, *Cell* **34**:597–607.
38. Patchinsky, T., Hunter, T., and Sefton, B. M., 1986, Phosphorylation of the transforming protein of Rous sarcoma virus: Direct demonstration of phosphorylation of serine 17 and identification of an additional site of tyrosine phosphorylation in pp60^{v-src} of Prague Rous sarcoma virus, *J. Virol.* **59**:73–81.
39. Collett, M. S., Erikson, E., and Erikson, R. L., 1979, Structural analysis of the avian sarcoma virus-transforming protein: Sites of phosphorylation, *J. Virol.* **29**:770–781.
40. Roth, C. W., Richert, N. D., Pastan, I., and Gottesman, M. M., 1983, Cyclic AMP treatment of Rous sarcoma virus-transformed Chinese hamster ovary cells increases phosphorylation of pp60^{c-src} and increases pp60^{src} kinase activity, *J. Biol. Chem.* **258**:10768–10773.
41. Gentry, L. E., Chaffin, K. E., Shoyab, M., and Purchio, A. F., 1986, Novel serine phosphorylation of pp60^{c-src} in intact cells after tumor promoter treatment, *Mol. Cell. Biol.* **6**:735–738.
42. Gould, K. L., Woodgett, J. R., Cooper, J. A., Buss, J. E., Shalloway, D., and Hunter, T., 1985, Protein kinase C phosphorylates pp60^{src} at a novel site, *Cell* **42**:849–857.
43. Courtneidge, S. A., and Smith, A. E., 1984, The complex of polyoma virus middle T-antigen and pp60^{c-src}, *EMBO J.* **3**:585–591.
44. Bolen, J. B., Thiele, C. J., Israel, M. A., Yonemoto, W., Lipsich, L. A., and Brugge, J. S., 1984, Enhancement of cellular *src* gene product associated tyrosyl kinase activity following polyoma virus infection and transformation, *Cell* **38**:767–777.
45. Yonemoto, W., Jarvis-Morar, M., Brugge, J. S., Bolen, J. B., and Israel, M., 1985, Tyrosine phosphorylation within the amino-terminal domain of pp60^{src} molecules associated with polyoma virus middle-sized tumor antigen, *Proc. Natl. Acad. Sci. USA* **82**:4568–4572.
46. Swarup, G., Cohen, S., and Garbers, D. L., 1982, Inhibition of membrane phosphotyrosyl-protein phosphatase activity by vanadate, *Biochem. Biophys. Res. Commun.* **107**:1104–1109.
47. Collett, M. S., Belzer, S. K., and Purchio, A. F., 1984, Structurally and functionally modified forms of pp60^{v-src} in Rous sarcoma virus-transformed cell lysates, *Mol. Cell. Biol.* **4**:1213–1220.
48. Brown, D. G., and Gordon, J. A., 1984, The stimulation of pp60^{v-src} kinase activity by vanadate in intact cells accompanies a new phosphorylation state of the enzyme, *J. Biol. Chem.* **249**:9580–9586.

49. Yonemoto, W., Filson, A. J., Queral-Lustig, A. E., Wang, J., Brugge, J. S., 1987, Detection of phosphotyrosine-containing proteins in polyoma virus middle tumor antigen transformed cells after treatment with phosphotyrosine phosphatase inhibitor, *Mol. Cell. Biol.* **7:**905–913.
50. Radke, K., Gilmore, T., and Martin, G. S., 1980, Transformation by Rous sarcoma virus: A cellular substrate for transformation-specific protein phosphorylation contains phosphotyrosine, *Cell* **21:**821–282.
51. Erikson, E., and Erikson, R. L., 1980, Identification of a cellular protein substrate phosphorylated by the avian sarcoma virus-transforming gene product, *Cell* **21:**829–836.
52. Wang, J. Y., 1985, Isolation of antibodies for phosphotyrosine by immunization with a v-*abl* oncogene encoded protein, *Mol. Cell. Biol* **5:**3640–3643.
53. Jakobovits, E. B., Majors, J. E., and Varmus, H. E., 1984, Hormonal regulation of the Rous sarcoma virus *src* gene via a heterologous promoter defines a threshold dose for cellular transformation, *Cell* **38:**757–765.
54. Sorge, L. K., Levy, B. T., and Maness, P. F., 1984, pp60^{c-src} is developmentally regulated in the neural retina, *Cell* **36:**249–257.
55. Cotton, P. C., and Brugge, J. S., 1983, Neural tissues express high levels of the cellular *src* gene product pp60^{c-src}, *Mol. Cell. Biol.* **3:**1157–1162.
56. Fults, D. W., Towle, A. C., Lauder, J. M., and Maness, P. F., 1985, pp60^{c-src} in the developing cerebellum, *Mol. Cell. Biol.* **5:**27–32.
57. Brugge, J. S., Cotton, P. C., Queral, A. E., Barrett, J. N., Nonner, D., and Keane, R. W., 1985, Neurones express high levels of a structurally modified, activated form of pp60^{c-src}, *Nature* **316:**524–526.
58. Levy, B. T., Sorge, L. K., Meymandi, A., and Maness, P. F., 1984, pp60^{c-src} kinase in embryonic tissues of chick and human, *Dev. Biol.* **104:**9–17.
59. Barnekow, A., and Bauer, H., 1984, The differential expression of the cellular *src*-gene product pp60^{src} and its phosphokinase activity in normal chicken cells and tissues, *Biochim. Biophys. Acta* **782:**94–102.
60. Schartl, M., and Barnekow, A., 1984, Differential expression of the cellular *src* gene during vertebrate development, *Dev. Biol.* **105:**415–422.
61. Brugge, J., Cotton, P., Lustig, A., Yonemoto, W., Lipsich, L., Coussens, P., Barrett, J. N., Nonner, P., and Keane, R. W., 1987, Characterization of the altered form of the c-*src* gene product in neuronal cells, *Genes and Development*, **1:**287–296.
62. Simon, M. A., Drees, B., Kornberg, T., and Bishop, J. M., 1985, the nucleotide sequence and the tissue-specific expression of Drosophila c-*src*, *Cell* **42:**831–840.
63. Golden, A., Nemeth, S. P., and Brugge, J. S., 1986, Blood platelets express high levels of the pp60^{c-src}-specific tyrosine kinase activity, *Proc. Natl. Acad. Sci. USA* **83:**852–856.

11
Protein Kinase C as the Site of Action of the Phorbol Ester Tumor Promoters

Marie L. Dell'Aquila, Barbour S. Warren, David J. de Vries, and Peter M. Blumberg

1. Introduction

The phorbol esters are among the most potent tumor promoters in the two-stage model for mouse skin carcinogenesis.[1-3] They induce numerous biochemical and cellular responses in mouse skin, of which increased cellular proliferation and hyperplasia are particularly prominent. In *in vitro* systems likewise, the phorbol esters induce a wide spectrum of responses, including various alterations in cellular proliferation, differentiation, and intercellular communication.[4-6] The phorbol esters thus represent valuable tools to explore cellular function. The elucidation of their mechanism of action at the molecular level is relevant not only to the study of experimental chemical carcinogenesis but also to broader areas of research in cellular growth and differentiation.

The high potency of the phorbol esters in the induction of biological responses had suggested action at a specific receptor. However, the extremely lipophilic nature of phorbol esters such as PMA made it difficult to detect specific binding due to the high levels of nonspecific interaction which occurred. This problem was overcome by the introduction of [^{3}H]phorbol-12,13-dibutyrate (PDBu), a ligand that retains marked potency while being less lipophilic. Specific binding of phorbol

---

Marie L. Dell'Aquila, Barbour S. Warren, David J. de Vries, and Peter M. Blumberg • Molecular Mechanisms of Tumor Promotion Section, Laboratory of Cellular Carcinogenesis and Tumor Promotion, National Cancer Institute, Bethesda, Maryland 20892.

esters was initially demonstrated in chick embryo fibroblasts[7] and in the particulate fraction of mouse skin[8] but has since been characterized in many systems.[9]

Protein kinase C, the calcium- and phospholipid-dependent protein kinase, was initially characterized by Nishizuka and co-workers. Its properties showed marked similarities to those of the phorbol ester receptor,[10] and, indeed, Castagna et al.[11] were able to show that phorbol esters at nanomolar concentrations could stimulate protein kinase C enzymatic activity under conditions of limiting calcium and phospholipid. Direct evidence that phorbol ester binding activity and the protein kinase C enzymatic assays were detecting different domains of the same protein came from the demonstration that both activities copurified.[12–14] Protein kinase C activity purified to homogeneity from rat brain cytosol yielded a single protein with a molecular weight of 82,000 as determined by SDS–PAGE.[10] The enzyme is composed of a single polypeptide chain with no subunit structure. As expected, the stoichiometry of binding [^{3}H]-PDBu to the enzyme was found to be 1:1.[10,15]

Protein kinase C is an enzyme with broad substrate specificity that has been found in most mammalian cells with the highest concentrations being in brain tissue. The enzyme is usually inactive and requires the presence of both calcium and an anionic phospholipid such as phosphatidylserine for activity. Unsaturated diacylglycerols such as diolein increase the affinity of the kinase for calcium and thereby activate the enzyme at limiting calcium concentrations.[16] Diacylglycerols are generated by hormonally induced, receptor-mediated hydrolysis of inositol phospholipids. The addition of diglycerides to intact cells[16] or their generation in situ by treatment with phospholipase C[17,18] induces a number of phorbol ester responses. In addition, diglycerides competitively inhibit phorbol ester binding to protein kinase C in vitro.[19] The diacylglycerols are thus the postulated endogenous analogues of the phorbol esters. The identification of protein kinase C as the major receptor for phorbol esters provides a direct link between these compounds and the phosphatidylinositol (PI) turnover signaling pathway.

A specific increase in PI turnover has been documented in numerous tissues after exposure to a variety of biologically active compounds (reviewed in Refs. 20–22). Functions proposed for the elevated PI turnover include membrane fusion,[23,24] early events in cell proliferation,[25,26] elevation of intracellular calcium levels,[27] and arachidonic acid release.[28] The first step in PI turnover is initiated by a calcium-dependent phospholipase C-type phosphodiesterase that hydrolyzes inositol phospholipids to yield diacylglycerol and inositol phosphates.[29] The rapid disappearance of the phospholipid PI-4,5-diphosphate (PIP_2) observed in stimulated cells leading to the generation of the water-soluble product inositol-1,4,5-triphosphate (IP_3) has been proposed by Berridge and colleagues to be an intracellular messenger for the release of calcium from internal stores.[30] The other product of PIP_2 hydrolysis, diacylglycerol, serves as an intracellular mediator for the activation of protein kinase C.[16] The receptor-mediated induction of PI metabolism, therefore, appears to be a multifunctional second-messenger membrane-transducing mechanism.

For the activation of protein kinase C by diacylglycerol, it is important to note that diacylglycerol is largely absent from cellular membranes in unstimulated cells.

TABLE I
Some Biological Effects of Phorbol Esters

Response	References
Mouse skin tumor promotion	1, 2
Differentiation	
Mouse keratinocytes	33
HL-60 cells	3
Mitogenesis	
Fibroblasts	4
Lymphocytes	44
Inhibition of Friend cell differentiation	36, 37
Inhibition of EGF binding	67, 68
Activation of immunocompetent cells	4
Platelet activation	145, 146
Increased prostaglandin release	210
Increased arachidonic acid release	5
Increased hexose transport	2, 3
Increased phospholipid synthesis acitivity	4
Increased ornithine decarboxylase activity	4
Increased NADPH oxidase activity	34, 35
Increased protein kinase activity	210

After it is generated in the plasma membrane in response to hormonal stimulation, the diacylglycerol is rapidly degraded, permitting tight regulation of physiological processes. The short half-life of diacylglycerol in cells[21] compared to that of phorbol esters, which are not readily degraded by cellular metabolism,[31] may explain some of their differences in cellular action.

The responses induced by phorbol esters in different biological systems are numerous and varied (Table I). Both suppression and promotion of cell differentiation have been found with the ultimate biological response to these agents thus being determined by the cell type rather than simply stimulation of the signaling system. A similar conclusion holds true for responses mediated by the cAMP pathway,[32] a second major signal-transducing system in cells. In many cell types, the phorbol esters have proven to be valuable tools in the investigation of the role of protein kinase C in mitogenesis.

2. Mitogenic Activity of Phorbol Esters

2.1. Induction of Mitogenesis by Phorbol Ester Treatment

Induction of epidermal hyperplasia is one of the characteristic actions of the phorbol esters. This early observation led to investigations of the *in vitro* mitogenic activity of phorbol esters in mouse fibroblasts. Stimulation of cell division by PMA was described in mouse fibroblasts,[38-40] human diploid fibroblasts (WI-38),[41] chick embryo fibroblasts,[42] mouse epidermal cells,[43] and lymphocytes.[44,45] These original studies demonstrated that phorbol esters activated mitogenic pathways at extremely low concentrations, suggesting that their action was mediated by a high-affinity receptor.

2.2. Down-Regulation of Protein Kinase C after Phorbol Ester Treatment

In many cell systems, a prolonged exposure to phorbol esters has been shown to produce a loss of available [^{3}H]-PDBu binding sites.[46,47] This down-modulation of phorbol ester binding sites resulted in loss of cellular responsiveness to subsequent phorbol ester treatment. The molecular nature of down-modulation of phorbol ester receptors was characterized in a rat pituitary cell line, GH_4C_1.[48] Prolonged exposure to either phorbol ester (homologous down-regulation) or thyrotropin-releasing hormone (heterologous down-regulation) resulted in a loss of available [^{3}H]-PDBu binding sites to approximately 20% of control levels after 24 hr of treatment. Scatchard analysis indicated this decrease in binding was due to a loss of receptors with no alteration in binding affinity.

Phorbol ester down-modulation of receptors has also been characterized in Swiss 3T3 cells by Rozengurt and co-workers. In this system, phorbol ester pretreatment for 48 hr caused loss of both phorbol ester binding[49] and protein kinase C enzymatic activity measured in cell-free, detergent-solubilized extracts.[50] The effect was dose-dependent and reversible, with protein kinase C activity reappearing after phorbol ester removal. The phorbol ester pretreatment abolished the mitogenic response of the cells to phorbol esters; however, cells retained responsiveness to a wide variety of other mitogens.[51] Microinjection of purified protein kinase C into quiescent cultures of Swiss 3T3 cells pretreated with the phorbol ester PDBu restored the mitogenic response of the cells to PDBu.[52] These findings directly establish the involvement of protein kinase C in this biological response.

2.3. Phosphorylation of an 80,000-Dalton Protein

Understanding the molecular basis of regulation of cell growth requires the identification of early signals that mediate the initiation of the mitogenic response. In efforts to characterize initial cellular events during mitogenesis, Rozengurt and colleagues investigated protein phosphorylation induced by biologically active phorbol esters in intact quiescent mouse fibroblasts arrested in the G0/G1 phase of the cell cycle by serum deprivation.[53] PDBu enhanced the phosphorylation of an 80,000-dalton cellular protein (termed 80k) in a dose-dependent manner in Swiss 3T3 cells with half-maximal response obtained at 32 nM. The involvement of protein kinase C in this phosphorylation event was further indicated by the fact that phospholipid breakdown induced by exogenous treatment of the cells with phospholipase C also caused rapid increases in 80k phosphorylation. If cells were subjected to prolonged pretreatment with PDBu, leading to a marked decrease in phorbol ester binding sites, phorphorylation of 80k by phorbol ester or phospholipase C was prevented. Further studies verified a similar action by endogenously added diacylglycerols in the stimulation of 80k phosphorylation and blockage of this phosphorylation by PDBu pretreatment.[54] The same 80k phosphorylation can be generated in cell-free extracts by activation of protein kinase C via the addition of phosphatidylserine, calcium, and PDBu in the presence of [^{32}P]-ATP.[55] Though the function of this phosphoprotein remains to be established, these studies demonstrate that 80k phosphorylation is protein kinase C-mediated and analysis of changes in 80k phosphorylation offers a means to assess the *in vivo* activation of protein kinase C by various mitogenic agents in these cells.

3. Activation of Protein Kinase C by Growth Factors

3.1. Mitogens for Fibroblasts

Using the rapid phosphorylation of 80k as a specific marker for protein kinase C activity, Rozengurt and colleagues investigated the activation of protein kinase C by certain mitogens in Swiss 3T3 cells. Treatment of cells with the mitogen platelet-derived growth factor (PDGF), which has been shown to induce inositol phospholipid turnover,[56] caused a rapid increase in 80k phosphorylation.[53] Further, this effect was lost in cells following prolonged pretreatment with a phorbol ester. In contrast, insulin (1–10 μg/ml) or epidermal growth factor (EGF) (5 ng/ml), which do not stimulate endogenous phospholipase C activity in 3T3 cells, failed to stimulate 80k phosphorylation.[57]

Bombesin, a regulatory peptide originally isolated from frog skin,[58] together with several bombesin-related mammalian peptides [e.g., gastrin-releasing peptide (GRP)] have also been identified as potent mitogens in Swiss 3T3 cells.[59] The mitogenic response to these peptides is mediated by specific, high-affinity receptors that are distinguishable from those of other growth factors.[60] On the addition of bombesin to quiescent cultures of Swiss 3T3 cells, a rapid increase in 80k phosphorylation was observed,[61] indicating a functional role for protein kinase C. These peptides are of special interest due to recent reports demonstrating the presence of high levels of bombesinlike peptides in human pulmonary[62-64] and thyroid carcinomas.[65]

Recent studies by Blackshear *et al.*[66] have examined the extent of the importance of protein kinase C in the transduction of mitogenic signals in mitogen-treated fibroblasts. Treatment of serum-deprived 3T3-L1 fibroblasts with PMA, diacylglycerol, PDGF, or fibroblast growth factor (FGF) stimulated 80k phosphorylation. This response was induced only slightly by EGF and not at all by insulin. In contrast, proteins of 22,000 and 31,000 daltons were phosphorylated in response to insulin as well as PMA, diacylglycerols, EGF, PDGF, and FGF. To assess the involvement of protein kinase C, phosphorylation of these proteins was compared in control cells and cells pretreated with phorbol ester to down-regulate protein kinase C. Such pretreatment prevented stimulation of 80k phosphorylation by PMA, diacylglycerol, PDGF, or FGF, indicating this effect was protein kinase C-mediated. The increased phosphorylation of the 22k and 31k proteins in response to PMA was also abolished. However, increased phosphorylation of these proteins in response to insulin, PDGF, and FGF was still observed in protein kinase C down-regulated cells. These findings indicate that although PDGF and FGF may exert some of their effects through protein kinase C (as evidenced by 80k phosphorylation), a second, non-protein kinase C pathway of protein phosphorylation exists and is activated by these mitogens.

3.2. Transmodulation of Growth Factor Receptors by Protein Kinase C

Activation of protein kinase C is known to modulate a number of receptors for various growth factors and hormones including EGF,[67,68] thyrotropin-releasing hormone,[69] insulin,[70] somatomedin C,[70] transferrin,[71] and somatostatin.[72] This phenomenon has been best characterized in the EGF system. The original observation was that treatment of intact cells with tumor-promoting phorbol esters caused a marked inhibition of [^{125}I]-EGF binding to specific cell surface receptors.[67,68] The response was extremely rapid and highly temperature-dependent, unlike the down-regulation of EGF receptors seen after exposure to EGF. This transmodulation of the EGF receptor was subsequently observed in a variety of cell types in response to protein kinase C activation not only by tumor promoters such as phorbol esters and teleocidin, but also by various mitogens including vasopressin, fibroblast-derived growth factor (FDGF), PDGF, transforming growth factor β (TGF-β), and bombesin (reviewed in Ref. 73). The inhibition of EGF binding by these transmodulating ligands results from an apparent decrease in the apparent affinity of the EGF receptor and not from a decrease in the number of binding sites. The decrease in EGF binding affinity was time-dependent after phorbol ester treatment and binding returned to control values by 24 hr.[67]

The transmodulation of the EGF receptor by these heterologous ligands may be through a common mechanism involving the activation of protein kinase C. The finding that a synthetic diacylglycerol also induces a rapid decrease in affinity of the EGF receptor in intact 3T3 cells further strengthens the involvement of protein kinase C in this transmodulation.[73,74] Further, if cells were pretreated with PDBu for 40 hr, the inhibition of EGF binding by the diacylglycerol was prevented. Such pretreatment did not affect [^{125}I]-EGF or [^{125}I]-GRP binding either in apparent affinity or in total receptor number and both EGF and bombesin retained mitogenic activity in these pretreated cells.[75] Thus, EGF and bombesin receptors were present and functional in those PDBu-desensitized cells. Even though bombesin receptors remained functional in PDBu-pretreated cells, desensitization caused a marked decrease in the ability of bombesin to inhibit [^{125}I]-EGF binding.[61] The inability of bombesin to inhibit [^{125}I]-EGF binding in cells in which protein kinase C has been down-regulated by prolonged treatment with phorbol ester indicates an important role for the kinase in this response.

A number of recent reports have provided evidence that the EGF receptor is a substrate for protein kinase C. In A431, a human epidermoid carcinoma cell line, treatment with phorbol ester resulted in an increase in the phosphorylation state of the EGF receptor accompanied by a decrease in both its tyrosine-specific kinase activity and its EGF binding ability.[76,77] The major site of phosphorylation was a specific threonine residue (Thr-654) located 10 residues internal to the postulated transmembrane domain of the receptor.[78] Phosphorylation at this unique site can also be induced in fibroblasts by treatment with PDGF,[79] which activates protein kinase C via PI turnover. Phosphorylation of the EGF receptor may thus regulate its activity: EGF-stimulated autophosphorylation of the EGF receptor enhances its enzymatic activity while protein kinase C-catalyzed phosphorylation of the EGF receptor depresses its tyrosine kinase activity.[80] In this fashion, protein kinase C could exert negative control on EGF-induced responses. It should be noted that the biochemical studies are in apparent conflict with biological data suggesting that phorbol esters can synergize with EGF in stimulating mitogenesis.[81]

3.3. Mitogens for Lymphoid Cells

A major functional role has also been established for protein kinase C in proliferative responses in cells of the immune system. Interleukin-2 (IL-2) is a regulatory peptide that promotes the growth and activation of antigen-specific T lymphocytes.[82,83] The biological activity of IL-2 involves interaction with specific high-affinity receptors acquired after antigen or lectin stimulation. This interaction initiates a proliferative signal, promoting S phase progression and an ensuing clonal expansion. The binding of IL-2 to its respective high-affinity receptor has been well characterized,[84] and a monoclonal antibody, designated anti-Tac, to an antigenic determinant of the human IL-2 receptor has been developed.[85,86]

Farrar and Anderson have shown that IL-2 interaction with its receptor produces a rapid and transient redistribution of protein kinase C from cytosol to the plasma membrane, presumably reflecting protein kinase C activation.[87] PMA, which has been shown to mimic several actions of IL-2, also induced a similar protein kinase C transposition to the plasma membrane but for a longer duration. Likewise, interleukin-3 (IL-3), a colony-stimulating factor important in the regulation of hematopoiesis,[88] stimulated protein kinase C translocation to the plasma membrane after receptor interaction in a manner analogous to IL-2.[89] These findings suggest a role for protein kinase C in the signal transduction for these extracellular messengers which control growth and differentiation in cells of the immune system.

The involvement of protein kinase C with IL-2-mediated regulation of proliferation in lymphoid cells has also been demonstrated by its phosphorylation of the IL-2 receptor. The IL-2 receptor is not expressed constitutively in resting lymphocytes. Thus, the transient expression of IL-2 receptors constitutes another regulatory element in proliferative control. Phorbol esters have been shown not only to regulate IL-2 receptor expression,[90] but also to stimulate the phosphorylation of the antigenic epitope of the human IL-2 receptor, Tac, on lymphocytes.[91,92] Further studies confirmed the stimulation of phosphorylation of the IL-2 receptor in normal human T lymphocytes by phorbol ester or diacylglycerol.[92]

In addition, immunoprecipitated IL-2 receptor from lymphocyte membranes was phosphorylated by purified protein kinase C. Tryptic peptide analysis verified that the *in vitro* and *in vivo* phosphorylation occurred on the same sites, Ser-247 and Thr-250 in the carboxy-terminal cytoplasmic tail of the IL-2 receptor.[93]

4. Interaction between Protein Kinase C and Oncogenes

4.1. Induction of PI Turnover by Oncogenes

Many oncogene products possess tyrosine-specific kinase activity. Receptors for several growth factors, including the EGF receptor, also display tyrosine-specific kinase activity, suggesting a functional role for this enzymatic activity in cell proliferation. The oncogenic protein kinases are thought to induce transformation through either inappropriate or excessive protein phosphorylation on tyrosine.

Several oncogenes have been shown to enhance PI turnover in intact cells. In chick embryo fibroblasts, transformation by Rous sarcoma virus induced PI kinase activity resulting in a 50–100% increase in the labeling of PI-4-phosphate and PIP_2 as compared to uninfected cells.[94] Both of these phosphorylations are steps in the conversion of PI to diacylglycerol and the subsequent activation of protein kinase C. Likewise, Macara and colleagues have shown that cells transformed by v-*ros*, the transforming gene of the avian sarcoma virus UR-2, displayed elevated levels of [^{3}H]glycerol-labeled diacylglycerol.[95] In both these instances, the PI kinase activity was proven not to be intrinsic to the products of these transforming genes, as initially thought.[96,97] Instead, the PI kinase activities were found to be associated with the transforming gene products $pp60^{v-src}$ and $pp68^{v-ros}$ and thus coimmunoprecipitated upon purification. The activation of the PI cycle by these tyrosine-specific kinases provides a mechanism by which these oncogenes could induce protein kinase C activity in transformed cells.

PI turnover is also induced by oncogenes that do not encode a tyrosine-specific kinase. Similar to $pp60^{v-src}$, PI kinase activity was found in cells infected with polyoma virus in immunoprecipitates made with antiserum specific for middle T antigen, the transforming protein of polyoma virus.[98] Further, a study with middle T-defective mutants showed a close correlation between this PI kinase activity and transformation. In intact cells infected with transformation-competent polyoma virus, elevated levels of PI turnover were observed.[99] v-*sis*, the transforming gene of simian sarcoma virus, encodes a protein analogous to the β chain of PDGF which also binds PDGF receptors. The binding of PDGF to its receptor in intact cells raises diacylglycerol levels; similarly, elevation in diacylglycerol levels was observed in *sis*-transformed NRK cells.[100] The *ras* oncogenes (K-*ras*, H-*ras*, N-*ras*) encode membrane-associated GTP-binding proteins with GTPase activity.[101] In neutrophils and mast cells, receptors for certain hormones are coupled to phospholipase C via a GTP-binding protein.[101,102] Nonhydrolyzable analogues of GTP were able to substitute for external ligands, suggesting that guanine nucleotide regulatory proteins may act by stimulating phospholipase C activity.[103] Elevated diacylglycerol levels have been found in K-*ras*-transformed NRK cells, implying that this oncogene may induce protein kinase C activation via phospholipase C.[100] Preliminary evidence also indicates that K-*ras* transformation of 3T3 cells induces translocation of protein kinase C activity from the cytosol to the particulate fraction.[104]

4.2. Phosphorylation of Oncogene Products by Protein Kinase C

Much evidence has been presented recently that the transforming protein of Rous sarcoma virus is a substrate for protein kinase C. Gould and co-workers investigated $pp60^{v-src}$ phosphorylation induced by PMA treatment in a variety of cell types.[105] Treatment with PMA, teleocidin, or diacylglycerol caused nearly complete phosphorylation of $pp60^{v-src}$ at a novel site (Ser-12) located with the amino-terminus of the molecule. This same effect was noted for $pp60^{c-src}$, the normal cellular homologue of $pp60^{v-src}$, in cells treated with PMA or PDGF.[106] Purified protein kinase C also phosphorylated $pp60^{v-src}$ *in vitro* at this same unique site, based on comparative tryptic mapping.[107] Further, the site of phosphorylation in $pp60^{c-src}$ displayed marked homology to the site phosphorylated in the EGF receptor by protein kinase C,[105] suggesting that phosphorylation at this site has some physiological function.

The transforming protein of polyoma virus, middle T antigen, is known to bind and activate the tyrosine kinase $pp60^{c-src}$.[108] This complex is thought to be important for transformation by polyoma. In polyoma virus-transformed cells, stimulation of protein kinase C increased the phosphorylation of tyrosine residues of the viral middle T antigen in complex with $pp60^{c-src}$.[109,110] Thus, cellular protein kinase C is clearly implicated in cellular transformation by polyoma virus.

4.3. Enhancement of Oncogene-Induced Transformation by Phorbol Esters

At the biological level, evidence implicating protein kinase C in the action of oncogenes comes from the many reports that phorbol esters enhance oncogene-induced transformation in various cell lines. The tumor promoters PMA or teleocidin markedly enhanced the transformation of C3H 10T1/2 or NIH 3T3 mouse fibroblasts when these cells were transfected with the cloned human bladder cancer c-*ras*^{H} oncogene.[111] This observation was extended to a rat embryo fibroblast cell line where a 6- to 14-fold increase in transformed foci was observed.[112] Such systems could serve as useful models to analyze synergistic interactions between the protein kinase C-mediated actions of tumor promoters and activated oncogenes during multistage carcinogenesis.

5. Relation between Phorbol Esters and Diacylglycerols as Activators of Protein Kinase C

As discussed earlier, diacylglycerols are generated by the signal-linked breakdown of inositol phospholipids and are thought to be the endogenous activators of protein kinase C. Three lines of evidence suggest that the diacylglycerols interact with protein kinase C analogously to the phorbol esters. First, diacylglycerols competitively inhibited [^{3}H]-PDBu binding.[19] Not only was the binding affinity of [^{3}H]-PDBu decreased with no change in the number of binding sites, but the magnitude of the decrease in affinity as a function of the amount of diglyceride quantitatively fitted the predicted relationship for a competitive inhibitor. Second, the off-rate of [^{3}H]-PDBu from protein kinase C was independent of the concentration of diacylglycerol.[113] Competitive inhibitors only affect the on-rate for binding, not the off-rate. In contrast, were the diacylglycerols to have been affecting the [^{3}H]-PDBu binding affinity through perturbation of the phospholipid environment, then both the on-rate and the off-rate would have been affected. Third, the stoichiometry of binding of diacylglycerol to protein kinase C was 1:1, as demonstrated by two entirely independent techniques. This laboratory quantitated the perturbation in apparent diacylglycerol binding affinity as a function of the ratio of ligand to receptor, from which the stoichiometry could be derived.[114] Bell and co-workers measured the amount of diacylglycerol required for protein kinase C activation in Triton X-100 mixed micelles at limiting dilutions of diacylglycerol.[115] Although formal demonstration that the diacylglycerol is interacting with protein kinase C at the identical site as phorbol esters will await affinity labeling, the competition data are most simply explained by such a model.

Structure–activity analysis indicates that the diacylglycerols require adequate lipophilicity for activity. On the other hand, activity is also lost for derivatives with very long, saturated side chains. The critical features may be the ability of the diacylglycerol to pack in the membrane, in that activity is retained for derivatives either with both a long and a short side chain or if there is a *cis*-double bond in one chain. Both conditions would cause disorder in the hydrophobic domain, preventing efficient packing. Other required features are that the diacylglycerol be in the sn-1,2 configuration, possess an ester linkage at the 1-position and an ester or amide at the 2-position. The free hydroxyl group is essential. These features are consistent with computer modeling of the phorbol ester pharmacophore (Section 10).

Comparison of the potencies of phorbol esters and similarly substituted diacylglycerols revealed that the phorbol esters were in all cases more potent. The magnitude of the difference varied dramatically, however; whereas phorbol 12-myristate 13-acetate was 30,000-fold more potent than glycerol 1-myristate 2-acetate,[116,117] the difference between the dilaurate derivatives was only 20-fold. The phorbol esters and diacylglycerols thus show different side chain dependencies.

As expected from their analogous *in vitro* behavior, the diacylglycerols show many of the same *in vivo* actions as the phorbol esters. Since the normal diacylglycerols are too lipophilic to equilibrate with the cellular membranes if added to the culture medium, more hydrophilic derivatives need to be used, such as 1-oleoyl 2-acetylglycerol,[118] 1,2-dioctanoylglycerol,[119] or glycerol 1-myristate 2-acetate.[120]

Numerous examples of similar responses exist such as superoxide production[120] and lysosomal enzyme release[121] in neutrophils, mitogenesis in Swiss 3T3 cells,[54] A431 cells,[122] and lymphocytes,[123] release of histamine from mast cells,[124] platelet activation,[125] decreased EGF binding,[126] inhibition of prolactin release from decidual cells,[127] inhibition of EGF and vanadate activation of

Ca^{2+} influx and Na^+/H^+ exchange in A431 cells,[128] differentiation of Friend cells,[129] activation of sea urchin eggs,[130] secretion of mucin from rat submandibular glands,[131] release of LH, FSH, PRL, and TSH from dispersed rat pituitary cells,[132] and inhibition of cellular communication in 3T3 cells.[133]

Examples of cases in which the responses diverge are few; incorporation of choline into lysophosphatidylcholine in HL-60 cells,[173] production of cAMP in maturing granulosa cells,[134] and secretion of arginine esterase from rat submandibular glands.[131] In addition, not all diglycerides give comparable effects.

Because the diglycerides are rapidly metabolized[21] and of low potency, so that effective concentrations may be difficult to achieve, interpretation of negative results is problematic. In particular, it does not justify the conclusion that the phorbol esters therefore act through other targets.

Another approach to examining the relationship between the phorbol esters and the diacylglycerols has been through the use of phospholipase C, which produces diglycerides from membrane phospholipids. The results with phospholipase C have also generally agreed with those of the phorbol esters. Phospholipase C, like the phorbol esters, causes LHRH release from the immature rat hypothalamus,[135] progesterone production in granulosa cells,[136] release of GH, LH, and TSH from cultured pituitary cells,[137] activation of B cells,[138] inhibition of [^{3}H]-PDBu binding in GH_4C_1 cells,[139] induction of ornithine decarboxylase and a decrease in EGF binding in tracheal epithelial cells,[140] activation of neutrophils[141] and induction of ornithine decarboxylase and transglutaminase in primary mouse epidermal cells.[142]

In contrast, phospholipase C treatment was not found to induce the differentiation of HL-60 cells.[143] Again the interpretation is difficult since phospholipase C treatment will cause more widespread membrane changes which might lead to toxicity before adequate diglyceride levels were achieved.

6. Synergism between Phorbol Esters and Calcium

Receptor-mediated inositol phospholipid breakdown initiates two arms of signal transduction, namely protein kinase C activation by endogenously generated diacylglycerol and IP_3-induced calcium mobilization. The independent induction of these two branches has been described in platelets by the addition of either exogenous synthetic diacylglycerol or phorbol ester to activate protein kinase C and the use of the calcium ionophore A23187 to mobilize calcium.[144] The independent activation of these two pathways was monitored by measuring the phosphorylation of two cellular proteins with molecular weights of 20k and 40k. When platelets were stimulated by agonists that elicit a full biological response (i.e., thrombin, collagen, or platelet-activating factor), a concomitant phosphorylation of both a 40k protein and a 20k protein, myosin light chain, was observed.[144,145] Treatment of platelets with synthetic diacylglycerol alone mediated 40k phosphorylation, indicative of protein kinase C activation. Alternatively, if calcium ionophore alone was added, the 20k protein was selectively phosphorylated, presumably by a calmodulin-mediated phosphorylation. In neither case, however, was serotonin secretion observed.[146] The concentration of agonist in these studies was reported to be critical, as specificity between the 20k and 40k phosphorylation responses was lost at greater than 0.5 μM A23187 or 50 μg/ml diacylglycerol (reviewed in Ref. 147).

Though neither agent alone fully activated platelets, when these cells were treated with permeable diacylglycerol or PMA in the presence of a low concentration of A23187, the release reactions were dramatically enhanced.[146] A similar synergistic action has been demonstrated for release reactions and secretion in a variety of other cell types, including mast cells, neutrophils, and many endocrine tissues (reviewed in Ref. 148). Thus, under appropriate conditions there exists a synergism between protein kinase C activation and Ca^{2+} mobilization.

A different type of synergistic interaction between Ca^{2+} and phorbol esters has been described by Cuatrecasas and co-workers. Unlike the platelet system, these results indicate a direct effect of Ca^{2+} on protein kinase C. In reconstituted erythrocyte membranes, synergism between Ca^{2+} and protein kinase C activators was observed for intracellular translocation of protein kinase C.[149] The binding of purified protein kinase C to these membranes was stimulated by Ca^{2+} concentrations between 50 nM and 5 μM with the sharpest increase in binding observed between 100 and 500 nM free Ca^{2+}. Thus, physiological increases of intracellular Ca^{2+} might cause a translocation of protein kinase C from a soluble to a membrane-bound compartment, facilitating activation of the kinase. Studies in a second system revealed that calcium ionophore indeed synergized protein phosphorylation induced by protein kinase C in whole cells. Raising intracellular Ca^{2+} levels, though ineffective itself, increased the efficacy of phorbol ester-induced activation of protein kinase C as measured by transferrin receptor phosphorylation and down-regulation in HL-60 cells.[150] These studies suggested a direct effect of Ca^{2+} on protein kinase C. The authors proposed a molecular model in which intracellular Ca^{2+} recruits protein kinase C to the plasma membrane, thus "priming" the system for activation by phorbol ester. Such a model implies that protein kinase C mediates at least some of the effects of Ca^{2+} in these systems.

7. The Protein Kinase C Pathway and Other Signal Transduction Systems

7.1. The cAMP Second Messenger System

A second major signal transduction pathway in many cells involves a receptor-regulatory protein–adenylate cyclase–cAMP–protein kinase A cascade. Numerous reports indicate an interaction between this second messenger pathway and that involving protein kinase C. Again, phorbol ester-induced effects have provided much of the information on the role of the protein kinase C. Several accounts of phorbol esters eliciting both inhibition[151–154] and stimulation[155–158] of the cAMP-generating mechanism have been published. In addition, stimulation of the cAMP-mediated pathway has been reported to exert at least a functional inhibition[159] as well as facilitation[160] of protein kinase C-mediated processes. In order to explain the heterogeneity of these interactions, Nishizuka[97] has proposed an attractive scheme in which the specific cell types studied are divided into classes displaying discrete combinations of inhibition or stimulation between the two signal transduction pathways. Although evidence exists for such trends, the level of understanding of these pathways does not allow assignment of a particular mechanism of action.

Determination of the interactions between these signal transduction pathways is difficult due to the complexity of these systems. For the protein kinase C pathway, diacylglycerol generation is subject to regulation at many sites. Also, the down-regulation of available protein kinase C subsequent to prolonged phorbol ester treatment (Section 2.2) has numerous effects on the biochemical makeup of the treated cells. Many studies dealing with the interactions of the signal transduction systems do not address the question of duration and dose of phorbol ester treatment. Taken to an extreme, unjustified comparisons are made concerning the characteristics of cAMP-mediated processes of whole cells versus cell lysates, micromolar versus nanomolar concentrations of phorbol esters, and over periods of hours versus minutes. Another source of complexity in these interacting systems is the presence of dual inhibitory and stimulatory systems for the regulation of adenylate cyclase. Thus, up- and down-regulation of cAMP-dependent protein kinase-mediated events do not follow a simple turn-on/turn-off mechanism.

Studies dealing with the molecular site of action of protein kinase C in regulating cAMP generation underline the complexity of this system. Protein kinase C has been reported to promote phosphorylation of the adenylate cyclase-linked β-adrenergic receptor[151,153] and the inhibitory nucleotide regulatory protein.[158] Feasibly, protein kinase C-mediated phosphorylation of other components of the cAMP second messenger will also be demonstrated.

7.2. Calcium Mobilization

In addition to its role as a coactivator of protein kinase C, Ca^{2+} has an important role in signal transduction and cellular dynamics. Not surprisingly, many investigators have looked for means by which protein kinase C may interact with Ca^{2+}-related processes.

The action of phorbol esters in platelet activation is similar to that produced by hormones that activate protein kinase C but, in contrast to the hormone effect, phorbol esters are thought to exert their effects independently of a rise in intracellular Ca^{2+} concentration.[161] However, more recently Ware et al.[162] have employed a more sensitive Ca^{2+} fluorophore and detected a phorbol ester-induced rise in Ca^{2+} levels which possibly was not detected in earlier studies due to its discrete localization within the cell.

Hormones that act through the PI pathway and stimulate Ca^{2+} mobilization appear to have these effects blocked by phorbol esters over a short (minutes) time scale.[163–165] The mechanism by which this inhibition occurs is not known.

Voltage-dependent Ca^{2+} channels are also sensitive to the actions of protein kinase C. While enhanced Ca^{2+} current was reported in a molluscan system,[166] in several mammalian systems protein kinase C reduced Ca^{2+} current through an action that inhibited opening of the voltage-gated Ca^{2+} channel.[167-169]

Thus, while Ca^{2+} and protein kinase C appear to have synergistic actions in many systems (Section 6), there also appears to be a specific effect of protein kinase C activation to oppose those processes that increase intracellular Ca^{2+}.

Another important intracellular divalent cation whose intracellular levels are affected by phorbol esters is magnesium. Grubbs and Maguire[170] found that while PMA exerts little effect on Ca^{2+} transport in resting murine S49 lymphoma cells, it produced a severalfold increase in Mg^{2+} influx. The role of Mg^{2+} in processes such as receptor-coupled modulation of adenylate cyclase indicates this action of protein kinase C may also be of importance in signal transduction.

8. Phosphorylation and Substrate Specificity

Protein phosphorylation represents a major mechanism of cellular regulation. A critical question concerning the response of various cell types to the phorbol esters has been to determine which proteins are phosphorylated after phorbol ester treatment. Two approaches have been used.

The first approach has been the *in vitro* screening of various proteins as substrates for protein kinase C and the explanation of various effects of protein kinase C agonists by this phosphorylation/activation. Several reviews have compiled lists of proteins identified in this manner as substrates for phosphorylation by protein kinase C.[118,171] A critical issue is whether comparable phosphorylation events occur *in vivo*.

A second method, indirect but physiologically more relevant, has been to examine cellular protein phosphorylation by gel electrophoresis and subsequently characterize and hopefully identify the critical proteins affected. Several studies have used one-dimensional gel electrophoresis to characterize proteins whose phosphorylation is indicative of protein kinase C activation. One-dimensional gels have the advantage that phosphorylation can be monitored in large numbers of samples. The principal disadvantage is that only major changes in phosphorylation can be detected over the background. As discussed in the preceding sections, prominent among the changes observed on one-dimensional gels are phosphorylation of a 40k protein in platelets[125] and phosphorylation of an 80k protein in Swiss 3T3 cells.[53]

Due to its greatly increased resolving ability, two-dimensional gel electrophoresis is particularly suited for characterizing the variety of proteins whose phosphorylation is affected upon protein kinase C activation. Systems studied using this approach include HL-60 cells,[172,173] neutrophils,[174,175] and liver.[176] Typically, phorbol ester treatment induces enhanced phosphorylation of 5–15 proteins detectable in two-dimensional gels, depending somewhat on the system. Interpretation of the results is complicated because changes may reflect not only direct phosphorylation by protein kinase C but also secondary changes due to the action of protein kinase C on other kinases or phosphatases. For example, Feuerstein and

Cooper characterized a 17k/5.5 pI protein in HL-60 cells which is rapidly phosphorylated in response to active but not inactive phorbol ester derivatives and has its phosphorylation blocked by the protein kinase C inhibitor trifluoroperazine.[172] It is rapidly phosphorylated in A431 human epidermal carcinoma cells whose growth is also inhibited by the phorbol esters.[177] Its phosphorylation is minimally affected in three cell types that do not differentiate in response to phorbol esters. The protein is not a substrate for protein kinase C, however, but rather for the cAMP-dependent protein kinase.

Characterization of interactions between the protein kinase C and cAMP systems at the level of phosphorylation has been studied in greatest detail in the S49 mouse lymphoma line.[178,179] Two mutant sublines of S49 exist which are lacking two elements of the cAMP-dependent kinase system: the catalytic subunit of the kinase itself and one of the guanyl nucleotide binding factors of adenylate cyclase, "N_S." This study compared the protein phosphorylation patterns of wild-type cells and these two mutants in response to PMA and dibutyryl cAMP, both alone and in combination. As expected, PMA treatment led to the phosphorylation of substrates in the absence of cAMP-dependent protein kinase. Additional substrates were identified, however, whose phosphorylation relied on stimulation of the adenylate cyclase system, or whose phosphorylation decreased in the presence of cAMP.

A second kinase system indirectly activated by phorbol ester treatment is that of the S6 kinase. S6 is a ribosomal protein involved in protein synthesis. It is phosphorylated in response to PMA treatment as well as expression of $pp60^{v-src}$ and treatment of quiescent cells with serum.[180] Although initial studies suggested that its phosphorylation might be mediated by protein kinase C (it is a substrate for protein kinase C *in vitro*), subsequent analysis suggested that its phosphorylation reflected activation of a different kinase, S6 kinase. The partially purified kinase proved to be distinct from protein kinase C because of a lack of activation by Ca^{2+} and phosphatidylserine. Further, in protein kinase C down-regulated cells, S6 kinase activity was not stimulated by PMA. However, serum and $pp60^{v-src}$ enhanced S6 phosphorylation.[181] The point at which the pathways of activation of the S6 kinase by phorbol esters and growth factors converge is not known.

Induction of elevated phosphorylation of some proteins at tyrosine residues is an additional, presumably indirect effect of protein kinase C activation. Although *in vitro* protein kinase C phosphorylates protein only at threonine and serine, treatment of chicken embryo fibroblasts with PMA caused phosphorylation on tyrosine of a 42k protein.[182] Elevated phosphorylation at tyrosine of this protein had previously been observed in response to transformation by avian sarcoma virus and by growth factors.[183] Inhibition of the tyrosine phosphorylation activity of the EGF receptor as a consequence of EGF receptor phosphorylation by protein kinase C has been discussed above.

Two-dimensional protein phosphorylation patterns have proven valuable for comparison of PMA action with that of other compounds believed to be acting through protein kinase C, e.g., diacylglycerols, or through induction of PI turnover. Examples of such studies include the activation of protein kinase C by TRH in GH_3 primary cells[184] and stimulation of hepatocytes with angiotensin II and vasopressin.[176]

9. Mechanisms of Heterogeneity of Responses Seen after Protein Kinase C Activation

Although the major phorbol ester receptor is protein kinase C, considerable evidence suggests heterogeneity in phorbol ester interaction in different systems. This heterogeneity could be generated from protein kinase C itself. Alternatively, it could reflect multiple targets in addition to protein kinase C. Biological evidence for heterogeneity of phorbol ester receptors is suggested by different dose–response curves and structure–activity requirements for different biological responses (reviewed in Ref. 185). The second-stage tumor promoter mezerein provides a noteworthy illustration. In most systems examined, mezerein shows a markedly lower binding affinity (20- to 136-fold) than does PMA.[7,48,186,187] As these findings would predict, mezerein is itself of low potency as a tumor promoter.[188] However, PMA and mezerein have similar potencies for inflammation, hyperplasia, and second-stage tumor promotion in skin,[185] suggesting the possible existence of an additional binding target. In most systems, Scatchard analysis of [^{3}H]-PDBu binding yielded linear plots, consistent with a homogeneous class of binding sites. However, in certain notable exceptions, curved plots were obtained. In mouse skin particulate preparations, Scatchard analysis of [^{3}H]-PDBu binding yielded curvilinear plots, consistent with three classes of binding sites.[189] In intact primary mouse keratinocytes, heterogeneous binding was also observed, with the degree of heterogeneity depending on the state of differentiation of the target cells.[190]

Another source of heterogeneity could be generated by the association of protein kinase C with different phospholipid environments. As shown by Konig et al., there is a 30-fold difference in the binding affinity of protein kinase C associated with phosphatidylserine as compared to a mixture of phospholipids corresponding to that in human erythrocytes.[191] Likewise, Bell and co-workers have suggested that the amounts of cationic lipids, in particular sphingosine, might regulate binding affinity.[192] Indeed, reconstitution of receptor into liposomes of two different compositions was shown to generate curved Scatchard plots.[193] A critical issue is whether heterogeneity in lipid domains actually occurs in intact cells. The use of photoaffinity probes specific for the phospholipids associated with protein kinase C might provide an approach to resolve this problem.[194]

The altered subcellular distribution of protein kinase C in response to phorbol esters or other activators of protein kinase C is a third possible mechanism for the generation of heterogeneity. Protein kinase C is found in both the cytosol and bound to membranes (reviewed in Ref. 104). Phorbol ester treatment of intact cells induces the redistribution of protein kinase C from the soluble to the particulate fraction as typically measured by the lysis of cells in the presence of chelators. A similar redistribution of protein kinase C in PMA-treated GH$_3$ cells was measured in immunoprecipitates, indicating that the disappearance of binding activity in phorbol ester-treated cells is associated with loss of the enzyme protein.[195] In different cell types, translocation of protein kinase C to different subcellular membranes has been documented immunocytochemically. Polyclonal antisera to protein kinase C indicated that the kinase was localized in the plasma membrane of phorbol ester-treated HL-60 cells[196] while in certain neural cells the enzyme was detected inside the nucleus, concentrated in a region adjacent to the inner nuclear membrane.[197] This subcellular translocation of protein kinase C could represent an important regulatory event by influencing substrate accessibility. More detailed assessment of protein kinase C subcellular localization after activation may give insight into pathways conveying the impact of protein kinase C activation into other cellular compartments.

Another mechanism for altering protein kinase C activity is treatment with proteases. *In vitro*, mild proteolysis can activate protein kinase C to produce a catalytically active fragment of about 51k.[198] This catalytic fragment has been cleaved from the regulatory domain of the enzyme so that its phosphotransferase activity is now independent of Ca^{2+}, phospholipid, and diacylglycerol. The susceptibility of protein kinase C to proteolysis is enhanced *in vitro* when the enzyme is in the phospholipid-associated, activated state.[199] *In vivo*, generation of the catalytic fragment in response to phorbol ester treatment has been confirmed in both neutrophils[200] and platelets.[201]

Recently, protein kinase C cloning efforts led to the identification of a family of genes that encode polypeptides closely related to protein kinase C.[202,203] Several genes of this family have been cloned, and analysis by Southern hybridization suggests that even more protein kinase C genes may exist. The significance of the existence of a family of closely related but distinct protein kinase C molecules is considerable. Cellular responses may be effected through the activation of different members of the protein kinase C family. Thus, the heterogeneity in biological responses to phorbol esters may in part be explained by different expression in individual cell types. Studies investigating the patterns of protein kinase C expression in individual cell lines might reveal lineage-specific differences which could account for observed cellular responses to phorbol esters.

Although phorbol esters appear to be diacylglycerol analogues, in certain systems they do not function identically. Diacylglycerols do not mimic all phorbol ester-induced responses and in some studies yielded different patterns of protein phosphorylation (Section 8). These differences could be explained in several ways. (1) Protein kinase C may not be the sole functional phorbol ester receptor but rather the quantitatively major form in most tissues. (2) The stability of phorbol ester relative to the rapidly metabolized diacylglycerols may lead to abnormal, chronic stimulation of protein kinase C. (3) Phorbol esters, in contrast to diacylglycerols, should be able to equilibrate with internal cell membranes and may cause abnormal distribution of activated kinase. (4) The endogenous generation of diacylglycerol, unlike phorbol ester treatment, can elevate intracellular Ca^{2+} and thereby activate protein kinases other than protein kinase C.

A divergence in structure/activity is seen for the bryostatins, a group of non-phorbol compounds isolated from the marine bryozoan *Bugula neritina*. These compounds were intially isolated on the basis of their antineoplastic activity,[204] and more recently were found to mimic a number of phorbol ester-induced responses. Bryostatin 1 at nanomolar concentrations induced superoxide generation and competed for [^{3}H]-PDBu binding in polymorphonuclear leukocytes.[205] Like phorbol esters, bryostatin 1 stimulated DNA synthesis in Swiss 3T3 cells and this stimulation was abolished in cells desensitized by prolonged pretreatment with

PMA.[206] These findings suggested that bryostatins and phorbol esters act via the same receptor, protein kinase C. In HL-60 cells, the action of the bryostatins was found to be more complex. Although bryostatin 1 induced protein kinase C translocation to the plasma membrane and bound to this receptor in intact cells, it was unable to induce differentiation in HL-60 cells.[207] Furthermore, bryostatin blocked phorbol ester-induced differentiation in these cells in a dose-dependent fashion. A similar anti-phorbol ester action has been characterized by our laboratory in Friend cells where bryostatin 1 specifically blocked the phorbol ester-induced inhibition of Friend cell differentiation. Thus, the bryostatins may represent an important probe to investigate the heterogeneity of phorbol ester-induced responses and a possible general mechanism for the inhibition of certain protein kinase C-mediated events.

10. Modeling Studies to Elucidate Stucture–Function Relationships for Ligands of Protein Kinase C

The precise structural requirements for the biological activities exhibited by phorbol esters and structurally unrelated compounds that mimic their action have not been established. However, since these agents appear to interact at the same modulatory site on protein kinase C (as indicated by the inhibition of [^{3}H]-PDBu binding), they would be expected to possess a similar three-dimensional arrangement of homologous functional groups in their bound or active form. Efforts to identify this common structural basis through synthetic modification of PMA have been limited due to the complexity and instability of these compounds. Nevertheless, comparison of different classes of natural products has proved informative.

In a recent study, an explanation for the similar biological activities of four structurally different classes of irritants and tumor promoters was derived from a comparison of the three-dimensional array of their heteroatoms and other groups that would influence binding. Extensive correlation of these pharmacophoric features for phorbol, ingenol, gnidimacrin, and teleocidin suggested that the important functional groups were the C-4, C-9, and C-20 hydroxyls of phorbol.[208] Based on earlier reports that indicated that a long-chain acid esterified to either the C-12 or C-13 hydroxyl group of phorbol was important for activity,[209] Nishizuka has suggested a structural analogy between the 12,13-diester of phorbol and the 1,2-diester of diglycerides[16] (Fig. 1A). However, the comparable activity of phorbol 12,13-diesters and ingenol 3-monoesters (which lack hydroxyl groups at 12 and 13) argues against this suggestion. A significantly better correlation (rms deviation = 0.03 Å) was found when the heteroatoms of diacylglycerol were compared instead with the C-4, C-9, and C-20 hydroxyls[208] (Fig. 1B).

Compounds synthesized with the essential structural features predicted by this model possessed certain phorbol ester activities.[208] These compounds competitively inhibited phorbol ester binding, although at a somewhat lower potency than the diglycerides. In intact cells the compounds also induced several responses characteristic of phorbol esters, including inhibition of EGF binding in 3T3 cells and

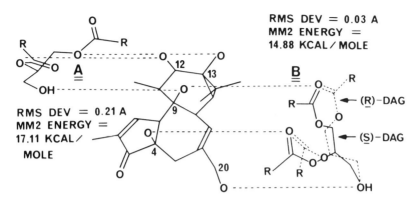

FIGURE 1. Computer (MM2)-optimized geometries of diacylglycerols (DAG) constrained so that three oxygen atoms of DAG overlay with the corresponding oxygens of phorbol 20-bromofuroate.[208] (MM2 = molecular mechanics, program 2.)

phosphorylation of the 40k protein in platelets. In some other systems, toxicity prevented the observation of specific cellular responses.

Studies such as these confirm the usefulness of natural product chemistry in the understanding of biochemical and cellular functions. The phorbol esters have played a key role in the investigation of the action of protein kinase C. Although a specific antagonist of protein kinase C activity has not yet been found, knowledge gained from the study of phorbol esters may make possible the development of pharmacological agents for intervention in the biological actions of protein kinase C.

ACKNOWLEDGMENT. We thank Dr. Stuart H. Yuspa for his critical reading of the manuscript.

References

1. Diamond, L., O'Brien, T. G., and Baird, W. M., 1980, Tumor promotors and the mechanism of tumor promotion, *Adv. Cancer Res.* **32:**1–74.
2. Hecker, E., Fusenig, N. E., Kunz, W., Marks, F., and Thielmann, H. W. (eds.), 1982, *Carcinogenesis—A Comprehensive Survey,* Vol. 7, Raven Press, New York.
3. Slaga, T. J. (ed.), 1984, *Mechanism of Tumor Promotion,* Vol. 2, CRC Press, Boca Raton, Fla.
4. Blumberg, P. M., 1980, In vitro studies on the mode of action of the phorbol esters, potent tumor promoters: Part 1, *CRC Crit. Rev. Toxicol.* **8:**153–197.
5. Blumberg, P. M., 1981, In vitro studies on the mode of action of the phorbol esters, potent tumor promoters: Part 2, *CRC Crit. Rev. Toxicol.* **8:**199–234.
6. Diamond, L., 1984, Tumor promotors and cell transformation, *Pharmacol. Ther.* **26:**89–145.
7. Driedger, P. E., and Blumberg, P. M., 1980, Specific binding of phorbol ester tumor promotors, *Proc. Natl. Acad. Sci. USA* **77:**567–571.
8. Delclos, K. B., Nagle, D. S., and Blumberg, P. M., 1980, Specific binding of phorbol ester tumor promotors to mouse skin, *Cell* **19:**1025–1032.

9. Blumberg, P. M., Dunn, J. A., Jaken, S., Jeng, A. Y., Leach, K. L., Sharkey, N. A., and Yeh, E., 1984, Specific receptors for phorbol ester tumor promotors and their involvement in biological responses, in: *Mechanisms of Tumor Promotion* (T. J. Slaga, ed.), Vol. 3, CRC Press, Boca Raton, Fla., pp. 143–186.
10. Kikkawa, U., Takai, Y., Tanaka, Y., Miyake, R., and Nishizuka, Y., 1983, Protein kinase C as a possible receptor protein of tumor-promoting phorbol esters, *J. Biol. Chem.* **258**:11442–11445.
11. Castagna, M., Takai, Y., Kaibuchi, K., Sano, K., Kikkawa, U., and Nishizuka, Y., 1982, Direct activation of calcium-activated, phospholipid-dependent protein kinase by tumor-promoting phorbol esters, *J. Biol. Chem.* **257**:7847–7851.
12. Niedel, J. E., Kuhn, L. J., and Vandenbark, G. R., 1983, Phorbol diester receptor copurifies with protein kinase C, *Proc. Natl. Acad. Sci. USA* **80**:36–40.
13. Leach, K. L., James, M. L., and Blumberg, P. M., 1983, Characterization of a specific phorbol ester aporeceptor in mouse brain cytosol, *Proc. Natl. Acad. Sci. USA* **80**:4208–4212.
14. Ashendel, C. L., Staller, J. M., and Boutwell, R. K., 1983, Protein kinase activity associated with a phorbol ester receptor purified from mouse brain, *Cancer Res.* **43**:4333–4337.
15. Parker, P. J., Stabel, S., and Waterfield, M. D., 1984, Purification to homogeneity of protein kinase C from bovine brain—Identity with the phorbol ester receptor, *EMBO J.* **3**:953–959.
16. Nishizuka, Y., 1984, The role of protein kinase C in cell surface signal transduction and tumor promotion, *Nature* **308**:693–698.
17. Jetten, A. M., Ganong, B. R., Vandenbark, G. R., Shirley, J. E., and Bell, R. M., 1985, Role of protein kinase C in diacylglycerol-mediated induction of ornithine decarboxylase and reduction of EGF binding, *Proc. Natl. Acad. Sci. USA* **82**:1941–1945.
18. Jeng, A. Y., Lichti, U., Strickland, J. E., and Blumberg, P. M., 1985, Similar effects of phospholipase C and phorbol ester tumor promotors, *Cancer Res.* **45**:5714–5721.
19. Sharkey, N. A., Leach, K. L., and Blumberg, P. M., 1984, Competitive inhibition by diacylglycerol of specific phorbol ester binding, *Proc. Natl. Acad. Sci. USA* **81**:607–610.
20. Michell, R. H., 1975, Inositol phospholipids and cell surface receptor function, *Biochim. Biophys. Acta* **415**:81–147.
21. Berridge, M. J., 1981, Phosphatidylinositol hydrolysis: A multifunctional transducing mechanism, *Mol. Cell. Endocrinol.* **24**:115–140.
22. Berridge, M. J., 1986, Cell signalling through phospholipid metabolism, *J. Cell Sci.* **4** (Suppl.):137–153.
23. Hokin, L. E., 1968, Dynamic aspects of phospholipids during protein secretion, *Int. Rev. Cytol.* **23**:187–208.
24. Pickard, M. R., and Hawthorne, J. N., 1978, The labelling of nerve ending phospholipids in guinea pig brain *in vivo* and the effect of electrical stimulation on phosphatidylinositol metabolism in prelabelled synaptosomes, *J. Neurochem.* **30**:145–155.
25. Fisher, D. B., and Mueller, G. C., 1971, Studies on the mechanism by which phytohemagglutinin rapidly stimulates phospholipid metabolism in human lymphocytes, *Biochim. Biophys. Acta* **248**:434–438.
26. Masuzawa, Y., Osawa, T., Inoue, K., and Nojima, S., 1973, Effects of various mitogens on the phospholipid metabolism of human peripheral lymphocytes, *Biochim. Biophys. Acta* **326**:339–344.
27. Dawson, R. M. C., and Irvine, R. F., 1978, Possible role of lysosomal phospholipases in inducing tissue prostaglandin synthesis, *Adv. Prostagl. Thromb. Res.* **3**:47–54.
28. Bell, R. M., Kennerly, D. A. Stanford, N., and Majerus, P. W., 1979, Diglyceride lipase, a pathway for arachidonate release from human platelets, *Proc. Natl. Acad. Sci. USA* **76**:3238–3241.
29. Dawson, R. M. C., Frienkel, N., Jungawala, F. B., and Clarke, N., 1971, The enzymatic formation of myoinositor 1:2-cyclic phosphate from phosphatidylinositol, *Biochem. J.* **122**:605–607.
30. Streb, H., Irvine, R. F., Berridge, M. J., and Schulz, I., 1983, Release of Ca^{2+} from a nonmitochondrial intracellular store in pancreatic acinar cells: Defined characteristics and unanswered questions, *Philos. Trans. R. Soc. London Ser. B* **296**:123–137.
31. Berry, D. L., Lieber, M. R., Fischer, S. M., and Slaga, T. J., 1977, Qualitative and quantitative separation of a series of tumor promoters by high pressure liquid chromatography, *Cancer Lett.* **3**:125–132.

32. Hemmings, B. A., 1985, Regulation of cAMP-dependent protein kinase in cultured cells, *Curr. Top. Cell. Regul.* **27**:117–132.
33. Yuspa, S. H., 1984, Tumor promotion in epidermal cells in culture, in: *Mechanisms of Tumor Promotion* (T. J. Slaga, ed.), Vol. 3, CRC Press, Boca Raton, Fla., pp. 1–11.
34. Rovera, G., O'Brien, T. G., and Diamond, L., 1979, Induction of differentiation in human promyelocytic leukemia cells by tumor promotors, *Science* **204**:868–870.
35. Feuerstein, N., and Cooper, H. L., 1984, Rapid phosphorylation–dephosphorylation of specific proteins induced by phorbol esters in HL-60 cells: Further characterization of the 17 kilodalton and 27 kilodalton proteins in myeloid leukemic cells and human monocytes, *J. Biol. Chem.* **259**:2782–2788.
36. Yamasaki, H., Fibach, E., Nudel, U., Weinstein, I. B., Rifkind, R. A., and Marks, P. A., 1977, Tumor promotors inhibit spontaneous and induced differentiation of murine erythroleukemia cells in culture, *Proc. Natl. Acad. Sci. USA* **74**:3451–3455.
37. Rovera, G., O'Brien, T. G., and Diamond, L., 1977, Tumor promotors inhibit spontaneous differentiation of Friend erythroleukemia cells in culture, *Proc. Natl. Acad. Sci. USA* **74**:2894–2899.
38. Sivak, A., 1977, Induction of cell division in Balb/c 3T3 cells by phorbol myristate acetate or bovine serum: Effects of inhibition of cyclic AMP phosphodiesterase and Na^+/K^+-ATPase, *In Vitro* **13**:337–343.
39. Sivak, A., 1972, Induction of cell division: Role of cell membrane sites, *J. Cell. Physiol.* **80**:167–173.
40. Sivak, A., 1977, Comparison of the biological activity of the tumor promoter phorbol myristate acetate and a metabolite, phorbol myristate acetate in cell culture, *Cancer Lett.* **2**:285–290.
41. Diamond, L., O'Brien, S., Donaldson, C., and Shimizu, Y., 1974, Growth stimulation of human diploid fibroblasts by the tumor promotor 12-O tetradecanoylphorbol-13-acetate, *Int. J. Cancer* **13**:721–730.
42. Driedger, P. E., and Blumberg, P. M., 1977, The effect of phorbol diesters on chicken embryo fibroblasts, *Cancer Res.* **37**:3257–3265.
43. Yuspa, S. H., Lichti, U., Ben, T., Patterson, E., Hennings, H., Slaga, T. J., Colburn, N., and Kelsey, W., 1976, Phorbol esters stimulate DNA synthesis and ornithine decarboxylase activity in mouse epidermal cell cultures, *Nature* **262**:4062–4068.
44. Mastro, A. M., and Mueller, G. C., 1974, Synergistic action of phorbol esters in mitogen-activated bovine lymphocytes, *Exp. Cell Res.* **88**:40–47.
45. Whitfield, J. F., MacManus, J. P., and Gillan, D. J., 1974, Calcium-dependent stimulation by a phorbol ester (PMA) of thymic lymphoblast DNA synthesis and proliferation, *J. Cell. Physiol.* **82**:151–155.
46. Solanki, V., Slaga, T. J., Callahan, M., and Huberman, E., 1981, Down regulation of specific binding of [20-^{3}H]phorbol 12,13-dibutyrate and phorbol ester-induced differentiation of human promyelocytic leukemia cells, *Proc. Natl. Acad. Sci. USA* **78**:1722–1725.
47. Solanki, V., and Slaga, T. J., 1981, Specific binding of phorbol ester tumor promotors to intact primary epidermal cells from Sencar mice, *Proc. Natl. Acad. Sci. USA* **78**:2549–2553.
48. Jaken, S., Tashjian, A. H., and Blumberg, P. M., 1981, Characterization of phorbol ester receptors and their down modulation in GH_4C_1 rat pituitary cells, *Cancer Res.* **41**:2175–2181.
49. Collins, M. K. L., and Rozengurt, E., 1982, Binding of phorbol esters to high-affinity sites on murine fibroblastic cells elicits a mitogenic response, *J. Cell. Physiol.* **112**:42–50.
50. Rodriguez-Pena, A., and Rozengurt, E., 1984, Disappearance of Ca^{2+}-sensitive, phospholipid-dependent protein kinase activity in phorbol ester treated 3T3 cells, *Biochem. Biophys. Res. Commun.* **120**:1053–1059.
51. Collins, M. K. L., and Rozengurt, E., 1984, Homologous and heterologous mitogenic desensitization of Swiss 3T3 cells to phorbol esters and vasopressin: Role of receptor and postreceptor steps, *J. Cell. Physiol.* **118**:133–142.
52. Pasti, G., Lacal, J. C., Warren, B. S. Aaronson, S., and Blumberg, P. M., 1986, Restoration of phorbol ester responsiveness in down regulation of Swiss 3T3 cells by microinjection of protein kinase C, Nature **324**:375–377.
53. Rozengurt, E., Rodriguez-Pena, M., and Smith, K. A., 1983, Phorbol esters, phospholipase C, and

growth factors rapidly stimulate the phosphorylation of a M_r 80,000 protein in intact quiescent 3T3 cells, *Proc. Natl. Acad. Sci. USA* **80:**7244–7248.
54. Rozengurt, E., Rodriguez-Pena, A., Coombs, M., and Sinnett-Smith, J., 1984, Diacylglycerol stimulates DNA synthesis and cell division in mouse 3T3 cells: Role of calcium-sensitive, phospholipid-dependent protein kinase, *Proc. Natl. Acad. Sci. USA* **81:**5748–5752.
55. Rodriguez-Pena, A., and Rozengurt, E., 1986, Phosphorylation of an acidic M_r 80,000 cellular protein in cell-free system and intact Swiss 3T3 cells: Specific marker of protein kinase C activity, *EMBO J.* **5:**77–83.
56. Habenicht, A. J., Glomset, J. A., and King, W. C., 1981, Early changes in phosphatidylinositol and arachidonic acid metabolism in quiescent Swiss 3T3 cells stimulated to divide by platelet-derived growth factor, *J. Biol. Chem.* **256:**12329–12335.
57. Vara, F., and Rozengurt, E., 1985, Stimulation of Na^+/H^+ antiport activity by epidermal growth factor and insulin occurs without activation of protein kinase C, *Biochem. Biophys. Res. Commun.* **130:**646–653.
58. Anastasi, A., Erpsamer, V., and Bucci, M., 1971, Isolation and structure of bombesin and alytsin, two analogous active peptides from the skin of the European amphibians *Bombina* and *Alytes, Experientia* **27:**166–167.
59. Rozengurt, E., and Sinnett-Smith, J., 1983, Bombesin stimulation of DNA synthesis and cell division in cultures of Swiss 3T3 cells, *Proc. Natl. Acad. Sci. USA* **80:**2936–2940.
60. Zachary, I., and Rozengurt, E., 1985, High affinity receptors from peptides of the bombesin family in Swiss 3T3 cells, *Proc. Natl. Acad. Sci. USA* **82:**7616–7620.
61. Zachary, I., Sinnett-Smith, J. W., and Rozengurt, E., 1986, Early events elicited by bombesin and structurally related peptides in quiescent Swiss 3T3 cells. I. Activation of protein kinase C, and inhibition of epidermal growth factor binding, *J. Cell Biol.* **102:**2211–2222.
62. Erisman, M. D., Linoila, R. I., Hernandez, O., DiAugustine, R. P., and Lazarus, L. H., 1982, Human lung small-cell carcinoma contains bombesin, *Proc. Natl. Acad. Sci. USA* **79:**2379–2383.
63. Moody, T. W., Pert, C. B., Gazdar, A. F., Carney, D. N., and Minna, J. B., 1981, High levels of intracellular bombesin characterize human small cell lung carcinoma, *Science* **214:**1246–1248.
64. Roth, K. A., Evans, C. J., Weber, L. E., Barchas, J. D., Bostwick, D. G., and Bensch, K. G., 1983, GRP-related peptides in a human malignant lung carcinoid tumour, *Cancer Res.* **43:**5411–5415.
65. Matsubayashi, S., Yanaihara, C., Ohkubo, M., Fukata, S., Hayashi, Y., Tamai, H., Nakagawa, T., Miyauchi, A., Kuma, K., Abe, K., Suzuki, T., and Yanaihara, N., 1984, Gastrin-releasing peptide immunoreactivity in medullary thyroid carcinoma, *Cancer* **53:**2472–2477.
66. Blackshear, P. J., Witters, L. A., Girard, P. R., Kuo, J. F., and Quamo, S. N., 1985, Growth factor-stimulated protein phosphorylation in 3T3-L1 cells: Evidence for protein kinase C-dependent and -independent pathways, *J. Biol. Chem.* **260:**13304–13315.
67. Brown, K. D., Dicker, P., and Rozengurt, E., 1979, Inhibition of epidermal growth factor binding to surface receptors by tumor promotors, *Biochem. Biophys. Res. Commun.* **86:**1037–1043.
68. Lee, L. S., and Weinstein, I. B., 1979, Mechanism of tumor promoter inhibition of cellular binding of epidermal growth factor, *Proc. Natl. Acad. Sci. USA* **76:**5168–5172.
69. Osbourne, R., and Tashjian, A., 1982, Modulation of peptide binding to specific receptors on rat pituitary cells by tumor promoting phorbol esters: Decreased binding of thyrotropin-releasing hormone and somatostatin as well as epidermal growth factor, *Cancer Res.* **42:**4375–4381.
70. Jacobs, S., Sahyoun, N. E., Saltier, A. R., and Cuatrecasas, P., 1983, Phorbol esters stimulate the phosphorylation of receptors for insulin and somatomedin C, *Proc. Natl. Acad. Sci. USA* **80:**6211–6213.
71. May, W. S., Jacobs, S., and Cuatrecasas, P., 1984, Association of phorbol ester-induced hyperphosphorylation and reversible regulation of transferrin membrane receptors in HL-60 cells, *Proc. Natl. Acad. Sci. USA* **81:**2016–2020.
72. Zeggari, M., Susini, C., Viguerie, N. Esteve, J. P., Vaysse, N., and Ribert, A., 1985, Tumor promoter inhibition of cellular binding of somatostatin, *Biochem. Biophys. Res. Commun.* **128:**850–857.
73. Zachary, I., and Rozengurt, E., 1985, Modulation of the epidermal growth factor receptor by mitogenic ligands: Effects of bombesin and role of protein kinase C, *Cancer Surv.* **4:**729–765.
74. McCaffrey, P. G., Friedman, B., and Rosner, M. R., 1984, Diacylglycerol modulates binding and phosphorylation of the epidermal growth factor receptor, *J. Biol. Chem.* **259:**12502–12507.

75. Sinnett-Smith, J., and Rozengurt, E., 1985, Diacylglycerol treatment rapidly decreases the affinity of the epidermal growth factor receptors of Swiss 3T3 cells, *J. Cell. Physiol.* **124**:81–86.
76. Cochet, C., Gill, G. N., Meisenhelder, J., Cooper, J. A., and Hunter, A., 1984, C-kinase phosphorylates the EGF receptor and reduces its EGF-stimulated protein tyrosine kinase activity, *J. Biol. Chem.* **259**:2553–2558.
77. Davis, R. J. and Czech, M., 1984, Tumor promoting phorbol diesters cause the phosphorylation of epidermal growth factor receptor, *J. Biol. Chem.* **259**:8545–8549.
78. Davis, R. J., and Czech, M., 1985, Tumor promoting phorbol diesters cause the phosphorylation of epidermal growth factor receptors in normal human fibroblasts at threonine 654, *Proc. Natl. Acad. Sci. USA* **82**:1974–1978.
79. Davis, R. J., and Czech, M., 1985, Platelet-derived growth factor mimics phorbol diester action on epidermal growth factor receptor phosphorylation at threonine 654, *Proc. Natl. Acad. Sci. USA* **82**:4080–4085.
80. Bertics, P. J., Weber, W., Cochet, C., and Gill, G. N., 1985, Regulation of the epidermal growth factor receptor by phosphorylation, *J. Cell. Biochem.* **29**:195–208.
81. Matrisian, L. M., Bowden, G. T., and Magun, B. E., 1981, Mechanism of synergistic induction of DNA synthesis by epidermal growth factor and tumor promoters, *J. Cell. Physiol.* **108**:417–425.
82. Ruscetti, F. W., Morgan D., and Gallo, R. C., 1977, Functional and morphologic characterization of human T cells continuously grown in vitro, J. Immunol. **119**:131–138.
83. Smith, K. A., 1984, Interleukin-2, *Annu. Rev. Immunol.* **2**:319–333.
84. Robb, R., Munck, A., and Smith, K. A., 1981, T-cell growth factor receptors, *J. Exp. Med.* **154**:1455–1474.
85. Leonard, W., Depper, J., Uchiyama, T., Smith, K., Waldmann, T., and Greene, W. 1982, A monoclonal antibody that appears to recognize the receptor for human T-cell growth factor: Partial characterization of the receptor, *Nature* **300**:267–269.
86. Ross, R. J., and Greene, W. C., 1983, Direct demonstration of the identity of T-cell growth factor binding protein and Tac antigen, *J. Exp. Med.* **158**:1332–1337.
87. Farrar, W. L., and Anderson, W. B., 1985, Interleukin-2 stimulates association of protein kinase C with plasma membrane, *Nature* **315**:233–235.
88. Howard, M., Burgess, A. W., McPhee, D., and Metcalf, D., 1979, T-cell hybridoma secreting hemopoietic regulatory molecules: Granulocyte macrophage and eosinophil colony-stimulating factors, *Cell* **18**:993–999.
89. Farrar, W. L., Thomas, T. P., and Anderson, W. B., 1985, Altered cytosol/membrane enzyme redistribution on interleukin-3 activation of protein kinase C, *Nature* **315**:235–237.
90. Depper, J. M., Leonard, W. J., Kronke, M., Noguchi, P., Cunningham, R., Waldmann, T. A., and Greene, W. C., 1984, Regulation of IL-2 receptor expression: Effects of phorbol esters, phospholipase C and reexposure to lectin or antigen, *J. Immunol.* **133**:3054–3058.
91. Shackelford, D. A., and Trowbridge, I. S., 1984, Induction of expression and phosphorylation of the human IL-2 receptor by a phorbol diester, *J. Biol. Chem.* **259**:11706–11712.
92. Taguchi, M., Thomas, T. P., Anderson, W. B., and Farrar, W. L., 1986, Direct phosphorylation of the IL-2 receptor Tac antigen epitope by protein kinase C, *Biochem. Biophys. Res. Commun.* **135**:239–246.
93. Shackelford, D. A., and Trowbridge, I. S., 1986, Identification of lymphocyte integral membrane proteins as substrates for protein kinase C: Phosphorylation of the IL-2 receptor, class I HLA antigens, and T200 glycoprotein, *J. Biol. Chem.* **261**:8334–8341.
94. Sugimoto, Y., Whitman, M., Cantley, L. C., and Erikson, R. L., 1984, Evidence that Rous sarcoma virus transforming gene product phosphorylates phosphatidylinositol and diacylglycerol, *Proc. Natl. Acad. Sci. USA* **81**:2117–2121.
95. Macara, I. G., Marinetti, G. V., Livingston, J. N., and Balduzzi, P. C., 1985, Lipid phosphorylating activities and tyrosine kinases: a possible role for phosphatidyl inositol turnover in transformation, in: *Cancer Cells* (J. Feramisco, B. Ozanne, and C. Stiles, eds.), Vol. 3, Cold Spring Harbor Laboratory, Cold Spring Harbor, N. Y., pp. 365–382.
96. Macara, I. G., Marinetti, G. V., and Balduzzi, P. C., 1984, Transforming protein of avian sarcoma virus UR2 is associated with phosphatidylinositol kinase activity: Possible role in tumorigenesis, *Proc. Natl. Acad. Sci. USA* **81**:2728–2732.

97. Nishizuka, Y., 1986, Perspectives on the role of protein kinase C in stimulus–response coupling, *J. Natl. Cancer Inst.* **76:**363–370.
98. Whitman, M., Kaplan, D. R., Schaffhausen, B., Cantley, L., and Roberts, T. M., 1985, Association of phosphatidylinositol kinase activity with polyoma middle-T competent for transformation, *Nature* **315:**239–242.
99. Kaplan, D. R., Whitman, M., Schaffhausen, B., Raptis, L., Garcea, R. L., Pallas, D., Roberts, T. M., and Cantley, L., 1986, Phosphatidylinositol metabolism and polyoma-mediated transformation, *Proc. Natl. Acad. Sci. USA* **83:**3624–3628.
100. Preiss, J. P., Loomis, C. R., Bishop. W. R., Stein, R. B., Niedel, J. E., and Bell, R. M., 1986, Quantitative measurement of sn-1,2-diacylglycerols present in platelets, hepatocytes, and *ras* and *sis* transformed normal rat kidney cells, *J. Biol. Chem.* **261:**8597–8600.
101. Berridge, M. J., and Irvine, R. F., 1984, Inositol trisphosphate, a novel second messenger intracellular signal transduction, *Nature* **312:**315–321.
102. Der, C. J., Finkel, T., and Cooper, G. M., 1986, Biological and biochemical properties of human H-*ras* genes mutated at codon 61, *Cell* **44:**167–176.
103. Cockroft, S., and Gomperts, B. D., 1985, Role of guanine nucleotide binding protein in the activation of polyphosphoinositide phosphodiesterase, *Nature* **314:**534–536.
104. Anderson, W. B., Estival, A., Tapiovaara, H., and Gopalakrishna, R., 1985, Altered subcellular distribution of protein kinase C (a phorbol ester receptor). Possible role in tumor promotion and the regulation of cell growth: Relationship to changes in adenylate cyclase activity, *Adv. Cyclic Nucleotide Protein Phosphorylation Res.* **19:**287–306.
105. Gould, K. L., Woodgett, J. R., Cooper, J. A., Buss, J. E., Shalloway, D., and Hunter, T., 1985, Protein kinase C phosphorylates pp60src at a novel site, *Cell* **42:**849–857.
106. Purchio, A. F., Gentry, L., and Shoyab, M., 1986, Phosphorylation of pp60v-src by the TPA receptor kinase (protein kinase C), *Virology* **150:**524–529.
107. Tamura, T., Friis, R. R., and Bauer, H., 1984, pp60c-src is a substrate for phosphorylation when cells are stimulated to enter cycle, *FEBS Lett.* **177:**151–156.
108. Courtneidge, S. A., 1985, Activation of the pp60$^{c\text{-}src}$ kinase by middle T antigen binding or dephosphorylation, *EMBO J.* **4:**1471–1477.
109. Raptis, L., Boynton, A. L., and Whitfield, J. F., 1986, Protein kinase C promotes the phosphorylation of immunoprecipitated middle T antigen from polyoma virus-transformed cells, *Biochem. Biophys. Res. Commun.* **136:**995–1000.
110. Balmer-Hofer, K., and Benjamin, T. L., 1985, Phosphorylation of middle T antigen and cellular proteins in purified plasma membranes of polyoma virus-infected cells, *EMBO J.* **4:**2321–2327.
111. Hsiao, W. L., Gattoni-Celli, S., and Weinstein, I. B., 1986, Oncogene-induced transformation of C3H 10T1/2 cells is enhanced by tumor promoters, *Science* **226:**552–555.
112. Hsiao, W. L., Wu, T., and Weinstein, I. B., 1986, Oncogene-induced transformation of a rat embryo fibroblast cell line is enhanced by tumor promoters, *Mol. Cell. Biol.* **6:**1943–1950.
113. Sharkey, N. A., and Blumberg, P. M., 1985, Kinetic evidence that 1,2-diolein inhibits phorbol ester binding to protein kinase C via a competitive mechanism, *Biochem. Biophys. Commun.* **133:**1051–1056.
114. Konig, B. DiNitto, P. A., and Blumberg, P. M., 1985, Stoichiometric binding of diacylglycerol to the phorbol ester receptor, *J. Cell. Biochem.* **29:**37–44.
115. Hannun, Y. A., Loomis, C. R., and Bell, R. M., 1986, Protein kinase activation in mixed micelles: Mechanistic implications of phospholipid, diacylglycerol, and calcium, *J. Biol. Chem.* **261:**7184–7190.
116. Sharkey, N. A., and Blumberg, P. M., 1985, Highly lipophilic phorbol esters as inhibitors of specific [^{3}H]phorbol 12,13-dibutyrate binding, *Cancer Res.* **45:**19–24.
117. Sharkey, N. A., and Blumberg, P. M., 1986, Comparison of the activity of phorbol 12-myristate 13-acetate and the diglyceride glycerol 1-myristate 2-acetate, *Carcinogenesis* **7:**677–679.
118. Nishizuka, Y., 1984, The role of protein kinase C in cell surface signal transduction and tumor promotion, *Nature* **308:**693–698.
119. Ebeling, J. C., Vandenbark, G. R., Kuhn, L. J., Ganong, B. R., Bell, R. M., and Niedel, J. E., 1985, Diacylglycerols mimic phorbol diester induction of leukemic cell differentiation, *Proc. Natl. Acad. Sci. USA* **82:**815–819.
120. O'Flaherty, J. R., Schmitt, J. D., McCall, C. E., and Wykle, R. L., 1984, Diacylglycerols enhance

human neutrophil degranulation responses: Relevancy to a multiple mediator hypothesis of cell function, *Biochem. Biophys. Res. Commun.* **123**:64–70.

121. Fujita, I., Irita, K., Takeshige, K., and Minakami, S., 1984, Diacylglycerol, 1-oleoyl-acetyl-glycerol, stimulates superoxide-generation from human neutrophils, *Biochem. Biophys. Res. Commun.* **120**:318–324.

122. Davis, R. J. Ganong, B. R., Bell, R. M., and Czech, M. P., 1985, sn-1,2-Dioctanoylglycerol: A cell-permeable diacylglycerol that mimics phorbol diester action on the epidermal growth factor receptor and mitogenesis, *J. Biol. Chem.* **260**:1562–1566.

123. Kaibuchi, K., Takai, Y., and Nishizuka, Y., 1985, Protein kinase C and calcium ion in mitogenic response of macrophage-depleted human peripheral lymphocytes, *J. Biol. Chem.* **260**:1366–1369.

124. Katakami, Y., Kaibuchi, K., Sawamura, M., Takai, Y., and Nishizuka, Y., 1984, Synergistic action of protein kinase C and calcium for histamine release from rat periotoneal mast cells, *Biochem. Biophys. Res. Commun.* **121**:573–578.

125. Kaibuchi, K., Sano, K., Hoshijima, M., Takai, Y., and Nishizuka, Y., 1982, Phosphatidylinositol turnover in platelet activation; calcium mobilization and protein phosphorylation, *Cell Calcium* **3**:323–325.

126. Brown, K. D., Blay, J., Irvine, R. F., Heslop, J. P., and Berridge, M. J., 1984, Reduction of epidermal growth factor receptor affinity by heterologous ligands: Evidence for a mechanism involving the breakdown of phosphoinositides and the activation of protein kinase C, *Biochem. Biophys. Res. Commun.* **123**:377–384.

127. Harman, I., Costello, A., Ganong, B., Bell, R. M., and Handwerger, S., 1986, Activation of protein kinase C inhibits synthesis and release of decidual prolactin, *Am. J. Physiol.* **251**:E172–177.

128. Macara, I. G., 1986, Activation of $^{45}Ca^{2+}$ influx and $^{22}Na^+/H^+$ exchange by epidermal growth factor and vanadate in A431 cells is independent of phosphatidylinositol turnover and is inhibited by phorbol ester and diacylglycerol, *J. Biol. Chem.* **261**:9321–9327.

129. Giroldi, J., Hamel, E., and Yamasaki, H., 1986, 1-Oleoyl 2-acetyl glycerol inhibits differentiation of TPA-sensitive but not of TPA-resistant Friend cells, *Carcinogenesis* **7**:1183–1186.

130. Shen, S. S., and Burgart, L. J., 1986, 1,2-Diacylglycerols mimic phorbol 12-myristate 13-acetate activation of the sea urchin egg, *J. Cell. Physiol.* **127**:330–340.

131. Fleming, N., Bilan, P. T., and Sliwinski-Lis, E., 1986, Effects of a phorbol ester and diacylglycerols on secretion of mucin and arginine esterase by rat submandibular gland cells, *Pfluegers. Arch.* **406**:6–11.

132. Negro-Vilar, A., and Lapetina, E. G., 1985, 1,2-Didecanoylglycerol and phorbol 12,13-dibutyrate enhance anterior pituitary hormone secretion *in vitro*, *Endocrinology* **117**:1559–1564.

133. Enomoto, T., and Yamasaki, H., 1985, Rapid inhibition of intercellular communication between BALB/c 3T3 cells by diacylglycerol, a possible endogenous functional analogue of phorbol esters, *Cancer Res.* **45**:3706–3710.

134. Shinohara, O., Knecht, M., and Catt, K. J., 1985, Differential actions of phorbol ester and diacylglycerol on inhibition of granulosa cell maturation, *Biochem. Biophys. Res. Commun.* **133**:468–474.

135. Ojeda, S. R., Urbanski, H. F., Katz, K. H., Costa, M. E., and Conn, P. M., 1986, Activation of two different but complementary biochemical pathways stimulates release of hypothalamic luteinizing hormone-releasing hormone *Proc. Natl. Acad. Sci. USA* **83**:4932–4936.

136. Shinohara, O., Knecht, M., Feng, P., and Catt, K. J., 1986, Activation of protein kinase C potentiates cyclic AMP production and stimulates steroidogenesis in differentiated ovarian granulosa cells, *J. Steroid Biochem.* **24**:161–168.

137. Judd, A. M., Koike, K., Yasumoto, T., and MacLeod, R. M., 1986, Protein kinase C activators and calcium-mobilizing agents synergistically increase GH, LH, and TSH secretion from anterior pituitary cells, *Neuroendocrinology* **42**:197–202.

138. Ransom, J. T., and Cambier, J. C., 1986, B cell activation. VII. Independent and synergistic effects of mobilized calcium and diacylglycerol on membrane potential and I-A expression, *J. Immunol.* **136**:66–72.

139. Jaken, S., 1985, Increased diacylglycerol content with phospholipase C or hormone treatment: Inhibition of phorbol ester binding and induction of phorbol ester-like biological responses, *Endocrinology* **117**:2301–2306.

140. Jetten, A. M., Ganong, B. R., Vandenbark, G. R., Shirley, J. E., and Bell, R. M., 1985, Role of

protein kinase C in diacylglycerol-mediated induction of ornithine decarboxylase and reduction of epidermal growth factor binding, *Proc. Natl. Acad. Sci. USA* **82**:1941–1945.
141. Grzeskowiak, M., Della Bianca, V., De Togni, P., Papini, E., and Rossi, F., 1985, Independence with respect to Ca^{2+} changes of the neutrophil respiratory and secretory response to exogenous phospholipase C and possible involvement of diacylglycerol and protein kinase C, *Biochim. Biophys. Acta* **844**:81–90.
142. Jeng, A. Y., Lichti, U., Strickland, J. E., and Blumberg, P. M., 1985, Similar effects of phospholipase C and phorbol ester tumor promoters on primary mouse epidermal cells, *Cancer Res.* **45**:5714–5721.
143. Yamamoto, S., Gotoh, H., Aizu, E., and Kato, R., 1985, Failure of 1-oleoyl-acetylglycerol to mimic the cell-differentiating action of 12-O-tetradecanoylphorbol 13-acetate in HL-60 cells, *J. Biol. Chem.* **260**:14230–14234.
144. Kawahara, Y., Takai, Y., Minakuchi, R., Sano, K., and Nishizuka, Y., 1980, Phospholipid turnover as a possible transmembrane signal for protein phosphorylation during human platelet activation by thrombin, *Biochem. Biophys. Res. Commun.* **97**:309–317.
145. Ieyasu, H., Takai, Y., Kaibuchi, K., Sawamura, M., and Nishizuka, Y., 1982, A role of calcium-activated, phospholipid-dependent protein kinase in platelet-activating factor-induced serotonin release from rabbit platelets, *Biochem. Biophys. Res. Commun.* **108**:1701–1708.
146. Kaibuchi, K., Takai, Y., Sawamura, M., Hoshijima, M., Fujikura, T., and Nishizuka, Y., 1983, Synergistic functions of protein phosphorylation and calcium mobilization in platelet activation, *J. Biol. Chem.* **258**:6701–6704.
147. Nishizuka, Y., 1984, The role of protein kinase C in cell surface signal transduction and tumor promotion, *Nature* **308**:693–698.
148. Takai, Y., Kikkawa, U., Kaibuchi, K., and Nishizuka, Y., 1984, Membrane phospholipid metabolism and signal transduction for protein phosphorylation, *Adv. Cyclic Nucleotide Protein Phosphorylation Res.* **8**:119–157.
149. Wolf, M., LeVine, H., May, W. S., Cuatrecasas, P., and Sahyoun, N., 1985, A model for intracellular translocation of protein kinase C involving synergism between Ca^{2+} and phorbol esters, *Nature* **317**:546–549.
150. May, W. S., Sahyoun, N., Wolf, M., and Cuatrecasas, P., 1985, Role of intracellular calcium mobilization in the regulation of protein kinase C-mediated membrane processes, *Nature* **317**:549–551.
151. Sibley, D. R., Nambi, P., Peters, J. R., and Lefkowitz, R. J., 1984, Phorbol diesters promote β-adrenergic receptor phosphorylation and adenylate cyclase desensitization in duck erythrocytes, *Biochem. Biophys. Res. Commun.* **121**:973–979.
152. Mukhopadhyat, A. K., and Schumaker, M., 1985, Inhibition of HCG-stimulated adenylate cyclase in purified mouse Leydig cells by the phorbol ester PMA, *FEBS Lett.* **187**:56–60.
153. Kelleher, D. J., Pessin, J. E., Ruoho, A. E., and Johnson, G. L., 1984, Phorbol ester induces desensitization of adenylate cyclase and phosphorylation of the β-adrenergic receptor in turkey erythrocytes, *Proc. Natl. Acad. Sci. USA* **81**:4316–4320.
154. Kassis, S., Zaremba, T., Patel, J., and Fishman, P. H., 1985, Phorbol esters and β-adrenergic agonists mediate desensitization of adenylate cyclase in rat glioma C6 cells by distinct mechanisms, *J. Biol. Chem.* **260**:8911–8917.
155. Bell, J. D., Buxton, I. L., and Brunton, L. L., 1985, Enhancement of adenylate cyclase activity in S49 lymphoma cells by phorbol esters: Putative effect of C kinase on alpha s-GTP-catalytic subunit interaction, *J. Biol. Chem.* **260**:2625–2628.
156. Summers, S. T., and Cronin, M. J., 1986, Phorbol esters enhance basal and stimulated adenylate cyclase activity in a pituitary cell line, *Biochem. Biophys. Res. Commun.* **135**:276–281.
157. Olianas, M. C., and Onali, P., 1986, Phorbol esters increase GTP-dependent adenylate cyclase activity in rat brain striatal membranes, *J. Neurochem.* **47**:890–897.
158. Katada, T., Gilman, A. G., Watanabe, Y., Bauer, S., and Jakobs, K. H., 1985, Protein kinase C phosphorylates the inhibitory guanine-nucleotide-binding regulatory component and apparently suppresses its function in hormonal inhibition of adenylate cyclase, *Eur. J. Biochem.* **151**:431–437.
159. Takai, Y., Kaibuchi, K., Sano, K., and Nishizuka, Y., 1982, Counteraction of calcium-activated phospholipid-dependent protein kinase activation by adenosine 3′,5′-monophosphate in platelets, *J. Biochem.* **91**:403–406.
160. Tamagawa, T., Niki, H., and Niki, A., 1985, Insulin release independent of a rise in cytosolic free Ca^{2+} by forskolin and phorbol ester, *FEBS Lett.* **183**:430–432.

161. Rink, R. J. Sanchez, A., and Hallam, T. J., 1983, Diacylglycerol and phorbol ester stimulate secretion without raising cytoplasmic free calcium in human platelets, *Nature* **305**:317–319.
162. Ware, J. A., Johnson, P. C., Smith, M., and Salzman, E. W., 1985, Aequorin detects increased cytoplasmic calcium in platelets stimulated with phorbol ester or diacylglycerol, *Biochem. Biophys. Res. Commun.* **133**:98–104.
163. Zavoico, G. B., Halenda, S. P. Shaafi, R. I., and Feinstein, M. B., 1985, Phorbol myristate acetate inhibits thrombin-stimulated Ca^{2+} mobilization and phosphatidylinositol 4,5-bisphosphate hydrolysis in human platelets, *Proc. Natl. Acad. Sci. USA* **82**:3859–3862.
164. Orellana, S. A., Solski, P. A. and Brown, J. H., 1985, Phorbol ester inhibits phosphoinositide hydrolysis and calcium mobilization in cultured astrocytoma cells, *J. Biol. Chem.* **260**:5236–5239.
165. Tohmatsu, T., Hattori, H., Nagao, S., Ohki, K., and Nozawa, Y., 1986, Reversal by protein kinase C inhibitor of suppressive actions of phorbol-12-myristate-13-acetate on polyphosphoinositide metabolism and cytosolic Ca^{2+} mobilization in thrombin-stimulated human platelets, *Biochem. Biophys. Res. Commun.* **134**:868–875.
166. DeRiemer, S. A., Strong, J. A., Albert, K. A., Greengard, P., and Kaczmarek, L. K., 1985, Enhancement of calcium current in Aplysia neurones by phorbol ester and protein kinase C, *Nature* **313**:313–316.
167. Rane, S. G., and Dunlap, K., 1986, Kinase C activator 1,2-oleoylacetylglycerol attenuates voltage-dependent calcium current in sensory neurons, *Proc. Natl. Acad. Sci. USA* **83**:184–188.
168. Messing, R. O., Carpenter, C. L., and Greenberg, D. A., 1986, Inhibition of calcium flux and calcium channel antagonist binding in the PC12 neural cell line by phorbol esters and protein kinase C, *Biochem. Biophys. Res. Commun.* **136**:1049–1056.
169. Di Virgilio, F. Pozzan, T., Wollheim, C. B., Vicentini, L. M., and Meldolesi, J., 1986, Tumor promoter phorbol myristate acetate inhibits Ca^{2+} influx through voltage-gated Ca^{2+} channels in two secretory cell lines, PC12 and RINm5F, *J. Biol. Chem.* **261**:32–35.
170. Grubbs, R. D., and Maguire, M. E., 1986, Regulation of magnesium but not calcium transport by phorbol ester, *J. Biol. Chem.* **261**:12550–12554.
171. Ashendel, C. L. 1985, The phorbol ester receptor: A phospholipid-regulated protein kinase, *Biochim. Biophys. Acta* **822**:219–242.
172. Feuerstein, N., and Cooper, H. L., 1983, Rapid protein phosphorylation induced by phorbol ester in HL-60 cells: Unique alkali stable phosphorylation of a 17,000 dalton protein detected by two-dimensional gel electrophoresis, *Cancer Res.* **258**:10786–10793.
173. Kreutter, D., Caldwell, A. B., and Morin, M. J., 1985, Dissociation of protein kinase C activation from phorbol ester-induced maturation of HL-60 leukemia cells, *J. Biol. Chem.* **260**:5979–5984.
174. Hayakawa, T., Suzuki, K., Suzuki, S., Andrews, P. C., and Babior, B. M., 1986, A possible role for protein phosphorylation in the activation of the respiratory burst in human neutrophils, *J. Biol. Chem.* **261**:9109–9115.
175. White, J. R., Huang, C. K., Hill, J. M., Naccache, P. H., Becker, E. L., and Sha'afi, R. I., 1984, Effect of phorbol 12-myristate 13-acetate and its analog 4-alpha-phorbol 12,13-didecanoate on protein phosphorylation and lysosomal enzyme release in rabbit neutrophils, *J. Biol. Chem.* **259**:8605–8611.
176. Garrison, J. C., Johnsen, D. E., and Campanile, C. P., 1984, Evidence for the role of phosphorylase kinase, protein kinase C and other Ca^{2+}-sensitive protein kinases in the response of hepatocytes to angiotensin II and vasopressin, *J. Biol. Chem.* **259**:3283–3292.
177. Feuerstein, N., Sahai, A., Anderson, W. B., Salomon, D. S., and Cooper, H. L., 1984, Differential phosphorylation events associated with phorbol ester effects on acceleration versus inhibition of cell growth, *Cancer Res.* **44**:5227–5233.
178. Kiss, Z., and Steinberg, R. A., 1985, Phorbol ester-mediated protein phosphorylations in S49 mouse lymphoma cells, *Cancer Res.* **45**:2732–2740.
179. Kiss, Z., and Steinberg, R. A., 1985, Interactions between cyclic AMP- and phorbol ester-dependent phosphorylation systems in S49 mouse lymphoma cells, *J. Cell. Physiol.* **125**:200–206.
180. Blenis, J., and Erikson, R. L. 1985, Regulation of a ribosomal protein S6 kinase activity by the Rous sarcoma virus transforming protein, serum, or phorbol ester, *Proc. Natl. Acad. Sci. USA* **82**:7621–7625.
181. Blenis, J., and Erikson, R. L., 1986, Stimulation of ribosomal protein S6 kinase activity by $pp60^{v-src}$ or by serum: Dissociation from phorbol ester-stimulated activity, *Proc. Natl. Acad. Sci. USA* **83**:1733–1737.

182. Gilmore, T., and Martin, G. S. 1983, Phorbol ester and diacylglycerol induce protein phosphorylation at tyrosine, *Nature* **306**:487–490.
183. Cooper, J. A., Sefton, B. M., and Hunter, T., 1984, Diverse mitogenic agents induce phosphorylation of two related 42,000-dalton proteins on tyrosine in quiescent chick cells, *Mol. Cell. Biol.* **4**:30–37.
184. Drust, D. S., and Martin, T. F., 1984, Thyrotropin-releasing hormone rapidly activates protein phosphorylation in GH3 pituitary cells by a lipid-linked, protein kinase C-mediated pathway, *J. Biol. Chem.* **259**:14520–14530.
185. Blumberg, P. M., Delclos, K. B., Dunn, J. A., Jaken, S., Leach, K. L., and Yeh, E., 1983, Phorbol ester receptors and the *in vitro* effects of tumor promotors, *Ann. N.Y. Acad. Sci.* **407**:303–315.
186. Dunphy, W. G., Delclos, K. B., and Blumberg, P. M., 1980, Characterization of specific binding of [^{3}H]phorbol 12,13-dibutyrate and [^{3}H]phorbol 12-myristate 13-acetate to mouse brain, *Cancer Res.* **40**:3635–3641.
187. Lew, K. K., Chritton, S., and Blumberg, P. M., 1982, Biological responsiveness to the phorbol esters and specific binding of a manipulable genetic system, *Teratogen. Carcinogen. Mutagen.* **2**:19–30.
188. Mufson, R. A., Fischer, S. M., Verma, A. K., Gleason, G. L., Slaga, T. J., and Boutwell, R. K., 1979, Effects of 12-O-tetradecanoylphorbol-13-acetate and mezerein on epidermal ornithine decarboxylase activity, isoproterenol-stimulated levels of cyclic adenonisine 3′,5′-monophosphate, and induction of mouse skin, *Cancer Res.* **39**:4791–4795.
189. Dunn, J. A., and Blumberg, P. M., 1983, Specific binding of [20-^{3}H]12-deoxyphorbol 13-isobutyrate to phorbol ester receptor subclasses in mouse skin particulate preparations, *Cancer Res.* **43**:4632–4637.
190. Dunn, J. A., Jeng, A. Y., Yuspa, S. H., and Blumbeg, P. M., 1985, Heterogeneity of [^{3}H]phorbol 12,13-dibutyrate binding in primary mouse keratinocytes at different stages of maturation, *Cancer Res.* **45**:5540–5546.
191. Konig, B., DiNitto, P. A., and Blumberg, P. M., 1985, Phospholipid and Ca^{2+} dependency of phorbol ester receptors, *J. Cell. Biochem.* **27**:255–256.
192. Hannun, Y. A., Loomis, C. R., Merrill, A. H., and Bell, R. M., 1986, Sphingosine inhibition of protein kinase C activity and of phorbol dibutyrate binding *in vitro* and in human platelets, *J. Biol. Chem.* **261**:12604–12609.
193. Blumberg, P. M., Sharkey, N. A., Konig, B., Jaken, S., Leach, K. L., and Jeng, A. Y., 1983, Phorbol ester receptor—Insight into the initial events in the mechanism of action of the phorbol esters, *Int. Symp. Princess Takamatsu Cancer Res.* **14**:75–87.
194. Delclos, K. B., Yeh, E., and Blumberg, P. M., 1983, Specific labelling of mouse brain membrane phospholipids with [20-^{3}H]phorbol 12-p-azidobenzoate 13-benzoate, a photolabile phorbol ester, *Proc. Natl. Acad. Sci. USA* **80**:3054–3058.
195. Ballester, R., and Rosen, O. M., 1985, Fate of immunoprecipitable protein kinase C in GH$_3$ cells treated with phorbol 12-myristate 13-acetate, *J. Biol. Chem.* **260**:15194–15199.
196. Shoji, M., Girard, P. R., Mazzei, G. J., Vogler, W. R., and Kuo, J. F., 1986, Immunocytochemical evidence for phorbol ester-induced protein kinase C translocation in HL-60 cells, *Biochem. Biophys. Res. Commun.* **135**:1144–1149.
197. Wood, J. G., Girard, P. R., Mazzei, G. J., and Kuo, J. F., 1986, Immunocytochemical localization of protein kinase C in identified neuronal compartments of rat brain, *J. Neurosci.* **6**:2571–2577.
198. Yamamoto, M., Takai, Y., Inoue, M., Kishimoto, A., and Nishizuka, Y., 1978, Characterization of cyclic nucleotide-independent protein kinase produced enzymatically from its proenzyme by calcium-dependent neutral protease from rat liver, *J. Biochem.* **83**:207–213.
199. Kishimoto, A., Kajedawa, N., Shiota, M., and Nishizuka, Y., 1983, Proteolytic activation of calcium-activated, phospholipid-dependent protein kinase by calcium-dependent neutral protease, *J. Biol. Chem.* **258**:1156–1164.
200. Melloni, E., Pontremoli, S., Michetti, M., Sacco, O., Spartore, B., and Horecher, B. L., 1986, The involvement of calpain in the activation of protein kinase C in neutrophils stimulated by phorbol myristic acid, *J. Biol. Chem.* **261**:4101–4105.
201. Tapley, P. M., and Murray, A. W., 1985, Evidence that treatment of platelets with phorbol ester causes proteolytic activation of Ca^{2+}-activated, phospholipid-dependent protein kinase, *Eur. J. Biochem.* **151**:419–423.

202. Coussens, L., Parker, P. J., Rhee, L., Yang-Feng, T. L., Chen, E., Waterfield, M. D., Francke, U., and Ullrich, A., 1986, Multiple, distinct forms of bovine and human protein kinase C suggest diversity in cellular signaling pathways, *Science* **233**:859–866.
203. Knopf, J. L., Lee, M. H., Sultzman, L. A., Kriz, R. W., Loomis, C. R., Hewick, R. M., and Bell, R. M., 1986, Cloning and expression of multiple protein kinase C cDNA's, *Cell* **46**:491–502.
204. Pettit, G. R., Day, J. F., Hartwell, J. L., and Wood, H. B., 1970, Antineoplastic components of marine animals, *Nature* **227**:962–963.
205. Berkow, R. L., and Kraft, A. S., 1985, Bryostatin, a nonphorbol macrocyclic lactone, activates intact human polymorphonuclear leukocytes and binds to the phorbol ester receptor, *Biochem. Biophys. Res. Commun.* **131**:1109–1116.
206. Smith, J. B., Smith, L., and Pettit, B. R., 1985, Bryostatins: Potent new mitogens that mimic phorbol ester tumor promotors, *Biochem. Biophys. Res. Commun.* **137**:939–945.
207. Kraft, A. S., Smith, J. B., and Berkow, R. L., 1986, Bryostatin, an activator of the calcium-, phospholipid-dependent protein kinase, blocks phorbol ester-induced differentiation of human promyelocytic leukemia cells HL-60, *Proc. Natl. Acad. Sci. USA* **83**:1334–1338.
208. Wender, P. A., Koehler, K. F., Sharkey, N. A., Dell'Aquila, M. L., and Blumberg, P. M., 1986, Analysis of the phorbol ester pharmacophore on protein kinase C as a guide to the rational design of new classes of analogs, *Proc. Natl. Acad. Sci. USA* **83**:4214–4218.
209. Hecker, E., 1978, Structure activity relationships in diterpene esters irritant and cocarcinogenic to mouse skin, in: *Carcinogenesis—A Comprehensive Survey* (T. J. Slaga, A. Sivak, and R. K. Boutwell, eds.), Vol. 2, Raven Press, New York, pp. 11–48.
210. Hecker, E., 1985, Cell membrane associated protein kinase C as receptor of diterpene ester cocarcinogens of the tumor promotor type and the phenotypic expression of tumors, *Arzneim. Forsch./Drug Res.* **35**:1890–1903.

12
Involvement of Human Retrovirus in Specific T-Cell Leukemia

Mitsuaki Yoshida, Motoharu Seiki, Junichi Fujisawa, and Junichiro Inoue

1. Introduction

Animal retroviruses have provided very useful systems for studies on the molecular mechanisms of carcinogenesis.[1,2] Information on oncogenes and their cellular counterparts, and their significance in tumor development has been obtained from studies on retroviruses. Moreover, interest in oncogenes is now apparent in other fields of science, including developmental biology.

Human T-cell leukemia virus type 1 (HTLV-1), which is the first human retrovirus to be well characterized,[3,4] was shown to be associated with a specific human malignancy,[4–8] adult T-cell leukemia (ATL).[9] Therefore, human retrovirus, HTLV-1, associated with specific leukemia provides a useful system to define features of molecular mechanisms of human tumorigenesis. Since HTLV-1 was shown to be the etiological agent of the specific T-cell malignancy ATL,[7] these studies may lead to the development of methods for diagnosis, prevention, and treatment of ATL. As discussed in this chapter, HTLV-1 belongs to a group of retroviruses different from other chronic leukosis viruses and has a unique mechanism of leukemogenesis. Our current major interest in HTLV-1 is in how virus infection induces ATL. Since this virus is associated with a human disease, it is being studied in many laboratories, and there are many reports on various aspects of the virus and of ATL.

This chapter summarizes our studies on molecular aspects of gene expression of HTLV-1 and its association with ATL. Then, based on these results, the possible mechanism of development of ATL is discussed.

Mitsuaki Yoshida, Motoharu Seiki, Junichi Fujisawa, and Junichiro Inoue • Department of Viral Oncology, Japanese Foundation for Cancer Research, Tokyo 170, Japan.

Studies on HTLV-1 over the last 5 years have established the following facts:

1. The HTLV-1 genome consists of LTR-*gag-pol-env-pX*-LTR.[10] However, the *pX* sequence is not a typical oncogene derived from a cellular sequence.[10]
2. HTLV-1 is an exogenous virus for humans[4,11] and a member of the so-called "HTLV family," which includes HTLV type 2,[12] STLV type 1 (simian T-cell leukemia virus type 1),[13] and BLV (bovine leukemia virus),[14] all of which have an extra sequence "*pX*" between the *env* gene and 3' LTR. The pathogenicities of HTLV-2 and STLV-1 are not yet clear.
3. All ATL patients have antibodies against HTLV-1 proteins and a certain subpopulation of healthy adults living in the endemic area also possess antibodies. The clustering of the antibody-positive population in southwestern Japan,[6,7] the West Indies,[15] and central Africa[16] is coincident with ATL, demonstrating the association of the virus with ATL.
4. Antibody-positive individuals are HTLV-1 carriers. Almost all carriers are over 20 years of age and the incidence of carriers increases with age.[18] Only a small proportion of the carriers develop leukemia.
5. Virus infection shows familial aggregation.[17] The virus is transmitted from mother to child,[18] possibly through milk,[19] and husband to wife,[18] but not vice versa, and by blood transfusion.[20] All these transmissions seem to require transfer of living infected T cells from HTLV-1-positive subjects to recipients.
6. Virus infection *in vitro* by cocultivation with virus-producing cells frequently immortalizes T cells.[21-23] The immortalized cells always have helper-type surface markers, and overexpress receptors of interleukin-2.

2. Genomic Structure of HTLV-1

A provirus clone of HTLV-1 integrated into primary leukemic cells of an ATL patient was isolated[24] and its total nucleotide sequence determined.[10] A characteristic finding was the presence of an extra sequence, termed *pX*, in addition to the three genes required for virus replication (Fig. 1). The viral proteins encoded by these genes are summarized in Fig. 1.[25-27] The general strategy for identification of gene products was use of antisera against the synthetic peptides predicted from the DNA sequence of the genomic clone. The extra sequence *pX* is about 1.6 kbp and contains four possible open reading frames I–IV. However, the *pX* sequence did not appear to be a typical oncogene derived from cellular sequences, because it showed no significant homology with uninfected human cell DNA.[10] All known retroviral oncogenes are derived from cellular DNA sequences, and thus show high homologies to normal cell DNA even of different species. From this sequence analysis, HTLV-1 was concluded to be a replication-competent virus that carries no oncogene. These conclusions were consistent with the facts that there is no apparent association of a helperlike viral genome and that there is a long latency for development of leukemia after virus infection.

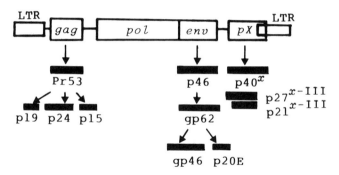

FIGURE 1. Schematic illustration of the gene arrangement of the HTLV-1 genome determined from the nucleotide sequence and proteins identified with specific antipeptide sera.

3. Etiological Agent of Leukemogenesis

The association of HTLV-1 with ATL was clearly demonstrated by extensive surveys of antibodies against the viral proteins in the sera of patients[5,6] and by detection of the provirus genome in leukemic cells from patients.[4,7,8] Antibodies to the viral proteins are present in subjects infected with HTLV-1, but their presence does not provide any insight into the mode of viral involvement in leukemogenesis. For determination of whether leukemic cells are directly infected with HTLV-1, the proviral genomes integrated into primary tumor cells were analyzed. Leukemic cells from all cases of ATL tested were found to contain the integrated proviral genome. Furthermore, in all cases the leukemic cells were found to be monoclonal with respect to the integration site of the proviral genome[7]: no exceptions were found among Japanese ATL patients in the endemic area. These results strongly suggest that HTLV-1 interacts directly with the target cells, which then become malignantly transformed cells.[28,29] If virus infection were associated with ATL indirectly—for example, by inducing a factor that promotes abnormal growth of particular cells or by reducing the immunocompetency of the host, or increasing the sensitivity of leukemic cells to virus infection—some cases should show leukemic cells that have multiple provirus integration sites or no such sites. In fact, such cases have been observed in some patients other than those with ATL: Some patients had antibodies against HTLV-1 proteins, but no detectable proviral genome in their tumor cells.[7] Such cases can be explained either by indirect involvement of virus infection with the tumors or by the occurrence of infection and tumor development as two independent phenomena. The existence of such cases was expected, because 10–20% of the healthy adults living in the endemic area in Japan are infected with HTLV-1. However, no such cases were found among ATL patients. Because no exceptions were found, the results on proviral integration strongly suggested that HTLV-1 interacts directly with target cells, which eventually become tumor cells, and also that infection of the target cells with HTLV-1 is a prerequisite for development of ATL.

4. *Trans*-Acting Viral Function for Leukemogenesis

As discussed in Section 3, HTLV-1 infection is involved in leukemogenesis at the level of infected cells. What is the mechanism of leukemogenesis induced by HTLV-1? Two mechanisms have been demonstrated for induction of animal tumors by retroviruses: One is that the viruses have their own viral oncogenes, and the other that the proviral genome is integrated into a specific locus on the chromosomal DNA and then activates an adjacent cellular oncogene (insertional mutagenesis). The former mechanism was demonstrated for many acute leukemia or sarcoma viruses,[8,9] and the latter for chicken lymphoma[30] and erythroblastosis[31] induced by avian leukosis viruses and mammary tumors[32] induced by mouse mammary tumor viruses. Since HTLV-1 has no typical oncogene in its genome,[10] the second model, insertional mutagenesis, was suspected to be the mechanism of ATL development. For examination of the integration of the specific provirus, a patient whose leukemic cells had a single copy of the HTLV-1 provirus was selected and the cellular DNA sequence flanking the integrated provirus was isolated. Using this cellular flanking region as a probe, the DNAs of other ATL patients were surveyed to determine the rearrangements of cellular DNA sequences induced by provirus integration. However, no rearrangement was detected in the DNAs of 34 ATL patients.[33] By use of combinations of other probes and different restriction enzymes, a region of about 25 kb was examined, which corresponded to the sites around the provirus integration site in the DNA which was used for preparation of the probes. The results were confirmed using another set of probes isolated from a different patient.[33] The results showed that there was no common region for provirus integration in leukemic cells. This finding suggested that a *cis*-acting function of LTR in the insertional mutagenesis model was not applicable to the HTLV-1 involved in ATL. This conclusion was further supported by the finding that the proviruses were integrated into different chromosomes in the ATL patients.[33]

Since the viral function involved in leukemogenesis must therefore be active irrespective of the site of provirus integration, a *trans*-acting viral function, possibly mediated by viral proteins, was suspected. Infection with HTLV-1 frequently immortalizes helper T cells *in vitro*.[21–23] No similar *in vitro* immortalization or transformation has been observed with chronic leukemia virus, so that a unique

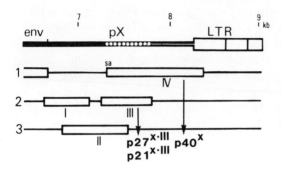

FIGURE 2. Arrangements of the extended open reading frames I–IV in the *pX* sequences and their gene products. sa, splicing acceptor site for expression of frame IV to code for p40^x. Lines 1–3 represent three possible translational registers. The region in the genome marked with asterisks is highly conserved in HTLV-1 and -2.

factor, possibly a *trans*-acting factor, was also predicted to be associated with *in vitro* immortalization. However, neither observation supported a *cis*-acting function of the LTR for activating adjacent cellular oncogenes.

5. Three Proteins Encoded by the *pX* Sequence

Studies on provirus integration into primary leukemic cells strongly suggested the involvement of *trans*-acting viral function in leukemogenesis. The unique sequence *pX* was assumed to be associated with the unique properties of the HTLV-1, because (1) all proviruses in leukemic cells retained the *pX* sequence, even when they were defective, and (2) the *pX* sequences in HTLV-1 and -2 were the most highly conserved, suggesting that they were important for viral replication or stimulation of growth of infected cells.

The *pX* sequence in the proviral genome contains four possible open reading frames I–IV.[10] The strategy used for identification of the gene product was synthesis of a small peptide predicted from the nucleotide sequence of each of these frames and subsequent preparation of antiserum against the peptide. Antiserum against the C-terminal peptide from frame IV precipitated a 40kD protein in HTLV-1-producing cell lines and this protein specifically competed with the corresponding peptide.[26] Thus, it was concluded that the open reading frame IV in the *pX* sequence encodes a 40kD protein, $p40^x$ (Fig. 2). This protein was also found in three other laboratories,[34–36] but as a 41kD[36] or 42kD protein.[34] As described later, the molecular weight of $p40^x$ calculated from the cDNA sequence is 39,842 daltons.

For a while after its identification, $p40^x$ was thought to be the only product of the *pX* sequence, and thus the properties of *pX* were thought to be mediated by $p40^x$. However, by comparison of the sequences of the *pX* regions of HTLV-1 and -2, a second protein encoded by an overlapping gene in the *pX* sequence was predicted. The *pX* sequences of HTLV-1 and -2 were conserved more than other parts of the viral genomes. Moreover, the 5' half of frame IV in the *pX* sequence was highly conserved, and in this region even the third bases in codons are well conserved (Fig. 2). Conservation of this sort in two homologous, but diverged genes is unusual. Frame III corresponds to this highly conserved region. Careful reexamination of the antipeptide sera used previously led us to find two new proteins encoded by the *pX* sequence. Testing with the antiserum against the peptide from frame III by an immunoblot assay revealed two proteins with molecular weights of 27kD and 21kD in all HTLV-1-producing cell lines.[27] Formation of bands of these two proteins was completely blocked by the specific peptide and these proteins were not detected in uninfected T-cell lines. Thus, these two proteins were considered to be encoded by frame III in the *pX* sequence, and were thus termed $p27^{x-III}$ and $p21^{x-III}$ (Fig. 2). These two proteins share the same antigenicity as recognized by antiserum against the C-terminal peptide, but there is no evidence for a precursor–product relationship between the two. They are both phosphorylated proteins,[27] but not glycoproteins.[27] The unusual character of these proteins $p27^{x-III}$

and p21^{x-III} is their high content of proline (21 and 24%, respectively) and serine (15 and 19%).[27]

An important problem is whether these *pX* proteins are expressed in primary leukemic cells of ATL patients. Neither *gag* proteins nor *env* gene products are expressed in primary leukemic cells, but in most cases their expression can be induced by *in vitro* culture for 2 or 3 days.[37] Like these viral proteins, the three *pX* gene products p40^x, p27^{x-III}, and p21^{x-III} were not detected in freshly isolated leukemic cells from ATL patients, but after culture for 3 days with fetal calf serum, the cells expressed anomalous amounts of p40^x and p27^{x-III}.[27] These results strongly suggest that the *pX* proteins, p40^x and p27^{x-III}, are regulated similarly to other viral proteins, thus suggesting that they are also expressed in certain cells of HTLV-1 carriers and ATL patients, although they are not expressed significantly *in vivo* by leukemic cells or by most infected cells. In fact, about one-quarter to one-third of the HTLV-1-positive population was found to have antibodies against p40^x. But, surprisingly, no significant level of antibodies against p27^{x-III} and/or p21^{x-III} was detected in the 50 independent sera from healthy carriers and ATL patients so far tested.[27] Therefore, p27^{x-III} and/or p21^{x-III} may be expressed only at a low level or have very limited antigenicity.

6. Mechanism of *pX* Gene Expression

If open reading frames III and IV in the *pX* sequence are translated from the first ATG in each frame, frames III and IV would code for 12kD and 27kD proteins, respectively.[10] However, 27kD, 21kD, and 40kD proteins were identified[26,27] (Section 5). For understanding the mechanisms of their expression, RNA splicings were studied by structural analysis of a cDNA clone.

A cDNA clone was isolated from 2.1-kb mRNA coding for p40^x.[38] The coding capacity of this cDNA clone was demonstrated by inserting it under the control of the SV40 promoter and transfecting it into COS cells. Expression of p40^x was demonstrated in transfected COS cells. Sequence analysis of the cDNA clone showed that it consisted of three exons, that is, the mRNA was formed by double splicing.[38] The first exon was derived from the 5' half of the R sequence in the LTR, the second from the 3' region of the *pol*, and the third from the *pX* sequence (Fig. 3). As a result of the second splicing between *pol* and *pX*, the ATG used for *env* gene translation is joined to a *pX* reading frame containing the original frame IV, and used for initiation of p40^x translation. Thus, the original frame IV is extended to the 5' side and only the first methionine is brought onto p40^x from the *env* domain. The molecular weight of this product of frame IV was calculated to be 39,842.[38] Similar results were reported by another group.[39]

The key sequences for the splicings for expression of p40^x were also found at similar positions in HTLV-2, BLV, and STLV-1 (Fig. 3),[38] suggesting that the unusual splicing mechanism producing the *pX* mRNA is common to all four members of the HTLV family, and thus strongly suggesting the biological significance of *pX* gene expression.

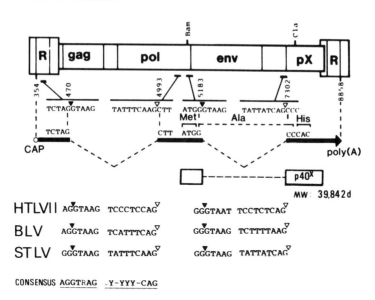

FIGURE 3. Double splicing for expression of p40^x and signals for similar splicings conserved in other members of the HTLV family.

With regard to expression of frame III, the mechanisms of translation of p27^{x-III} and p21^{x-III} are not clear. No uniquely spliced mRNA other than 2.1-kb mRNA was detected by blotting analysis of RNA from cells carrying a single complete copy of the HTLV-1 proviral genome. Moreover, attempts to isolate a cDNA clone of putative mRNA that is spliced differently from the 2.1-kb mRNA were unsuccessful. These negative observations and, more specifically, the nucleotide sequence of the 2.1-kb mRNA containing open reading frame III, suggest that the mRNA species can code for proteins p40^x, p27^{x-III}, and p21^{x-III}.

7. *Trans*-Activation of the LTR Mediated by *pX*

Although some retroviruses such as avian acute leukemia virus MC29 can induce many types of tumors, many viruses have target specificity for pathogenicity. The LTR sequences of some viruses are known to exert this tissue specificity. For example, the tissue specificities of two retroviruses can be modified by interchange of their LTRs.[40] Since ATL associated with HTLV-1 is a malignancy of almost exclusively helper-type T cells, the LTR of HTLV-1 might have T cell specificity. This possibility was tested using a plasmid pLTR-CAT containing a gene for CAT (chloramphenicol acetyltransferase) under the control of the LTR. After transfection of the pLTR-CAT into various cell lines, the CATase activity, which reflects the transcription rate from the LTR, was measured by acetylation of chloramphenicol. Taking the SV40 promotor activity of pSV-CAT as the standard, the

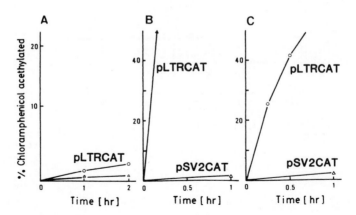

FIGURE 4. CATase expression directed by pLTR-CAT and pSV2-CAT plasmids in various human T-cell [MT-1(A) and MT-2(B)] and rat T-cell [TARL-2(C)] lines infected with HTLV-1. MT-2 and TARL-2 express the viral antigens, but MT-1 does not.

LTR activity was found to be almost equally active in epithelial, fibroblastic, B- and T-cell lines, clearly demonstrating no significant T-cell specificity.[41,42]

More surprisingly, the CATase activity directed by the LTR was stimulated more than 100-fold in human T-cell lines producing HTLV-1, and in some cell lines even 250-fold[41,42] (Fig. 4). On the other hand, similar stimulation was not observed in a human T-cell line that did not express the viral antigens although it was infected with HTLV-1.[42] These results strongly suggest the involvement of the viral antigen in activating the LTR function. In this assay, the expression of the CAT gene was directed by the LTR in unintegrated plasmid. Therefore, the high expression in HTLV-1-producing cells was concluded to be mediated by a *trans*-acting factor. In further tests in other cell lines, the augmented activity was also observed in cell lines that were thought to express only p40^x.[41,42] These observations strongly suggested that viral protein p40^x is involved in *trans*-activation of the LTR function. However, other explanations were still possible, such that HTLV-1 selectively induced transformation of some particular cells that originally had the *trans*-activating function or induced a cellular *trans*-activating factor irrespective of the function of the *pX* sequence.

8. The p40^x as *trans*-Activator of the LTR

To demonstrate more directly that p40^x is responsible for *trans*-activation of the LTR, the *pX* sequence was cloned into expression plasmid pMTPX (Fig. 5) and the p40^x was transiently expressed in cells by DNA transfection. The activation of LTR function, measured with pLTR-CAT, was dependent on cotransfection with the *pX* expression plasmid.[43-45] At the time of these experiments, p40^x was the only *pX* product known, and it was thought that the results directly demonstrated the association of p40^x with *trans*-activation of LTR function. However, later the

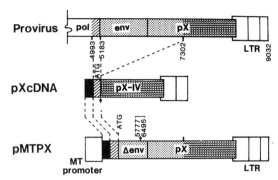

FIGURE 5. Construction of plasmid pMTPX expressing *pX*, which can express all possible frames in the *pX* sequence by alternative splicings. Numbers represent positions from the 5' end of the integrated provirus genome.

second and third proteins p27^{x-III} and p21^{x-III} were also found to be encoded by the *pX* sequence, and these open reading frames were found to mostly overlap. Therefore, it became evident that all previous experiments in which the *pX* sequence was expressed were not sufficient to distinguish the effects of these three products.

Accordingly, for determination of which product is primarily responsible for *trans*-activation of the LTR, each frame was introduced by a site-directed mutation that generated a termination codon, but did not induce gross alteration of the amino acids in the other frames.[46,47] Introduction of a termination codon into frame III, which encodes p27^{x-III} and p21^{x-III}, did not affect the *trans*-activating function. Introduction of a termination codon into frame II, which also overlaps frames III and IV, also did not reduce the activity. On the other hand, generation of a termination codon in frame IV, which codes for p40^x, completely abolished the activity. These careful studies clearly established that p40^x is a *trans*-activator of the LTR in unintegrated plasmid DNA, and that neither p27^{x-III} nor p21^{x-III} was apparently involved in this activation in the transient assay.

Consistent with this conclusion that p40^x is a *trans*-activator of transcription, p40^x was found to be a nuclear protein.[48-50] A fraction of p40^x was easily released from the nuclei during nuclear isolation, probably because of its presence in the nucleoplasm, but another fraction was tightly bound to nuclear components. This subcellular localization is very similar to those of T antigens of papova viruses,[51] Ela protein of adenoviruses,[52] and oncogene products[53] of *myc* and *myb*, all of which are associated with control of cell growth.

Like p40^x, p27^{x-III} was also found to be a nuclear protein but p21^{x-III} was not. Some of the p27^{x-III} was bound as tightly as p40^x to some nuclear structures. These similarities in its localization to that of p40^x suggest that p27^{x-III} is also involved in regulation of viral gene expression, although it was not thought to be involved in stimulation of the unintegrated LTR. It may be required for activation or regulation of expression of the viral genome integrated in the cellular DNA sequence. In fact, the *pX* sequence was shown to activate the integrated proviral genome,[45] although, in the system used, it was not clear whether p40^x alone was sufficient for the activation.

9. Elements Responsible for the *trans*-Activation

Trans-activation of transcription from the unintegrated LTR was shown to be mediated by p40^x. Although it is not known whether p40^x interacts with the LTR directly or indirectly, it was of interest to identify the sequence in the LTR responsible for the *trans*-acting transcriptional activation.

Serial deletions were introduced into a region upstream of the initiation site (Fig. 6), where many transcriptional units have regulatory elements. On increasing the size of the deletion from the 5' side, the LTR activity supporting CAT gene expression was gradually reduced and the largest deletion from −322 to −62 bp resulted in almost complete inactivation of the LTR responding to *trans*-activation.[54] Similar progressive inactivation was observed on increasing the size of deletions on the 3' side. These observations led us to conclude that the region from −322 to −62 bp contains an element responsible for the *trans*-activation mediated by p40^x and that the element is not a single unit, but two or more units. In this

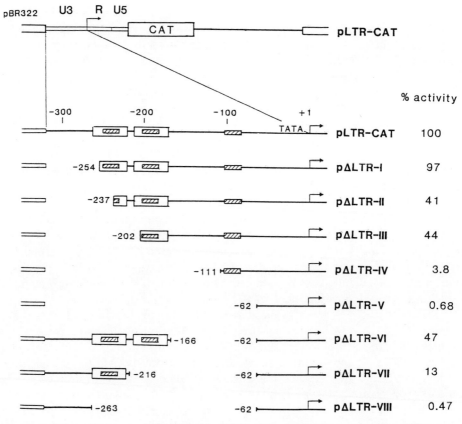

FIGURE 6. Deletion mutations in pLTR-CAT and activities of the mutants *trans*-activated in p40^x-expressing cells. Numbers are positions relative to the mRNA cap site. 42-bp (open boxes) and 21-bp (hatched boxes) repeats are shown.

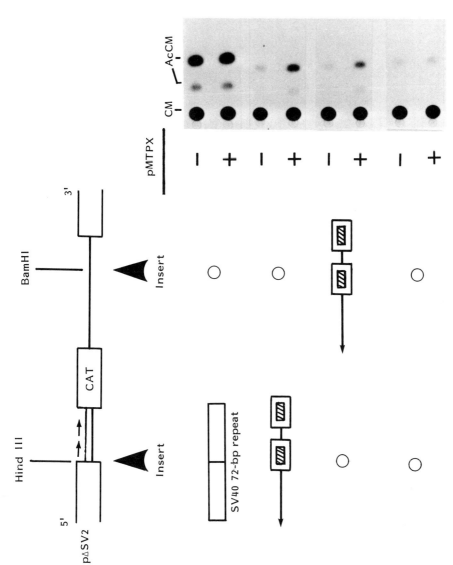

FIGURE 7. CATase activities with SV40 promoter deleted of the original enhancer (pΔSV2) but containing the LTR fragment. A 151-bp fragment containing two direct repeats of 21 or 42 bp was inserted and cotransfected with or without a pX expression plasmid (pMTPX). The arrows on the LTR fragments represent the 5'-to-3' direction and thick arrows indicate the sites in the pΔSV2 for the insertion. CM and AcCM are chloramphenicol and its acetylated form on thin-layer chromatograms, which were developed from left to right.

region, there are three repeats of 21 bp and two repeats of 42 bp, as shown in Fig. 6. These repeating sequences are conserved in the LTRs of HTLV-1 and -2, which are distantly related, suggesting the functional importance of these repeats. For examination of the effect of these repeating units on *trans*-activation, one or two units of 21 bp were cut out from the LTR, and inserted into pSV-CAT, from which the original SV40 enhancer sequences were removed. The SV40 promotor without the original enhancer was not activated by expression of p40^x. However, this SV40 promoter could be activated by p40^x upon insertion of the direct repeats of 21 bp isolated from the HTLV-1 LTR[55] (Fig. 7). Furthermore, the activations were also observed upon insertion of these direct repeats in the antiparallel direction and even upon their insertion on the 3' side of the CAT gene. These results clearly indicate that direct repeats of 21 bp are an enhancer of the LTR and that the enhancer sequence is responsible for the *trans*-activation mediated by p40^x. Similarly, enhancer sequences were shown to be responsible for *trans*-activation in HTLV-2[41] and BLV.[55] So far, *trans*-activation of the LTR has been assayed only *in vivo* by transfection of cells. Therefore, it is not known whether p40^x interacts directly with the enhancer sequence or indirectly through a cellular factor.

Deletions in the U5 and R regions in the LTR also reduced CAT gene expression, but these sequences were not responsible for the activation by p40^x. Nevertheless, it is evident that these sequences in the LTR are required for maximum expression of the LTR function, although their mode of action is not clear.

10. Possible Involvement of *pX* Function in Leukemogenesis

As discussed in the previous sections, the random provirus integration in *in vitro* immortalization of infected T cells strongly suggest that *trans*-acting viral function is associated with leukemogenes induced by HTLV-1 infection. On the other hand, a unique sequence (*pX*) in the HTLV-1 genome was found to encode three proteins and one of these, p40^x, was clearly shown to activate expression of the viral genome at the transcriptional level. These two independent findings suggest the simple idea that transcriptional *trans*-activation of the LTR mediated by p40^x may also activate cellular genes that include those associated with growth control of T cells (Fig. 8). Since enhancer sequences of 21 bp are responsible for the p40^x-mediated *trans*-activation, a cellular sequence containing a similar enhancer would be transcriptionally activated. The other two *pX* proteins, p27^{x-III} and p21^{x-III}, were not shown in transient assays to be directly involved in the LTR activation. But the fact that like p40^x, p27^{x-III} is localized in nuclei strongly suggests its involvement in activation of expression of the viral gene integrated into cellular DNA, and also of the cellular gene. Assuming that these putative genes are regulated in a T-cell-specific manner, this hypothetical mechanism would explain that pathogenic specificity for helper-type T cells. HTLV-1 can be introduced into many types of cells, such as B cells and fibroblasts, but almost invariably, only helper T cells become malignant *in vivo* or immortalized *in vitro*. Therefore, an intracellular event is expected to determine the cell type specificity of HTLV-1 pathogenicity rather than specificity of the receptor for the virus infection.

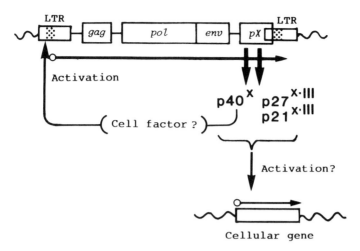

FIGURE 8. Summary of the function of the *pX* proteins in viral replication and its possible involvement in activation of cellular genes. Dotted areas in the LTR represent an enhancer sequence containing 21-bp direct repeats.

If *pX* proteins are directly involved in cellular gene activation, their continuous expression in leukemic cells may be expected. However, no viral proteins, including the *pX* products, have been detected in primary tumors or infected cells, although they are produced after *in vitro* culture of the cells for a few days. Therefore, these *pX* proteins may be involved in an early stage of ATL development, but not in the maintenance of the leukemic state. Alternatively, trace amounts of *pX* proteins, that are too low to detect, may be sufficient to maintain cells in the leukemic state. The latter possibility seems unlikely because their expression could not be detected by a sensitive blotting assay for viral mRNA in primary tumor cells from peripheral blood and lymph nodes (unpublished data). Thus, the former hypothesis is preferable. Here, the expression of *pX* proteins induces abnormal growth of infected T cells, but only one or a few cells acquire other abnormal growth properties that are independent of *pX* expression before being rejected by the host immunological defense system. This second alteration of primarily stimulated cells may not occur frequently, and so could explain the long latency and low frequency of development of ATL.

References

1. Vogt, P. K., 1977, Genetics of RNA tumor viruses, *Compr. Virol.* **10**:341–455.
2. Bishop, J. M., 1982, Oncogenes, *Sci. Am.* **246**:68–78.
3. Poiesz, B. J., Ruscetti, F. W., Gazdar, A. F., Bunn, P. A., Minna, J. D., and Gallo, R. C., 1980, Detection and isolation of type C retrovirus particles from fresh and cultured lymphocytes of a patient with cutaneous T cell lymphoma, *Proc. Natl. Acad. Sci. USA* **77**:7415–7419.
4. Yoshida, M., Myoshi, I., and Hinuma, Y., 1982, Isolation and characterization of retrovirus from cell lines of human adult T cell leukemia and its implication in the disease, *Proc. Natl. Acad. Sci. USA* **79**:2031–2035.

5. Kalyanaraman, V. S., Sarngadharan, M. G., Nakao, Y., Ito, Y., Aoki, T., and Gallo, R. C., 1982, Natural antibodies to the structural core protein (p24) of the human T cell leukemia (lymphoma) retrovirus found in sera of leukemia patients in Japan, *Proc. Natl. Acad. Sci. USA* **79:**1653–1657.
6. Hinuma, Y., Nagata, K., Misoka, M., Nakai, M., Matsumoto, T., Kinoshita, K., Shirakawa, S., and Miyoshi, I., 1981, Adult T cell leukemia: Antigen in an ATL cell line and detection of antibodies to the antigen in human sera, *Proc. Natl. Acad. Sci. USA* **78:**6476–6480.
7. Yoshida, M., Seiki, M., Yamaguchi, K., and Takatsuki, K., 1984, Monoclonal integration of HTLV in all primary tumors of adult T-cell leukemia suggests causative role of HTLV in the disease, *Proc. Natl. Acad. Sci. USA* **81:**2534–2537.
8. Wong-Staal, F., Hahn, H., Manzari, V., Colonbini, S., Franchini, G., Gelman, E. P., and Gallo, R. C., 1983, A survey of human leukemias for sequences of a human retrovirus, *Nature* **302:**626–628.
9. Uchiyama, T., Yodoi, J., Sagawa, K., Takatsuki, K., and Uchino, H., 1977, Adult T cell leukemia: Clinical and hematological features of 16 cases, *Blood* **50:**481–491.
10. Seiki, M., Hattori, S., Hirayama, Y., and Yoshida, M., 1983, Human adult T cell leukemia virus: Complete nucleotide sequence of the provirus genome integrated in leukemia cell DNA, *Proc. Natl. Acad. Sci. USA* **80:**3618–3622.
11. Reitz, M. S., Poiesz, B. J., Ruscetti, F. W., and Gallo, R. C., 1981, Characterization and distribution of nucleic acid sequences of a novel type C retrovirus isolated from neoplastic human T lymphocytes, *Proc. Natl. Acad. Sci. USA* **78:**1887–1891.
12. Shimotohno, K., Takahashi, Y., Shimizu, N., Gojobori, T., Golde, D. W., Chen, I. S. Y., Miwa, M., and Sugimura, T., 1985, Complete nucleotide sequence of an infectious clone of human T-cell leukemia virus type II: An open reading frame for the protease gene, *Proc. Natl. Acad. Sci. USA* **82:**3101–3105.
13. Watanabe, T., Seiki, M., Tsujimoto, H., Miyoshi, I., Hayami, M., and Yoshida, M., 1985, Sequence homology of the simian retrovirus (STLV) genome with human T-cell leukemia virus type I (HTLV-I), *Virology* **144:**59–65.
14. Sagata, N., Yasunaga, T., and Ikawa, Y., 1984, Identification of a potential protease-coding gene in the genomes of bovine leukemia and human T-cell leukemia viruses, *FEBS Lett.* **178:**79–82.
15. Blattner, W. A., Kalyanaraman, V. S., Robert-Guroff, M., Lister, A., Galton, D. A. G., Sarin, P. S., Crawford, M. H., Catovsky, D., Greaves, M., and Gallo, R. C., 1982, The human type C retrovirus, HTLV, in blacks from the Caribbean region, and relationship to adult T cell leukemia/lymphoma, *Int. J. Cancer* **30:**257–264.
16. Hunsmann, G., Schneider, J., Schmitt, J., and Yamamoto, N., 1983, Detection of serum antibodies to adult T-cell leukemia virus in non-human primates and in people from Africa, *Int. J. Cancer* **32:**329–332.
17. Ichimaru, M., Kinoshita, K., Kamihira, S., Ikeda, S., Yamada, Y., and Amagasaki, T., 1979, T cell malignant lymphoma in Nagasaki district and its problems, *Jpn. J. Clin. Oncol.* **9:**337–346.
18. Tajima, K., Tominaga, S., Suchi, T., Kawagoe, T., Komoda, H., Hinuma, Y., Oda, T., and Fujita, K., 1982, Epidemiological analysis of the distribution of antibody to adult T-cell leukemia virus, *Gann* **73:**893–901.
19. Kinoshita, K., Hino, S., Amagasaki, T., Ikeda, S., Yamada, Y., Suzuyama, J., Momita, S., Toriya, K., Kamihira, S., and Ichimaru, M., 1984, Demonstration of adult T-cell leukemia virus antigen in milk from three sero-positive mothers, *Gann* **75:**103–105.
20. Okochi, K., Sato, H., and Hinuma, Y., 1983, A retrospective study on transmission of adult T cell leukemia virus by blood transfusion; sero-conversion in recipients, *Vox Sang.* **46:**245–253.
21. Miyoshi, I., Kubonishi, I., Yoshimoto, S., Akagi, T., Ohtsuki, Y., Shiraishi, Y., Nagata, K., and Hinuma, Y., 1981, Type C virus particles in a cord T cell line derived by cocultivating normal human cord leukocytes and human leukemic T cells, *Nature* **294:**770–771.
22. Yamamoto, N., Okada, M., Koyanagi, Y., Kannagi, M., and Hinuma, Y., 1982, Transformation of human leukocytes by cocultivation with an adult T cell leukemia virus producer cell line, *Science* **217:**737–739.
23. Popovic, M., Lange-Wantzin, G., Sarin, P. S., Mann, D., and Gallo, R. C., 1983, Transformation of human umbilical cord blood T cells by human T cell leukemia/lymphoma virus, *Proc. Natl. Acad. Sci. USA* **80:**5402–5406.

24. Seiki, M., Hattori, S., and Yoshida, M., 1982, Human adult T cell leukemia virus: Molecular cloning of the provirus DNA and the unique terminal structure, *Proc. Natl. Acad. Sci. USA* **79**:6899–6902.
25. Hattori, S., Kiyokawa, T., Imagawa, K., Shimizu, F., Hashimura, E., Seiki, M., and Yoshida, M., 1984, Identification of gag and env gene products of human T-cell leukemia virus (HTLV), *Virology* **136**:338–347.
26. Kiyokawa, T., Seiki, M., Imagawa, K., Shimizu, F., and Yoshida, M., 1984, Identification of a protein (p40^x) encoded by a unique sequence *pX* of human T-cell leukemia virus type I, *Gann* **75**:747–751.
27. Kiyokawa, T., Seiki, M., Iwashita, S., Imagawa, K., Shimizu, F., and Yoshida, M., 1985, p27^{x-III} and p21^{x-III}, proteins encoded by the *pX* sequence of human T-cell leukemia virus type I, *Proc. Natl. Acad. Sci. USA* **82**:8359–8363.
28. Yoshida, M., Seiki, M., Hattori, S., and Watanabe, T., 1984, Genome structure of human T-cell leukemia virus and its involvement in the development of adult T-cell leukemia, in: *Human T-cell Leukemia/Lymphoma Virus* (R. C. Gallo, M. Essex, and L. Gross, eds.), Cold Spring Harbor Laboratory, Cold Spring Harbor, N.Y., pp. 141–148.
29. Yoshida, M., Hattori, S., and Seiki, M., 1985, Molecular biology of human T-cell leukemia virus associated with adult T-cell leukemia, *Curr. Top. Microbiol. Immunol.* **115**:157–175.
30. Hayward, W. S., Neel, B. G., and Astrin, S. M., 1981, Activation of a cellular onc gene by promoter insertion in ALV-induced lymphoid leukosis, *Nature* **290**:475–480.
31. Lewis, W. G., Crittenden, L. B., and Kung, H.-J., 1983, Activation of the cellular oncogene c-erbB by LTR insertion: Molecular basis for induction of erythroblastosis by avian leukosis virus, *Cell* **33**:357–368.
32. Nusse, R., and Varmus, H. E., 1982, Many tumors induced by the mouse mammary tumor virus contain a provirus integrated in the same region of the host genome, *Cell* **31**:99–109.
33. Seiki, M., Eddy, R., Shows, T. B., and Yoshida, M., 1984, Nonspecific integration of the HTLV provirus genome into adult T-cell leukemia cells, *Nature* **309**:640–642.
34. Lee, T. H., Coligan, J. E., Sodroski, J. G., Haseltine, W. A., Salahuddin, S. Z., Wong-Staal, F., Gallo, R. C., and Essex, M., 1984, Antigens encoded by the 3'-terminal region of human T-cell leukemia virus: Evidence for a functional gene, *Science* **226**:57–61.
35. Slamon, D. J., Shimotohno, K., Cline, M. J., Golde, D. W., and Chen, I. S. Y., 1984, Identification of the putative transforming protein of the human T-cell leukemia viruses HTLV-I and -II, *Science* **226**:61–65.
36. Miwa, M., Shimotohno, K., Hoshino, H., Fujino, M., and Sugimura, T., 1984, Detection of *pX* proteins in human T-cell leukemia virus (HTLV)-infected cells by using antibody against peptide deduced from sequences of X-IV DNA of HTLV-I and X-C DNA of HTLV-II proviruses, *Gann* **75**:752–755.
37. Hinuma, Y., Gotoh, Y., Sugamura, K., Nagata, K., Goto, T., Nakai, M., Kamada, N., Matsumoto, T., and Kinoshita, K., 1982, A retrovirus associated with human adult T-cell leukemia: *In vitro* activation, *Gann* **73**:341–344.
38. Seiki, M., Hikikoshi, A., Taniguchi, T., and Yoshida, M., 1985, Expression of the *pX* gene of HTLV-I: General splicing mechanism in the HTLV family, *Science* **228**:1532–1534.
39. Wachsman, W., Golde, D. W., Temple, P. A., Orr, E. C., Clark, S. C., and Chen, I. S. Y., 1985, HTLV x-gene product: Requirement for the *env* methionine initiation codon, *Science* **228**:1534–1536.
40. Chatis, P. A., Holland, C. A., Hartley, J. W., Rowe, W. P., and Hopkins, N., 1983, Role for the 3'end of the genome in determining disease specificity of Friend and Moloney murine leukemia viruses, *Proc. Natl. Acad. Sci. USA* **80**:4408–4411.
41. Sodroski, J. G., Rosen, C. A., and Haseltine, W. A., 1984, Trans-acting transcriptional activation of the long terminal repeat of human T lymphotropic viruses in infected cells, *Science* **225**:381–385.
42. Fujisawa, J., Seiki, M., Kiyokawa, T., and Yoshida, M., 1985, Functional activation of long terminal repeat of human T-cell leukemia virus type I by *trans*-acting factor, *Proc. Natl. Acad. Sci. USA* **82**:2277–2281.
43. Sodroski, J., Rosen, C., Goh, W. C., and Haseltine, W., 1985, A transcriptional activator protein encoded by the x-lor region of the human T-cell leukemia virus, *Science* **228**:1430–1434.

44. Felber, B. K., Paskalis, H., Kleinman-Ewing, C., Wong-Staal, F., and Pavlakis, G. N., 1985, The pX protein of HTLV-I is a transcriptional activator of its long terminal repeats, *Science* **229**:675–679.
45. Chen, I. S. Y., Slamon, D. J., Rosenblatt, J. D., Shah, N. P., Quan, S. G., and Wachsman, W., 1985, The x gene is essential for HTLV replication, *Science* **229**:54–58.
46. Seiki, M., Inoue, J., Takeda, T., Hikikoshi, A., Sato, M., and Yoshida, M., 1985, The p40^x of human T-cell leukemia virus type I is a trans-acting activator of viral gene transcription, *Gann* **76**:1127–1131.
47. Seiki, M., Inoue, J., Takeda, T., and Yoshida, M., 1986, Direct evidence that p40^x of human T-cell leukemia virus type I is a trans-acting transcriptional activator, *EMBO J.* **5**:561–565.
48. Goh, H. G., Sodroski, J., Rosen, C., Essex, M., and Haseltine, W. A., 1985, Subcellular localization of the product of the long open reading frame of human T-cell leukemia virus type I, *Science* **227**:1227–1229.
49. Slamon, D. J., Press, M. F., Souza, L. M., Murdock, D. C., Cline, M. J., Golde, D. W., Gasson, J. C., and Chen, I. S. Y., 1985, Studies of the putative transforming protein of the type I human T-cell leukemia virus, *Science* **228**:1427–1430.
50. Kiyokawa, T., Kawaguchi, T., Seiki, M., and Yoshida, M., 1985, Association with nucleus of pX gene product of human T-cell leukemia virus type I, *Virology* **147**:462–465.
51. Soule, H. R., and Butel, J. S., 1979, Subcellular localization of simian virus large tumor antigen, *J. Virol.* **30**:523–532.
52. Feldman, L. T., and Nevins, J. R., 1983, Localization of the adenovirus E1A protein, a positive-antigen transcriptional factor, in infected cells, *Mol. Cell. Biol.* **3**:829–838.
53. Rohrschneider, L. R., and Gentry, L. E., 1984, Subcellular locations of retroviral transforming proteins define multiple mechanisms of the transformation, in: *Advances in Viral Oncology*, Vol. 4 (G. Klein, ed.), Raven Press, New York, pp. 269–306.
54. Fujisawa, J., Seiki, M., Sato, M., and Yoshida, M., 1986, A transcriptional enhancer sequence of HTLV-I is responsible for *trans*-activation mediated by *p40* of HTLV-I, *EMBO J.* **5**:713–718.
55. Rosen, C., Sodroski, J., Kettman, R., and Haseltine, W. A., 1986, Activation of enhancer sequences in type II human T-cell leukemia virus and bovine leukemia virus long terminal repeats by virus-associated *trans*-acting regulatory factors, *J. Virol.* **57**:738–744.

13
Mechanisms of Virus-Induced Alterations of Expression of Class I Genes and Their Role on Tumorigenesis

DANIEL MERUELO

1. Alterations of Class I Gene Expression in Virus-Induced Tumors

A number of years ago my collaborators and I[1] first documented the striking role that virus-induced alterations in H-2 (class I) antigen expression played in resistance to neoplasia. This fact has also been recently documented by numerous investigators.[2-6] Our early studies showed that resistance to RadLV-induced leukemia is mediated by a gene(s) in the *H-2D* region of the major murine histocompatibility complex (MHC).[7] These experiments indicated a role in disease resistance for changes in H-2 expression which occur immediately after virus inoculation and after tumorigenesis.[1,8] For example, elevated H-2D antigen expression occurs immediately after virus inoculation on thymocytes of resistant but not of susceptible mice.[1] Furthermore, an inverse relationship exists between expression of H-2 and viral antigens.[1] After RadLV infection viral antigen expression is greater in susceptible animals, which show little if any changes in H-2 antigen expression, whereas H-2 antigen expression is maximal for virus-infected resistant mice which are subjected to rapid elimination of viral antigen-expressing cells.[1] Finally, H-2 antigens usually disappear from the surface of RadLV-transformed cells when overt leukemia develops.[1] Thus, resistance to the disease is associated with

---

DANIEL MERUELO • Department of Pathology and Kaplan Cancer Center, New York University Medical Center, New York, New York 10016.

increased H-2 antigen expression, and the onset of leukemia is associated with disappearance of these antigens. (Some of these findings are diagrammed in Fig. 1.)

Further experiments demonstrated that elevated H-2 antigen expression enhances the effectiveness of the host' immune response to virus-infected cells and functions in resistance to leukemogenesis.[8] For example, cell-mediated immunity against RadLV-transformed or -infected cells can be detected with ease when H-2-positive target cells are substituted in the cell-mediated lympholysis (CML) assay

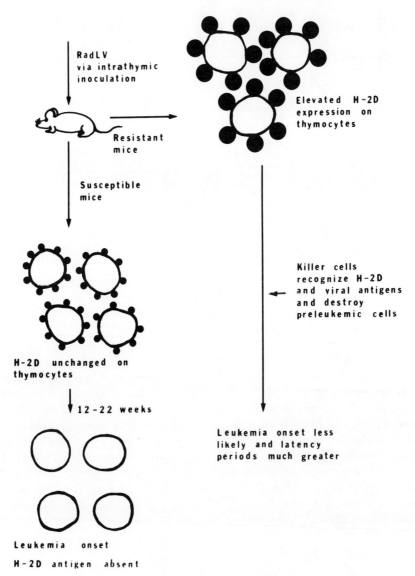

FIGURE 1. Schematic of RadLV effects on class I expression, highlighting the differences between strains of mice bearing the resistant H-2^d or susceptible H-2^s, H-2^q, H-2^k, or H-2^b haplotype.

for the normally H-2-nonexpressing, RadLV-transformed cells.[8] (The still unexplained observation was made in these studies that H-2-nonexpressing, RadLV-transformed cells could be induced to reexpress H-2 antigens by *in vivo* passage.)[8] Also, resistant mice were shown to have many more effector cells when infected with RadLV than did susceptible mice. In addition, injection of normal (uninfected) thymocytes into syngeneic recipients of resistant or susceptible *H-2* type does not stimulate a CML response.[8] However, injection of RadLV-infected thymocytes from resistant mice produces a vigorous CML response, and such thymocytes elicit the strongest response at a time when their cell surface H-2 and viral antigen expression is most elevated.[8] By contrast, injection of infected thymocytes from susceptible mice, which express viral antigens but low levels of H-2 antigens, does not stimulate a CML reaction.[8]

Furthermore, target recognition appears to be specific for certain H-2 antigens because cytolytic T lymphocytes (CTL) from F_1 mice challenged with infected, resistant thymocytes can lyse resistant target cells but not susceptible targets despite adequate amounts of H-2 antigens in both target cell types.[8]

Other experiments indicated that transfer of thymocytes infected with RadLV for a short time (e.g., 3 weeks), from resistant or susceptible mice, into irradiated, adult, syngeneic F_1 mice leads to death from tumorigenesis in a relatively short time.[8] It is unlikely that such death results from the induction of leukemia by virus present in transformed cells, because the progress of disease is too rapid and RadLV is not known to induce leukemia when injected into adult mice. Rather, preleukemic cells are probably generated quickly in the thymus by RadLV infection of either resistant or susceptible mice, and transfer of these cells to the irradiated host environment permits their escape from the immunological control that normally prevents their growth. These observations clearly indicated the importance of immunosurveillance in controlling tumorigenesis by RadLV, and the impact that altered H-2 antigen expression has in the effectiveness of the host's immune response to deal with virus-infected cells.

Similar observations have been made by others. For example, by contrast with normal lymphocytes, the AKR T-cell leukemia line K36.16 expresses normal levels of H-2D antigens but very low levels of H-2K molecules.[9] K36.16 tumor cells are resistant to H-2K-restricted killing by T cells and grow rapidly when transplanted into normal hosts.[9] The H-2K antigen, but not the H-2D antigen (as is the case for the RadLV in the studies cited in the preceding discussion), appears to be involved in CTL recognition of the Gross virus antigen expressed on these cells. If the low levels of H-2K antigens expressed by these cells contribute to their escape from CTL-mediated lysis, it would follow that transfection and expression of the normal AKR *H-2K* gene into K36.16 cells should lead to their rejection *in vivo*. Indeed, Hui *et al.*[2] have shown that transfected tumor cells expressing the transferred *H-2K* sequences at high levels are rejected when transplanted into normal, histocompatible mice. In support of the involvement of CTL-mediated rejection of the transfected cells is the observation that these grow rapidly when the hosts are immunosuppressed by moderate doses of radiation. Thus, H-2K expression on these cells correlated directly with their ability to grow in immunocompetent hosts.

Similar studies were performed by De Baetselier *et al.*[10] with variants of the methylcholanthrene (MCA)-induced sarcoma T10. MCA- and virus-induced sarcomas generally express tumor-associated antigens (TAA) which can be recognized in the context of self-MHC by antitumor CTLs. T10 cells express at least one such TAA. Since these cells are derived from an F_1 mouse hybrid, they should express both sets of parental H-2 antigens. However, a highly metastatic variant of this tumor, IE7, and a moderately metastatic variant, IC9, both of which express H-2D-encoded class I products, fail to express either *H-2K* allele.[11] By introducing each parental *H-2K* allele separately into IE7 and IC9, Wallich et al.[3] found that expression of either of the parental H-2K antigens markedly lowered the metastatic phenotype of the transplanted cells, as would be expected if the transfected H-2K molecules played an important role in tumor rejection. Furthermore, CTLs generated against T10 cells that expressed some of the parental alleles only recognized target cells expressing the same antigens.[11] Thus, H-2K loss appears to allow the variant tumor cells to escape immune destruction.

Another experimental system allowing a similar conclusion involves adenovirus-infected or -transformed cells. Cells transformed with the highly oncogenic adenovirus strain Ad12 are generally able to grow and metastasize, but those transformed by other adenovirus strains (e.g., Ad5) are not.[4] The first suggestion that class I gene expression could be involved in the different metastatic potential of these cells came from the observation that there is much lower expression of class I antigens on the Ad12- versus Ad2- or Ad5-transformed cells.[4] The adenovirus system is quite similar in many respects to the RadLV one we described earlier. For example, it is particularly interesting that, as is the case for RadLV infection,[1] Ad12 infection, in contrast to cellular transformation, results in a transient 10- to 15-fold increase in the expression of class I mRNA.[12] It should be emphasized, however, that unlike the situation for RadLV, residual levels of class I transcripts are present in Ad12-transformed cells.

In the adenovirus system, as contrasted with the RadLV model and the work with K36.16 and MCA variants, it is not absolutely clear that failure of the T-cell immune surveillance system is entirely responsible for the oncogenic character of the Ad12 transformants. Other studies have suggested that the neoplastic potential of hamster and rat cells transformed with the various adenovirus strains may result more from their inability to be lysed by macrophages and NK cells than from their lack of susceptibility to CTL lysis.[13,14] In general, all adenovirus transformants, regardless of the strain of virus used, seem to share similar sensitivity to killing by either alloreactive CTLs, or CTLs generated against adenovirus-transformed tumors.[6,13] However, the above explanation is problematic because nude mice, which are generally lacking in T cells such as CTLs but which express high levels of NK activity, readily succumb to growth of cells transformed by any strain of adenovirus.[15]

In any case, recent transfection experiments[16] similar to those of Hui et al.[2] and De Baetselier et al.[10] suggest that level of class I expression on these tumors determines, to some degree, whether they will grow or not. These experiments, however, are not on as solid ground as those discussed above because an *H-2L* gene from the BALB/c mouse strain was transfected into Ad12-transformed cells of the C57BL/6 strain and then transplanted into BALB/c mice. Since the cells used are derived from very dissimilar mouse strains, tumor rejection might have resulted from antigenic differences unrelated to the differences in class I gene expression between the transfected and parental Ad12-transformed cells.

Given the variety of systems that have been used, the phenomena of virus-induced alterations in class I gene expression appear to be on solid footing. The next step is to elucidate the mechanisms(s) by which virus infections or transformation can deregulate class I expression. It is probably reasonable to presume that no single mechanism can account for the numerous alterations in class I gene expression reported. Furthermore, the generality of the phenomena remains to be established. In what follows we shall review the progress made by our laboratory in approaching a molecular understanding of RadLV-induced changes in class I gene expression. We shall also allude to experiments by others designed to understand virus-induced alterations in class I gene expression.

2. RadLV Transformation Is Associated with Decreased Expression of H-2 Class I mRNA and Increased Methylation of Class I Genes

Class I mRNA is undetectable in RadLV tumors[17] (Fig. 2). This is consistent with our observations that transformation of thymocytes by RadLV leads to a loss of H-2 antigen expression as measured by two-dimensional gel electrophoresis, immunoflourescence, absorption analysis, and immunoprecipitation.[1] It further suggests that the loss of expression results either from decreased transcription of class I genes or from instability of the class I mRNA synthesized. The lack of detectable mRNA is particularly impressive in that the probe used in these experiments, pH-2II, is broadly cross-reactive hybridizing with all class I (K, D, L, Qa, and TL) genes,[17] suggesting all class I gene transcription is shut down by RadLV. Consistent with this notion is the observation that changes in methylation of MHC DNA are associated with RadLV transformation of murine thymocytes.[17] Every RadLV-induced tumor shows altered methylation patterns for class I DNA as compared to normal cells.[17] One result obtained in these studies is that every tumor appears to have unique alterations in methylation, even RadLV tumors derived from the same mouse strain. Since every tumor displays the same methylation pattern whenever it is assayed (in uncloned cell lines), this implies that alterations in methylation occur before the tumorigenic cells expand from a limited number of transformed cells; i.e., a clonal origin for most RadLV tumors. It also confirms our previous observations that virus interactions with class I genes occur early after virus infection.

RadLV-association changes in *H-2* DNA methylation are not related to normal regulatory influences that affect *H-2* expression in various organs. On average, thymocytes express about five- to tenfold less H-2 antigens than splenocytes,[18] yet no differences in methylation can be detected between splenocytes and thymocytes.[17] Thus, if altered methylation of *H-2* DNA associated with RadLV transformation is responsible for decreased H-2 antigen expression, such a mechanism is different

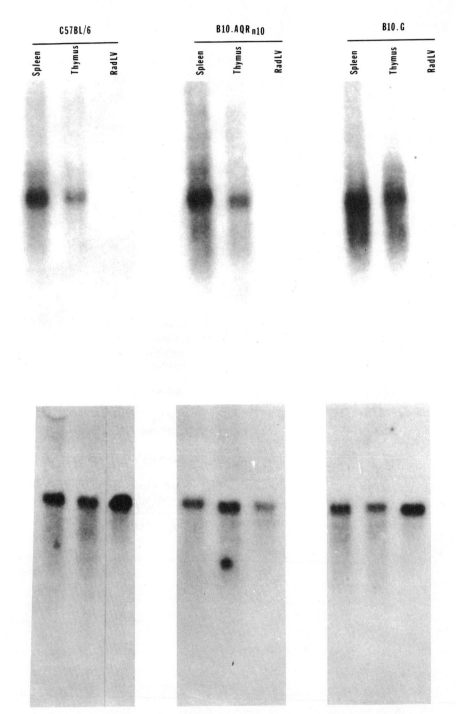

FIGURE 2. Comparison of class I MHC mRNA levels in normal mouse thymocytes and splenocytes with those of RadLV-induced thymomas. (A) Expression of H-2, shown with the pH-2II probe, is observed in both normal cell types of each of the three mouse strains, while there is considerable reduction of H-2 in the corresponding thymoma. (B) Expression of β-actin mRNA is unaffected by RadLV transformation.

from normal regulatory influences affecting *H-2* expression in distinct populations of lymphocytes.

An important question we raised is whether the alterations in methylation associated with RadLV are restricted to class I genes or affect other genes (non-MHC) in the genome. We examined some genes such as *Thy-1*, *γ2b*, and *β2-microglobulin (β2M)* which are not encoded in the MHC. DNA from at least 20 *in vivo* and *in vitro* derived tumors was analyzed with probes specific for these genes.[17] No differences were observed concerning *Msp*I versus *Hpa*II digestion of *Thy-1* DNA from RadLV-transformed and uninfected/transformed thymocytes. Similar results were obtained for β2-microglobulin. In the case of γ2b, no changes were observed except that one of the tumors examined appeared to be *less* methylated than normal DNA.[17] Although this preliminary examination of the other genes is quite limited, and many more coding sequences need to be examined, it suggests that alterations in methylation induced by RadLV might be specific for class I genes.

Others[16,19,20] have also shown that changes in the levels of DNA methylation surrounding class I genes might be relevant to the issue of class I gene expression. For example, both metastatic and nonmetastatic phenotypes could be imparted to subclones of the Lewis lung carcinoma by treatment with 5-azacytidine, which is thought to specifically inhibit the enzymes involved in the maintenance of DNA methylation patterns. The changes in metastatic potential of the cells thus achieved were accompanied by altered expression of specific class I genes.

3. Rearrangement of Class I DNA Regions Is Associated with RadLV Tumorigenesis

Rearrangements in MHC DNA also appear to accompany RadLV transformation as detected by digestion with restriction enzymes.[17] Novel restriction fragments of length polymorphisms (RFLPs), which are often seen with *H-2* probes, are not detected with probes for other genes such as *γ2b* and *β2M*.[17] While tumors with class I RFLPs are seen less frequently than tumors with changed DNA methylation patterns, it is difficult to ascertain the true incidence of novel class I RFLPs because of the multiplicity of fragments hybridizing with class I probes. A displacement of a fragment to a different location in a gel may be masked by a fragment of the same size already present in normal DNA.

While studies to determine more precisely which class I genes were methylated and rearranged are still preliminary, it is possible to conclude that alterations in methylation associated with RadLV appear to affect numerous class I genes, but when novel RFLPs arise they involve a more limited number of class I genes.[17] Thus, rearrangements might play a role in loss of H-2 antigen expression but it would be difficult to envision a mechanism by which such limited rearrangements would affect expression of all class I genes, and this is one of the issues we still need to understand.

4. Interactions of Viruses with Class I Promoter and Enhancerlike Sequences and Their Role in the Regulation of Class I Gene Expression

Certain DNA sequences have been shown to play key roles in the accuracy and efficiency of transcription by RNA polymerase II.[21,26] The cap site has been identified as the point at which transcription starts. The Goldberg–Hogness or TATA box usually located some 30 base pairs (bp) upstream, serves as a strong promoter. In the -50 to -90 region ($+1$ refers to the cap site), sequences of the CAAT type[22,26] or GC-rich regions[23,26] have also been defined to be important for transcription. Finally, sequences located even farther upstream, or sometimes within the genes, and known as enhancer sequences, have been shown to stimulate transcription from homologous or heterologous promoters in a manner relatively independent of their distance and orientation with regard to the promoters.[24,26] In addition, enhancers have been implicated in the tissue-specific expression of the gene with which they are associated.[25,26]

Kimura et al.[26] have recently demonstrated that enhancer sequences associated with the H-$2K^b$ gene are not the target of the inhibition of H-2 expression observed in Ad12-transformed primary cells. They also found that the H-$2K^b$ promoter is repressed about eight to ten times more efficiently by Ad12 than by Ad5. These experiments reproduce in a transient assay, and with the H-$2K^b$ promoter alone, what is observed *in vivo* with the endogenous H-2 gene. In the latter case, the E1a region of the virus has been involved. However, because as was mentioned earlier, during acute infection by Ad5 and Ad12 H-$2K^b$ transcription is stimulated rather than repressed, it has been pointed out that the relationship between H-2 promoters, enhancers, and viral elements is likely to be complex.[26]

The possibility that RadLV alters class I expression by interacting with class I enhancers and promoters is currently under investigation.

5. Evidence for Insertion of Viral Sequences in the MHC

One mechanism by which class I gene expression can be altered by viruses is direct integration in the MHC. Clearly, such integrations, if they occur, can significantly impact on the arrangement, structure, and expression of class I genes.

While we have as yet not been able to demonstrate directly that viral insertions in the MHC result in alterations of MHC antigen expression, we would take exception with the notions that such viral integrations in the MHC occur randomly, infrequently, or can be observed because of the selective advantage they impart on the cells in which the event takes place.[27] We have noted an association between viruses and histocompatibility genes which has too high a frequency[28,29] to be considered the result of random insertions followed by selection, unless it is argued that random insertions in the germline MHC region confer a selective advantage in the evolution of mouse strains. This argument is flawed for several reasons. Numerous mouse strains can be shown to have viruses inserted adjacent to histocompatibility genes even when normal, non-tumor-bearing cells are examined. Furthermore, mouse strains that lack some or all of the specific class I viral sequences so far defined seem to fare equally well in most regards as those that possess the MHC-linked retroviral genomes in the MHC.

As an example of a mouse strain which carries MHC-linked viruses, we can discuss our studies of C57BL/10 mice. Examination of a battery of 38 class I gene-containing cosmids, isolated from two large genomic libraries constructed from C57BL/10 spleen DNA, led to the localization of viral sequences within the murine MHC.[30] The viral probes used hybridized with only four cosmids, containing overlapping mouse sequences that defined four class I gene-related sequences in a region of 90 kb of DNA in the *TLa* region[30] (Fig. 3). The data showed that at least two distinct viral sequences were contained in the cluster. Extensive analysis, including cloning and sequencing of these viruses (*TLev1* and *TLev2*), suggested that at least one of the two viral genomes *TLev1* is complete, i.e., it contains the typical LTR-*gag-pol-env*-LTR configuration.[31] *TLev-2* has so far been shown to encode *gag-pol* sequences.[31]

It has been shown that hybridization of the viral probes with the *H-2* cosmid clones does not occur because homology exists between the viral and *H-2* sequences, but rather because both types of sequences are tightly linked in the same cosmid DNA.[30] Detailed restriction analyses of each of the two viral sequences within the *H-2*-containing cosmids have shown that there is a minimum distance of between 600 and 1000 bp between viral and *H-2* sequences. In addition, there is no cross-hybridization between any of multiple viral probes tested and *H-2* probes.[30] A computer comparison of *H-2* nucleotide sequences for K^b, D^b, and L^d genes with the available viral sequences indicates no obvious homology.[30]

A potential caveat initially considered with respect to these findings was that the particular mice used to derive the cosmid libraries examined may coincidentally have had an inserted viral genome at *H-2*. However, three sets of *H-2*-positive cosmid clones, derived from three independently created cosmid libraries, were exam-

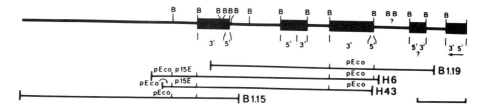

FIGURE 3. Cosmid map of the *H-2* cluster encompassed by cosmid clones H6, H43, B1.15, and B1.19, which contain overlapping sequences. Regions hybridizing to the 5′ or 3′ end of *H-2* genes, detected with probes pH-2III and pH-2II, respectively, are indicated by dark boxes. cDNA clone pH-2II codes for amino acids 167–352 and 600 nucleotides of the 3′ untranslated region of *H-2*, which have been removed from our probe. Clone pH-2III codes for the 5′ region of *H-2*, encoding amino acids 63–120. The extent of the hybridizing gene sequence is defined only by the restriction sites shown [for endonuclease *Bam*HI (B)]. The orientation was determined from additional restriction analysis with *Sac*II, *Kpn*I, *Hpa*I, and *Cla*I (data not shown). Viral sequences were detected using a pAKR-Eco 2.7-kb *Sal*I/*Bam*HI fragment within the envelope *(env)* gene (4.3–7.0 kb) of complete infectious AKR provirus cloned probe λAKR 623. Bar = 10 Kb.

ined and the results were internally consistent, making such a possibility highly unlikely. In addition, use of congenic mouse strains has now shown that all mice carrying the TLa^b haplotype contain similar viral sequences (C. Pampeno and D. Meruelo, unpublished observation).

In collaboration with Dr. L. Hood and colleagues, we have now conducted an extensive analysis to determine whether viral sequences are associated with MHC class I genes in BALB/c mouse genomes. Preliminary results to date have been published.[32] These studies indicate that viral sequences appear to flank a substantial number of class I genes.[32] It is also particularly interesting that we have not yet detected viral sequences in the Qa region of BALB/c mice.[32] If this observation holds up, it would be considered significant in that the Qa cosmid clusters examined encompass at least 260 kb of DNA (or approximately one-half the total DNA tested) and 8 of the 36 class I genes defined. An explanation for this observation may relate to the possible role of retroviruses in the generation of class I polymorphisms. The Qa molecules appear unique among known class I molecules in that they lack the high degree of polymorphisms exhibited by other class I molecules. In fact, no structural polymorphism has been detected even among *Mus* subspecies for any Qa molecule.[33] The only polymorphism observed in the Q subregion is with regard to the expression of the various Qa family antigens. Therefore, the generation of diversity in this latter subregion appears to be governed by deletion mechanisms rather than by several mechanisms postulated to be involved in generation of MHC polymorphism such as gene conversion.[33,34] Alternatively, selective pressures prevent the accumulation of mutations in these genes. Thus, Qa subregion gene mutations might not be tolerated by the organisms, although one Qa subregion gene may substitute for another. In support of this notion, Flaherty *et al.*[33] cite the situation for certain other multigene families such as hemoglobin, where the persistence of fetal hemoglobin production can compensate for extensive deletions in the adult hemoglobin δ- and β-chain genes, yet certain mutations in these same hemoglobin genes can cause severe anemia.

6. Integration of Viruses Adjacent to *H* and *Ly* Genes Appears to Occur at a High Frequency

The above studies are complemented by other studies from our laboratory[28,29] and at least two other groups.[35,36] These studies demonstrate that viral sequences are also found adjacent to non-MHC histocompatibility loci and genes encoding several Ly (H/Ly) antigens. Southern gel electrophoresis and hybridization have shown that RFLPs hybridizing with xenotropic and ecotropic envelope virus probes map adjacent to minor H or Ly-antigen-encoding loci. For example, RFLPs hybridizing with viral probes are associated with *Ly-17* on chromosome 1; *H-30, H-3*, and *H-13* on chromosome 2; *Ly-21* on chromosome 7; *H-28* on chromosome 3; *H-16* on chromosome 4; *Thy-1* on chromosome 9; and *H-38* on chromosome 12.[28] Although some viral loci map to locations where no *H/Ly* has yet been mapped, the frequency and tightness of linkages observed in the many instances studied, coupled with the large number of as yet unmapped H/ly loci, suggest that the associations found are significant.[29,30]

This issue has been studied further by cloning and characterizing one of the endogenous viral sequences, B65.5PV2, mapping near the minor H locus, H-30.[29] A subfragment (358-bp probe) of this endogenous viral sequence has been found to cross-hybridize with DNA sequences adjacent to 15 additional minor H loci.[29] This observation demonstrates that a group of related retroviruses is associated with DNA regions coding for minor H loci and may have evolutionary and functional connotations and raises several important questions. For example, are the linkages between the 358-bp hybridizing RFLPs and minor H loci significant? Could the simplest explanation to these findings be that minor H genes are viral genomes? Two observations argue circumstantially against this possibility. First, viral sequences in the MHC are clearly distinct from class I (H) genes. While they are physically very close (600–3000 bp), they do not overlap.[30] Second, studies similar to ours, in which DNAs of minor H mutants were compared to those from the parental strains, failed to reveal any viral RFLPs (P. Wettstein, personal communication). Mutations of minor H genes evidently do not alter the restriction map of the viral genome.

Another question raised: is it merely that there are large numbers of both minor H loci and endogenous viral genomes which, purely on a random basis, are bound to be close to each other a significant number of times? Given the large number of known endogenous retroviral sequences, random integrations might be expected to lead to a few instances were a virus is found close to an H or Ly gene. However, the fact that 15 out of 31 RFLPs detected with the 358-bp probe could be mapped to a region with a minor H locus nearby, suggests this association is not random. Furthermore, along this line of reasoning, examination of a cosmid library of total C57BL/6By genomic DNA suggests that the probe hybridizes with sequences in the genome once every 20,000–40,000 kb. Since most of the H loci closely associated with retroviruses show a recombination frequency less than one centimorgan (or about 2000 kb), it is unlikely that the associations observed are random. Further support for this notion comes from work of Blatt et al. (personal communication) that has linked the preferential retroviral integration site Pim-1 to the H-2 complex. However, the argument can be made in this instance that we are looking at a selected population of insertion sites, since the preferred integrations were determined in tumor cells.

Numerous recent studies[37–48] have documented the existence of *non*random, preferential integration sites for murine retroviruses. The most recent at the time of writing is provided by the insertional mutagenesis studies of King et al.[48] These investigators started out from the premise that retroviruses were ideal tools for insertional mutagenesis studies because their insertion may be expected to be random. However, at least in this study, the premise of random viral insertion proved unfounded. Theoretically, the screening of a library of 2–5×10^9 independent insertion sites should have covered the entire mouse genome (3×10^9 nucleotides) and yielded 1800 to 34,000 insertions leading to the hprt$^-$ phenotype sought.[48] Instead, only 17 hprt$^-$ lines were obtained, clearly indicating that proviral insertion into cellular DNA is not always random.[48] This experiment was done *in vitro*, with an xeogenous virus carrier, and under artificial selection pressures. The *in vivo* selection for insertion sites that will allow stable endogenous retroviruses to propagate in the germ line is likely to be even less random.

The available evidence suggests that the multiple related copies of endogenous retroviruses found in the genome are not the result of tandem duplications of virogenes but arise from new and independent integrations.[29] When somatic amplification of endogenous retroviruses has been studied, it has become apparent that the viral gene duplications and reintegrations accounting for the amplification occur at specific genomic sites under *in vivo* conditions. It has been postulated that insertion at specific sites occurs in the germ line as well.[37]

Specificity of integration in terms of flanking genes is not inconsistent with what is known about the integrative mechanism of viruses, nor the conclusion that proviruses are inserted in many different regions of host genomes. Thus, *H* and *Ly* genes are dispersed throughout the genome.[29] Insertion of viruses adjacent to these loci would give the appearance of random integrations, if no note was made of the fact that insertions are near *H/Ly* loci.

Given our findings and those of others to date, it might be reasonable to propose that there is specificity in the observed location of endogenous retroviral sequences and that there is a significant reason for the association between MuLV sequences and *H* or *Ly* loci. Some evidence has suggested that the high mutation rate associated with *H* genes of the MHC complex results from the incorporation of viral genomes into germinal cells.[49,50] In addition, the mechanisms by which *H*-2 mutants arise appear to involve gene conversion events[51,52] which might occur via transposonlike structures such as murine leukemia viruses or *Alu*-like repeat elements.

A significant role for retroviruses in *H* gene polymorphism and evolution is suggested as well by studies from Atherton *et al.*[53] These authors have proposed that if viruses play a role in the generation of *H* gene polymorphism, one would expect that whenever the retroviral family of genes is highly polymorphic in one species, the *H* genes would be expected to display significant polymorphism. In the absence of polymorphism in the retroviral gene family, *H* genes would be expected to show little diversity. This is precisely what has been observed thus far in the Syrian hamster which lacks or shows minimal polymorphism for class I gene products as detected serologically, biochemically, and by DNA restriction enzyme analysis. The hybridization data from genetically disparate hamster strains reveal no evidence of polymorphism of retroviral sequences.[53] Atherton *et al.*[53] performed their studies with the inbred hamster strains plus other partially inbred hamster strains developed from a dozen wild hamsters captured in Syria in 1970 at locations many kilometers apart.[54,55] Similar observations with squirrel populations that separated over a million years ago have been made by P. Wettstein (personal communication).

If it is evolutionarily advantageous to host and virus to link viral and *H* sequences and if specificity of integration is the norm, an important question becomes, what is the molecular mechanism by which the actual preferential site location is achieved? As stated earlier, the 3' *pol* region of retroviruses plays an important role in viral integration by encoding a site-specific endonuclease which nicks the host DNA at the insertion site.[56-58] Since the 358 bp is derived from this endonuclease-coding segment of the virus, and this sequence is maintained in all *H*-linked viral sequences, one possibility is that minor loci share specific sites in their flanking DNA sequences, which the 3′ *pol*-encoded nuclease of these conserved viruses recognize. Many other possibilities exist, but until more is known, it is premature to speculate further.

Even if we could prove specificity of integration next to MHC genes, this might only provide one clue to understanding the mechanism by which viruses alter class I gene transcription. Clearly, it is unlikely that a single integration event adjacent to a class I gene would account for shutdown of all class I gene transcription.

In fact, in most systems studied, virus-induced transformation does not lead to shutdown of all class I genes, but rather of specific class I genes. This might also be true even for RadLV, despite the lack of any detectable class I mRNA. We have clearly shown that after RadLV infection, the effect on transcription is selective, with H-2D antigen expression being most dramatically affected. In addition, many *Qa* and *TL* genes are known *not* to be expressed on thymocytes, which are the target cells for RadLV transformation. Therefore, the actual extent of class I genes affected by the virus is unknown. More needs to be done to understand possible mechanisms of virus-induced suppression of specific and/or numerous class I genes.

7. Molecular Mechanisms by Which Flanking Viral Sequences May Alter Class I Gene Expression

Recent studies from our laboratory suggest that at least in some cases, retroviruses flanking *H* genes insert themselves within repetitive sequences. These studies suggest that part of the answer as to the mechanism by which viral insertion adjacent to MHC genes affects transcription levels might be found in studying the relationship between viruses, class I or *H* genes, and repetitive sequences. Studies from other laboratories also seem to lend support to this notion. For example, Goodenow *et al.*[27] report that analysis of SV40-transformed 3T3 cells reveals that a variety of class I genes, as identified by specific oligonucleotide probes, can be activated in these cells and remark that many of the class I genes activated by SV40 appear to contain repetitive elements known as B2 repeats, whose transcription is known to be enhanced in SV40-transformed cells.[59]

Two types of repetitive elements are generally recognized: the *short interspersed repeats* of which more than 10^5 copies are found scattered throughout the mouse genome[62] and the *long transposable elements,* of which more will be said in a moment. The original isolation of the short repeats from the genome library was achieved by means of a dsRNA B probe[60] and hence they are known as B1 and B2 sequences. According to some,[61] short interspersed repeats represent a class of transposable elements in eukaryotic genomes transcribed by RNA polymerase III.[61] It is postulated that their primary transcript serves as a template for reverse transcriptase, and the resultant double-standard DNA inserts into new positions in the genome at the sites of occasional chromosome cleavages.[61,62]

The long transposable elements, or the so-called mobile dispersed genes (mdg) which are also widespread in eukaryotic genomes,[63] differ from the short repetitive elements not only in length (larger than 5–10 kb) but also are transcribed by RNA polymerase II rather than polymerase III.[62] Their insertion also creates a duplication of the target sequence, but only 4–5 bp long. They are less abundant in the genome than B-type repeats, and are significantly less conserved.[62–69]

It has been suggested that because of the conservative character of ubiquitous repeats and the presence of signal sequences within them, they may play some functional role,[62] particularly in regulation of transcription. For example, they may be involved in splicing, a hypothesis that has found strong experimental support, at least in the case of snRNA U1.[63,70] The existence of a B1 repeat within an intron could allow multistep splicing, and influence gene expression by modulating the level of processing and splicing that occurs.[71] The apparent homology between B1 sequence and the replication origins of papova viruses[62] has suggested that another possible role of repeats of the B type is to serve as sites of the initiation of replication.[62] However, no direct evidence for their function as origins of replication has been obtained.

B1 and B2 sequences may be involved in the regulation of gene expression by yet other mechanisms.[62] For example, they may serve as elements of a modulator type influencing the transcription rate, as structural elements determining RNA interaction with the nuclear skeleton during RNA transport, and so on.[62] Regulation through transposition is also a possibility. The transposition of B copies into new sites of the genome could have generated many cryptic differences among individuals, and, in this way, accelerated the evolutionary process.[62] (Clearly, this could be most relevant for evolution of H genes.) For example, sequencing studies of goat globin genes[72] have shown that these genes are virtually identical, except for differences in their short inserted repeated sequences within the introns. Nonetheless, these genes are developmentally regulated. Since their only difference is in the repetitive sequences, a role for them is almost certain in the regulation of expression observed.

As should be apparent from what has been stated thus far, a role of repetitive DNA sequences in the regulation of gene expression seems quite plausible. However, to date, it has not been possible to correlate the presence of specific repetitive DNA element adjacent to a coding sequence with the expression of that coding sequence. Furthermore, until recently there has been no demonstration that specific repeated elements are associated with various members of a multigene family. However, a missing element for such a hypothesis to be taken more seriously has now been demonstrated. The nonrandom association of certain repetitive sequences with different MHC genes has recently been documented[73]. Class I mRNAs have also been shown to contain B2 repetitive sequences.[74] In conjunction with this finding, Lelanne et al.[74] postulated that the B2 element next to several class I genes became inserted after gene duplication. It now remains to be shown whether these repetitive sequence elements can play a role in the regulation of expression and/or generation of polymorphism of the MHC multigene family, and whether their role in this capacity is impacted on by the insertion of viral sequences in these regions. Whatever the answer might be, it is clear that only further studies of retrovirus-induced alterations in class I gene expression will shed light on which mechanisms and interactions might be relevant and ultimately help elucidate the molecular events that account for these virus-induced alterations.

ACKNOWLEDGMENT. Supported by NIH Grants CA22247, CA31346, CA35482, and CA41420. The author is an Established Investigator of the American Heart Association.

References

1. Meruelo, D., Nimelstein, S., Jones, P., Lieberman, M., and McDevitt, H. O., 1978, Increased synthesis and expression of H-2 antigens as a result of radiation leukemia virus infection: A possible mechanism for *H-2* linked control of virus-induced neoplasia, *J. Exp. Med.* **147**:470–479.
2. Hui, K., Grosveld, H., and Festenstein, H., 1984, Rejection of transplantable AKR leukemia cells following MHC DNA-mediated cell transformation, *Nature* **311**:750–752.
3. Wallich, R., Bulbuc, N., Hammerling, G. J., Katzav, S., Segal, S., and Feldman, M., 1985, Abrogation of metastatic properties of tumor cell by *de novo* expression of H-2K antigens following *H-2* gene transfection, *Nature* **315**:301–305.
4. Schrier, P. I., Bernards, R., Vaessen, R. T. M. J., Houweling, A., and van der Erb, A. J., 1983, Expression of class I major histocompatibility antigens switched off by highly oncogenic adenovirus-12 in transformed cells, *Nature* **305**:771–775.
5. Eager, K. B., Williams, J., Breiding, D., Pan, S., Knowles, B., Appella, E., and Ricciardi, R. P., 1985, Expression of histocompatibility antigens, H-2K, -D, and -L is reduced in adenovirus-12-transformed mouse cells and is restored by interferon γ, *Proc. Natl. Acad. Sci. USA* **82**:5525–5529.
6. Mellow, G. H., Fohring, B., Dougherty, J., Gallimor, P. H., and Raska, K., 1984, Tumorigenicity of adenovirus-transformed rat cells and expression of class I histocompatibility antigen, *Virology* **134**:460–465.
7. Meruelo, D., Lieberman, M., Ginzton, N., Deak, B., and McDevitt, H. O., 1977, Genetic control of radiation leukemia virus-induced tumorigenesis. I. Role of the murine major histocompatibility complex, *H-2*, *J. Exp. Med.* **146**:1079–1088.
8. Meruelo, D., 1979, A role for elevated H-2 antigen expression in resistance to RadLV-induced leukemogenesis: Enhancement of effective tumor surveillance by killer lymphocytes, *J. Exp. Med.* **149**:898–909.
9. Schmidt, W., and Festenstein, H., 1982, Resistance to cell-mediated cytotoxicity is correlated with reduction of H-2K gene products in AKR leukemia, *Immunogenetics* **16**:257–264.
10. De Baetselier, P., Katzav, S., Gorelik, E., Feldman, M., and Segal, S., 1980, Differential expression of H-2 gene products in tumor cells is associated with their metastogenic properties, *Nature* **288**:179–181.
11. Katzav, S., Taitakov, B., De Baetselier, P., Isakov, N., Feldman, M., and Segal, S., 1983, Role of MHC-encoded glycoproteins in tumor dissemination, *Transplant. Proc.* **15**:162–170.
12. Rosentahal, A., Wright, S., Quade, K., Gallimore, P., Cedar, H., and Grosveld, F., 1985, Increased MHC H-2K gene transcription in cultured mouse embryo cells after adenovirus infection, *Nature* **315**:579–581.
13. Raska, K., Jr., and Gallimore, P. H., 1982, An inverse relation of the oncogenic potential of adenovirus-transformed cells and their sensitivity to killing by syngeneic natural killer cells, *Virology* **123**:8–18.
14. Lewis, A. M., Jr., and Cook, J. L., 1985, A new role for DNA virus early proteins in viral carcinogenesis, *Science* **227**:15–19.
15. Herberman, R. B., and Holden, H. T., 1978, Natural cell-mediated immunity, *Adv. Cancer Res.* **27**:305–377.

16. Tanaka, K., Isselbacher, K. J., Khoury, G., and Jay, G., 1985, Reversals of oncogenesis by the expression of a major histocompatibility complex class I gene, *Science* **228**:26–30.
17. Meruelo, D., Kornreich, R., Rossomando, A., Pampeno, C., Boral, A., Silver, J. L., Buxbaum, J., Weiss, E. H., Devlin, J. J., Mellor, A. L. Flavell, R. A., and Pellicer, A., 1985, Lack of class I *H-2* antigens in cells transformed by radiation leukemia virus is associated with methylation and rearrangement of H-2 DNA, *Proc. Natl. Acad. Sci. USA* **83**:4504–4508.
18. Edidin, M., 1972, The tissue distribution and cellular location of transplantation antigens, in: *Transplantation Antigens* (B.D. Kahan and R.A. Reisfeld, eds.), Academic Press, New York, pp. 125–140.
19. Olsson, L., and Forchhamer, J., 1984, Induction of the metastatic phenotype in a mouse tumor by 5-azacytidine and characterization of an antigen associated with metastatic activity, *Proc. Natl. Acad. Sci. USA* **81**:3389–3393.
20. Olsson, L., 1985, Identification of a gene and its product associated with metastatic activity, *Fed. Proc. Fed. Am. Soc. Exp. Biol.* **44**:1336.
21. Breathnach, R., and Chambon, P., 1981, Organization and expression of eukaryotic split genes coding for proteins, *Annu. Rev. Biochem.* **50**:349–383.
22. Dierks, P., Van Ooyen, A., Cochran, M. D., Dobkin, C., Reiser, J., and Weissman, C., 1983, Three regions upstream from the cap site are required for effective and accurate transcription of the rabbit β-globin gene in mouse 3T6 cells, *Cell* **32**:695–706.
23. McKnight, S. L., and Kingsbury, R., 1982, Transcriptional control signals of a eukaryotic protein-coding gene, *Science* **217**:316–325.
24. Gluzman, Y., and Shenk, T. (eds.), 1983, *Enhancers and Eukaryotic Gene Expression*, Cold Spring Harbor Laboratory, Cold Spring Harbor, N.Y.
25. Banerji, J., Olsson, L., and Shaffner, W., 1983, A lymphocyte specific cellular enhancer is located downstream of the joining region in immunoglobulin heavy chains, *Cell* **33**:729–740.
26. Kimura, A., Israel, A., Lebaile, O., and Kourilsky, P., 1986, Detailed analysis of the mouse H-2K^b promoter: Enhancer-like sequences and their role in the regulation of class I gene expression, *Cell* **44**:261–272.
27. Goodenow, R. S., Vogel, J. M., and Linsk, R., 1985, Histocompatibility antigens on murine tumors, *Science* **230**:777–783.
28. Meruelo, D., Rossomando, A., Offer, M., Buxbaum, J., and Pellicer, A., 1983, Association of endogenous viral loci with genes encoding murine histocompatibility and lymphocyte differentiation antigens, *Proc. Natl. Acad. Sci. USA* **80**:5032–5036.
29. Rossomando, A., and Meruelo, D., 1986, Viral sequences are associated with many histocompatibility genes, *Immunogenetics* **23**:233–245.
30. Meruelo, D., Kornreich, R., Rossomando, A., Pampeno, C., Mellor, A. L., Weiss, E. H., Flavell, R. A., and Pellicer, A., 1984, Murine leukemia virus sequences are encoded in the murine major histocompatibility complex, *Proc. Natl. Acad. Sci. USA* **81**:1804–1808.
31. Pampeno, C., and Meruelo, D., 1986, Isolation of retroviral-like sequences from the *TL* locus of the C57BL/10 murine major histocompatibility complex, *J. Virol.* **58**:296–306.
32. Meruelo, D., 1985, Retroviral sequences flanking histocompatibility genes alter their expression, function, and evolution: A hypothesis, in: *Genetic Control of Host Resistance to Infection and Immunity* (E. Skamene, ed.), Liss, New York, pp. 655–669.
33. Flaherty, L., DiBiase, K., Lynes, M. A., Seidman, J. B., Weinberger, O., and Rinchik, E. M., 1985, Characteristics of a Q subregion gene in the murine major histocompatibility complex, *Proc. Natl. Acad. Sci. USA* **82**:1503–1507.
34. Mellor, A. L., Weiss, E. H., Kress, M., Jay, G., and Flavell, R. A., 1984, A nonpolymorphic class I gene in the murine major histocompatibility complex, *Cell* **36**:139–144.
35. Blatt, C., Mileham, K., Haas, M., Nesbitt, M., Harper, M., and Simon, M., 1983, Chromosomal mapping of the mink cell focus-forming and xenotropic *env* gene family in the mouse, *Proc. Natl. Acad. Sci. USA* **80**:6298–6302.
36. Wejman, J., Taylor, B., Jenkins, N., and Copeland, N. G., 1984, Endogenous zenotropic murine leukemia virus related sequences map to chromosomal regions encoding mouse lymphocyte antigens, *J. Virol.* **50**:237–249.
37. Jaenisch, R., 1980, Germ line and Mendelian transmission of exogenous type C viruses, in: *Molecular Biology of RNA Tumor Viruses* (J. R. Stephenson, ed.), Academic Press, New York, pp. 131–162.

38. Lemay, G., and Jolicoeur, P., 1984, Rearrangement of a DNA sequence homologous to cell–virus junction fragment in several Moloney murine leukemia virus-induced rat thymomas, *Proc. Natl. Acad. Sci. USA* **81**:38–42.
39. Nusse, R., and Varmus, H. E., 1982, Many tumors induced by the mouse mammary tumor virus contain a provirus integrated in the same region of the host genome, *Cell* **31**:99–109.
40. Nusse, R., van Ooyery, A., Cox, D., Fung, Y. K. T., and Varmus, H., 1984, Mode of proviral activation of a putative mammary oncogene (int-1) on mouse chromosome 15, *Nature* **307**:131–136.
41. Hayward, W. S., Neel, B. G., and Astrin, S., 1981, Activation of a cellular *onc* gene by promoter insertion in ALV-induced lymphoid leukosis, *Nature* **209**:475–480.
42. Jenkins, N. A. Copeland, M. G., Taylor, B. A., and Lee, B., 1981, Dilute(d) coat colour mutation of DBA/2J mice is associated with the site of integration of an ecotropic MuLV genome, *Nature* **283**:370–374.
43. Neel, B. G., Hayward, W. S., Robinson, H. L., Farey, J., and Astrin, S. M., 1981, Avian leukosis virus-induced tumors have common proviral integration sites and synthesize discrete new RNAs: Oncogenesis by promoter insertion, *Cell* **23**:323–344.
44. Noori-Daloii, M., Swift, R. A., Kung, H. J., Crittender, L. B., and Witter, R. L., 1981, Specific integration of REV proviruses in avian bursal lymphomas, *Nature* **294**:574–576.
45. Li, Y., Holland, C. A., Hartley, J. W., and Hopkins, N., 1984, Viral integration near *c-myc* in 10–20% of MCF 247-induced AKR lymphomas, *Proc. Natl. Acad. Sci. USA* **81**:6808–6811.
46. Tsichlis, P. N., Strauss, P. G., and Fu Hu, L., 1983, A common region for proviral DNA integration of MoMuLV-induced rat thymic lymphomas, *Nature* **302**:445–449.
47. Van der Putten, H., Quint, W., Verma, I. M., and Berns, A., 1982, Moloney murine leukemia virus-induced tumors: Recombinant proviruses in active chromatin regions, *Nucleic Acids Res.* **10**:577–592.
48. King, W., Patel, M. D., Lobel, L. I., Goff, S. P., and Nguyen-Huu, M. C., 1985, Insertion mutagenesis of embryonal carcinoma cells by retroviruses, *Science* **228**:554–558.
49. Bailey, D., 1966, Heritable histocompatibility changes: Lysogeny in mice, *Transplantation* **4**:482–487.
50. Melvold, R., W., and Kohn, H. I., 1975, Histocompatibility gene mutation rates: H-2 and non-H-2, *Mutat. Res.* **27**:415–418.
51. Pease, L., Schulze, D., Pfaffenback, G., and Nathenson, S., 1983, Spontaneous H-2 mutants provide evidence that a copy mechanism analgous to gene conversion generates polymorphism in the major histocompatibility complex, *Proc. Natl. Acad. Sci. USA* **80**:242–246.
52. Weiss, E., Golden, L., Zakut, R., Mellor, A., Fahrner, K., Kvist, S., and Flavell, R., 1983, The DNA sequence of the $H-2K^b$ gene: Evidence for gene conversion as a mechanism for the generation of polymorphism in histocompatibility antigens, *EMBO J.* **2**:453–462.
53. Atherton, S. S., Streilein, R. D., and Streilein, J. W., 1984, Lack of polymorphism for *C*-type retrovirus sequences in the Syrian hamster, in: *Advances in Gene Technology: Human Genetic Disorders* (F. S. Ahmad, J. Block, W. A., Schultz, D. Scott, and W. J. Whelan, eds.), ICSU Press, Cambridge, pp. 128–129.
54. Streilein, J. W., Duncan, W. R., and Homburger, F., 1980, Immunogenetic relationship among genetically defined inbred domestic Syrian hamster strain, *Transplantation* **30**:358–361.
55. Murphy, M. R., 1971, Natural history of Syrian golden-hamster: Reconnaissance expedition, *Am. Zool.* **11**:632–635.
56. Schwartzberg, P., Colicelli, J., and Goff, S. P., 1984, Construction and analysis of deletion mutations in the *pol* gene of Moloney murine leukemia virus: A new viral function required for productive infection, *Cell* **37**:1043–1052.
57. Panganiba, A., and Temin, H. M., 1984, Sequences required for integration of spleen necrosis virus DNA, in: *RNA Tumor Viruses* (T. Hunter and S. Martin, eds.), Cold Spring Harbor Laboratory, Cold Spring Harbor, N.Y., p. 22.
58. Donehawer, L. A., and Varmus, H. E., 1984, A mutant murine leukemic virus with a single missense codon in *pol* is defective in a function affecting integration, *Proc. Natl. Acad. Sci. USA* **81**:6461.
59. Singh, K., Casey, M., Saragosti, S., and Botchan, M., 1985, Expression of enhanced RNA polymerase III transcripts encoded by the B2 repeats in simian virus 40-transformed mouse cells, *Nature* **314**:553–556.
60. Kramerov, D. A., Grigoryan, A. A., Ryskov, A. P., and Georgiev, G. P. 1979, Long double-stranded

sequences (DSRNA-B) of nuclear pre-messenger RNA consist of a few highly abundant classes of sequences—Evidence from DNA cloning experiments, *Nucleic Acids Res.* **6:**697–713.
61. Jagadeeswaran, P., Forget, B. G., and Weissman, S. M., 1981, Short interspersed repetitive DNA elements in eukaryotes—Transposable DNA elements generated by reverse transcription of RNA Pol III transcripts, *Cell* **26:**141–142.
62. Krayev, A. S., Markusheva, T. V., Kramerov, D. A., Ryskov, A. P., Skryabin, K. G., Bayev, A. A. and Georgiev, G. P., 1982, Ubiquitous transposon-like repeats B1 and B2 of the mouse genome: B2 sequencing, *Nucleic Acids Res.* **10:**7461–7475.
63. Georgiev, G. P., Ilyin, Y. U., Chmeliauskaite, V. G., Ryskov, A. P., Kramerov, D. A., Skryabin, K. G., Krayev, A. S., Leukanidir, E. M., and Grigorkyan, M. S., 1981, Mobile dispersed genetic elements and other middle repetitive DNA sequences in the genomes of Drosophila and mouse—Transcription and biological significance, *Cold Spring Harbor Symp. Quant. Biol.* **45:**641–654.
64. Deininger, P. L., Jolly, D. J., Rubin, C. M., Friedmann, T., and Schmid, C. W., 1981, Base sequence studies of 300 nucleotide renatured repeated human DNA clones, *J. Mol. Biol.* **151:**17–33.
65. Grimaldi, G., Queen, C., and Singer, M. F., 1981, Interspersed repeated sequences in the African-green monkey genome that are homologous to the human Alu family, *Nucleic Acids Res.* **9:**5553–5568.
66. Krayev, A. S., Kramerov, D. A., Skryabin, K. G., Ryskov, A. P., Bagev, A. A., and Georgiev, G. P., 1980, The nucleotide sequence of the ubiquitous repetitive DNA sequence B1 complementary to the most abundant class of mouse fold-back RNA, *Nucleic Acids Res.* **8:**1201–1215.
67. Ryskov, A. P., Saunders, G. F., Farashyan, V. R., and Georgiev, G. P., 1973, Double-helical regions in nuclear precursors of messenger-RNA (pre-messenger RNA), *Biochim. Biophys. Acta* **312:**152–164.
68. Haynes, S. R., Toomey, T. P., Leinwald, L., and Jelinek, W., 1981, The Chinese-hamster Alu-equivalent sequence-A conserved, highly repetitious, interspersed deoxyribonucleic acid sequence in mammals a structure suggestive of a transposable element, *Mol. Cell. Biol.* **1:**573–583.
69. Stumph, W. E., Kristo, P., Tsai, M.-J., and O'Mally, W., 1981, A chicken middle-repetitive DNA sequence which shares homology with mammalian ubiquitous repeats, *Nucleic Acids Res.* **9:**5383–5397.
70. Lerner, M. R., and Steitz, J. A., 1981, Snurps and scyrps, *Cell* **25:**298–300.
71. Davidson, E. H., and Bruther, R. J., 1979, Regulation of gene expression: Possible role of repetitive sequences, *Science* **204:**1052–1059.
72. Schon, E. A., Clealy, M. L., Haynes, J. R., and Lingel, J. B., 1981, Structure and evolution of goat γ-, β^c- and β^a-globin genes: Three developmentally regulated genes contain inserted elements, *Cell* **27:**359–369.
73. Singer, D. H., Lifshitz, R., Abelson, L., Nyirjisy, P., and Rudekoff, S., 1983, Specific association of repetitive DNA sequences with major histocompatibility genes, *Mol. Cell. Biol.* **3:**903–913.
74. Lelanne, J. L., Transy, C., Guerain, S., Darche, S., Meulien, P., and Kourilsky, P., 1985, Expression of class I gene in the major histocompatibility complex: Identification of eight distinct mRNAs in DBA/2 mouse liver, *Cell* **41:**469–478.

14
Antigenic Requirements for the Recognition of Moloney Murine Leukemia Virus-Induced Tumors by Cytotoxic T Lymphocytes

DAVID C. FLYER, DOUGLAS V. FALLER, AND
STEVEN J. BURAKOFF

1. Introduction

During viral transformation and/or infection, antigenic changes occur on the surface of transformed and infected cells that initiate a variety of cellular as well as humoral immune responses. A major component of this cellular immune response is the generation of cytotoxic T lymphocytes (CTL) which are capable of specifically recognizing and then lysing transformed or infected cells. Analysis of the specificity of these CTL has shown that immune recognition of virus-encoded antigens expressed as integral membrane proteins on the surface of these cells is in association with the class I major histocompatibility antigens H-2K, D, and L.[1-3] Target cells must not only express the appropriate viral antigen but must also express the same class I major histocompatibility antigen (H-2K, D, and L) as the effector CTL. This phenomenon is referred to as H-2-restricted, CTL recognition. This associative recognition (viral + H-2 antigens) required for CTL recognition contrasts with the recognition of antigen alone by antibody and has made it difficult to precisely define the antigenic determinants recognized by H-2-restricted, virus-specific CTL.

One system in which the antigenic requirements for immune recognition by CTL have been extensively studied is the CTL response in mice to tumors induced by the Moloney murine leukemia virus: Moloney sarcoma virus (M-MuLV:MSV) complex. Inoculation of the M-MuLV:MSV complex into immunocompetent mice results in the rapid development of tumors at the site of inoculation followed by spontaneous tumor regression. This spontaneous tumor regression is the result of the development of a significant cellular immune response of which CTL specific for M-MuLV:MSV-induced tumors are a readily detectable component.[4] These CTL appear before and after tumor regression in the lymphoid organs and the blood of the tumor-bearing mice as well as in the tumor itself.[5,6] Highly reactive populations of M-MuLV:MSV-specific CTL capable of lysing virus-induced tumors and infected cells can be generated when spleen cells from mice which have rejected an M-MuLV:MSV-induced tumor are specifically stimulated *in vitro* in mixed lymphocyte–tumor cell cultures.[7]

Analysis of the specificity of M-MuLV:MSV-induced CTL has shown that immune lysis of tumor cells is dependent upon the recognition of viral antigens expressed on the cell surface in association with class I antigens. The reactivity of these CTL appears to be directed against the antigens encoded by the M-MuLV helper virus as evidenced by the cross-reactivity between M-MuLV:MSV-induced sarcomas and M-MuLV-induced lymphomas as well as by the ability of M-MuLV-induced lymphomas but not MSV nonproducer transformed cells to immunize mice against M-MuLV:MSV-induced tumors in transplantation assays.[8–10]

For the past several years, our laboratories have been interested in defining the role of the various M-MuLV antigens and the role of the class I MHC glycoproteins in the CTL recognition of M-MuLV-induced lymphomas and M-MuLV-infected cells. This review will focus on our efforts and the efforts of others to address this issue.

2. M-MuLV Antigens as CTL Target Recognition Structures

2.1. Cell Surface Expression of Viral Antigens

The identification of virus-encoded antigens which act as CTL target antigens for M-MuLV-specific CTL has been difficult since M-MuLV-induced lymphomas synthesize numerous viral antigens, some of which are expressed to a greater or lesser degree on the cell surface. Two viral proteins present on the surface of MuLV-induced tumors which are candidates for virus-encoded CTL target antigens are the envelope glycoprotein and a glycosylated form of the *gag* precursor.

The envelope glycoprotein, which forms the knoblike spikes on the surface of the virus and which is found on the surface of M-MuLV-induced lymphomas, is a complex comprised of two subunits, gp70 and p15E. These two proteins are produced by posttranscriptional cleavage of a precursor polyprotein, pr80^{env}, with carbohydrate subsequently attached to several sites on gp70 but not p15E. Studies have shown that p15E is an integral membrane protein which traverses the membrane at its carboxy-terminal hydrophobic sequence.[11,12] The gp70 complexes with p15E via disulfide linkages and possibly also by noncovalent association.[11,13,14] Monoclonal antibodies have identified a number of different antigenic epitopes on gp70 and p15E but virus neutralizing and cytotoxic activity are associated only with antibody directed to gp70.[15] gp70 is also responsible for the binding of the virus to specific cell surface receptors necessary for viral infection to take place.[16]

The *gag* gene of M-MuLV encodes a precursor polyprotein, pr65^{gag}, which is processed to yield four individual structural proteins, p15, 30, 10, and 12, which comprise the viral capsid. This processing is dependent upon a virus-encoded protease which has been mapped to the 5' portion of the polymerase gene.[17] In addition to pr65^{gag}, a glycosylated form of the gag precursor, pr85^{gag}, is also synthesized.[18–21] The initiation site of the pr85^{gag} mRNA is located upstream of the normal pr65^{gag} AUG initiation codon with the additional sequences targeting the protein for insertion into the plasma membrane as well as glycosylation.[18,20–23]

The expression of both the envelope glycoprotein and the glycosylated gag precursor, pr85^{gag}, on the surface of M-MuLV-infected cells is easily detected by staining with FITC-conjugated anti-envelope glycoprotein and anti-gag antibodies. Analysis of the level of cell surface expression of the two viral antigens as determined by flow cytometry indicates that the level of envelope glycoprotein expressed on the surface of M-MuLV-infected cells appears to be almost fourfold greater than the level of gag (Flyer, unpublished data).

2.2. CTL Recognition of the Envelope Glycoprotein

Recognition of the viral envelope glycoprotein has been suggested by a number of studies. Purified gp70 has been reported to be able to successfully block the recognition and subsequent lysis of M-MuLV-induced tumor cells by M-MuLV-specific CTL.[24] Such blocking activity could only be attributed to gp70 and not to any of the other viral antigens. As the T-cell receptor recognizes antigen in association with class I antigens, it is unlikely that purified gp70 was able to block CTL-mediated lysis by receptor blockade. A more likely explanation would be that purified gp70 added to the mixture of tumor cells and CTL bound to the surface of the CTL and in effect created competitive "cold target" inhibitors. Studies done with Friend MuLV, a closely related MuLV, has shown that Friend MuLV-induced tumor variants which are gp70-negative are resistant to lysis by Friend MuLV-specific CTL while the gp70-positive parental tumor cells are not.[25] Despite these studies which indicate a role for envelope glycoprotein in CTL recognition, the blocking of CTL recognition with gp70-specific antibodies has never been successful.[26]

The approach that our laboratories have taken to evaluate the role of the M-MuLV envelope glycoprotein in CTL recognition has been to construct cell lines that express the envelope glycoprotein in the absence of the M-MuLV antigens. This was accomplished by transfecting a molecular clone of the M-MuLV *env* gene into murine fibroblasts. These cells synthesize all the *env* gene products (pr80^{env}, gp70, and p15E) that are observed in M-MuLV-infected cells with significant amounts of these proteins expressed at the cell surface.[27,28] The transfected *env* gene product has retained its biological activity as evidenced by its ability to inhibit super infection of the transfected cells by pseudotyped MSV.

In order to determine whether or not the envelope glycoprotein acts as a recognition antigen for M-MuLV-specific CTL, *env*-transfected cells were used as targets in an *in vitro* lymphocyte-mediated cytotoxicity assay. The results indicated that expression of the envelope glycoprotein on the cell surface of the transfected fibroblasts rendered them susceptible to CTL-mediated lysis. In addition, the lysis of transfected cells expressing different levels of the envelope glycoprotein correlated with the amount present on the cell curface as detected by FACS analysis.[27,28] The lysis of *env*-transfected fibroblasts by M-MuLV-specific CTL indicates a role for *env* gene products in CTL recognition. The *env*-transfected firboblasts, like M-MuLV-infected cells, express both gp70 and p15E at the cell surface. While it appears that the expression of gp70 on the cell surface is dependent upon p15E, whether the CTL recognition sites are present on gp70, p15E, or both remains to be determined.

2.3. CTL Recognition of *gag* Gene Products

While early data suggested a role for *env* gene products in CTL recognition of M-MuLV-infected cells and tumor cells, there has been no evidence that *gag* gene products act as CTL target antigens. To further investigate the role of *gag* gene products, murine fibroblasts were transfected with a cloned M-MuLV *gag* gene. Analysis of *gag*-transfected fibroblasts indicated an expression pattern of gag proteins different from that observed in M-MuLV-infected cells. While M-MuLV-infected cells synthesized the gag precursor $pr65^{gag}$, the glycosylated form of the gag precursor $pr85^{gag}$, and the four virion core proteins p15, 30, 12, and 10, only $pr65^{gag}$ and $pr85^{gag}$ were expressed in the *gag*-transfected cells.[28] The failure of $pr65^{gag}$ to be processed into the four virion core proteins can be attributed to the absence of the viral protease responsible for this cleavage in the transfected cells. This protease is encoded in the 5' portion of the *pol* gene which was not included in the expression vector.[17] The failure of the $pr65^{gag}$ to undergo the appropriate cleavage should not alter the expression of gag proteins on the surface of *gag*-transfected cells as the only gag protein known to be expressed on the cell surface of infected cells is the glycosylated $pr85^{gag}$.

Although the cell surface expression of *gag* on the transfected fibroblasts was equal to or greater than that detected on the surface of M-MuLV-infected cells, when *gag*-transfected fibroblasts are used as targets in *in vitro* lymphocyte-mediated cytotoxicity assays, no lysis by M-MuLV-specific CTL is observed. The failure of *gag*-transfected fibroblasts to be lysed would appear to suggest that gag determinants do not function as CTL target recognition structures for M-MuLV CTL. However, the recent characterization of three CTL clones which appear to be specific for cells expressing the glycosylated gag precursor, $pr85^{gag}$, would indicate that this is not always the case.[29] Our inability to detect lysis of *gag*-transfected fibroblasts may instead be due to the failure to generate gag-specific CTL in our secondary *in vitro* stimulation protocol or the fact that gag-reactive CTL are in such low numbers that they cannot be detected in our cytotoxicity assay. While gag proteins are expressed on the cell surface of M-MuLV-infected cells, the level of expression is severalfold lower than the expression of the envelope glycoprotein. As the envelope glycoprotein is the major cell surface viral antigen, it may also serve as the major CTL target antigen. This could result in a lower frequency of gag-reactive CTL and may explain the discrepancy between our results with bulk CTL populations and the clonal results of van der Hoorn et al.[29]

In contrast to the M-MuLV system, gag proteins appear to play a major role in CTL recognition of AKV/Gross MuLV-infected and tumor cells. Plata et al. have reported that certain anti-p30 monoclonal antibodies are capable of blocking CTL-mediated lysis of Gross MuLV tumors by as much as 90%.[30] In addition, murine L-cell fibroblasts expressing the transfected AKV *gag* and H-$2K^b$ genes are susceptible to lysis by H-2^b-restricted, AKV/Gross MuLV-specific CTL.[31] Although the hybridomas producing the anti-p30 monoclonal antibodies generated by Plata et al. were derived from mice immunized with Friend MuLV, the antibodies were unable to block the lysis of Friend MuLV-induced tumor cells by Friend-specific CTL. The differences in the CTL response to the various retroviruses have yet to be clearly defined.

3. Requirement for Class I MHC in CTL Recognition

3.1. MHC-Restricted Recognition of M-MuLV-Induced Tumors

Equally important in CTL recognition are the class I glycoproteins coded for by genes in the murine histocompatibility (H-2) complex. As discussed earlier, these glycoproteins act as restriction elements in that CTL generated against syngeneic cells that have been virally infected or chemically modified only recognize those target cells which express the same class I (H-$2K$, D, and L) gene products as the effector cell. In addition, quantitative variations in the levels of class I antigen expression correlated with the level of lysis of virus-infected[32] and chemically modified cells[33] by H-2-restricted CTL.

In the two most studied strains of mice, C57BL/6 (H-2^b) and BALB/c (H-2^d), CTL recognition of M-MuLV tumors is found to be associated with only one class I antigen.[34,35] These are D^b in C57BL/6 mice and K^d in BALB/c mice. No recognition is observed in association with either D^d or K^b, respectively, in these two mouse strains. The data suggest that the complete immunodominance of some viral antigen–class I associations precludes the recognition of the association of viral antigens with other class I antigens by CTL precursors.

While the secondary *in vitro* CTL responses of C57BL/6 mice against M-MuLV is restricted by the D^b gene product, K^b-associated responses can be generated in recombinant B10.A(5R) mice (H-$2K^b$, D^d.)[36] Mouse strains which possess a mutation in the D^b gene have been shown to exhibit an altered CTL response to M-MuLV. B6.C-H-2^{bm14} mice do not generate CTL against M-MuLV whereas B6.C-H-2^{bm13} mice exhibit an increased K^b component of the CTL response.[37] The data indicate that the D^b region of the MHC may act as an immune response gene by determining the restriction specificity and by regulating the magnitude of the M-MuLV CTL response.

3.2. Regulation of Class I Antigens by M-MuLV

Because of the role class I MHC antigens play in CTL recognition, the regulation of MHC antigens by viruses in virus-infected and transformed cells may significantly alter the host's ability to respond to viral infection and virus-induced neoplasia. This point is clearly illustrated by the recent reports in which cells infected with the oncogenic strain of human adenovirus Ad12 were found to express reduced levels of class I antigens on the cell surface[38] and were poorly recognized by Ad12-specific CTL.[39] While Ad12-transformed mouse cells were highly tumorigenic *in vivo,* introduction of a functional class I antigen into these cells abrogated their tumorigenicity.[40]

In our studies we have observed that following the infection of murine BALB/c-3T3 fibroblasts with M-MuLV, an enhanced level of class I antigen expression can be detected by FACS analysis.[41] Fibroblasts infected with M-MuLV exhibit up to a tenfold increase in cell surface expression of all three class I antigens (K, D, and L). Our data indicate that M-MuLV exerts its control of class I expression at the genomic level in that the level of class I mRNA, as well as the level of β_2-microglobulin mRNA, are increased.[41] The mechanism by which M-MuLV enhances MHC expression has yet to be identified. It is known, however, that this is not an interferon effect as (1) no antiviral interferon activity can be detected in the culture supernatants of infected fibroblasts and (2) there is no difference in the level of interferon-β mRNA present in infected and uninfected fibroblasts (D. V. Faller, unpublished data).

When BALB/c-3T3 fibroblasts are coinfected with M-MuLV and MSV, no increase in the level either of class I antigen expression or of class I mRNA is observed, suggesting that MSV exerts an inhibitory effect on the M-MuLV-induced enhancement of class I antigens. The inhibition appears specific for M-MuLV-enhanced class I expression as MSV infection does not inhibit the enhancement of class I antigens by interferon-γ.[41]

3.3. Effect of Enhanced MHC Expression on CTL Recognition

Experiments using allospecific CTL indicate that the level of lysis on infected and uninfected fibroblasts parallels the level of class I expression seen by FACS analysis. The lysis of M-MuLV-infected fibroblasts by allospecific CTL was found to be approximately fourfold greater than that observed on uninfected fibroblasts while the level of lysis of fibroblasts coinfected with M-MuLV and MSV was identical to the uninfected control.[41] Similar results were found when the susceptibility of infected and uninfected fibroblasts to lysis by class I-restricted M-MuLV-specific CTL was examined in that the level of lysis on M-MuLV-infected fibroblasts was approximately fourfold greater than that observed on fibroblasts coinfected with

M-MuLV and MSV even though the two sets of infected fibroblasts express the same level of viral antigen.

The reduced susceptibility of fibroblasts coinfected with M-MuLV and MSV appears to be the result of insufficient levels of class I expression necessary for optimal CTL recognition. To increase the expression of class I antigens expressed by M-MuLV:MSV-infected fibroblasts, the fibroblasts were treated with 100 units/ml of murine interferon-γ for 48 hr. Treatment of M-MuLV:MSV-infected fibroblasts with interferon-γ increased their susceptibility to lysis by M-MuLV-specific CTL to levels comparable to those observed with M-MuLV-infected fibroblasts.[40] Although treatment of M-MuLV-infected fibroblasts with interferon-γ also results in an increase in their expression of class I antigens, treatment with interferon-γ does not increase the susceptibility of M-MuLV-infected fibroblasts to lysis by M-MuLV-specific CTL, suggesting that once an optimal level of MHC expression is reached, additional increases do not alter CTL recognition.

When fibroblasts coinfected with M-MuLV and MSV were used as targets for class I-restricted, M-MuLV-specific CTL clones, we observed that several CTL clones were unable to lyse these targets even though the cells are lysed, albeit at reduced levels, by bulk CTL populations. The inability of these clones to recognize and lyse these fibroblasts suggests that individual CTL clones may have different quantitative requirements for class I antigens on target cells. The reduced activity of the bulk CTL populations would therefore reflect the presence of some CTL clones that require less class I expression for recognition and lysis but were not isolated in our limited dilution cloning protocol. Interferon-γ treatment of fibroblasts coinfected with M-MuLV and MSV renders them susceptible to lysis by the CTL clones, indicating that the required viral antigens are present and that once the level of class I antigen is made sufficiently high, recognition and lysis will occur.

4. Concluding Remarks

The identification of viral antigens on the surface of M-MuLV-induced tumors that are recognized by CTL has been difficult due to the inability to analyze the contribution of the individual viral antigens independently and to the requirement for the recognition of viral genes in association with class I MHC antigens. Using transfected cells expressing cloned viral genes, we have been able to examine the role of the M-MuLV envelope glycoprotein and *gag* gene product in CTL recognition. The lysis of *env*-transfected cells but not of *gag*-transfected cells by M-MuLV-specific CTL suggests that the envelope glycoprotein serves as the major target antigen for CTL recognition. Although others have isolated *gag*-reactive CTL clones,[28] our CTL populations appear to have only *env*-reactivity, suggesting that *gag*-reactive CTL may be present only in very low numbers. The difference in the frequency of *env*- and *gag*-reactive CTL may be due in part to the relative levels of the two viral antigens on the surface of M-MuLV-infected and transformed cells.

The regulation of class I MHC gene product expression by retroviruses represents a significant immunologic phenomenon given the central role class I antigens play in the CTL recognition process. While we have studied the effect of retrovirus infection of MHC expression in fibroblasts, other groups have reported similar findings in lymphoid cells.[42,43] Increased class I expression on M-MuLV-infected cells enhances CTL recognition and may thereby promote a more rapid elimination of virus-infected cells *in vivo*. The recognition of preleukemic, virus-infected cells may also be enhanced, influencing the susceptibility of the host to retrovirus-induced leukemia. Differential enhancement of class I antigen expression by radiation leukemia virus has been reported for resistant versus susceptible strains of mice,[42] suggesting that virus-enhanced MHC expression does play a role in the host's resistance to neoplasia by retroviruses.

From the studies presented here, it is apparent that it is not only the presence of the appropriate viral and class I antigens that dictates the recognition of target cells by CTL, but the level of expression of the two antigens as well. The use of cloned viral and class I genes will hopefully allow a more detailed analysis of the role of the different antigens in CTL recognition. The use of promoters of varied strengths will enable the construction of target cells expressing varied levels of antigen while the use of site-specific mutagenesis and hybrid genes will enable the alteration of antigens in a desired fashion to explore the fine specificity of T-cell recognition. With this approach it should be possible to develop an experimental system in which one can manipulate the expression of the specific antigens one wishes to study.

References

1. Shearer, G. M., Rehn, T. G., and Schmitt-Verhulst, A. M., 1976, Role of the murine major histocompatibility complex in the specificity of in vitro T-cell mediated lympholysis against chemically modified autologous lymphocytes, *Transplant. Rev.* **29**:222–248.
2. Doherty, P. C., Blanden, R. O., and Zinkernagel, R. M., 1976, Specificity of virus immune effector cells for H-2K and H-2D compatible interactions: Implications for H-2 antigen diversity, *Transplant. Rev.* **29**:89–124.
3. Zinkernagel, R. M., 1978, Thymus and lymphohemopoietic cells: Their role in T cell maturation in selection of T cells, H-2 restriction specificity and in H-2 linked Ir-gene control, *Immunol. Rev.* **42**:224–270.
4. Levy, J. P., and Leclerc, J. C., 1977, The murine sarcoma virus-induced tumor: Exception or general model in tumor immunology? *Adv. Cancer Res.* **24**:1–66.
5. Plata, F., MacDonald, H. R., and Sordat, B., 1975, Studies on the distribution and origin of cytolytic T lymphocytes present in mice bearing Moloney murine sarcoma (MSV)-induced tumors, *Bibl. Haematol.* **43**:274–281.
6. Plata, F., and Sordat, B., 1977, Murine sarcoma virus (MSV)-induced tumors in mice. I. Distribution of MSV-immune cytolytic T lymphocytes in vivo, *Int. J. Cancer* **19**:205–214.
7. Plata, F., Cerottini, J.-C., and Brunner, K. T., 1975, Primary and secondary in vitro generation of cytotoxic T lymphocytes in the murine sarcoma virus system, *Eur. J. Immunol.* **5**:227–233.
8. Fefer, A., McCoy, J. L., and Glynn, J. P., 1967, Antigenicity of virus-induced murine sarcoma (Moloney), *Cancer Res.* **27**:962–971.
9. Stephenson, J. R., and Aronson, S. A., 1972, Oncogenic properties of murine sarcoma virus-transformed BALB/c-3T3 nonproducer cells, *J. Exp. Med.* **135**:503–514.
10. Law, L. W., and Ting, R. C., 1970, Antigenic properties of a nonreleaser neoplasm induced in the mouse by mature sarcoma virus, *J. Natl. Cancer Inst.* **44**:615–623.
11. Pinter, A., and Honnen, W. J., 1983, Topography of murine leukemia virus envelope proteins: Characterization of transmembrane components, *J. Virol.* **46**:1056–1060.

12. Lentz, J., Crowther, R., Straceski, A., and Haseltine, A., 1982, Nucleotide sequence of the AKV env gene, *J. Virol.* **42:**519–529.
13. Pinter, A., and Fleissner, E., 1977, The presence of disulfide-linked gp70–p15(E) complexes in AKV-MuLV, *Virology* **83:**417–422.
14. Pinter, A., Lieman-Hurwitz, J., and Fleissner, E., 1978, The nature of the association between the murine leukemia virus envelope proteins, *Virology* **91:**345–351.
15. Cicurel, L., Lee, J. C., Enjuanes, L. E., and Ihle, J. N., 1980, Monoclonal antibodies to the envelope protein of Moloney leukemia virus: Characterization of recombinant viruses, *Transplant. Proc.* **12:**394–397.
16. DeLarco, J., and Todaro, G. J., 1976, Membrane receptors for murine leukemia viruses: Characterization using purified viral envelope glycoprotein, gp71, *Cell* **8:**365–371.
17. Crawford, S., and Goff, S. P., 1985, A deletion mutation in the 5′ part of the *pol* gene on Moloney murine leukemia virus blocks proteolytic processing of the *gag* gene and *pol* polyproteins, *J. Virol.* **53:**899–907.
18. Edwards, S. A., and Fan, H., 1979, *gag*-related polyproteins of Moloney murine leukemia virus: Evidence for independent synthesis of glycosylated and unglycosylated forms, *J. Virol.* **30:**551–563.
19. Edwards, S. A., and Fan, H., 1980, Sequence relationship of glycosylated and unglycosylated *gag* polyproteins of Moloney murine leukemia virus, *J. Virol.* **35:**41–51.
20. Ledbetter, J. A., Nowinski, R., and Eisenman, R. N., 1978, Biosynthesis and metabolism of viral proteins expressed on the surface of murine leukemia virus infected cells, *Virology* **91:**116–129.
21. Arcement, J. L., Karshin, W. I., Naso, R. B., and Arlinghaus, R. B., 1977, "Gag" polyprotein precursors of Rauscher murine leukemia virus, *Virology* **81:**284–297.
22. Schultz, A. M., Rabin, E. H., and Oroszlan, S., 1979, Post-translational modification of Rauscher leukemia virus precursor polyproteins encoded by the *gag* gene, *J. Virol.* **30:**255–266.
23. Schwartzenberg, P., Colicelli, J., and Goff, S. P., 1983, Deletion mutants of Moloney leukemia virus which lack glycosylated *gag* protein are replication competent, *J. Virol.* **46:**538–546.
24. Enjuanes, L., Lee, J. C., and Ihle, J. N., 1979, Antigenic specificities of the cellular immune response of C56BL/6 mice to the Moloney leukemia sarcoma virus complex, *J. Immunol.* **122:**665–674.
25. Collins, J. K., Britt, W. J., and Chesebro, B., 1980, Cytotoxic T Lymphocyte recognition of gp70 on Friend virus-induced erythroleukemia cell clones, *J. Immunol.* **125:**1318–1324.
26. Gomard, E., Levy, J. P., Plata, F., Henin, Y., Duprez, V., Bismuth, A., and Reme, T., 1978, Studies on the nature of the cell surface antigen reacting with cytolytic T lymphocytes in murine on cornavirus-induced tumor cells, *Eur. J. Immunol.* **8:**228–236.
27. Flyer, D. C., Burakoff, S. J., and Faller, D. V., 1983, Cytotoxic T lymphocyte recognition of transfected cells expressing a cloned retroviral gene, *Nature (Lond.)* **305:**815–818.
28. Flyer, D. C., Burakoff, S. J., and Faller, D. V., 1985, Expression and CTL recognition of cloned subgenomic fragments of Moloney murine leukemia virus in murine cells, *Surv. Immunol. Res.* **4:**168–172.
29. van der Hoorn, F. A., Lahaye, T., Muller, V., Ogle, M. A., and Engers, H. D., 1985, Characterization of gp85^{gag} as an antigen recognized by Moloney leukemia virus-specific cytolytic T cell clones that function in vivo, *J. Exp. Med.* **162:**128–144.
30. Plata, F., Kalil, J., Zilber, M-T., Fellous, M., and Levy, D., 1984, Identification of a viral antigen recognized by H-2 restricted cytolytic T lymphocytes on a murine leukemia virus-induced tumor, *J. Immunol.* **131:**2551–2556.
31. Abastado, J.-P., Plata, F., Morello, D., Daniel-Vedele, F., and Kourilsky, P., 1985, H-2 restricted cytolysis of L cells doubly transformed with a cloned H-2K^b gene and cloned retroviral DNA, *J. Immunol.* **135:**3512–3519.
32. Plata, F., Tilkin, A. F., Levy, J. P., and Lilly., F., 1981, Quantitative variations in the expression of H-2 antigens on murine leukemia virus-induced tumor cells can effect the H-2 restriction patterns of tumor-specific cytolytic T lymphocytes, *J. Exp. Med.* **154:**1795–1810.
33. Kuppers, R. C., Ballas, Z. K., Green, W. R., and Henney, C. S., 1981, Quantitative appraisal of H-2 products in T cell-mediated lysis by allogeneic and syngeneic effector cells, *J. Immunol.* **127:**500–504.
34. Gomard, E., Duprez, V., Hemin, Y., and Levy, J. P., 1976, H-2 region products are determinant

in immune cytolysis of syngeneic tumor cells by anti-MSV T lymphocytes. *Nature (Lond.)* **260**:707–710.
35. Gomard, E., Duprez, V., Reme, T., Colombani, J. M., and Levy, J. P., 1977, Exclusive involvement of H-2D^b or H-2K^d product in the interaction between T-killer lymphocytes and syngeneic H-2^b or H-2^d viral lymphomas, *J. Exp. Med.* **146**:909–922.
36. Gomard, E., Henin, Y., Colombani, M. J., and Levy, J. P., 1980, Immune response genes control T-killer-cell response against Moloney tumor-antigen cytolysis regulating reactions against the best available H-2 + viral antigen association, *J. Exp. Med.* **151**:1468–1476.
37. Stukart, M. J., Vos, A., Boes, J., Melvold, R. W., Bailey, D. W., and Melief, C. J. M., 1982, A crucial role of the H-2D locus in the regulation of both the D- and K-associated cytotoxic T lymphocyte response against Moloney leukemia virus, demonstrated with two D^b mutants, *J. Immunol.* **128**:1360–1364.
38. Schrier, P. I., Benards, R., Vaessen, R. T. M. J., Houweling, A., and van der Eb, A. J., 1983, Expression of class I major histocompatibility antigens switch off by highly oncogenic Adenovirus 12 in transformed rat cells, *Nature* **305**:771–776.
39. Bernards, R., Schrier, P. I., Houweling, A., Bos, J. L., van der Eb, A. J., Zijlstra, M., and Melief, C. J. M., 1983, Tumorigenicity of cells transformed by Adenovirus type 12 by evasion of T-cell immunity, *Nature* **305**:776–779.
40. Tanaka, K., Isselbacher, K. J., Khoury, G., and Jay, G., 1985, Reversal of oncogenesis by the expression of a major histocompatibility complex class I gene, *Science* **228**:26–30.
41. Flyer, D. C., Burakoff, S. J., and Faller, D. V., 1985, Retrovirus-induced changes in major histocompatability complex antigen expression influence susceptibility to lysis by cytotoxic T lymphocytes, *J. Immunol.* **135**:2287–2292.
42. Meruelo, D., Nimelstein, S. H., Jones, P. P., Lieberman, M., and McDevitt, H., 1978, Increased synthesis and expression of H-2 antigens on the thymocytes as a result of radiation leukemia virus infection: A possible mechanism for H-2 linked control of virus-induced neoplasia, *J. Exp. Med.* **147**:470–487.
43. Henley, S. L., Weise, K. S., and Acton, R. T., 1984, Productive murine leukemia virus (MuLV) infection of EL-4 lymphoblastoid cells: Selective elevation of H-2 surface expression and possible association of Thy-1 antigen with viruses, *Adv. Exp. Med. Biol.* **172**:365–377.

15

SV40 Tumor Antigen

Importance of Cell Surface Localization in Transformation and Immunological Control of Neoplasia

SATVIR S. TEVETHIA AND JANET S. BUTEL

1. Introduction

The concept of immunosurveillance is nowhere as readily demonstrated as in the SV40 tumor system. The age-related resistance to tumor induction by SV40, abrogation of that resistance by thymectomy and x irradiation, immunological intervention during the latent period resulting in the abrogation of virus carcinogenesis, high rate of spontaneous regression of primary SV40 tumors in the host, prevention of SV40 tumor transplantation in the preimmunized host, and involvement of thymus derived lymphocytes in the immunologically mediated rejection of SV40 tumor cells are all indicative of the fact that SV40-tranformed cells derived either SV40-induced tumors or, transformed *in vitro*, possess strong antigens at the cell surface. In this article, we will review the nature and viral origin of the transplantation rejection antigen and will discuss evidence that the protein involved in the initiation and maintenance of transformation by SV40 tumor or T antigen also provides a target for the cellular immune response by virtue of its location at the surface of SV40-transformed cells.

2. Multifunctional Nature of SV40 T-Antigen

T-ag is a remarkably multifunctional protein.[35,51] A variety of activities have been defined by *in vitro* and *in vivo* assays and assigned to T-ag. Some of the prop-

---

SATVIR S. TEVETHIA • Department of Microbiology, The Pennsylvania State University College of Medicine, Hershey, Pennsylvania 17033. JANET S. BUTEL • Department of Virology and Epidemiology, Baylor College of Medicine, Houston, Texas 77030.

erties are directly related to viral genome expression and replication. T-ag is required for initiation of viral DNA synthesis in infected cells, and in the process exhibits sequence-specific binding to the origin of replication on the viral genome. T-ag also autoregulates the transcription of early sequences (i.e., its own synthesis) and is involved in the induction of viral late transcription. It exhibits an intrinsic ATPase activity and is associated with an unidentified protein kinase activity. A domain near the carboxy-terminus of T-ag helps determine the host range for productive infection.

Numerous effects also are exerted on the host cell by T-ag. It binds to cellular DNA (but with lower affinity than to viral origin sequences), initiates cellular DNA synthesis, induces cellular enzyme synthesis, activates ribosomal DNA transcription, and expresses the adenovirus helper function. It also forms a tight complex with a cellular protein, p53,[23] thought to be involved in normal cell proliferation. A functional T-ag is required for both initiation and maintenance of cellular transformation.[35] Finally, T-ag induces immunity against SV40 tumor cells.[46] Many of these cellular effects reflect the fact that SV40 infection stimulates the host cell to enter S phase. SV40 forces cell DNA synthesis because a papovavirus does not encode all the enzymes necessary for replication and is dependent upon the host cell for their provision. This strategy for replication undoubtedly explains the oncogenic potential of the virus, i.e., the ability to promote the unregulated growth of cells. Transformation would occur as an accidental by-product of functions that have evolved to assure virus replication *in vivo*.

T-ag is extensively chemically modified, not suprising in light of its functional diversity. It is phosphorylated, *N*-acetylated, glycosylated, ADP-ribosylated, acylated, and adenylated. Extensive modification may be the basis for the aberrant migration of T-ag in polyacrylamide gels. Although T-ag has a calculated molecular weight of 82,000, it migrates with an apparent molecular weight of 90,000–100,000. T-ag is localized in both the nucleus and plasma membrane of SV40-infected and transformed cells. This unusual subcellular distribution for a transforming protein also may contribute to the ability of T-ag to perform different functions in infected and transformed cells.

3. Cell Surface Localization of SV40

The multiple biochemical and biological functions attributed to T-ag suggest that the protein must interact with the host cell in different ways. Some of the recognized T-ag functions are clearly nucleus-based (e.g., viral DNA synthesis), whereas the involvement of T-ag in the SV40 tumor-specific transplantation antigen (TSTA) system suggests a membrane function. T-ag has been recognized to accumulate in the nuclei of infected and transformed cells since its original detection by immunofluorescence.[30,32] The role of surface T-ag in inducing tumor immunity will be considered in detail below.

In view of the functional involvement of nuclear T-ag in TSTA activity, surface membranes were examined for polypeptides sharing antigenic determinants with nuclear T-ag. There is now a large body of biochemical and serological evidence demonstrating the existence of a membrane-associated form of SV40 T-ag. Subcellular fractionation techniques, coupled with immunoprecipitation, revealed the presence of a virus-coded protein in the plasma membranes of SV40-infected and -transformed cells that was indistinguishable from nuclear T-ag.[26,41,42,49] The membrane form of T-ag represented only a minor fraction (2%) of the total cellular T-ag. Rigorous controls established that the surface-associated T-ag fraction was not an artifact due to copurification of nuclear T-ag during cell fractionation procedures.

Several variations of cell surface immunological assays have confirmed the presence of T-ag in the plasma membrane. Sera from SV40 tumor-bearing animals and antisera prepared against purified T-ag react with the surface of infected or transformed cells by indirect immunofluorescence.[4,14,24,42] It is easier to detect surface T-ag by immunofluorescence if the cell surface architecture has been modified by paraformaldehyde fixation or EDTA treatment, prompting suggestions that many of the membrane associated T-ag molecules may be cryptic in adherent cells on plastic.[4] Iodinated protein A binding assays have confirmed the surface reactivity of anti-T sera,[14,15,25,42] and blocking tests involving antibody directed against purified T-ag established with the specificity of the surface reactivity displayed by the anti-T sera. Enzyme-catalyzed cell surface radioiodination can label surface T-ag,[36,40,43,55] under labeling conditions documented to iodinate only surface proteins. Finally a differential immunoprecipitation technique has been used to distinguish metabolically labeled, surface-associated T-ag from nuclear T-ag.[36]

The small amounts of surface T-ag can be difficult to detect. High-titered antisera are necessary, labeling conditions must be carefully optimized in iodination studies, and adequate numbers of cells must be analyzed. The growth state of the cells also is an important factor, as the amount of detectable surface T-ag is markedly reduced when the cells are not actively dividing.[39]

The disposition of T-ag in the plasma membrane is not yet completely deciphered. However, it is clear that the molecules assume a specific arrangement in the membrane with both the amino- and carboxy-termini exposed on the extracellular surface of the cell and with more internal sequences embedded in the membrane. This conclusion is based on the binding patterns of an antiserum against a carboxy-terminal synthetic peptide[5] and of a panel of monoclonal antibodies,[2,37,53] as well as recognition by CTL. A specific orientation argues for a functional role for the membrane-associated molecules.

There are subtle differences between the orientation of surface T-ag in SV40-infected and -transformed cells. In infected cells, amino-terminal epitopes and carboxy-terminal epitopes are equally accessible to antibody binding, whereas in transformed cells, carboxy-terminal epitopes appear to be relatively more accessible.[37] Further, two monoclonal antibodies that reacted well with surface T-ag on transformed cells reacted poorly with the surface of infected cells.[37] Membrane T-ag of transformed cells is complexed with cellular protein p53, whereas surface T-ag of infected cells is not.[37] One interesting possibility is that the association with p53 might cause conformational changes in the surface T-ag in transformed cells, thereby accounting for the subtle changes in antigenic reactivity and perhaps prompting unique functional activities as well.

There appear to be two solubility subclasses of surface T-ag in transformed cells.[21] One subclass, representing about one-third of the total surface T-ag population, is extractable with a nonionic detergent, NP40, whereas extraction of the second subclass requires the use of a zwitterionic detergent, Empigen BB. It was suggested that these solubility characteristics are of the latter subclass of surface T-ag, which is associated with the plasma membrane lamina, a structure that presumably underlies the lipid bilayer of the plasma membrane and connects it to the cytoskeleton. The Empigen-extractable surface T-ag was shown to be modified by fatty acid acylation, in contrast to the NP40-soluble subclass.

The state of association of surface T-ag with the plasma membrane appears to be influenced by cell shape, which is determined, in turn, by culture conditions. The exposure of the NP40-soluble class may vary among different cells, being less exposed on cells grown in suspension.[71] Differences also have been noted in the stability of metabolically labeled surface T-ag in cells cultured in different ways.[71,38]

4. Comparative Analysis of Nuclear and Surface Forms of SV40 T-ag

The forms of T-ag localized in different subcellular compartments are strikingly similar in biochemical characteristics. The apparent molecular weights of nuclear membrane T-ag are the same.[16,26,41] Both forms are phosphorylated[41] and both label with galactose,[17] although Schmidt-Ullrich et al.[40] have reported that only membrane T-ag is labeled with glucosamine. The possibility of differential glycosylation of the two forms awaits further study.

There are multiple phosphorylation sites (> 8) that are clustered in two separate regions of T-ag, one near each end of the polypeptide.[56,57,58] A distinct subset of sites is initially phosphorylated in the cytoplasm, with additional sites subsequently being phosphorylated in the nucleus of infected cells.[59] The rate of phosphate turnover is greater than the rate of turnover of the protein,[60] suggesting that phosphorylation and dephosphorylation may be a mechanism by which T-ag exerts regulator effects. The extent of phosphorylation does influence the ability of T-ag to oligomerize and to bind to DNA.[61,62,63] Detailed phosphorylation patterns for membrane T-ag are not yet available.

Surface T-ag is modified by fatty acid acylation[21]; the nuclear T-ag population is not. Acylation is the only structural feature of T-ag found to be unique to the surface T-ag fraction. However, only a subpopulation of the total surface T-ag is palmitylated, corresponding to those molecules postulated to be associated with the plasma membrane lamina.[21] This modification may occur at the plasma membrane by exchange between T-ag and membrane lipids. It is known that palmitylation can occur as long as 48 hr after some proteins are synthesized,[64] and it has been reported that exogenously added T-ag can undergo palmitylation by association with the exterior cell surface.[65] The modification of T-ag by fatty acid acylation provides firm biochemical evidence for the presence of SV40 T-ag in the cell membranes.

The amount of T-ag and T-ag/p53 complex in the plasma membrane is related to cell growth, with much higher levels detectable when the cells are actively dividing.[39] Interestingly, p53 has been found at the plasma membrane of normal cells, but only during mitosis.[28]

The mechanism by which SV40 T-ag gets sorted and transported to the plasma membrane is unknown,[16] as the molecule does not possess structural features typical of membrane proteins. The early region contains an alternate open reading frame near the 3′ end, which, if used, would result in a product, designated T*-ag, that would be the same as authentic T-ag except for the carboxy-terminal 70 amino acids.[66] The unique carboxy-terminus of T*-ag would be distinctly more hydrophobic and could provide a membrane insertion sequence. Despite the attractiveness of a model in which T*-ag represents mT-ag, recent data rule out that possibility.[16,71,44] Peptide mapping studies showed that the carboxy-terminal, methionine-containing tryptic peptides were identical in nuclear and membrane T-ag and different from that of a T*-ag encoded by a deletion mutant constructed to utilize the alternate reading frame.[16]

5. Involvement of Surface T-ag as a Target for Cellular Immune Reactions

The evidence cited in the above sections for the presence of T-ag at the surface of SV40-transformed and -infected permissive cells was made possible by the prediction from immunological studies which first indirectly showed that T-ag must be present at the surface of transformed cells. Initial studies (reviewed in Ref. 46) demonstrated the existence of SV40-specific antigens at the surface of SV40-transformed hamster cells by the classical transplantation rejection test in syngeneic hamsters. In this assay, the animals were immunized with either SV40 or irradiated SV40-transformed cells and were later challenged with syngeneic virus-free SV40 tumor cells. The immunized animals specifically rejected the transplantation of these tumor cells. That antigen designated TSTA did not cross-react with any of the known oncogenic viruses but was common to both *in vitro* transformed cells as well as cells derived from a tumor induced by SV40. The TSTA was not related to any of the virion proteins as tumor cells were found to be free of these proteins. Later studies using SV40-transformed BALB/c mouse cells indicated that the resistance to tumor transplantation was mediated by lymphoid cells[54] and specifically by T cells.[47]

Immunogenetic studies using recombinant mouse strains have indicated that the cellular immune response against SV40 TSTA is H-2 restricted. In the case of H-2^b mice, the response is restricted to both H-2K^b and H-2D^b MHC class I antigens.[9,22,52] Different mouse strains differ in their ability to respond to the generation of SV40-specific CTL. The CTL which specifically recognize SV40-specific antigen at the cell surface are of Lyt-1^-2^+ phenotype.[7] The involvement of CTL in tumor rejection *in vivo* was demonstrated in C3H mice in which SV40-transformed cells which had lost the expression of K^k antigen grew progressively in mice preimmunized with SV40. In addition, these cells were not lysed *in vitro* by the SV40-specific CTL.[11] In a very interesting study, *in vivo* tumorigenic potential of

SV40 correlated with the ability of different mouse strains to mount a CTL response,[1] indicating that class I-restricted CTL which lyse syngeneic SV40-transformed cells also recognize SV40-specific antigen *in vivo* leading to the rejection of emerging tumors.

Since SV40-transformed cells synthesize only two proteins coded by the early region of the viral genome, the role of these proteins in the induction of tumor rejection response was considered questionable due to the nuclear location of T-ag in transformed cells and it was considered unlikely that a nuclear protein would provide a target at the cell membrane as most cellular immune reactions leading to the lysis of tumor cells take place at the membrane level. The first studies linking the SV40 early region to the TSTA were carried out by Rapp *et al.*[33] who showed that an adenovirus 7–SV40 hybrid virus in which part of the adenovirus genome was replaced by the SV40 early region coding for T-ag was able to immunize hamsters against SV40 tumor challenge. Studies by Lewis and Rowe[27] narrowed the SV40 early region which induced the synthesis of TSTA to SV40 map units between 0.48 and 0.17. Additional evidence[8,45] showed that the TSTA was synthesized during the viral permissive cycle in monkey cells. The conditions for the synthesis of TSTA were similar to the ones under which only T-ag synthesis occurred. Similar evidence was obtained in nonpermissive mouse cells. When infected with SV40, the cells became susceptible to lysis by the CTL specifically generated to SV40 TSTA.[13,31] Temperature-sensitive mutants of SV40 with a lesion in T-ag showed a defect in the expression of TSTA.[72] The evidence cited above clearly indicated that the TSTA antigenicity is either part of T-ag or requires T-ag for its synthesis. Anderson *et al.*[67] were the first to show that SV40 T-ag isolated from the nucleus of SV40-transformed human cells was able to immunize BALB/c mice against the transplantation of syngeneic SV40 tumor cells, thus showing that the SV40 T-ag located mainly in the nucleus was able to induce a rejection response *in vivo*. Later studies using T-ag purified to homogeneity confirmed this finding.[48,68] In addition, the immunizing activity of T-ag could be removed by absorption of T-ag by anti-T antibody. And finally, the denatured T-ag eluted from SDS–polyacrylamide gels also immunized mice against SV40 tumor cell challenge,[7] indicating that the immunogenic sites are denaturation resistant. These immunological studies predicted that T-ag must be present at the cell membranes of transformed cells since it induced cellular immune responses resulting in tumor rejection.

6. Surface T Provides a Target for SV40-Specific CTL

Both indirect as well as some direct evidence have indicated that surface T-ag in association with class I MHC antigens interacts with the T-cell receptor on SV40-specific CTL. It has been demonstrated that mice, upon immunization with allogeneic or xenogeneic SV40-transformed cells, will generate pre-CTL which upon antigenic stimulation *in vitro* differentiated into CTL, suggesting that T-ag is processed *in vivo* by the host antigen-presenting cells.[12] In addition, T-ag purified to homogeneity not only immunized mice against SV40 tumor transplantation but also induced the generation of SV40-specific pre-CTL *in vivo*.[48,68] A correlation between the loss of susceptibility to CTL lysis by SV40-transformed polyoma tumor cells and the simultaneous loss of T-ag from these transformed cells indicated that the continuous synthesis of T-ag was required to provide a target for CTL and that it is the T-ag at the cell surface which interacts with the CTL.[73]

Pan and Knowles[69] provided the most compelling evidence in favor of surface T-ag's recognition by CTL by showing that the lysis of SV40-transformed cells by a cloned line of SV40-specific CTL could be blocked by a monoclonal antibody to SV40 T-ag. However, these authors have not mapped the region of T-ag which binds this monoclonal antibody. This finding is interesting since others have failed to block CTL lysis by anti-T antibodies.[10,49]

7. Localization of CTL-Reactive Sites on Surface T-ag

Initial studies had indicated that the antigens responsible for inducing a tumor rejection response in syngeneic mice were localized to the 28k carboxy-terminal fragment of T-ag synthesized by cells infected with adeno 2–SV40 hybrid virus Ad2.ND1.[18] The same authors also showed that the 42k T-ag sharing the common carboxy-terminal region with the 28k T-ag synthesized by Ad2.ND2 also immunized mice against tumor transplantation.[19] It could not be concluded, however, whether there is only one TSTA site on both the 28-k and 42-k proteins or whether there are additional sites on the 42k T-ag. Tevethia et al.[50] were the first to show that CTL-reactive sites exist on the 33k amino-terminal half of T-ag. This truncated T-ag was expressed in mouse cells transfected with a plasmid containing SV40 DNA of 0.72–0.42 map units and was found to be susceptible to lysis by SV40-specific CTL generated in C3H mice (H-2^k) using syngeneic SV40-transformed cells. Although the cells synthesizing the 33k amino-terminal T-ag fragment provided a target for the CTL, these cells failed to immunize mice against the tumor transplantation. This failure to immunize can probably be explained by the short half-life of the truncated protein of approximately 40 min.[34] Later studies confirmed these findings and also localized the CTL-reactive antigenic sites in the distal half of T-ag.[13] The amino-terminal half of 48k T-ag expressed in stable form in H-2^b transformed cells was shown to be able to immunize mice against SV40 tumor transplantation,[70] suggesting that the antigenic sites in the amino-terminal half of T-ag could independently induce a rejection response as was observed with the 28k carboxy-terminal T-ag.[18]

The first evidence for the existence of more than one distinct antigenic site at the surface of SV40-transformed cells was provided by Campbell et al.[3] who isolated two independent CTL clones designated K11 and K19, both of which recognized SV40 T-ag in association with class I H-2D^b antigen. These two CTL clones have been maintained in tissue culture in vitro without losing specificity for SV40. CTL clone K11 recognized SV40 T-ag only whereas CTL clone K19 showed cross-reactivity between SV40 and a T-ag coded by a related human papovavirus, BK. The BK virus T-ag serologically cross-reacts with SV40 T-ag. The approximate location of these two antigenic sites on T-ag recognized by CTL clones K19 and K11 was determined by the use of a plasmid that contained SV40 early region encoding an amino-terminal fragment of 48k. B6 cells transformed by this plasmid and expressing the 48k T-ag were found to be susceptible to lysis by both of these CTL clones.[70] Thus there are two distinct CTL-reactive antigenic sites which map in the amino-terminal half of T-ag. The results of O'Connell and Gooding[29] localized one CTL-reactive site in the amino-terminal half of the molecule and another in the caboxy-terminal half. These results show that more than one antigenic site, recognized by the CTL, exists on surface T-ag and can be distinguished by the use of CTL clones.

8. Fine Mapping of CTL-Reactive Sites for CTL Clones K19 and K11

Finer mapping studies using mutants carrying deletions of various sizes have indicated that the antigenic site for CTL clone K19 maps to amino acids 220–223 and for K11 to amino acids 189–211 (J. L. Anderson and S. S. Tevethia, unpublished). B6 cell lines expressing T-ag deleted in these amino acids are not susceptible to lysis by these CTL clones. In addition to these two distinct antigenic sites, we have also identified an additional antigenic site in the amino-terminal region of T-ag and which is defined by amino acids 224–228. While the deletion of these amino acids from T-ag eliminates the activity of CTL clones, it does not indicate that these amino acids are directly involved in the recognition of the T-cell receptor. Attempts are being made to identify other antigenic sites in T-ag.

9. Role of Surface T in Immunosurveillance

T-ag is a transforming protein and is largely located in the nucleus of SV40-infected and -transformed cells. The finding that mutants which are defective in the transport of T-ag to the nucleus transform cells efficiently *in vitro* appears to implicate surface T-ag in cell transformation. So far attempts to identify an SV40-transformed cell that is missing surface T-ag have been largely unsuccessful. One consequence of surface T-ag is its association with MHC class I antigen and its recognition by T cells. This recognition mechanism provides a powerful immunosurveillance to the host reacting to a developing tumor. Our recent findings indicate that an SV40 CTL clone (K11) is capable of abrogating the transformation of primary mouse embryo cells by SV40 DNA *in vitro*.[20]

SV40 causes latent infection in monkeys and replicates in monkey kidney cells. In an effort to determine whether a CTL-mediated immune response is operational in the natural host, we have utilized the SV40-infected monkey cells which have been transfected with mouse *H-2K^b* or *H-2D^b* genes as targets for the SV40-specific *H-2^b* CTL clones. Our results (M. Bates and S. S. Tevethia, unpublished) have shown that mouse CTL clones with specificity to T-ag are capable of killing SV40-infected monkey kidney cells if the appropriate recognition elements are present. These findings suggest that a T-cell-mediated immune response may exist in the natural hosts and may modify these infections. Whether a breakdown in T-cell-mediated immune response to surface T-ag results in activation of papovavirus from latency remains to be investigated.

ACKNOWLEDGMENT. This study was supported by research grants CA 22555 to J.S.B. and CA 25000 to S.S.T. from the National Cancer Institute, National Institutes of Health.

References

1. Abramczuk, J., Pan, S., Maul, G., and Knowles, B., 1984, Tumor induction by simian virus 40 in mice is controlled by long-term persistence of the viral genome and the immune response of the host, *J. Virol.* **49:**540–548.
2. Ball, R. K., Siegl, B., Quellhorst, S., Brandner, G., and Braun, D. G., 1984, Monoclonal antibodies against simian virus 40 nuclear large T tumour antigen: Epitope mapping, papova virus cross-reaction and cell surface staining, *EMBO J.* **3:**1485–1491.
3. Campbell, A. E., Foley, L. F., and Tevethia, S. S., 1983, Demonstration of multiple antigenic sites of the SV40 transplantation rejection antigen by using cytotoxic lymphocyte clones, *J. Immunol.* **130:**490–492.
4. Deppert, W., Hanke, K., and Henning, R., 1980, Simian virus 40 T antigen-related cell surface antigen: Serological demonstration on simian virus 40-transformed monolayer cells in situ, *J. Virol.* **35:**505–518.
5. Deppert, W., and Walter, G., 1982, Domains of simian virus 40 large T-antigen exposed on the cell surface, *Virology* **122:**56–70.
6. Flyer, D. C., and Tevethia, S. S., 1982, Biology of simian virus 40 (SV40) transplantation antigen (TrAg). VIII. Retention of SV40 TrAg sites on purified SV40 large T-antigen following denaturation with sodium dodecyl sulfate, *Virology* **117:**267–270.
7. Flyer, D. C., Anderson, R. W., and Tevethia, S. S., 1982, Lyt phenotype of H-2^b CTL effectors and precursors specific for the SV40 transplantation rejection antigen, *J. Immunol.* **129:**2368–2371.
8. Girardi, A. J., and Defendi, V., 1970, Induction of SV40 transplantation antigen (TrAg) during the lytic cycle, *Virology* **42:**688–698.
9. Gooding, L. R., 1979a, Specificities of killing by T lymphocytes generated against SV40 transformants: Studies employing recombinants within the H-2 complex, *J. Immunol.* **122:**1002–1008.
10. Gooding, L. R., 1979b, Antibody blockade of lysis by T lymphocyte effectors generated against syngeneic SV40 transformed cells, *J. Immunol.* **122:**2328–2336.
11. Gooding, L. R., 1982, Characterization of a progressive tumor from C3H fibroblasts transformed in vitro with SV40 virus: Immunoresistance in vivo correlates with phenotypic loss of H-2K^k, *J. Immunol.* **129:**1306–1312.
12. Gooding, L. R., and Edwards, C. B., 1980, H-2 antigen requirements in the in vitro induction of SV40-specific cytotoxic T lymphocytes, *J. Immunol.* **124:**1258–1262.
13. Gooding, L. R., and O'Connell, K. A., 1983, Recognition by cytotoxic T lymphocytes of cells expressing fragments of the SV40 tumor antigen, *J. Immunol.* **131:**2580–2586.
14. Henning, R., Lange-Mutschler, J., and Deppert, W., 1981, SV40-transformed cells express SV40 T antigen-related antigens on the cell surface, *Virology* **108:**325–337.
15. Ismail, A., Baumann, E. A., and Hand, R., 1981, Cell surface T antigen in cells infected with simian virus 40 or an adenovirus–simian virus 40 hybrid, Ad2+D2, *J. Virol.* **40:**615–619.
16. Jarvis, D. L., Cole, C. N., and Butel, J. S., 1986, Absence of a structural basis for the intracellular recognition and differential localization of the nuclear and plasma membrane-associated forms of SV40 large tumor antigen, *Mol. Cell. Biol.* **6:**758–767.
17. Jarvis, D. L., and Butel, J. S., 1985, Modification of simian virus 40 large tumor antigen by glycosylation, *Virology* **141:**173–189.
18. Jay, G., Jay, F. T., Chang, C., Friedman, R. M., and Levine, A. S., 1978, Tumor specific transplantation antigen: Use of the Ad2+ND1, hybrid virus to identify the protein responsible for simian virus 40 tumor rejection and its genetic origin, *Proc. Natl. Acad. Sci. USA* **75:**3055–3059.

19. Jay, G., Jay, F. T., Chang, C., Levine, A. S., and Friedman, R. M., 1979, Induction and simian virus 40-specific tumor rejection by the Ad2+ND2 hybrid virus, *J. Gen. Virol.* **44**:287–296.
20. Karjalainen, H. E., Tevethia, M. J., and Tevethia, S. S., 1985, Abrogation of simian virus 40 DNA-mediated transformation of primary C57Bl/6 mouse embryo fibroblasts by exposure to a simian virus 40-specific cytotoxic T-lymphocyte clone, *J. Virol.* **56**:373–377.
21. Klockmann, U., and Deppert, W., 1983, Acylated simian virus 40 large T antigen: A new subclass associated with a detergent-resistant lamina of the plasma membrane, *EMBO J.* **2**:1151–1157.
22. Knowles, B. B., Koncar, M., Pfizenmaier, K., Solter, D., Aden, D. P., and Trinchieri, G., 1979, Genetic control of the cytotoxic T cell response to SV40 tumor-associated specific antigen, *J. Immunol.* **122**:1798–1806.
23. Lane, D. P., and Crawford, L. V., 1979, T antigen is bound to a host protein in SV40-transformed cells, *Nature* **278**:261–263.
24. Lanford, R. E., and Butel, J. S., 1979, Antigenic relationship of SV40 early proteins to purified large T polypeptide, *Virology* **97**:295–306.
25. Lange-Mutschler, J., Deppert, W., Hanke, K., and Henning, R., 1981, Detection of simian virus 40 T-antigen-related antigens by a ^{125}I-protein A-binding assay and by immunofluorescence microscopy on the surface of SV40-transformed monolayer cells, *J. Gen. Virol.* **52**:301–312.
26. Luborsky, S. W., and Chandrasekaran, K., 1980, Subcellular distribution of simian virus 40 T antigen species in various cell lines: The 56K protein, *Int. J. Cancer* **25**:517–527.
27. Lewis, A. M., Jr., and Rowe, W. P., 1973, Studies of nondefective adenovirus 2–simian virus 40 hybrid viruses. VIII. Association of simian virus 40 transplantation antigen with a specific region of the early viral genome, *J. Virol.* **112**:836–840.
28. Milner, J., and Cook, A., 1986, Visualization by immunocytochemistry of p53 at the plasma membrane of both non-transformed and SV40-transformed cells, *Virology* **150**:265–269.
29. O'Connell, K. A., and Gooding, L. R., 1984, Cloned cytotoxic T cells recognize cells expressing discrete fragments of SV40 tumor antigen, *J. Immunol.* **132**:953–958.
30. Pope, J. H., and Rowe, W. P., 1964, Detection of a specific antigen in SV40-transformed cells by immunofluorescence, *J. Exp. Med.* **120**:121–128.
31. Pretell, J., Greenfield, R. S., and Tevethia, S. S., 1979, Biology of simian virus 40 (SV40) transplantation rejection antigen (TrAg). V. In vitro demonstration of SV40 TrAg in SV40 infected nonpermissive mouse cells by the lymphocyte mediated cytotoxicity assay, *Virology* **97**:32–41.
32. Rapp, F., Butel, J. S., and Melnick, J. L., 1964, Virus-induced intranuclear antigen in cells transformed by papovavirus SV40, *Proc. Soc. Exp. Biol. Med.* **116**:1131–1135.
33. Rapp, F., Tevethia, S. S., and Melnick, J. L., 1966, Papovavirus SV40 transplantation immunity conferred by an adenovirus–SV40 hybrid, *J. Natl. Cancer Inst.* **36**:707–708.
34. Reddy, V. B., Tevethia, S. S., Tevethia, M. J., and Weissman, S. M., 1982, Nonselective expression of simian virus 40 large tumor antigen fragments in mouse cells, *Proc. Natl. Acad. Sci. USA* **79**:2064–2067.
35. Rigby, P. W. J., and Lane, D. P., 1983, Structure and function of simian virus 40 large T antigen, in: *Advances in Viral Oncology.* Vol. 3 (G. Klein, ed.), Raven Press, New York, pp. 31–57.
36. Santos, M., and Butel, J. S., 1982, Association of SV40 large tumor antigen and cellular proteins on the surface of SV40-transformed mouse cells, *Virology* **120**:1–17.
37. Santos, M., and Butel, J. S., 1984a, Dynamic nature of the association of large tumor antigen and p53 cellular protein with the surfaces of simian virus 40-transformed cells, *J. Virol.* **49**:50–56.
38. Santos, M., and Butel, J. S., 1984b, Antigenic structure of simian virus 40 large tumor antigen and association with cellular protein p53 on the surfaces of simian virus 40-infected and -transformed cells, *J. Virol.* **51**:376–383.
39. Santos, M., and Butel, J. S., 1985, Surface T-antigen expression in simian virus 40-transformed mouse cells: Correlation with cell growth rate, *Mol. Cell. Biol.* **5**:1051–1057.
40. Schmidt-Ullrich, R., Thompson, W. S., Kahn, S. J., Monroe, M. T., and Wallch, D. F. H., 1982, Simian virus 40 (SV40)-specific isoelectric point-4.7–94,000-M_r, membrane glycoprotein: Major peptide homology exhibited with the nuclear and membrane-associated 94,000-M_r SV40 T-antigen in hamsters, *J. Natl. Cancer Inst.* **69**:839–849.
41. Soule, H. R., and Butel, J. S., 1979, Subcellular localization of simian virus 40 large tumor antigen, *J. Virol.* **30**:523–532.

42. Soule, H. R., Lanford, R. E., and Butel, J. S., 1980, Antigenic and immunogenic characteristics of nuclear and membrane-associated simian virus 40 tumor antigen, *J. Virol.* **33**:887–901.
43. Soule, H. R., Lanford, R. E., and Butel, J. S., 1982, Detection of simian virus 40 surface-associated large tumor antigen by enzyme-catalyzed radioiodination, *Int. J. Cancer* **29**:337–344.
44. Tevethia, M. J., Anderson, R. W., Tevethia, S. S., Simmons, D., Feunteun, J., and Cole, C., 1986, Influence of amino acids encoded in the 3′ open reading frame of the SV40 early region on transformation and antigenicity of large T antigen, *Virology* **150**:361–372.
45. Tevethia, M. J., and Tevethia, S. S., 1976, Biology of SV40 transplantation antigen (TrAg). I. Demonstration of SV40 TrAg on glutaraldehyde-fixed SV40-infected African green monkey kidney cells, *Virology* **69**:474–489.
46. Tevethia, S. S., 1980, Immunology of simian virus 40, in: *Viral Oncology* (G. Klein, ed.), Raven Press, New York, pp. 581–601.
47. Tevethia, S. S., Blasecki, J. W., Waneck, G., and Goldstein, A., 1974, Requirement of thymus-derived ∅ positive lymphocytes for rejection of DNA virus (SV40) tumors in mice, *J. Immunol.* **113**:1417–1423.
48. Tevethia, S. S., Flyer, D. C., and Tjian, R., 1980a, Biology of simian virus 40 (SV40) transplantation antigen (TrAg). VI. Mechanism of induction of SV40 transplantation immunity in mice by purified SV40 T antigen (D2 protein), *Virology* **107**:13–23.
49. Tevethia, S. S., Greenfield, R. S., Flyer, D. C., and Tevethia, M. J., 1980b, Simian virus 40 (SV40) transplantation rejection antigen: Relationship to SV40 specific proteins, *Cold Spring Harbor Symp. Quant. Biol.* **44**:235–242.
50. Tevethia, S. S., Tevethia, M. J., Lewis, A. J., Reddy, V. B., and Weissman, S. M., 1983., Biology of simian virus 40 (SV40) transplantation antigen (TrAg). IX. Analysis of TrAg in mouse cells synthesizing truncated large T antigen, *Virology* **128**:319–330.
51. Tooze, J., (ed.), 1980, *Molecular Biology of Tumor Viruses, Part 2*, pp. 61–338, Cold Spring Harbor Laboratory, Cold Spring Harbor, N.Y.
52. Trinchieri, G., Aden, D. P., and Knowles, B. B., 1976, Cell mediated cytotoxicity to SV40 specific tumor associated antigens, *Nature* **261**:312–314.
53. Whittaker, L., Fuks, A., and Hand, R., 1985, Plasma membrane orientation of simian virus 40 T antigen in three transformed cell lines mapped with monoclonal antibodies, *J. Virol.* **53**:366–373.
54. Zarling, J. M., and Tevethia, S. S., 1973, Transplantation immunity to SV40-transformed cells in tumor-bearing mice. I. Development of cellular immunity to simian virus 40 tumor-specific transplantation antigens during tumorigenesis by transplanted cells, *J. Natl. Cancer Inst.* **50**:137–147.
55. Chandrasekaran, K., Winterbourne, D. J., Luborsky, S. W., and Mora, P. T., 1981, Surface proteins of simian virus 40 transformed cells, *Int. J. Cancer* **27**:397–407.
56. Scheidtmann, K.-H., Eschle, B., and Walter, G., 1982, Simian virus 40 large T antigen is phosphorylated at multiple sites clustered in two separate regions, *J. Virol.* **44**:116–133.
57. Schwyzer, M., Weil, R., Frank, G., and Zuber, H., 1980, Amino acid sequence analysis of fragments generated by partial proteolysis from large simian virus 40 tumor antigen, *J. Biol. Chem.* **255**:5627–5634.
58. Van Roy, F., Fransen, L. and Fiers, W., 1983, Improved localization of phosphorylation sites in simian virus 40 large T antigen, *J. Virol.* **45**:315–331.
59. Scheidtmann, K.-H., Hardung, M., Eschle, B., and Walter, G., 1984b, DNA-binding activity of simian virus 40 large T antigen correlates with a distinct phosphorylation state, *J. Virol.* **50**:1–12.
60. Edwards, C. A. F., Khoury, G., and Martin, R. G., 1979, Phosphorylation of T-antigen and control of T-antigen expression in cells transformed by wild-type and tsA mutants of simian virus 40, *J. Virol.* **29**:753–762.
61. Fanning, E., Nowak, B., and Burger, C., 1981, Detection and characterization of multiple forms of simian virus 40 large T-antigen, *J. Virol.* **37**:92–102.
62. Shaw, S. B., and Tegtmeyer, P., 1981, Binding of phosphorylated A protein to SV40 DNA, *Virology* **115**:88–96.
63. Scheidtmann, K.-H., Schickddanz, J., Walter, G., Lanford, R. E., and Butel, J. S., 1984a, Differential phosphorylation of cytoplasmic and nuclear variants of simian virus 40 large T antigen encoded by simian virus 40-adenovirus 7 hybrid viruses, *J. Virol.* **50**:636–640.

64. Omary, M. B., and Trowbridge, I. S., 1981, Biosynthesis of human transferrin receptor in cultured cells, *J. Biol. Chem.* **256:**12888–12892.
65. Lange-Mutschler, J., and Henning, R., 1984, Cell surface binding simian virus 40 large T antigen becomes anchored and stably linked to lipids of the target cells, *Virology* **136:**404–413.
66. Mark, D. F., and Berg, P., 1980, A third splice site in SV40 early mRNA, *Cold Spring Harbor Symp. Quant. Biol.* **44:**55–62.
67. Anderson, J. L., Martin, R. G., Chang, C., Mora, P. T., and Livingston, D. M., 1977, Nuclear preparations of SV40 transformed cells contain tumor-specific transplantation antigen activity, *Virology* **76:**420–425.
68. Chang, C., Chang, R., Mora, P. T., and Hu, C. -P., 1982, Generation of cytotoxic lymphocytes by SV40-induced antigens, *J. Immunol.* **128:**2160–2163.
69. Pan, S., and Knowles, B. B., 1983, Monoclonal antibody to SV40 T antigen block lysis of cloned cytotoxic cell line specific for SV40 TASA, *Virology* **125:**1–71.
70. Tevethia, S. S., Lewis, A. J., Campbell, A. E., Tevethia, M. J., and Rigby, P. W. J., 1984, Simian virus 40 specific cytotoxic lymphocyte clones localize two distinct TSTA sites on cells synthesizing a 48kD SV40 T antigen, *Virology* **133:**443–447.
71. Klockman, U., and Deppert, W., 1985, Evidence for transmembrane orientation of acylated simian virus 40 large T antigen, *J. Virol.* **56:**541–548.
72. Tevethia, M., and Tevethia, S. S., 1977, Biology of SV40 transplantation antigen (TrAg). III. Involvement of SV40 gene A in the expression of TrAg in permissive cells, *Virology* **81:**212–223.
73. Flyer, D. C., Pretell, J., Campbell A. E., Liao, W. S., Tevethia, M. J., Taylor, J. M. and Tevethia S. S., 1983, Analysis of TrAg in mouse cells synthesizing truncated SV40 large T antigen. X. Tumorigenic potential of mouse cells transferred by SV40 in high responder C57BL/G mice and correlation with the persistence of SV40 TrAg, early proteins and viral sequences, *Virology* **131:**207–220.
74. Chang, C., Martin, R. G., Livingston, D. M., Luborsky, S. W., Hu, C. P., and Mora, P. T., 1979, Relationship between T-antigen and tumor specific transplantation antigen in simian virus 40-transformed cells, *J. Virol.* **29:**69–75.

16

Correlation of Natural Killer Cell Recognition with *ras* Oncogene Expression

PAUL W. JOHNSON, WILLIAM S. TRIMBLE,
NOBUMICHI HOZUMI, ANTHONY J. PAWSON, AND
JOHN C. RODER

1. Introduction

Recognition and immune surveillance of spontaneous tumors has been attributed to widely diverse populations of lymphocytes, which include macrophages and natural killer (NK) cells.[1] NK or NK-like cells have been proposed as a first line of defense against *in vivo* tumor growth because they possess attributes that are distinct from cytotoxic T lymphocytes (CTL) and are ideally suited to immediate recognition and killing of transformed targets. They exhibit neither major histocompatibility complex (MHC) restriction nor immunologic memory and, hence, appear more primitive than their CTL couunterparts. However, *in vitro* studies have demonstrated that NK cells kill a wide variety of neoplastic cells without requiring prior sensitization and can differentiate between tumor and nontumor targets.[2] In addition, *in vivo* studies have also indicated that NK surveillance may play an important role in regulation of spontaneous tumors.[3]

The advent of gene transfer has facilitated experiments designed to investigate changes associated with expression of specific cloned genes. This technology has played a major role in the study of oncogenes since transfection of suitable recip-

---

PAUL W. JOHNSON, WILLIAM S. TRIMBLE, NOBUMICHI HOZUMI, AND JOHN C. RODER • Division of Molecular Immunology and Neurobiology, Mount Sinai Hospital Research Institute, Toronto, Ontario M5G 1X5, Canada.   ANTHONY J. PAWSON • Division of Molecular and Developmental Biology, Mount Sinai Hospital Research Institute, Toronto, Ontario M5G 1X5, Canada.

ient cells has allowed the identification and characterization of transforming DNA sequences. In addition, these techniques have also established cell lines that permit the study of early neoplastic changes related to oncogene expression.

Since many experiments correlating NK activity with resistance to tumor growth have used long-passaged transplantable tumor cell lines, data from these investigations may not be directly applicable to the concept of immune surveillance. Host surveillance mechanisms must recognize and destroy neoplastic cells in early stages of transformation. The experiments described below were designed to determine whether NK cells can recognize and kill targets that have been newly transformed by specific oncogenes. This attribute would play an important role in control of spontaneous tumors.

2. Results

The data have been grouped into three major categories. The first section deals with studies on rat cell lines transfected with either the v-Ki-*ras* or v-*fps* oncogene. The second section describes experiments conducted with C3H 10T1/2 mouse fibroblasts transfected with *EJ*, the mutated c-Ha-*ras*. This oncogene either was under the control of its own promoter or was part of a metallothionein–*ras* construct which allowed inducible *ras* expression. The third section presents evidence which indicates that effects observed with the metallothionein–*ras* construct are confined to NK cells and are not part of other killing pathways.

2.1. Enhanced NK Killing of Fibroblasts Transfected with v-Ki-*ras*

Rat-1, a cell line derived from Fisher rat F2408 fibroblasts,[4] was transfected as described[5] with plasmid DNA containing the entire 7.0-kb v-Ki-*ras* oncogene ligated into the *Eco*RI site of pBR322.[6] Transformants were selected by anchorage-independent growth in 1% methyl cellulose[7] and formed colonies at 3–6 weeks. Transfected cells were injected into NIH-II *nu/nu* mice (2×10^6 cells per mouse, subcutaneously) and tumors were explanted after 3 weeks and reestablished as cell lines. Similar injections of untransfected rat-1 cells resulted in a benign, barely palpable nodule which was also established as a cell line and used as a control in most experiments. Transfected, untransfected, and nude mouse-passaged cells were then labeled with ^{51}Cr and analyzed in 6-hr chromium release assays with a variety of effectors.

Cytotoxicity assays performed on transformed and untransformed lines demonstrated that transfection of the v-Ki-*ras* oncogene enhanced NK-mediated killing. As shown in Table I, the transformed cells were much more sensitive ($p < 0.01$) than the rat-1 parent to lysis by nylon-wool-passed spleen cells from poly(I·C)-boosted CBA/J mice. This difference was > 20-fold in terms of lytic units calculated at 10% lysis. Since the parent cells are essentially resistant to NK lysis, they required > 10^6 lymphocytes to even approach this level of killing. The difference was even more striking when the LH49 NK clone was used as an effector.

TABLE I
NK Sensitivity in a Panel of Transfected Targets and Nontransfected Controls[a]

Target cells	Number of experiments	Cytolysis	
		LU/10^6	Percent lysis (100:1 E/T)
YAC 1.2	7	332 ± 122	42 ± 2
Rat-1	5	< 0.5 ± 0.2	3 ± 2
Nodule	6	1 ± 0.9	7 ± 2
K-ras-A	7	26 ± 12	23 ± 10
K-ras-A tumor	7	58 ± 20	32 ± 6
K-ras-B	2	6 ± 1	10 ± 4
K-ras-B tumor	1	10	17
K-ras-C	2	8 ± 2	18 ± 2
K-ras-C tumor	4	11 ± 3	21 ± 5

[a]Fresh spleen cells from CBA/J mice boosted 24 hr previously with 100 µg of poly(I·C) were titrated in serial dilutions at effector/target (E/T) ratios between 200/1 and 3/1. Lytic units (LU) were calculated from titration curves; 1 LU is defined as the number of lymphocytes required to lyse 10% of ^{51}Cr-labeled targets in a 6-hr assay; the values shown represent LU per 10^6 effector lymphocytes. Percent lysis is indicated at the 100:1 effector/target ratio. Data represent the mean and standard error of replicate experiments.

Transformed cells passaged through nude mice were also highly NK-sensitive. Data from seven replicate experiments demonstrate that the newly transformed and tumor-derived v-Ki-ras-A cells are 26- to 58-fold more NK-sensitive than rat-1 and 6- to 12-fold less sensitive than the standard NK-sensitive tumor, YAC 1.2. Both the parental rat-1 cells and a line established from the nodule at the site of injection of untransfected rat-1 cells were NK-resistant (< 1 lytic unit per 10^6). Two additional, independently derived transfectants (Ki-ras-B and Ki-ras-C) were also converted to NK sensitivity after transfection with v-KI-ras.

The characteristics of effector cells that caused cytolysis of transfected targets are summarized in Table II. Although normal, unfractionated CBA/J mouse spleen cells killed the Ki-ras tumor, boosting of the mice with poly(I·C) resulted in a fourfold enchancement of the effect. The poly(I·C) augmentation was less apparent when using nylon-wool-passed spleen cells depleted of Lyt-1$^+$ and Lyt-2$^+$ effectors. In addition to being Lyt-1$^-$, 2$^-$ phenotype, the antitumor effector cells were partially depleted by treatment with anti-asialo GM$_1$ and totally eliminated by treatment with anti-NK-1.2 or anti-NK-2.1. Experiments with cloned NK cell lines of varying cytolytic potential, kindly provided by C. Brooks,[(8)] demonstrated that the clones lysed YAC 1.2 and the transformed tumor line in parallel, but had negligible effects on the rat-1 parent. The phenotype of the effector cell that mediates cytolysis of rat-1 transformants is Lyt-1$^-$, 2$^-$, NK-1.2$^+$, NK-2.1$^+$, asialo GM$_1$$^+$, nylon wool nonadherent, and poly(I·C) boostable. Therefore, the effect is most likely mediated by NK cells.

The nature of the presumptive target structures on the transformants was investigated using cold-target competition studies to ascertain if determinants were different from those expressed on YAC 1.2. As shown in Table III, unlabeled Ki-ras-A tumor cells competed with NK cells for lysis of ^{51}Cr-labeled YAC. The degree

TABLE II
Characterization of the NK Effector Cell[a]

Strain	Assay time (hr)	Poly(I·C)[c]	Treatment	Cytolysis[b] YAC Percent lysis	YAC LU/10^6	Ki-ras-A tumor Percent lysis	Ki-ras-A tumor LU/10^6
CBA/J	6	−	None	33	ND	11	8
	6	+	None	42	ND	26	30
	10	−	NWP, anti-Lyt-1.1,2.1[d]	80	160	55	45
	10	+	NWP, anti-Lyt-1.1,2.1	82	250	52	50
	10	−	NWP, anti-Lyt-1.1,2.1, asialo GM$_1$[e]	63	125	28	25
BALB/c	10	+−	NWP, anti-Lyt-1.1,2.1	74	100	35	41
	10	+−	NWP, anti-Lyt-1.1,2.1, NK-1.2[e]	44	34	1	<1
	10	+	NWP, anti-Lyt-1.1,2.1, NK-2.1	47	31	2	<1
Wistar–Firth rat	4	+	None	17	25	33	115
C57BL/6	6	−	NK clone LH49[f]	21	500	41	1000
	6	−	NK clone L250-A9	7	62	9	110

[a] The experiments summarized in this table were repeated three times with similar results.
[b] Percent cytolysis of ^{51}Cr-labeled target cells is indicated at a 100:1 effector/target ratio. A lytic unit (LU) is the number of lymphocytes required to lyse 10% of targets. The values shown represent LU per 10^6 effector lymphocytes.
[c] Mice were injected intraperitoneally with 100 μg (1 unit) of poly(I·C) 24 hr before assay.
[d] Spleen cells were pooled from 4 to 18 animals per group and passed over nylon wool columns. Spleen cells (10^7/ml) were treated for 1 hr at 4°C with a 1:20 dilution of monoclonal anti-Lyt-1.1 mixed with 1:20 anit-Lyt-2.1 (Cedarlane Laboratories), followed by washing and a 1-hr incubation at 37°C with 1:10 rabbit complement preabsorbed with mouse tissues. Effectors were passaged through nylon wool columns prior to depletion (NWP).
[e] After anti-Lyt-1.1,2.1 and complement treatment, spleen cells were treated with 1:100 rabbit anit-asialo GM$_1$ sera or 1:80 mouse anti-NK-1.2 or 2.1 sera followed by 1:10 rabbit complement.
[f] Percent lysis values for the NK clones LH49 and L250-A9 are at 4:1 effector/target ratios.

TABLE III
Competitive Inhibition of YAC and Ki-ras Transfectants[a]

Labeled target	Unlabeled competitor	Competitor (× 10^3) required for 50% inhibition of lysis
YAC 1.2	YAC 1.2	8
	K-ras-A tumor	4
	Rat-1	>64
	Nodule	>64
K-ras-A tumor	YAC 1.2	4
	K-ras-A tumor	4
	Rat-1	>64
	Nodule	>64

[a] YAC 1.2 and nude-mouse-passaged, v-Ki-ras-A-transfected, rat-1 tumor cells were labeled with ^{51}Cr and tested in a 6-hr cytotoxicity assay against nylon-wool-passed spleen cells from six CBA/J mice injected 24 hr previously with 100 μg (1 unit) of poly(I·C). Varying numbers of unlabeled target cells were added to a constant number of ^{51}Cr-labeled targets (2×10^3/well) at an effector/target ratio of 100:1. The mean percent lysis values from triplicate wells were plotted and the number of competitors causing 50% inhibition of cytolysis was calculated by interpolation. This experiment was repeated five times with similar results.

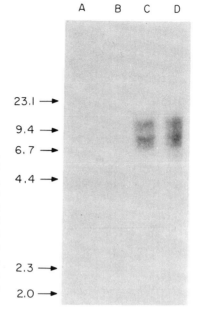

FIGURE 1. Southern blot of v-Ki-*ras*-A transfectants. DNA was extracted from transfected and untransfected cell lines and equal amounts were digested with *Eco*RI, electrophoresed on an agarose gel, and transferred to Gene Screen Plus by Southern blotting. Blots were probed with a 0.6-kb fragment of the transfected gene and autoradiographed. Tracks correspond to equal amounts of DNA from (A) rat-1; (B) benign nodule at site of injection of rat-1 into nude mice; (C) v-Ki-*ras*-A transfectants; (D) tumor from nude mice injected with v-Ki-*ras*-A transfectants. Size markers in kilobase pairs are shown on the left.

of inhibition was comparable to that observed between labeled and unlabeled YAC. Similar results were obtained in the reciprocal experiment in which Ki-*ras*-A tumor cells were labeled and killing was reduced by competition with either unlabeled Ki-*ras*-A tumor cells or unlabeled YAC 1.2. Nontransformed cells did not compete in either system.

Transfection was verified by Southern blot analysis[9] of genomic DNA hybridized to a 0.6-kb segment of the original v-Ki-*ras* transfected gene (see Fig. 1). This fragment, which corresponds to the viral long terminal repeats (LTR), hybridized strongly to DNA from the transfected cells but not control lines when blots were washed under high stringency and exposed overnight.

Experiments were also conducted with the v-*fps* oncogene which was transfected into rat-2 cells, a derivative of rat-1 lacking thymidine kinase activity.[10] These transfectants are similar to the v-Ki-*ras* transfectants in the following ways:

1. Both were transfected by calcium phosphate precipitation.
2. Both transfected oncogenes are controlled by retroviral LTR sequences.
3. Both were selected by growth in semisolid medium.
4. Both produce protein products which are associated with cell membrane structures.
5. Both are tumorigenic.

Cytotoxicity assays were performed with wtFSV (cells transfected with the v-*fps* oncogene from wild-type Fujinami sarcoma virus) as well as two subclones from this line which were reselected by growth in semisolid medium. In addition, assays

TABLE IV
NK Sensitivity of Rat-2 Cells Transfected with the v-fps Oncogene[a]

Target cells	Number of experiments	Cytolysis	
		LU/10^6	Percent lysis (100:1 E/T)
YAC 1.2	3	66 ± 3	33 ± 5
Rat-1	3	1 ± 0.5	3 ± 1
Rat-2	3	< 0.5 ± 0.1	1.5 ± 0.8
FSV-S(1073)	3	< 0.5 ± 0.1	<1 ± 0.5
wtFSV	3	< 0.5 ± 0.3	2 ± 2
Subclone 1	2	2.2 ± 1.5	6.4 ± 3
Subclone 2	2	< 0.5 ± 0.1	3 ± 2

[a]Spleen cells from CBA/J mice injected with 100 µg of poly(I·C) at 24 hr prior to assay were employed as effectors in 6-hr ^{51}Cr release assays. Data represent the mean and standard error of replicate experiments. Lytic units (LU) per 10^6 effector lymphocytes are calculated at the 10% lysis level.

were also performed on FSV-S(1073), a mutant constructed by oligonucleotide-directed mutagenesis in which tyrosine at position 1073 was substituted by serine.[11] This substitution reduces kinase activity, transforming ability, and tumorigenicity of the *fps* oncogene.

As indicated in Table IV, none of the *fps*-transfected lines exhibited enhanced sensitivity to lysis by poly(I·C)-stimulated effector spleen cells. The positive control, YAC 1.2, was lysed at high levels in all assays and demonstrated that these spleen cells were capable of killing a normally sensitive target. Comparison of Tables I and IV indicates that the enhanced NK sensitivity exhibited by cell lines transfected with v-Ki-*ras* could not be detected in comparable lines transfected with v-*fps*.

2.2. Increases in NK Sensitivity Mediated by Inducible c-Ha-*ras* Expression

Since the Harvey *ras* oncogene synthesizes protein products which are similar to those of the Kirsten *ras*, further experiments were conducted with fibroblasts transfected by Ha-*ras*, constructs. These vectors contained *EJ*, a highly transforming *ras* gene which is identical to c-Ha-*ras* except for one point mutation at codon 12.[12] Constitutive expression of this gene was obtained by ligating a 6.6-kb *Bam*HI fragment containing *EJ* and the normal *ras* promoter into the *Bam*HI site of pSV2 neo.[13] The resulting construct was designated pREJ.

Additional constructs were derived from pREJ in which the transforming gene was placed under control of the mouse metallothionein-I promoter.[14] This vector (pMTEJ) could be induced to express high levels of ras RNA and protein by culture of transfected lines in 50 µM zinc sulfate. Uninduced cells expressed a low basal level of ras which was six- to eightfold less than observed in cells cultured 36 hr in heavy metals.

All vectors were introduced into the murine fibroblast cell line C3H 10T1/2 and transformants were selected by culture in G418 since pSV2 neo confers resis-

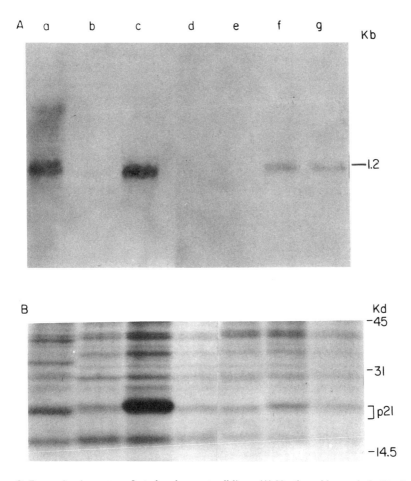

FIGURE 2. Expression in ras-transfected and parent cell lines. (A) Northern blot analysis. Total cellular RNA was isolated from cells grown in Dulbecco's MEM with 10% fetal calf serum in the presence or absence of 50 μM zinc sulfate. Samples of 20 μg RNA were electrophoresed, blotted, and hybridized to a 602-bp ras probe. Lanes contain RNA from (a) 5637 bladder carcinoma, (b) uninduced 212, (c) 212 plus 50 μM zinc, (d) 10T1/2, (e) 10T1/2 plus 50 μM zinc, (f) 245, and (g) 245 plus 50 μM zinc. (B) Immunoprecipitation of p21. Cultures labeled for 24 hr with 250 μCi/ml [^{35}S]methionine were washed, lysed, and immunoprecipitated with 5 μl YA6-172 rat monoclonal anti-p21 antibody.[20] Lysates contained 10^7 trichloroacetic acid-precipitable cpm in 300 μl solution. Following 4-hr incubation at 4°C, ras protein was precipitated with 25 μl of a 25% suspension of protein A–Sepharose beads coated with rabbit anti-rat IgG at 4°C overnight. Beads were then washed and electrophoresed on a 12.5% acrylamide slab gel. The gel was then soaked in Amplify (Amersham), dried, and exposed to XAR film (Kodak) at −70°C. Loading order of the gel is identical to that in panel A.

TABLE V
Zinc Induction of Transformation-Related Properties

	Doubling time[a] (hr)	Saturation density[b] ($\times 10^{-5}/cm^2$)	Anchorage independence[c] (percent plating efficiency)
10T1/2 − zinc	22	0.9	<0.001
10T1/2 + zinc	24	0.9	<0.001
212 − zinc	21	1.4	0.5
212 + zinc	22	3.6	8
245 − zinc	20	2.9	3
245 + zinc	24	3.0	2

[a]Cultures in the presence or absence of 50 µM zinc were counted during logarithmic growth phase.
[b]Confluent cultures were refed daily and counted following identical growth periods.
[c]Colonies of > 32 cells were counted 3 weeks after plating 10^4 cells in 0.3% agar over a 0.6% agar underlay.

tance to this antibiotic. Cell lines established by this protocol were designated 245, 246, and 247 (constitutive *ras* expressers derived from pREJ) as well as 212 (an inducible cell line transfected with pMTEJ). All constitutive *ras* cell lines as well as induced 212 exhibited focus formation and the lack of contact inhibition associated with transformed cells. Uninduced 212 was nearly normal in morphology and more closely resembled the parent C3H 10T1/2 cells.

Ras expression in constitutive and inducible cell lines was characterized by both Northern blot analysis and immunoprecipitation of p21 ras proteins. Total cellular RNA was isolated from cell lines 212, 245, and C3H 10T1/2 cultured with and without 50 µM zinc sulfate for 48 hr. Cell line 5637, a human bladder carcinoma cell line expressing constitutive *ras*, was included as a positive control. The Northern blot of uninduced and induced 212 (Fig. 2A, lanes b and c, respectively) demonstrates that culture in zinc and activation of the metallothionein promoter causes significant increases in *ras*-specific RNA. Uninduced 212 expresses a low level of ras RNA which is not observed in the parent C3H 10T1/2 cells (lanes d and e). The constitutively expressing 245 line exhibits an intermediate level of *ras* transcription (lanes f and g) which exceeds uninduced 212 but is less than 212 cultured in zinc. *Ras* expression in both the parent C3H 10T1/2 cell line and 245 was unaltered by zinc exposure.

Immunoprecipitation of p21 in the same cell lines demonstrated that ras protein synthesis correlated with expression of RNA (Fig. 2B). Induction of 212 resulted in large increases of *ras*-specific protein (lane c), which was present at low levels in the uninduced control (lane b). The constitutive 245 line exhibited intermediate p21 concentrations which were unaffected by zinc (lanes f and g) and were greater than levels observed in the 10T1/2 parent (lanes d and e). As expected, 5637 expressed high levels of ras protein.

Cellular properties associated with transformation were investigated for 10T1/2, 212, and 245 in the presence and absence of zinc (Table V). Growth rates were similar for all three lines although a small but consistent decrease in doubling time was observed in both 10T1/2 and 245 grown in zinc-supplemented medium.

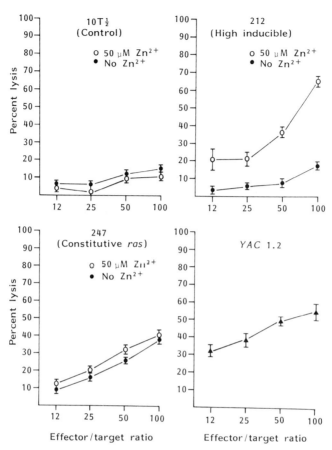

FIGURE 3. Comparison of cytolysis of *ras*-transfected cell lines. Cultures were induced for 36 hr in 50 μM zinc-supplemented medium and compared with unsupplemented controls. Effector spleen cells from CBA/J mice injected 24 hr prior to sacrifice with 1 unit of poly(I·C) were tested in 6-hr cytotoxicity assays. The mouse T-cell lymphoma line YAC 1.2 is included as a positive control. Experiments were repeated five times with similar results.

The 212 cells did not exhibit this change. Saturation density, a measurement for loss of contact inhibition of growth, was increased for 245 and induced 212 in which *ras* expression was elevated. These findings were substantiated by studies which compared anchorage-independent growth of each of the cell lines. Both the 245 and induced 212 exhibited anchorage-independent growth concomitant with loss of contact inhibition and increased *ras* expression.

Figure 3 depicts results of 6-hr cytotoxicity assays performed with spleen cells from CBA/J mice injected 24 hr prior to assay with 100 μg of poly(I·C). All target cell lines except for YAC 1.2 were tested in the presence and absence of 50 μM zinc which was added to the medium 36 hr prior to assay. The parent C3H 10T1/

TABLE VI
NK Susceptibility Using Spleen Cells as Effectors

Target cells[a]	Vector	Zn^{2+} induction	Number of experiments	Mean cytolysis ± S.E.		p value
				Percent lysis	$LU/10^6$	
				(200:1 E/T)		
10T1/2	—	—	4	10 ± 4	1.0 ± 0.3	
		+	3	17 ± 3	1.2 ± 0.3	N.S.
212	pMTEJ	—	15	18 ± 3	1.4 ± 0.4	
		+	17	57 ± 4	9.3 ± 1.2	< 0.001
245	pREJ	—	4	46 ± 10	5.7 ± 1.7	
246	pREJ	—	3	33 ± 13	3.5 ± 1.5	
247	pREJ	—	5	38 ± 2	6.0 ± 1.2	
		+	1	40 ± 3	8	N.S.
YAC 1.2	—	—		51 ± 4.5	19 3.3	

[a]Target cells were grown 24–48 hr in the presence or absence of 50 μM zinc sulfate added to the culture medium, and harvested by trypsinization. ^{51}Cr-labeled cells were tested in a 6-hr cytotoxicity assay against spleen cells pooled from 8- to 12-week-old CBA mice injected i.p. on day −1 with 100 μg of poly(I·C). Effector titration curves yielded maximum lysis values at a 200:1 effector/target ratio as shown and lowest values at 1.5:1. One lytic unit (LU) is the number of effector cells required to lyse 20% of the target cells; the values are mean % lysis ± S.E. and LU per 10^6 spleen cells in data pooled from 1 to 17 independent experiments. P values represent the probability (Student's t tests) that $LU/10^6$ values for zinc-induced compared to uninduced cells are not different.

2 transfection cell line (upper left panel) exhibited low levels of killing which was unaltered by the addition of zinc sulfate to the culture medium. Cytolysis of 247 cells, which were transfected with pREJ and therefore expressed constitutive *ras*, was also unchanged by culture in zinc. These cells were, however, more sensitive to lysis than the untransfected parent. This difference was observed in five separate experiments and when analyzed by Student's t test was found significant at $p <$ 0.005. The inducible 212 line (upper right panel) exhibited markedly increased killing in zinc-supplemented medium. Although uninduced 212 was killed at levels which were similar to the parent 10T1/2 cells, induction of *ras* expression enhanced killing so that 212 was not more sensitive to lysis than 247. Comparison of cytolysis curves of 212 and YAC 1.2 indicates that 212 was even more sensitive than YAC 1.2 at the high effector/target ratio.

Data from replicate experiments with both constitutive and inducible *ras* lines are presented in Table VI. In over a dozen separate tests, induction of *ras* expression in the 212 line was accompanied by a concomitant increase in killing. This change was unequivocally apparent in 6-hr assays with spleen cells sensitized by prior injection of poly(I·C). Overnight assays with unstimulated spleen cells also demonstrated comparable effects. The increase in cytolysis was approximately six-fold when calculated from lytic units averaged from over a dozen trials and was statistically significant at a P value < 0.001.

Similar analyses were performed on cell lines which expressed constitutive *ras*. Assays on 245, 246, and 247, lines established by transfection of pREJ into C3H 10T1/2, demonstrated that these cells were also more sensitive than the untransfected parent. None of these lines were, however, as sensitive as the induced 212

TABLE VII
Phenotype of NK Clones

Clone	Antibody[a]				
	Lyt-2.2	Lyt-1.2	NK-1.2	Asialo GM$_1$	Qa4
IIIF6	−	−	+++	++++	+++
1E4	−	−	+++	++++	+++
E4	+++	−	+++	−	++

[a]Aliquots of cells were suspended in RPMI 1640 medium supplemented with 5% fetal calf serum (5 × 10^5 cells in 500 μl medium). Cells were incubated with antibody for 1 hr at 4°C and then washed twice. F(ab')$_2$ fragments of FITC-conjugated goat anti-mouse (1/10 dilution) or goat anti-rabbit (1/20 dilution) IgG were then incubated with antibody-coupled cells for an additional hour at 4°C. Cells were then washed four times and analyzed with a FACS. The panel of first antibodies consisted of ascites mouse monoclonal (IgM) anti-Lyt-2.2 (1/500 dilution), ascites mouse monoclonal (IgG2b) anti-Lyt-1.2 (1/50 dilution), mouse serum anti-Nk-1.2 (1/50 dilution), rabbit anti-asialo GM$_1$ (1/100 dilution), and ascites mouse monoclonal (IgG) anti-Qa4 (1/100 dilution).

and they exhibited varying degrees of sensitivity which corresponded to approximate levels of *ras* expression.

Additional experiments were conducted with cloned NK cell lines, gifts of C. Brooks.[8] Although each line was cytolytic, the levels of killing varied between lines and was characteristic for each of the three clones. Table VII illustrates the relative phenotype of each clone as determined by a panel of antibodies analyzed with a fluorescence-activated cell sorter (FACS). Both IIIF6 (high cytolysis) and 1E4 (intermediate cytolysis) exhibited markers characteristic of NK cells (NK 1.2$^+$ and asialo GM$_1$$^+$) and not T cells (Lyt-2.2$^-$ and Lyt-1.2$^-$). The non-NK clone E4 (low cytolysis) displayed markers characteristic not only of NK cells (NK-1.2$^+$) but also of T cells (Lyt-2.2$^+$). In addition, this line was asialo GM$_1$$^-$. All lines reacted with Qa4 antibodies.

Cytolysis of induced 212, uninduced 212, and YAC 1.2 by the three cloned NK lines is illustrated in Table VIII. The rank order of killing by three effectors against the standard NK positive control YAC 1.2 was IIIF6 > 1E4 > non-NK clone E4. Killing of both induced and uninduced 212 also conformed to this pattern. As observed with poly(I·C)-stimulated spleen cell effectors (Table VI), lysis of

TABLE VIII
Cytolysis by NK Clones

	Cytolysis					
	212 + Zn[a]		212 − Zn		YAC 1.2	
Effector	Percent lysis[b]	LU/10^{6c}	Percent lysis	LU/10^6	Percent lysis	LU/10^6
NK clone IIIF6	97 ± 1	5 × 10^3	21 ± 1	1 × 10^2	51 ± 5	1 × 10^3
NK clone 1E4	83 ± 18	1 × 10^3	7 ± 1	< 2 × 10^2	34 ± 2	1.6 × 10^2
Non-NK clone E4	31 ± 3	8 × 10^1	4 ± 0.3	< 2 × 10^1	4 ± 0.4	< 2 × 10^1

[a]Cells were induced for 36 hr in medium supplemented with 50 μM zinc sulfate.
[b]Lysis values determined at 5:1 effector/target ratios.
[c]Lytic units are calculated at the 20% lysis level. These experiments were repeated three times with similar results.

TABLE IX
Characterization of Effector Cells

Effector	Treatment	Cytolysis (LU/10^6)^a		
		212 + Zn	212 − Zn	YAC 1.2
CBA spleen	Poly(I·C)^b	16	1.3	20
CBA spleen	None	13	<1.0	8
BALB/c nu/nu spleen	Poly(I·C)	40	8	17
C57BL/6 +/bg spleen	Poly(I·C)	12	ND	13
C57BL/6 bg/bg spleen	Poly(I·C)	4	ND	3
CBA spleen	Poly(I·C) + C'	10	ND	11
CBA spleen	Poly(I·C) + Lyt-1.1 + C'^{c}	13	ND	16
CBA spleen	Poly(I·C) + NK-1.2 + C'	<1	ND	<1
CBA spleen	Poly(I·C) + aGM$_1$ + C'	<1	ND	<1

^{a}A lytic unit (LU) is the number of effectors required to lyse 20% of targets. The values indicate LU per 10^6 lymphocytes.
^{b}Mice were injected intraperitoneally with 100 μg (IU) of poly(I·C) 24 hr before the assay.
^{c}A 1:20 dilution of monoclonal anti-Lyt-1.1, 1:100 mouse anti-NK-1.2, or 1:100 rabbit anti-asialo GM$_1$ was mixed with suspensions of 10^7 spleen cells/ml and incubated 1 hr at 4°C. Cells were then washed and incubated at 37°C for 1 hr with 1:10 rabbit complement preabsorbed with mouse tissues.

212 cells was markedly increased by culture in zinc. This effect was evident with all three effector lines. An additional observation is that killing of induced 212 by the clones surpassed lysis levels for YAC 1.2. As indicated in Table II, cytolysis of rat-1 cells transfected with the v-Ki-*ras* oncogene by cloned effectors also exceeded killing of the YAC positive control.

The data given in Table IX characterize effector mouse spleen cells which lyse both 212 and YAC 1.2 targets. Spleen cells from designated mouse strains were pretreated as indicated and employed as effectors in 6-hr chromium release assays. Killing levels are reported in lytic units per 10^6 lymphocytes at 20% percent lysis. Cytolysis of 212 cells (H-2^k haplotype) by spleen cells from CBA/J mice (also H-2^k) was compared for mice which were either injected 24 hr prior to assay with poly(I·C), or left untreated. As indicated, pretreatment with poly(I·C), an NK-activating agent, enhanced killing in both YAC 1.2 and induced or uninduced 212. Induction of 212 resulted in approximately tenfold increases in killing over uninduced 212 (as calculated from lytic units). Enhancement of killing by poly(I·C) was more apparent for YAC 1.2 targets than the 212.

Experiments were also conducted with poly(I·C)-treated BALB/c nude mice and C57BL/6 homozygous and heterozygous beige. Spleen cells from the T-cell-deficient nude mice were highly efficient at killing both YAC and 212 targets. This result is consistent with the observation that nude mice, although immune compromised in T-cell functions, retain levels of NK killing equal to or greater than totally immunocompetent strains.[15] Studies with C57BL/6 +/bg (heterozygous) and C57BL/6 bg/bg (homozygous) mice demonstrated that although the heterozygous beige strain efficiently killed both induced 212 and YAC 1.2, the homozygous strain was markedly impaired in killing ability. The beige mutation has been previously described[16] and confers a functional defect in NK killing to homozygous strains.

Depletion studies with CBA/J spleen cells from poly(I·C)-boosted mice indicated that the effector was not a T cell. Removal of cells exhibiting Lyt-1.1 (a T-cell marker) by antibody plus complement had no effect on cell-mediated lysis of either induced 212 or YAC 1.2. However, depletion of cells bearing either NK-1.2 or asialo GM_1 (markers associated with NK cells) totally abrogated with effect. In short, the results indicate that killing is mediated by NK or NK-like cells.

2.3. Specificity of the NK Effect

Alternative effectors for killing induced and uninduced 212 were generated to assess the specificity of the NK effect. These experiments were designed to determine whether lysis by antibody plus complement or cytotoxic T lymphocytes (CTL) was influenced by induction of *ras* expression.

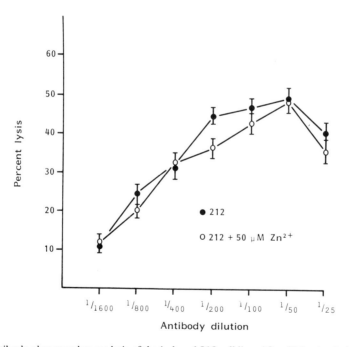

FIGURE 4. Antibody plus complement lysis of the induced 212 cell line. After 36-hr zinc induction, ^{51}Cr-labeled target cells were coupled to TNP and assayed with serial dilutions of antibody in 1/50 dilution of rabbit low-toxicity (M) complement. Release of ^{51}Cr was analyzed after 1-hr incubation for induced and control fibroblasts. Experiments were repeated twice with similar results.

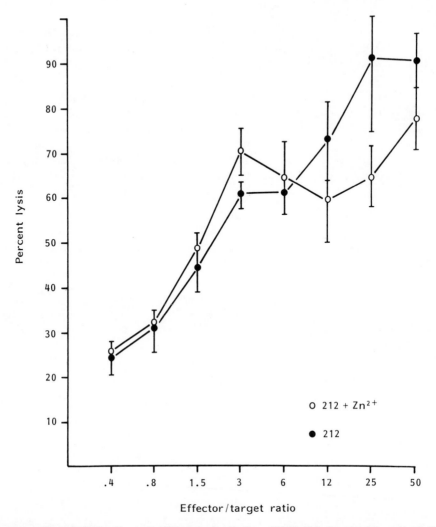

FIGURE 5. Killing of 212 by allogeneic CTL. Spleen cells from C57BL/6 *bg/bg* mice injected with C3H 10T1/2 cells (2 × 10^7 cells i.p.) and restimulated *in vitro* were tested against induced and uninduced 212 in 6-hr cytotoxicity assays. Data are expressed as mean ± standard error for triplicate wells.

Antibody plus complement lysis was determined for ^{51}Cr-labeled 212 cells coupled to trinitrophenol (TNP) hapten as previously described.[17] Zinc-induced and uninduced, TNP-coupled targets were incubated for 30 min at room temperature with serial dilutions of polyclonal anti-TNP antibodies. Rabbit low-toxicity complement (M) was then added at 1/50 and 1/25 dilutions, the mixture was incubated 1 hr at 37°C, cells were pelleted, and the supernatant was assayed for ^{51}Cr release. As depicted in Fig. 4, lysis by antibody plus 1/50 complement was unaffected by induction of *ras* expression. Similar results were obtained at 1/25 complement dilutions. FACS analysis of TNP binding indicated that coupling to the two cell types was equivalent and that differential binding of TNP was not a factor in these experiments (data not shown).

Figure 5 illustrates results of studies conducted with allogeneic CTL. Untransfected C3H 10T1/2 cells were injected into C57BL/6 bg/bg mice and spleen cells recovered in 2 weeks were subsequently restimulated *in vitro* as described.[18] CTL were raised in beige mice to ensure that killing of transfected targets could not be attributed to NK cells. Responder cells from the one-way mixed lymphocyte reactions were employed as effectors in killing assays with induced and uninduced 212. The data plotted in Fig. 5 indicate that although both targets were susceptible to killing, there was no differential lysis of the two cell lines.

3. Discussion

The results indicate that expression of at least two members of the *ras* oncogene family can induce sensitivity to NK-mediated killing. This effect was demonstrated both with a viral *ras* oncogene (v-Ki-*ras*) and with a mutated cellular *ras* oncogen (*EJ*) regulated by either a metallothionein gene or its own promoter. Transformed cells exhibited morphological changes associated with a malignant phenotype and acquired properties such as anchorage-independent growth. The induction of NK sensitivity was not, however, associated with either the process of transfection or transformation in general. Studies with the 212 cell line indicate that NK lysis was enhanced by elevated *ras* expression resulting from zinc activation of the metallothionein promoter. Since both induced and uninduced 212 contain the metallothionein–*ras* construct and have undergone transfection, there is little likelihood that differences in killing reflect changes resulting from tranfection per se.

This conclusion is supported by analyses of rat-2 cells transfected with v-*fps*. Although these transformed cells are highly tumorigenic and will form foci, colonies in semisolid medium, and tumors in syngeneic rats, they do not exhibit alterations in NK sensitivity. In addition, protein products of both *ras* and *fps* are associated with cell membranes so the effects do not appear related to a generalized perturbation of cell surface structures. Despite the similarity of their subcellular locations, the $p21^{ras}$ and $P130^{gag-fps}$ oncogene products have quite distinct biochemical activities; $p21^{ras}$ is a GTP-binding protein with associated GTPase activity[20] whereas $P130^{gag-fps}$ is a protein tyrosine kinase.[11] It would seem therefore that the reported changes may be related to a certain oncogene or class of oncogenes although more oncogenes remain to be tested before the generality of this phenomenon can be established.

Studies with methylcholanthrene (MCA) have demonstrated that chemical carcinogens can also alter sensitivity of fibroblasts to NK cells.[2] Although many diverse mechanisms may be responsible for this effect, it is of interest that MCA-transformed fibroblasts have been shown to contain activated *ras* oncogenes.[19] Tumor progression is likely related to a loss of NK sensitivity since selection of MCA-treated fibroblasts *in vivo* was correlated with emergence of NK-resistant variants.[2]

Characterization of the effectors killing Ki-*ras* and Ha-*ras* transfectants indicated that in both instances lysis was mediated by NK-like cells. Depletion of NK-1.2$^+$ or asialo GM$_1$$^+$ lymphocytes markedly reduced killing, whereas T-cell depletion had no effect. Effectors were stimulated by poly(I·C), an NK-activating agent, and were deficient in homozygous beige mice. NK cell clones were also demonstrated to kill *ras* transfectants at high levels.

The mechanism by which Ha-*ras* confers an NK-sensitive phenotype is not known but the changes induced in Ha-*ras*-transformed cells appear specific for NK cells. Lysis by alloimmune CTL or by antibody and complement was not altered in zinc-induced 212 targets expressing Ha-*ras*. Current investigations in our laboratory are being directed into further characterization of the influence of *ras* expression on NK killing and elucidating the mechanism for this effect.[21,22]

ACKNOWLEDGMENTS. This work was supported by grants from the Medical Research Council of Canada and the National Cancer Institute of Canada. W.S.T. is the recipient of an MRC Studentship and P.W.J. is the recipient of an MRC Fellowship.

References

1. Haller, O., Hansson, M., Kiessling, R., and Wigzell, H., 1977, Role of non-conventional natural killer cells in resistance against syngeneic tumor cells *in vivo*, *Nature* **270**:609–611.
2. Collins, J. L., Patek, P. Q., and Cohen, M., 1981, Tumorigenicity and lysis by natural killers, *J. Exp. Med.* **153**:89–106.
3. Haliotis, T., Ball, J. K., Dexter, D., and Roder, J. C., 1985, Spontaneous and induced primary oncogenesis in natural killer (NK) cell-deficient beige mutant mice, *Int. J. Cancer* **35**:505–513.
4. Botchan, M., Topp, W., and Sambrook, J., 1976, The arrangement of simian virus 40 sequences in the DNA of transformed cells, *Cell* **9**:269–287.
5. Wigler, M., Pellicer, A., Silverstein, S., and Axel, R., 1978, Biochemical transfer of single copy eukaryotic genes using total cellular DNA as donor, *Cell* **14**:725–731.
6. Tsuchida, N., and Uesugi, S., 1981, Structure and functions of the Kirsten murine sarcoma virus genome: Molecular cloning of biologically active Kirsten murine sarcoma virus DNA, *J. Virol.* **38**:720–727.
7. Stoker, M., O'Neill, C., and Waxman, V., 1968, Anchorage and growth regulation in normal and virus-transformed cells, *Int. J. Cancer* **3**:683–693.
8. Brooks, C. G., Kuribayashi, K., Sale, G. E., and Henney, C. S., 1982, Characterization of five cloned murine cell lines showing high cytolytic activity against YAC-1 cells, *J. Immunol.* **128**:2326–2335.
9. Southern, E. M., 1975, Detection of specific sequences among DNA fragments separated by gel electrophoresis, *J. Mol. Biol.* **98**:503–517.
10. Topp. W. C. 1981, Normal rat cell lines deficient in nuclear thymidine kinase, *Virology* **113**:408–411.
11. Weinmaster, G., and Pawson, T., 1986, Protein kinase activity of FSV P130$^{gag-fps}$ shows a strict specificity for tyrosine residues, *J. Biol. Chem.* **261**:328–333.
12. Shih, C., and Weinberg, R. A., 1982, Isolation of a transforming sequence from a human bladder carcinoma cell line, *Cell* **29**:161–169.
13. Southern, P. J., and Berg, P., 1982, Transformation of mammalian cells to antibiotic resistance with a bacterial gene under control of the SV40 early region promoter, *J. Mol. Appl. Genet.* **1**:327–341.
14. Durnam, D. M., Perrin, F., Gannon, F., and Palmiter, R. D., 1980, Isolation and characterization of the mouse metallothionein-I gene, *Proc. Natl. Acad. Sci. USA* **77**:6511–6515.

15. Herberman, R. B., Nunn, M. E., Holden, H. T., and Lavrin, D. H., 1975, Natural cytotoxic reactivity of mouse lymphoid cells against syngeneic and allogeneic tumors. II. Characterization of effector cells, *Int. J. Cancer* **16**:230–239.
16. Roder, J. C., and Duwe, A., 1979, The beige mutation in the mouse selectively impairs natural killer cell function, *Nature* **278**:451–453.
17. Fujiwara, H., Levy, R. B., Shearer, G. M., and Terry, W. D., 1979, Studies on *in vivo* priming of the TNP-reactive cytotoxic effector cell system. I. Comparison of the effects of intravenous inoculation with TNP conjugated cells on the development of contact sensitivity and cell-mediated lympholysis, *J. Immunol.* **123**:423–425.
18. Roder, J. C., Lohmann-Matthes, M.-L., Domiz, W., and Wigzell, H., 1979, The beige mutation in the mouse. II. Selectivity of the natural killer (NK) cell defect, *J. Immunol.* **123**:2174–2181.
19. Sukumar, S., Pulciani, S., Doniger, J., Di Paolo, J. A., Evans, C. H., Zbar, B., and Barbacid, M., 1984, A transforming *ras* gene in tumorigenic guinea pig cell lines initiated by diverse chemical carcinogens, *Science* **223**:1197–1199.
20. Furth, M. E., Davis, L. J., Fleurdelys, B., and Scolnick, E. M., 1982, Monoclonal antibodies to the p21 products of the transforming gene of Harvey murine sarcoma virus and of the cellular ras gene family, *J. Virol.* **43**:294–304.
21. Johnson, P. W., Baubock, C., and Roder, J. C., 1985, Transfection of a rat cell line with the v-Ki-ras oncogene is associated with enhanced susceptibility to natural killer cell lysis, *J. Exp. Med.* **162**:1732–1737.
22. Trimble, W., Johnson, P. W., Hozumi, N., and Roder, J. C., 1986, Inducible cellular transformation by a metallothionein–ras hybrid oncogene leads to natural killer cell susceptibility, *Nature* **321**:782–784.

17
A Regulatory Role of Natural Killer Cells (LGL) in T-Cell-Mediated Immune Response

KATSUO KUMAGAI, RYUJI SUZUKI, SATSUKI SUZUKI, TETSU TAKAHASHI, AND MINORU IGARASHI

1. Introduction

Accumulating evidence suggests that natural killer (NK) cells may play an important role in the regulation of differentiating normal cells, as well as being involved in elimination of developing malignant cells. NK cells recognize syngeneic or allogeneic bone marrow cells,[1,2] immature thymoctyes,[3] and granulocytic progenitor cells.[4] NK cells have also been implicated in the rejection of bone marrow transplants in mice.[5] Therefore, it is possible that NK cells contribute to regulation of specific humoral and cellular immunity mediated by T cells and B cells. An earlier study[6] showed that non-T, non-B human lymphocytes (L cells) mediated the enhancing effect on lymphocyte blastogenesis by virtue of having abundant Fc Ig receptors. Recently, we[7,8] and others[9,10] have demonstrated a suppressor property of human NK cells isolated from peripheral blood on pokeweed mitogen (PWM)-driven polyclonal Ig production of B cells. Several studies[8,11–13] have also shown that NK cells in mice, as well as in man, may exhibit a suppressive effect on murine B cells undergoing differentiation *in vivo* or being induced by mitogens or antigens *in vitro* or *in vivo*.

KATSUO KUMAGAI, RYUJI SUZUKI, SATSUKI SUZUKI, TETSU TAKAHASHI, AND MINORU IGARASHI • Department of Microbiology, Tohoku University School of Dentistry, Sendai 980, Japan.

Direct evidence linking NK cells with T-cell-mediated immune responses has been limited although a number of indirect observations have supported the concept. NK cells become activated during the early stages of tumor or allogeneic cell inoculation or viral infection.[14,15] Addition of interferon (IFN) into the mixed lymphocyte cultures, which can induce augmented NK activity, enhances generation of alloimmune cytotoxic T lymphocytes (CTL).[16] IFN administration prior to viral infection *in vivo* is also associated with effective induction of virus-specific CTL.[17] These observations led us to speculate that NK cells may display, together with macrophages, an initial interaction with certain cellular antigens, inducing CTL generation against specific cellular antigens. This hypothesis was reinforced by work that showed production of IFN when NK cells were incubated with tumor cells,[18] viruses,[19] interleukin-2 (IL-2).[20] IFN produced by NK cells may be involved in differentiation of CTL for which collaboration of IFN and IL-2 is required.[21] Recently, NK cells have also been recognized to be producers of other lymphokines or monokines.[22,23]

Based on this background, we[24] have recently demonstrated that murine NK cells recognize allogeneic cells and produce IL-2 and IFN in response to the cells, and through these soluble factors contribute to differentiation of alloimmune CTL. This chapter will review the immunological roles of NK cells as well as recent studies demonstrating the regulatory function of NK cells in cell-mediated immunity.

2. Experiments

2.1. Specific Depletion and Enrichment of Murine NK Cells Using Anti-Asialo GM_1 Serum

Murine large granular lymphocytes (LGL) with potent NK activity can be completely removed from mice by injecting a minute amount (5 μl) of rabbit antiserum to ganglio-*N*-tetraosylceramide (asialo GM_1), a neutral glycosphingolipid present in high quantities on the surface of murine NK cells.[24,25] This treatment resulted in almost complete elimination of NK activity and asialo GM_1^+ cells in the spleen 1 day after injection, and persisting for a week.[24] Throughout the experiments, it was shown that injection of anti-asialo GM_1 into mice was quite useful for selective depletion of LGL without any significant changes in the total number of mononuclear cells and the percentages of Thy-1$^+$ cells and their Lyt-1 and Lyt-2 subsets in the spleen of normal mice (Table I). There were no changes in the number of phagocytic cells although the possibility remained that a very small population of macrophages might be eliminated by the antibody.[26] Treatment of normal spleen cells with anti-asialo GM_1 and complement *in vitro* virtually depleted NK activity in the spleen cells without significant reduction in the percentages of T cells and their Lyt-1 and Lyt-2 subsets.[24,27-29]

To determine the direct effect of NK cells, NK (LGL)-enriched fractions were purified from the spleens by the combined methods at several successive steps. First, phagocytic cells and adherent cells were removed from these spleens by car-

TABLE I
Abrogation of NK Activity and Changes in Cell Populations in Mouse Spleen after Injection of Anti-Asialo GM_1

Mouse	Cell yield ($\times 10^7$) per mouse	Percentages of cells[a]					NK activity versus YAC-1 (LU/10^7)[a]
		Thy-1.2[b]	Lty-1[b]	Lyt-2[b]	Asialo GM_1[b]	Phagocyte[c]	
Control	2.05 ± 0.45	29.8 ± 2.4	20.3 ± 3.8	14.1 ± 2.9	11.2 ± 3.0	13.4 ± 3.4	7.0 ± 1.2
Anti-asialo GM_1-treated	2.11 ± 0.8	31.9 ± 2.2	21.1 ± 2.9	15.4 ± 1.1	0.2 ± 0.3	14.1 ± 3.6	<0.1

[a]Spleen mononuclear cells were isolated from mice 24 hr after i.v. injection of 10 μl of asialo GM_1 antiserum in 0.2 ml or saline containing 20% rabbit serum (control). All assays were carried out in triplicate. Mean and S.D. from three seperate experiments are shown.
[b]Detected by FACS analyzer.
[c]Phagocytosis of yeast particles.

TABLE II
Immunofluorescence Analysis of NK (Asialo GM_1^+) Cell Fractions Purified by Indirect Panning Methods

Spleen	Percent positive cells[a]						
	Asialo GM_1	Ly-5.1[b]	Thy-1.2	Lyt-1	Lyt-2	SIg	Phagocyte[c]
Purified NK fraction	93.8	95.5	0.2	0.3	0.3	0.3	<0.1

[a] All markers with the exception of Ly-5 were estimated by FACS analyzer. SIg, surface Ig. Values are the mean of one representative experiment.
[b] Estimated by indirect immunofluorescence.
[c] Estimated by yeast phagocytosis.

bonyl iron phagocytosis and nylon filtration, and these nonadherent mononuclear cells were subjected to a Percoll-density gradient to remove a majority of T cells in the high-density fractions. Subsequently, this partially purified NK population found in the low-density fractions on the Percoll gradient was treated with anti-Thy-1 monoclonal antibody and complement to remove T cells contaminating the fractions, and further purified by panning methods with anti-asialo GM_1.[26] The results (Table II) revealed that this final LGL-enriched preparation with potent NK activity generally consisted of greater than 90% asialo GM_1^+, Lyt-5$^+$ cells, and less than 0.5% of Thy-1$^+$, Lyt-1$^+$,2$^+$ cells.

2.2. Defective Generation of Alloimmune CTL in the Spleen Depleted of NK Cells *in Vivo* and *in Vitro*

C3H/He mice depleted of NK activity by administration of 10 μl of anti-asialo GM_1 serum were inoculated with C57BL/6 stimulator cells, and spleen cells were isolated from the mice at intervals after inoculation for 15 days, and then examined for CTL activity against an alloimmune-specific target, EL-4 (Fig. 1). In control mice (no antiserum treatment), cytotoxicity of alloimmune CTL against EL-4 was first detected 4 days after immunization, was highest for days 7 through 11, and returned to insignificant levels by day 15. In contrast, no significant levels of alloimmune CTL were generated in the anti-asialo GM_1-treated mice throughout the observation periods (Fig. 1). When these NK-depleted C3H/He mice were immunized with allogeneic BALB/c stimulator cells and then examined for generation of BALB/c-reactive CTL, no specific CTL were generated.

Spleen cells were isolated from the C3H/He mice that received anti-asialo GM_1 serum, incubated with C57BL/6 or BALB/c stimulator cells *in vitro*, and then assayed for generation of allospecific CTL for 6 days. No CTL activity against either C57BL/6 or BALB/c was detected in these NK-depleted spleen cells during the entire culture periods. These results indicate that the spleen cells depleted of NK cells by anti-asialo GM_1 administration respond poorly to allogeneic stimulation both *in vivo* and *in vitro*, resulting in limited generation of alloimmune CTL. When spleen cells isolated from mice injected with serial doses (0.6 to 20 μl) of anti-asialo GM_1 were examined for NK activity against YAC-1 and CTL generated

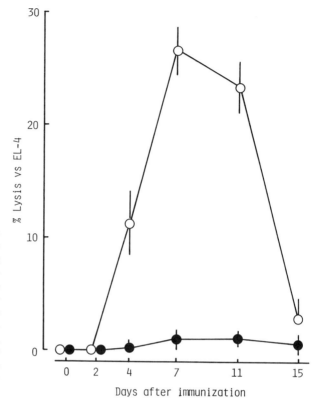

FIGURE 1. Failure to generate CTL against C57BL/6 (H-2^b) cell immunization in C3H/He (H-2^k) mice pretreated with i.v. injection of anti-asialo GM_1. Spleen mononuclear cells were isolated before immunization (day 0) and at the indicated times after immunization from control (O) and anti-asialo GM_1-treated mice (●), and were examined for cytotoxicity against EL-4 (H-2^b) cells at effector/target ratios of 50. Values are means and S.D. of three mice.

by C57BL/6 cell stimulation *in vitro*, a good correlation was found between decreased NK activity in the spleen and its reduced ability to generate alloimmune CTL *in vitro*.

We further examined the effect of NK depletion by anti-asialo GM_1 and complement *in vitro* on CTL generation of spleen cells in response to allogeneic stimulations *in vitro*. Responder spleen cells from C3H/He mice were subjected to complement lysis with anti-asialo GM_1. They were cultured with C57BL/6 and BALB/c stimulator cells, and 5 days later tested for cytolytic activity against EL-4 and Meth-A targets (Table III). Responder cells treated with anti-asialo GM_1, in which no NK activity was found, generated no allospecific CTL directed against the respective targets.

2.3. Restoration of CTL Generation in NK-Depleted Spleen Cells with Purified LGL

The foregoing *in vivo* and *in vitro* experiments suggest two possibilities: (1) that asialo GM_1^+ NK cells may be a cell population that plays an accessory role in CTL generation mediated by T–T cell interaction; the removal of which reduced

TABLE III
No in Vitro Generation of Alloimmune CTL in the Spleens of Anti-Asialo GM_1-Treated Mice[a]

Responder spleen	Stimulator	Percent cytotoxicity (E/T = 10)		
		EL-4	Meth-A	YAC-1
C3H/He mice	C57BL/6 spleen	26.9	2.1	20.3
	BALB/c spleen	1.4	24.3	26.9
Anti-asialo GM_1-treated C3H/HE mice	C57BL/6 spleen	4.0	0.9	5.2
	BALB/c spleen	2.0	1.1	3.4

[a]Spleen cells were isolated from C3H/He mice 24 hr after i.v. injection of 10 µl of anti-asialo GM_1 serum or saline containing 20% rabbit serum (control) and incubated with the indicated stimulator cells for 6 days. Values are the mean of three spleens.

the capability of CTL helpers and precursors to generate CTL effectors; and (2) that helpers or precursors of CTL or both bear asialo GM_1 antigent on the cell surface, the removal of which resulted in no generation of CTL effectors due to depletion of these populations. To examine these possibilities, LGL preparations that contained greater than 90% asialo GM_1^+ and Lyt-5$^+$ cells were purified by the methods described above. These highly purified LGL contained less than 0.5% Thy-1$^+$, Lyt-1$^+$, or Lyt-2$^+$ cells, as shown in Table II, suggesting that they contain neither helpers nor precursors of CTL. In fact, these purified LGL (2.5 or 1.25 × 10^6) generated no CTL when incubated with C57BL/6 cells in the absence of responder T cells (Table IV). The NK-depleted T cells, when reconstituted with

TABLE IV
Alloimmune CTL Response of NK-Depleted Spleen Cells When Cocultivated with Purified Asialo GM_1^+ and Lyt-1$^+$ Cells[a]

Addition		Responder NK-depleted spleen cells	CTL response (percent cytotoxicity, E/T = 10)		
Purified asialo GM_1^+ cells	Purified Lyt-1$^+$ cells		EL-4	Meth-A	YAC-1
2.5 × 10^6		−	1.8	−0.1	30.2
1.25 × 10^6		−	4.3	1.7	19.8
2.5 × 10^6		+	74.6	3.2	33.9
1.25 × 10^6		+	49.7	1.2	31.1
0.6 × 10^6		+	36.7	1.1	25.2
0.3 × 10^6		+	28.6	−2.0	20.3
	2.5 × 10^6	+	0.3	1.2	0.3
	1.25 × 10^6	+	−2.3	1.7	−0.9
−	−	Normal spleen	29.4	0.9	18.7

[a]Purified asialo GM_1^+ or Lyt-1$^+$ cells at the numbers indicated with or without 5 × 10^6 spleen cells from the anti-asialo GM_1-treated mice were incubated with 5 × 10^6 C57BL/6 stimulator cells at 37°C for 5 days and then assayed for the cytotoxicities. Values are the mean of triplicate cultures.

these purified LGL, generated a CTL response depending on the number of LGL added (Table IV). These results strongly suggest that an asialo GM_1^+ Thy-1$^-$ fraction contains cells that mediate an accessory effect on CTL generation from the T cells, but neither helpers nor precursors of CTL.

To further exclude the possibility that the apparent accessory function of asialo GM_1^+ fractions for CTL generation was mediated by Lyt-1$^+$ helper T cells that might contaminate the asialo GM_1^+ fractions at undetectable levels, providing an accessory function for CTL generation, we isolated Lyt-1$^+$ T cells from normal spleen mononuclear cells by panning methods, added them to NK-depleted T-cell fractions, and examined the CTL response to allogeneic cells from the mixed cultures. The results showed that the addition of Lyt-1$^+$ helper T cells to NK-depleted spleen cells resulted in no CTL response (Table IV). Thus, these results may indicate that NK (asialo GM_1^+) cells are required as an accessory cell population for CTL generation even in the presence of helper Lyt-1$^+$ T cells.

2.4. Production of IL-2 and IFN by Purified LGL in Response to Allogeneic Cells

We have found that restoration of CTL generation in NK-depleted T cells can be accomplished by soluble factors derived from NK cells induced by allogeneic cells. Significant levels of CTL were generated in the NK-depleted spleen cells when incubated with C57BL/6 stimulator cells in the presence of culture supernatants, which had been obtained from coculture of NK cells and allogeneic stimulator cells. We have also found that these NK-enriched fractions could produce IL-2 and IFN-γ in response to allogeneic cells. When asialo GM_1^+ cells were incubated with X-irradiated C57BL/6 spleen cells, IL-2 at a maximum of 2–5 U/ml and IFN at a maximum of 250–500 U/ml were produced in 2.5×10^6 purified asialo GM_1^+ cells. The amounts of IL-2 and IFN produced were dependent on the number of asialo GM_1^+ cells incubated. Even 0.6×10^6 asialo GM_1^+ cells produced 1.14 U/ml of IL-2 and 184 U/ml of IFN in response to C57BL/6 cells. We also found that the NK-depleted spleen cells were deficient in both IL-2 and IFN production.

Evidence suggests that Ir-gene-associated Ia$^+$ cells play a critical role in the antigenic stimulation of T lymphocytes. Cell proliferation and IL-2 production in primary antigen-specific responses or allogeneic mixed lymphocyte reaction are inhibited by anti-Ia sera.[30–33] IL-2 production by C3H/He NK cells in response to C57BL/6 stimulator cells or by C57BL/6 NK cells in response to C3H/He stimulators were, however, not inhibited by either anti-I-A^k or I-E^k antibodies (Table V). IFN production by these NK cells was also not inhibited by addition of anti-Ia sera. These results indicate that the IL-2 and IFN production by NK cells, unlike T cells, in response to allogeneic stimulator cells may not be linked to the interaction of surface Ia antigen expression by NK cells and allogeneic stimulator cells.

TABLE V
No Involvement of Ia Antigens in IL-2 Production by Purified Asialo GM_1^+ Cells Interacted with Allogeneic Stimulator Cells[a]

Purified asialo GM_1^+ (fraction 1)	Stimulator	Monoclonal antibody	IL-2 production (U/ml)
C3H/He	C57BL/6	—	1.79 ± 0.17
C3H/He	C57BL/6	1:10 I-A^k	2.14 ± 0.32
C3H/He	C57BL/6	1:100 I-A^k	1.61 ± 0.33
C3H/He	C57BL/6	1:10 I-E^k	1.52 ± 0.2
C3H/He	C57BL/6	1:100 I-E^k	1.71 ± 0.14
C57BL/6	C3H/He	—	2.41 ± 0.73
C57BL/6	C3H/He	1:20 I-A^k	1.82 ± 0.16
C57BL/6	C3H/He	1:20 I-E^k	2.07 ± 0.54
C3H/He	—	—	<0.02
C57BL/6	—	—	<0.02

[a] Partially purified NK (asialo GM_1^+) cells (2.5×10^6) from C3H/He or C57BL/6 mice were incubated with 5×10^6 mitomycin C-treated stimulator cells of C57BL/6 or C3H/He mice in the presence or absence of I-A^k or I-E^k monoclonal antibody at the indicated dilutions at 37°C for 4 days, and then examined for IL-2 activity in the culture supernatants by IL-2-dependent NK-7 cloned cells. Values are the mean of three cultures.

2.5. Restoration of Defective CTL Generation by NK-Depleted T Cells by IL-2 or IFN

As shown in Table VI, reconstitution with either 2 U/ml of human recombinant IL-2 (rIL-2) or 100 U/ml of human recombinant IFN-α A/D (rIFN-α-A/D) could restore generation of alloimmune CTL in the NK-depleted spleens to the levels of NK-containing control cultures. Addition of both IL-2 and IFN resulted in generation of CTL to much higher levels than in control cultures. NK-depleted T cells incubated with either IL-2 or IFN alone in the absence of stimulator cells generated no alloimmune CTL, although the augmented NK cytotoxicity against YAC-1 could be induced by IL-2.

We further examined whether injection of IL-2 or IFN into NK-depleted mice could restore their ability to generate alloimmune CTL *in vivo*. C3H/He mice were injected with 10 μl of anti-asialo GM_1, immunized with C57BL/6 stimulator cells, and then injected with 10^3 U of mouse rIFN-γ or 10^3 U of human rIL-2 at intervals after immunization. As shown in Fig. 2, defective CTL generation in NK-depleted mice, when injected with rIL-2 three or more days after immunization, was restored to the levels of NK-containing control mice. Injection of IFN-γ also restored CTL generation in NK-depleted mice. Injection of either IL-2 or IFN into NK-depleted mice resulted in no restoration of abrogated NK activity. These results indicate that IL-2 and IFN injected into mice, in place of NK cells, act on T–T cell interaction but not NK cells, and contribute to CTL generation.

2.6. Inhibition of Syngeneic Mixed Lymphocyte Reaction by NK Depletion of Responder T Cells

DNA replication in the spleen T lymphocytes stimulated by allogeneic stimulator cells, as well as generation of alloimmune CTL, is markedly depressed by NK depletion of responder spleen cells. However, mitogenic responses of the NK-depleted spleen cells to the T-cell mitogens Con A and PHA do not differ from responses seen with NK-containing spleen cells (Table VII). These results suggest a more significant contribution of the NK cells to the T-cell responses induced by cellular antigens, as compared to that induced with polyclonal mitogens.

The ability of murine and human T lymphocytes to proliferate *in vitro* in response to autologous or syngeneic non-T stimulator in the so-called autologous (AMLR) or syngeneic mixed lymphocyte reaction (SMLR) has been amply documented.[34] The stimulator cells are B cells, macrophages, and/or dendritic cells.[34,35] The responding cells in AMLR or SMLR have been shown to be in the helper/inducer class of Lyt-1^+23^- T cells in mice.[36] To test the possibility that NK cells contribute to the T-cell response induced by cellular antigens, we have examined the effect of NK depletion on murine SMLR using methods similar to those for allogeneic MLR. The responder cells were isolated from the spleen of C3H/He mice that had been administered anti-asialo GM_1. These NK-depleted responder cells and NK-containing control cells isolated from untreated mice were cultured with the nylon-adherent stimulator cells purified from C3H/He mice, and then the SMLR in both groups was examined. As shown in Fig. 3, DNA replication in NK-depleted responder T cells in the SMLR was markedly inhibited. These

TABLE VI
Effect of IL-2 and IFN on in Vitro Generation of Alloimmune CTL to C57BL/6 Cells in NK-Depleted Spleen Cells of C3H/He Mice[a]

Responder spleen	Stimulator	rIL-2 (2 U/ml)	rIFN-α A/D (100 U/ml)	Percent cytotoxicity (E/T = 10)		
				EL-4	Meth-A	YAC-1
Anti-asialo GM_1-treated	—	—	—	1	1	1
	—	+	—	0	−1	32
	—	—	+	0	−1	−1
	C57BL/6 spleen	—	—	4	0	10
	C57BL/6 spleen	+	—	38	0	29
	C57BL/6 spleen	—	+	42	1	20
	C57BL/6 spleen	+	+	64	2	54
Normal C3H/He mice	C57BL/6 spleen	—	—	34	1	20

[a] Pooled spleen cells from ten mice for each of anti-asialo GM_1-treated and control groups were incubated with X-irradiated C57BL/6 spleen cells in the presence or absence of rIL-2 and rIFN-α A/D at 37°C for 6 days. Values are the mean of three culture tubes.

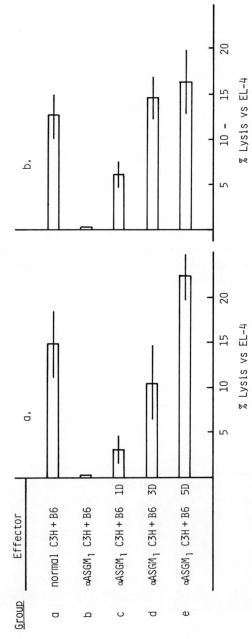

FIGURE 2. Effect of IL-2 and IFN-γ administration on defective CTL generation in anti-asialo GM_1-treated mice. C3H/He mice were injected with 10 μl of anti-asialo GM_1 (b–e) or saline containing 20% rabbit serum (a) and 24 hr later immunized with C57BL/6 stimulator cells. One, three, or five days later (c, d, and e), the mice were injected i.v. with 1000 U of recombinant human IL-2 (a) or 1000 U of recombinant mouse IFN-γ (b). All mice were assayed for CTL generation in the spleen on day 7. Values are mean and S.D. of three mice.

TABLE VII
Effect of NK-Depletion on Mitogenic Responses and Proliferative Responses in Allogeneic MLR

Sample	cpm [^{3}H]-TdR incorporation (mean ± S.D.)		
	Allogeneic MLR^a	Con A^b	PHA^b
Control	30794 ± 4562	94,826 ± 7413	25,675 ± 2716
Anti-asialo GM$_1$-treated	6972 ± 701	89,717 ± 5652	24,889 ± 3457

^{a}A million spleen mononuclear cells of C3H/He mice in 200 μl medium were incubated with 1 × 10^6 mitomycin C-treated spleen cells of C57BL/6 mice at 37°C for 5 days.
^{b}A million spleen mononuclear cells of C3H/He mice in 200 μl medium were incubated in the presence of 5 μg/ml Con A or 1% PHA-M at 37°C for 72 hr.

results indicate that the SMLR, as well as the allogeneic MLR, may require interaction of stimulator cells with NK cells, in addition to responder T cells. The SMLR is dependent on the expression of Ia antigens, as the addition of haplotype-specific anti-Ia sera to the cultures inhibits the reactions.[34] The interaction of syngeneic stimulator cells with NK cells was, however, not inhibited by anit-Ia sera (Igarashi *et al.*, in preparation).

3. Regulatory Features of NK Cells

A new aspect of immunoregulation mediated by NK cells is the observation that these cells may not only be killer cells but may also have the ability to secrete various cytokines. These cytokines are also secreted by T lymphocytes and macrophages. Earlier studies showed that NK cells had the ability to secrete IFN-α in response to tumor cells or viral infection.[18,19] Further studies[20,37,38] showed that highly purified murine and human LGL and their cloned cells could make IFN-γ as well as IFN-α. More recently, it has become apparent that a variety of other cytokines could also be produced by LGL. Highly purified human LGL, depleted of all detectable T cells or monocytes, could be stimulated to secrete substantial amounts of IL-2,[22] IL-1,[39] B-cell growth factor,[40] or colony-stimulating factor.[22] Different signals have been found to elicit a varying array of cytokines: mitogens, viruses, of NK-susceptible tumor target cells. Our study has shown that IL-2 and IFN can be produced by murine NK cells in response to allogeneic normal cells. Recently, we have also found that human LGL can endocytose gram-positive bacteria, accompanied by secretion of IL-1 and IFN.[23]

One might argue that the LGL with NK activity are different from cells that have cytokine-producing function. An approach to answer this question is to determine whether the various phenotypic subsets of LGL have parallel expression of NK activity and some of these other functions. Murine and human NK cells (LGL) are heterogeneous and include cytotoxic cells together with other subsets with different functions. A number of specialized LGL subsets can be identified through the expression of different surface markers. For example, studies in our laboratories have shown that a murine LGL population(s) with strong NK activity, and

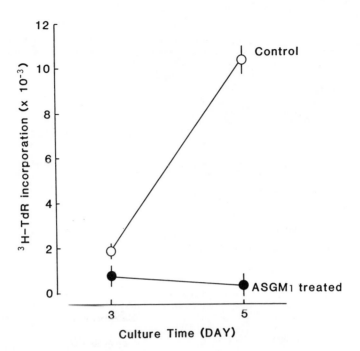

FIGURE 3. Suppression of DNA replication in the SMLR of NK-depleted (ASGM$_1^-$) T cells. Nonadherent T cells isolated from control mice (O) and those pretreated with anti-asialo GM$_1$ (●) were cocultured with mitomycin C-treated adherent cells isolated from control or treated mice, respectively. On day 3 and 5, [^{3}H]-TdR incorporation was assayed.

that expresses asialo GM$_1$ and Lyt-5 antigens, responded directly to IL-2 and produced IFN-γ, whereas another subset(s) of LGL, on which asialo GM$_1$ and Lyt-5 were expressed at lower quantities, responded to various stimuli and secreted IL-2 (Refs. 20, 41, and unpublished observations). In human LGL, the T3$^-$ Leu7$^-$Leu11$^+$ subset of high NK activity could produce IFN-γ in response to IL-2.[7] The same subset could also produce IL-1 and IFN in response to gram-positive bacteria.[23] On the other hand, a major population that produced IL-2 in response to mitogens was another Leu11$^-$ subset without NK activity (unpublished observations). Studies in other laboratories have also shown that IL-1-producing LGL are HLA-DR$^+$ M$_1^+$ B.73.1$^+$ (Leu11$^+$) with NK activity,[39] and those that produce IL-2 are HLA-DR$^+$ T11$^+$.[22] The BCGF-producing cells are LGL of phenotype Leu7$^+$ M$_1^+$ HLA-DR$^-$ E$^-$ Leu11$^-$ T3$^-$ and do not mediate NK activity.[42] These findings indicate that the LGL with the ability to produce cytokines may not be identical to the cells that have NK activity.

It has also been suggested that the activity of LGL to produce various cytokines may be subject to autoregulation. Thus, an LGL subset(s) responding to some stimuli, produces cytokines, which in turn stimulate another LGL subset(s) to become activated cells. Our recent investigations have revealed that the human Leu7$^+$Leu11$^-$ LGL subset responds to PHA and IL-1 and produces IL-2, which can, in turn, stimulate a T3⁻ Leu7⁻Leu11⁺ subset to become activated killer (AK) or lymphokine-activated killer (LAK) cells, which have the ability to kill a variety of freshly isolated tumor cells and tumor cell lines insensitive to NK lysis (Refs. 37, 43, and unpublished observations). Murine NK cells in response to IL-2 also develop into activated cells cytotoxic for tumor cell targets either sensitive or insensitive to NK lysis of syngeneic and allogeneic tumor origins. In these cases, IFN-γ produced by NK cells played a role as a triggering signal in differentiation of NK cells to AK or LAK cells.

Since the LGL can produce an array of cytokines, it is presumed that through these cytokines, LGL may have an immunoregulatory effect on antigen-specific immune responses, in addition to a regulatory effect on LGL themselves. In fact, accumulated evidence has demonstrated such an effect of murine and human LGL. A Leu7⁺ and/or Leu11⁺ subset of human LGL act as suppressor cells for Ig production by B cells induced by a polyclonal mitogen (PWM) or specific antigens.[7-11] Murine NK cells also had a suppressive activity on B-cell differentiation and Ig production.[8,12] In addition to this down-regulating effect of NK cells on immune responses, we have demonstrated that murine NK cells have an up-regulating effect on generation of alloimmune CTL. Human NK cells have also been shown to exhibit an up-regulating effect on CTL generation for influenza virus-infected cells[44] or PHA-induced T-cell colony formation.[45] Recently, others have also shown that human LGL exhibit accessory function of the T-lymphocyte proliferative response to the soluble stimulants *Staphylococcus* protein A or streptolysin O, or to surface antigens in the mixed lymphocyte reaction.[46] The human LGL may not only inhibit[1,4,47] but also promote[48] hematopoiesis, depending on the experimental conditions used. It is possible that cytokines such as IL-1, IL-2, IFN, other cytokines secreted from NK cells may be involved in the enhancing effect of NK cells in these T-cell-mediated immune responses.

On the other hand, activated NK (LAK) cells were found to have the ability to kill mitogen-induced activated T cells, in addition to tumor cells,[49] suggesting the possibility that they may have a suppressor activity on T-cell-mediated reactions. Recently, poly(I·C)-activated NK cells were also found to have the ability to suppress T-lymphocyte proliferation in MLR and AMLR cultures, acting on dendritic antigen-presenting cells.[50] Therefore, in addition to the up-regulation under the conditions demonstrated in the present studies, experiments examining selective activation of an LGL subset(s) with NK activity or selective depletion of another subset with accessory function, may result in a down-regulation of CTL generation.

In this chapter we have demonstrated that proliferation of T cells induced by syngeneic non-T stimulator cells (SMLR), as well as that by allogeneic stimulator cells (allogeneic MLR), was inhibited by depletion of NK cells in the responder cell population. The results have also shown that NK cells, when stimulated by syngeneic stimulators, produce soluble mediators, which contribute to the SMLR.

Both allogeneic MLR and SMLR induced by interaction between responder T cells and stimulator non-T cells are known to be dependent on the expression of syngeneic Ia antigenic determinants, since addition of haplotype-specific anti-Ia sera to the SMLR or allogeneic MLR cultures abolishes the reaction.[30,34] The structures on target cells that are recognized by NK cells have not yet been defined.

The recognition receptor on NK cells has also not been defined, although it is probably distinct from the T-cell receptor. NK cells kill NK-susceptible targets through undefined recognition receptors, but also may react to some cell surface structure without killing but with secretion of cytokines. The present study has shown that IL-2 and IFN production by murine NK cells in response to allogeneic or syngeneic cells are not inhibited by addition of anti-IE or -Ia sera, indicating that NK production of these factors induced by allogeneic or syngeneic cells may not be linked to the interaction of surface Ia antigen expression between responder NK cells and stimulator cells. These results indicate that NK cells may interact with either allogeneic or syngeneic cells independently of Ia determinants and produce soluble mediators, resulting in contribution to allogeneic MLR and SMLR mediated by T-cell recognition of Ia determinants. The identification of interaction between NK cells and allogeneic or syngeneic cells will require further detailed antigenic studies of NK cells. Our unpublished observations have, however, shown that an antiserum, LFA-1,[51] directed against an activated lymphocyte antigen, suppressed NK-driven lymphokine production in response to allogeneic and syngeneic cells (Suzuki et al., in preparation).

4. Summary and Conclusions

Figure 4 depicts a scheme for a role of NK cells (LGL) in allogeneic T-cell responses. In the allogeneic MLR exists a linear T–T cell interaction, in which helper T cells recognize MHC class II (Ia) antigen on the allogeneic stimulators such as macrophages, dendritic cells, or B cells, and are activated to secrete IL-2,

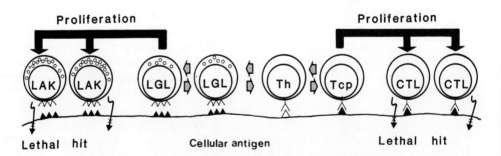

FIGURE 4. A scheme for NK cells (LGL) in allogeneic T-cell response. The killing of the target cells by activated killer (AK) or LAK cells is mediated by an MHC-independent recognition mechanism. It remains unknown, however, whether the recognition receptors on the LGL are the same as those on NK cells.

which can, in turn, stimulate the CTL precursors to proliferate and become CTL effectors, in collaboration with stimulation by MHC class I antigens. During this activation period, CTL produce other cytokines like IFN-γ, which may play a role as a differentiation signal in generation of helper and effector T cells. Our data have, however, suggested that allogeneic responses of T cells may require the interaction of NK cells (LGL) with allogeneic cells before T–T cell interaction. Thus, NK cells recognize allogeneic cells in some way but not through MHC antigens, and secrete cytokines such as IL-1, IL-2, or IFN. These cytokines secreted from LGL may play an accessory role in generation of alloimmune-specific CTL, acting on antigen–helper T cell interaction or helper T–cytotoxic T precursor cell interaction, or in the process of effector T-cell generation. On the other hand, the cytokines secreted from LGL may also have a role in autoregulation of NK cells. Thus, they can stimulate a subset(s) of LGL to proliferate and come to an activated state like LAK cells. LAK cells induced by the cytokines may play a role in elimination of antigenic cells through MHC-independent recognition mechanisms, resulting in a down-regulation of the T-cell immune reaction. Differentiation or activation of NK cells can also be mediated by the cytokines like IL-2 or IFN produced by T cells. It is possible therefore that development of T–T cell interaction in the allogeneic MLR participates in regulation of LGL or generation of LAK cells. In the AMLR or SMLR, as well as the allogeneic MLR, may exist a similar interaction between LGL and T cells.

References

1. Ritcardi, C., Santoni, A., Barlozzari, T., and Herberman, R. B., 1981, In vivo reactivity of mouse natural killer (NK) cells against normal bone marrow cells, *Cell. Immunol.* **60**:136–143.
2. Nunn, M. E., Herberman, R. B., and Holden, H. T., 1977, Natural cell-mediated cytotoxicity in mice against non-lymphoid tumor cells and some normal cells, *Int. J. Cancer* **20**:381–387.
3. Hansson, M., Kiessling, R., Anderson, B., Karre, K., and Roder, J., 1979, Natural killer (NK) cell sensitive T-cell subpopulation in the thymus: Inverse correlation to NK activity of the host, *Nature* **278**:174–176.
4. Hansson, M., Beran, M., Anderson, B., and Kiessling, R., 1982, Inhibition of in vitro granulopoiesis by autologous allogeneic human NK cells, *J. Immunol.* **129**:126–132.
5. Lotzova, E., Savary, C. A., and Pollack, S. B., 1983, Prevention of rejection of allogeneic bone marrow transplants by NK 1.1 antiserum, *Transplantation* **35**:490–494.
6. Carbalho, E. M., and Horwitz, D. A., 1980, Characterization of a non-T, non-B human blood lymphocyte that mediates the enhancing effects of immune complexes on lymphocyte blastogenesis, *J. Immunol.* **124**:1656–1661.
7. Arai, S., Yamamoto, H., Itoh, K., and Kumagai, K., 1983, Suppressive effect of human natural killer cells on pokeweed mitogen induced B cell differentiation, *J. Immunol.* **131**:651–657.
8. Kumagai, K., Suzuki, R., Suzuki, S., and Arai, S., 1985, Immunoregulatory effects of NK cells, in: *Mechanisms of Cytotoxicity by NK Cells* (R. B. Herberman and D. M. Callewaert, eds.), Academic Press, New York, pp. 489–498.
9. Tilden, A., Abo, T., and Balch, C. M., 1983, Suppressor cell function of human granular lymphocytes identified by the HNK-1 (Leu-7) monoclonal antibody, *J. Immunol.* **130**:1171–1175.
10. Brieva, J. A., Targan, S., and Stevens, R. H., 1984, NK and T cell subsets regulate antibody production by human *in vivo* antigen-induced lymphoblastoid B cells, *J. Immunol.* **132**:611–615.
11. Abruzzo, L. V., and Rowley, D. A., 1983, Homeostasis of the antibody response: Immunoregulation by NK cells, *Science* **222**:581–586.

12. Suzuki, S., Suzuki, R., Onta, T., and Kumagai, K., 1984, Suppression of B cell differentiation by NK cells, in: *Natural Killer Activity and Its Regulation* (T. Hoshino, H. Koren, and A. Uchida, eds.), Excerpta Medica, Amsterdam, pp. 296–300.
13. Robles, C. P., Pereira, P., Wortley, P., and Pollack, S. B., 1985, Regulation of the B cell response by NK cells, in: *Mechanisms of Cytotoxicity by NK Cells* (R. B. Herberman and D. M. Callewaert, eds.), Academic Press, New York, pp. 499–506.
14. Herberman, R. B., Nunn, M. E., Holden, H. T., Staal, S., and Djeu, J. Y., 1977, Augmentation of natural cytotoxic reactivity of mouse lymphoid cells against syngeneic and allogeneic target cells, *Int. J. Cancer* **19**:555–564.
15. Welsh, R. M., 1978, Cytotoxic cells induced during lymphocytic choriomeningitis virus infection of mice. I. Characterization of natural killer cell induction, *J. Exp. Med.* **148**:164–181.
16. Heron, I., Berg, K., and Cantell, K., 1976, Regulatory effect of interferon T cells in vitro, *J. Immunol.* **117**:1370–1373.
17. Kumagai, K., Itoh, K., Kurane, I., and Saitoh, F., 1981, Regulatory effect of interferon on effector cells in cell-mediated immune responses, in: *Self Defense Mechanisms: Role of Macrophages* (D. Mizuno, A. Z. Cohn, K. Takreya, and N. Ishida, eds.), University of Tokyo Press, Tokyo, Elsevier, Amsterdam, pp. 183–193.
18. Timonen, T., Saksela, E., Virtanen, I., and Cantell, K., 1980, Natural killer cells are responsible for the interferon production induced in human lymphocytes by tumor cell contact, *Eur. J. Immunol.* **10**:422–427.
19. Minato, N., Reid, L., Cantor, H., Lenzyel, P., and Bloom, B. R., 1980, Mode of regulation of natural killer cell activity of interferon, *J. Exp. Med.* **152**:124–137.
20. Handa, K., Suzuki, R., Matsui, H., Shimizu, Y., and Kumagai, K., 1983, Natural killer (NK) cells as a responder to interleukin 2 (IL-2). II. IL-2 induced interferon γ production, *J. Immunol.* **130**:988–992.
21. Farrar, W. L., Johnson, H. M., and Farrar, J. J., 1981, Regulation of the production of immune interferon and cytotoxic T lymphocytes by interleukin 2, *J. Immunol.* **126**:1120–1125.
22. Kasahara, T., Djeu, J. Y., Dongherty, S. F., and Oppenheim, J. J., 1983, Capacity of human large granular lymphocytes (LGL) to produce multiple lymphokines: Interleukin 2, interferon, and colony stimulating factor, *J. Immunol.* **131**:2379–2385.
23. Abo, T., Sugawara, S., Amenomori, A., Itoh, H., Rikiishi, H., Moro, I., and Kumagai, K., 1986, Selective endocytosis of gram positive bacteria and interleukin 1-like factor production by a subpopulation of large granular lymphocytes, *J. Immunol.* **136**:3189–3197.
24. Suzuki, R., Suzuki, S., Ebina, N., and Kumagai, K., 1985, Suppression of alloimmune cytotoxic T lymphocyte (CTL) generation by depletion of NK cells and restoration by interferon and/or interleukin 2 (IL2), *J. Immunol.* **134**:2139–2148.
25. Habu, S., Fukui, H., Shimamura, K., Kasai, M., Nagai, Y., and Tamaoki, N., 1981, *In vivo* effects of anti-asialo GM_1: Reduction of NK activity and enhancement of transplanted tumor growth in nude mice, *J. Immunol.* **127**:34–38.
26. Wiltrout, R. H., Santoni, A., Peterson, E. S., Knott, D. C., Uverton, W. R., Herberman, R. B., and Holden, H. T., 1985, Reactivity of anti-asialo GM_1 serum with tumoricidal and non-tumoricidal mouse macrophages, *J. Leukocyte Biol.* **37**:597–614.
27. Kasai, M., Iwamori, M., Nagai, Y., Okumura, K., and Tada, T., 1980, A glycolipid on the surface of mouse natural killer cells, *Eur. J. Immunol.* **10**:175–179.
28. Kumagai, K., Itoh, K., Suzuki, R., Hinuma, S., and Saitoh, F., 1982, Studies of murine large granular lymphocytes. I. Identification as effector cells in NK and K cytotoxicities, *J. Immunol.* **129**:388–394.
29. Itoh, K., Suzuki, R., Umezu, Y., Hanaumi, K., and Kumagai, K., 1982, Studies of murine large granular lymphocytes. II. Tissue, strain and age-distribution of LGL and LAL, *J. Immunol.* **129**:395–400.
30. Niederhuber, J. E., Frelinger, J. A., Dine, M. S., Shaffner, P., Dugan, E., and Shreffer, D. C., 1976, Effects of anti-Ia sera on mitogenic responses. II. Differential expression of the Ia marker on phytohemagglutinin- and concanavalin A-reactive T cells, *J. Exp. Med.* **143**:372–381.
31. Yamashita, U., and Shevach, E. M., 1977, The expression of Ia antigen immunocompetent cells in the guinea pig. II. Ia antigens on macrophages, *J. Immunol.* **119**:1584–1588.

32. Accolla, R., Moretta, A., and Cerottini, J. C., 1981, Allogeneic lymphocyte reactions in humans: Pretreatment of either the stimulator or the responder cell population with monoclonal anti-Ia antibodies leads to an inhibition of cell proliferation, *J. Immunol.* **127**:2438–2442.
33. Gilman, S. C., Rosenberg, J. S., and Feldman, J. D., 1980, Inhibition of interleukin 2 synthesis and T cell proliferation by a monoclonal-anti-Ia antibody, *J. Immunol.* **130**:1236–1240.
34. Weksler, M. E., Moody, C. E., Jr., and Kozak, R. W., 1981, The autologous mixed lymphocyte reaction, *Adv. Immunol.* **31**:271–283.
35. Nussenzweig, M. C., and Stenman, R. M., 1980, Contribution of dendritic cells to stimulation of the murine syngeneic mixed leukocyte reaction, *J. Exp. Med.* **151**:1196–1212.
36. Pastermak, R. D., Bocchieri, M. H., and Smith, J. B., 1980, Surface phenotype of responder cells in syngeneic mixed lymphocyte reaction in mice, *Cell. Immunol.* **49**:384–389.
37. Itoh, K., Shiiba, K., Shimizu, Y., Suzuki, R., and Kumagai, K., 1985, Generation of activated killer cells by recombinant interleukin 2 (rIL2) in collaboration with interferon γ (IFNγ), *J. Immunol.* **134**:3124–3129.
38. Munakata, T., Semba, N., Shibuya, Y., Kuwano, K., Akagi, M., and Arai, S., 1985, Induction of interferon-γ production by human natural killer cells stimulated by hydrogen peroxide, *J. Immunol.* **134**:2449–2455.
39. Scala, G., Allavena, P., Djeu, J. Y., Kasahara, T., Ortaldo, J. R., Herberman, R. B., and Oppenheim, J. J., 1984, Human large granular lymphocytes (LGL) are potent producers of interleukin 1, *Nature* **309**:56–59.
40. Procopio, A. D. G., Allavena, P., and Ortaldo, J. R., 1985, Noncytotoxic functions of natural killer (NK) cells: Large granular lymphocytes (LGL) produce a B cell growth factor (BCGF), *J. Immunol.* **135**:3264–3271.
41. Suzuki, R., Handa, K., Itoh, K., and Kumagai, K., 1983, Natural killer cells as a responder to interleukin 2 (IL2). I. Proliferative response and establishment of cloned cells, *J. Immunol.* **130**:981–987.
42. Pistoia, V., Cozzolino, F., Torcia, M., Castigli, E., and Ferrarini, M., 1985, Production of B cell growth factor by a Leu-7$^+$, OKM$_1$$^+$ non-T cell with the features of large granular lymphocytes (LGL), *J. Immunol.* **134**:3179–3184.
43. Itoh, K., Tilden, A. B., Kumagai, K., and Balch, C. M., 1985, Leu-11$^+$ lymphocytes with natural killer (NK) activity are precursors of recombinant interleukin 2 (rIL2)-induced activated killer (AK) cells, *J. Immunol.* **134**:802–807.
44. Burington, D. B., Djeu, J. Y., Wells, M. A., Killey, S. C., and Quinnan, G. V., 1984, Large granular lymphocytes provide an accessory function in the in vitro development of influenza A virus-specific cytotoxic T cells, *J. Immunol.* **132**:3154–3158.
45. Pistoia, V., Nocera, A., Ghio, R., Leprini, A., Perata, A., Pistone, M., and Ferrarini, M., 1983, PHA-induced human T-cell colony formation: Enhancing effect of large granular lymphocytes, *Exp. Hematol* **11**:249–259.
46. Scala, G., Allavena, P., Ortaldo, J. R., Herberman, R. B., and Oppenheim, J. J., 1985, Subsets of human large granular lymphocytes (LGL) exhibit accessory functions, *J. Immunol.* **134**:3049–3055.
47. Mangan, K. F., Hartnett, M. E., Matis, S. A., Winkelstein, A., and Abo, T., 1984, Natural killer cells suppress human erythroid stem cell proliferation in vitro, *Blood* **63**:260–269.
48. Pistoia, V., Ghio, R., Nocera, A., Ceprini, A., Perata, A., and Ferrarini, M., 1985, Large granular lymphocytes have a promoting activity on human peripheral blood erythroid burst forming units, *Blood* (in press).
49. Shiiba, K., Suzuki, R., Kawakami, K., Ohuchi, A., and Kumagai, K., 1986, Interleukin 2-activated killer cells: Generation in collaboration with interferon γ and its suppression in cancer patients, *Cancer Immunol. Immunother.* **21**:119–128.
50. Shah, P., Gilbortson, S. M., and Rowley, D. A., 1985, Dendritic cells that have interacted with antigen are targets for natural killer cells, *J. Exp. Med.* **162**:625–636.
51. Pavignon, D., Martz, E., Reynolds, T., Kurzinger, K., and Springer, T. A., 1981, Lymphocyte function associated antigen 1 (LFA-1): A surface antigen distinct from Lyt2,3 that participates in T lymphocyte mediated killing, *Proc. Natl. Acad. Sci. USA* **78**:4535–4539.

18

Immune Regulation in Neoplasia

Dominance of Suppressor Systems

J. Kevin Steele, Agnes Chan, Anthea T. Stammers,
Rakesh Singhai, and Julia G. Levy

1. Introduction

During the past decade, there has developed a consensus that neoplastically transformed cells (with a few notable exceptions) do not produce tumor-specific antigens (TSA), but rather may produce tumor markers better defined as tumor-associated antigens (TAA). TAA are those gene products expressed by the transformed cell constitutively which do not comprise part of the normal repertoire of its untransformed counterpart, either quantitatively or qualitatively. TAA can be the products of derepressed fetal genes such as carcinoembryonic antigen of α-fetoprotein, or of genes which are normally expressed at considerably lower levels in normal cells such as differentiation antigens, oncogene products, or cell surface receptor molecules. TAA as such must then be defined as part of the normal cell's genetic makeup and expression potential. Thus, the question as to whether or not malignant cells would be expected to be immunogenic is a legitimate one, which deserves considerable thought. We are not here discussing those tumor cells lines which express tumor-specific transplantation antigens (TSTA) or virally encoded antigens, since these clearly form a group of tumor types with well-recognized neoantigens and thus immunogenic potential. Rather, we will discuss here the majority of neoplasms, both murine and human, which do not express these types of antigens.

Despite the apparent lack of immunogenic properties of most tumors and tumor cell lines, the literature abounds in well-authenticated cases where immune

---

J. Kevin Steele • Department of Pathology, Harvard Medical School, Harvard University, Cambridge, Massachusetts. Agnes Chan, Anthea T. Stammers, and Julia G. Levy • Department of Microbiology, University of British Columbia, Vancouver, British Columbia V6T 1W5, Canada. Rakesh Singhai • Department of Pathology, Tufts University, Boston, Massachusetts.

responses to tumors, with respect to both tumor model systems and human neoplasms, have been noted. It is our contention, on the basis of the observations made above, that such responses should be considered as forms of autoimmune phenomena and that tumor immunity might possibly be viewed in this light. In this chapter we will discuss regulation of the immune response to tumors in this context, with emphasis on the effect of T suppressor cells (TsC), and present some recent results in the P815 murine mastocytoma system which are germane to this topic.

2. Immune Responsiveness to Neoplasms

There are numerous accounts in the literature on immune responses in humans to a variety of neoplasms. These cover a wide range of cancers including carcinoma of the breast,[1] lung cancer,[2,3] colorectal cancer,[4,5] and others.[6-8] Immune responsiveness to autologous tumors in human cancer has been assessed by T-cell-mediated killing of autologous or homologous tumor cells, or the detection of circulating antibody with specificity for homologous tumor cells or antigen extracts. There are, one should note, probably more instances in which specific immunity in human cancers could not be detected but many of these cases go unreported. At this time, in human cancers there has been no documented report on the isolation and characterization of a TSA associated with any tumor. Rather, tumor markers better defined as TAA have been detected and characterized. Is it therefore possible that in most instances in which a positive immune response to autologous tumor is noted, this response should be defined as a form of autoimmunity, and as such be subjected to possibly more rigorous immune regulation than other responses?

In murine tumor systems, it is the case that there does not appear to be any instance in which a syngeneic cell line has been shown to be devoid of some type of immunogenicity. The immunogenicity of murine tumors covers a broad range. There are those (usually methylcholanthrene-induced myosarcomas) which possess powerful immunogenic TSTA, unique for each tumor, which can bestow lasting T-cell-mediated antitumor immunity after a procedure of active immunization or tumor resection.[9,10] Similarly, the immunogenicity of UV-induced tumors is such that syngeneic animals will usually reject them, unless their immune systems have been manipulated in such a manner that TsC are dominant.[11,12] There are yet other tumor systems in mice in which no TSTA or potent antigens are present, but in which immunity, usually assessed by the development of cytotoxic T lymphocytes (CTL), can be measured.[13-17]

The P815 mastocytome of the DBA/2 mouse is at best poorly immunogenic. There are no recorded instances in which mice have been successfully immunized against this tumor, and spontaneous and permanent regressions during tumor growth are essentially nonexistent. However, it has been possible to establish that splenic CTL, specific for the P815 tumor, are present for a short period of time

(3–5 days) during the early phases of growth of subcutaneously implanted P815 cells.[18] We were able to show that these CTL were rapidly overridden by the development of tumor-specific TsC, and that the appearance of the TsC correlated with an acceleration of tumor growth. We also showed that CTL specific for P815 could be expanded and titrated *in vitro*.[19] It is our thesis that because of the "autoimmune" nature of the antitumor response, the generation of TsC as a means of control may be the strongest element of the response to tumors, and as such should be considered as a significant field of study, if immune modulation is to be seriously considered as a potential means of control of tumor growth.

3. T Suppressor Cells and T Suppressor Factors in Tumor Models

There are a number of well-defined models in which TsC and T suppressor factors (TsF) have been implicated in the modulation of the immune response to syngeneic tumors. The first observation of this kind was made by Greene and colleagues[20] who demonstrated clearly that a population of TsC and a soluble factor(s) obtained from them, were capable of significantly enhancing the growth of subcutaneously implanted tumors in tumor immune mice. Kripe and colleagues[11] using UV-induced tumors have also demonstrated the presence of apparently tumor-specific TsC which modulate the magnitude of response to these tumors. Fujimoto and collaborators showed similar development to TsC in mice bearing a variety of syngeneic tumors.[21] Other workers[22] showed that TsC in tumor-bearing mice developed from early thymocytes during the response to tumor. Bertschmann and Lüscher[23] and Nagarkatti and Kaplan[24] have reported results similar to ours in the P815 system, showing that tumor-bearing mice develop not only tumor-specific CTL but also specific TsC. There are also some indications that TsC are present in human patients with various neoplasms and that these may modulate responses to TAA.[25,26]

In the P815 system, it was shown a number of years ago that TsC developed in tumor-bearing mice and appeared to be able to abrogate the activity of existing tumor-specific CTL.[19] These TsC could be primed for by challenging mice intraperitoneally with P815 membrane extracts, and could suppress the *in vitro* generation of CTL specific for P815 if they were added to day 1 cultures of DBA/2 splenocytes with mitomycin-treated P815 cells.[27] An apparently tumor-specific TsF could be extracted from TsC-enriched populations and partially purified over affinity columns containing P815 membrane extracted material. The TsF thus isolated demonstrated specificity for the P815 tumor, had a molecular weight in the range of 70,000, and could be removed by passage over an anti-Ia^d column but not over an anti-mouse Ig column. The TsF was shown to originate from a T cell of the Lyt-1$^+$ phenotype and was able to replace TsC in the inhibition of *in vitro* generated anti-P815 CTL.[28] This phenotype for the TsC in the P815 system is analogous to that found by others in unrelated tumor systems[24] as well as in the P815 system itself.[29]

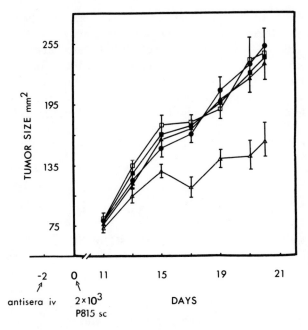

FIGURE 1. Effect of anti-P815 TsF *in vivo* on P815 tumor growth in syngeneic DBA/2 mice. Mice received 2 × 10³ P815 subcutaneously on day 0. On day −2, 50 μl of mouse antiserum was administered intravenously. Tumor size was measured in two dimensions with calipers. Data show the combined results of three separate experiments with a total of 21 mice per group. △, DBA/2 anti-P815-TsF (anti-idiotypic); □, C57BL/6 anti-P815-TsF; ▲, DBA/2 control antiserum; ■, C57BL/6 control antiserum.

4. Preparation of Anti-TsF Antisera and Monoclonal Antibodies

Attempts were made several years ago to raise antibodies in syngeneic and allogeneic mice to the P815 TsF. Affinity-enriched TsF (over P815 columns) was used, in complete Freund's adjuvant, to repeatedly immunize either DBA/2 or C57BL/6 mice. The antiserum thus obtained was tested *in vitro* and *in vivo*. It was found that both syngeneic and allogeneic antisera could neutralize the effect of the P815 TsF on the generation of CTL *in vitro*.[30] Perhaps more surprisingly, the syngeneic antiserum in particular was found to have a highly significant effect on the course of tumor growth in DBA/2 mice in that survival was increased by about 50% (Fig. 1)[31] in mice treated intravenously with small amounts of anti-TsF antiserum.

At the same time, a monoclonal antibody (MAb) with specificity for the P815 TsF was isolated. This MAb, B16G, was the product of a fusion between splenocytes of TsF-immunized BALB/c mice and NS-1 cells. DBA/2 mice injected intravenously on days −2 or 0 with 10 μg of B16G prior to administration of a tumorigenic dose of P815 cells showed significant slowing of tumor growth and increased survival. A few treated mice underwent complete regression of tumors.[32] Further investigation has shown that the B16G MAb is not directed to an idiotypic determinant of the P815 TsF and appears to react with a constant region of the suppressor molecule, since use of the MAb either *in vivo* or *in vitro* enhances a variety of immunological reactions in DBA/2 mice, presumably by blocking suppressor cell activity. The effects seen in the P815 system using either the conventional antisera or B16G are analogous to results reported earlier by Greene and collaborators[33,34] who showed that anti-I-J antiserum was able to slow the growth of syngeneic tumors in several systems, presumably by the inactivation of tumor-specific TsC or TsF.

5. Production of a P815-Specific TsF-Secreting T-Cell Hybridoma

We have been successful in producing a T-cell hybrid (A10) that secretes material with characteristics of a tumor-specific TsF: (1) it binds to B16G affinity columns and (2) P815 membrane extract columns; (3) it exerts both an *in vivo* and (4) *in vitro* effect on either tumor growth or the generation of P815 CTL, respectively.[35] Specifically, we have shown that as little as 20 ng of B16G affinity-purified A10 TsF administered intravenously will cause accelerated growth of P815 tumors (Fig. 2) if administered at the appropriate time. The effect appears to be specific since the growth of other unrelated tumors is not affected by equivalent treatment. Similarly, small amounts of affinity-purified A10 will suppress the generation of CTL specific for P815 in an *in vitro* assay system (Fig. 3).

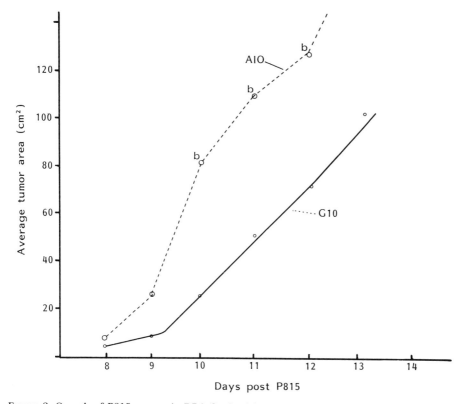

FIGURE 2. Growth of P815 tumors in DBA/2 mice injected intravenously with B16G-affinity-column-eluted ascites from A10 or an irrelevant hybridoma, G10. A total of 20 µg protein was injected per mouse on the same day that 3×10^3 P815 cells were administered subcutaneously. There were between 8 and 10 animals per group.

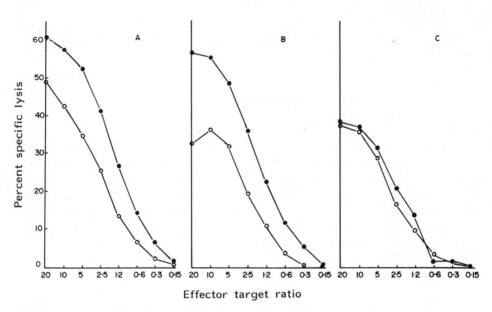

FIGURE 3. The generation of P815-specific CTL in splenocytes of DBA/2 mice cultured *in vitro* for 5 days with irradiated P815 cells. ^{51}Cr release assays were carried out for 18 hr at 37°C. Cultures contained either affinity-purified A10 at 50, 25, or 5 μg per culture (○) or equivalent material from BW 5147 ascites (●).

Preliminary biochemical characterization of the A10 TsF has been carried out. Affinity-purified material when run on SDS–PAGE displays certain characteristics regardless of whether the TsF is prepared from hybridoma cells grown in irradiated pristane-treated mice, media containing FCS, or serum-free medium (Fig. 4). The material runs with a major band of molecular weight of approximately 70,000, an apparent doublet running at between 44,000 and 47,000, and a low-molecular-weight component at approximately 25,000. The relative concentrations of these individual bands vary from one preparation to another. These findings are very similar to those described by Ferguson and co-workers[36] who have recently partially characterized a TsF responsible for suppressing the SRBC response. These workers claim that the reduced molecular weight of their TsF is about 70,000 and that lower-molecular-weight components constitute breakdown products of the 70,000 component. These results are also similar to the findings of Taniguchi and coworkers[37] in describing TsFs prepared from KLH-specific T-cell hybridomas. Analogous biochemical properties have been assigned to a TNP-binding TsF from sensitized animals.[38] Healy and co-workers[39] described a GAT-specific TsF isolated from a T-cell hybridoma. This molecule was identified on the membrane of the hybridoma, and in the cytosol. It occurred as a single polypeptide of 29,000–30,000 molecular weight which readily aggregated to form stable 65,000 polymers. Further biochemical analysis is necessary before definitive conclusions can be

6. Manipulation of the Immune Response to P815

We have been working on the assumption that the A10 molecule represents a soluble product of an id^+ Ts_1 $Lyt-1^+$ cell active in a suppressor cell circuit such as that described by Benacerraf.[40] This assumption is based not only on the properties of A10 itself, but also on earlier observations from this laboratory that the P815 TsF analogue to A10, isolated from TsC-enriched populations, was derived from $Lyt-1^+$ cells. If this assumption were correct, then the target for the A10 molecule would be an anti-idiotypic Ts_2 effector cell. Experiments were undertaken to try to manipulate the anti-P815 response in DBA/2 mice by way of administration of A10, either intravenously or as an active immunogen in complete Freund's

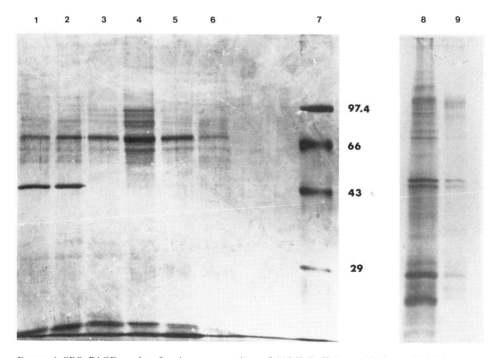

FIGURE 4. SDS–PAGE results of various preparations of A10 TsF affinity purified over B16G immunosorbent columns. Gels were stained with Coomassie blue. Lanes 1, 2, and 5, materials from ascites fluid; lanes 3, 4, and 6, materials purified from A10 cultures supernatants (DME + 10% FCS); lanes 8 and 9, materials purified from A10 culture supernatants from serum-free medium. Lane 7 contains molecular weight standards.

adjuvant (CFA), or in adoptive transfer experiments in which attempts were made to remove A10 analogue cells (Ts_1) by panning. It was hoped that various elements of the regulatory circuitry could be demonstrated in this way. Some of our results are presented here.

When animals are injected with A10 intravenously and simultaneously challenged with 10^4 P815 cells subcutaneously, there occasionally results a phenomenon that we have termed "suppressor-induced late regression" (SILR). In these mice, in which accelerated tumor growth was observed initially, marked regression occurred during the 14- to 21-day time period after tumor challenge such that a small number of A10-treated mice survived much longer than did controls. This pattern was observed in 10–15% of A10-treated animals. It was hypothesized that the massive input of Ts_1 factor on day 0 triggered the anti-P815 T-cell network much more rapidly, bypassing the antigenic activation and clonal expansion of the Ts_1 cell population which would normally be the case. After the activation of the anti-id Ts_2 effector, we reasoned, a contrasuppressor (id^+) might next come into

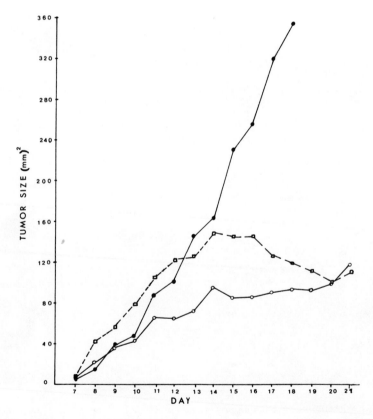

FIGURE 5. Effect of affinity-purified A10 on the growth of P815 tumors in DBA/2 mice. ●, PBS control; □, mice injected on day −7 and day 0 with 20 μg of A10 intravenously; ○, mice injected on day −14 and day −7 with 20 μg of A10 intravenously. All animals received 3×10^3 P815 cells subcutaneously on day 0.

TABLE I
Mean Survival Times of Mice Treated Intravenously Prior to Tumor Cell Challenge with Affinity-Purified A10 or Equivalent Material from BW 5147 Ascites[a]

Treatment	Mean survival time ± S.E.M.
20 μg A10 on day −14 and day −7	33.0 ± 2.3
20 μg A10 on day −7 and day 0	33.25 ± 1.8
20 μg BW 5147 ascites on day −7 and day 0	21.38 ± 1.49
50 μg PBS on day −7 and day 0	21.06 ± 2.04

[a]Each experimental group consisted of 16 animals.

play, thus inducing regression of the growing tumor by allowing antitumor immunity to dominate.

With the SILR phenomenon in mind, an attempt was made to trigger the control of regression (suppression of suppression) earlier. A10 (20 μg) was injected intravenously on day −7 and day 0. Animals were challenged with P815 on day 0. Control mice were injected intravenously on day −7 and day 0 with PBS or irrelevant material (BW 5147 ascites "purified" over B16G columns). This treatment (Fig. 5, Table I) caused an initial increase in the rate of tumor growth (presumably because of activated Ts_2 effector cells) followed by a marked, and sometimes total regression of tumor, with significantly enhanced survival rates over those of controls. Similar results were obtained when A10 was administered on day −14 or day −10 (Fig. 5).

These results support the hypothesis that A10 stimulates a Ts_2 effect or population, whose presence stimulates a contrasuppressive feedback, preventing further production of A10-analogue factors from Ts_1 inducers. Thus, when P815 cells are introduced, the suppressor circuitry is poorly functional, and contrasuppressive effects force a tumor regression early enough to significantly effect survival.

Currently, attempts are being made to identify idiotypically connected cells in the circuit proposed above, by adoptive transfer experiments in which each population is removed by panning. To date, we have demonstrated that reconstituting irradiated (600 rad) DBA/2 mice with a cell population depleted of Ts_1 cells by panning over plates coated with rabbit anti-A10 antiserum, results in enhanced immunity to P815 when animals are challenged 4 weeks later (Fig. 6). Cell populations panned over normal rabbit Ig served as a control. This result indicates that removal of a preexisting TsC population in the mouse spleen (Ts_1 id$^+$ inducers) prevents the formation of a normally functioning anti-P815 suppressor circuit. Attempts have been made to pan out a putative Ts_2 population over A10-coated plates, to date without success. This may be attributable to several causes: first, the A10 molecule may be too labile to serve as an effective panning reagent, and second, the Ts_2 id$^-$ population may need expansion and recruitment by Ts_1 inducers before it can be effectively panned away. These results are, however, suggestive of an operating T-suppressor network in DBA/2 mice challenged with P815 cells.

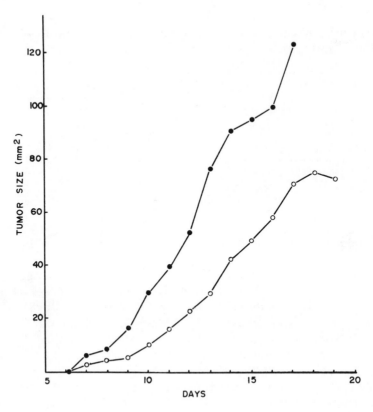

FIGURE 6. Growth of P815 tumors in adoptively transferred mice receiving syngeneic spleen cells panned over normal rabbit Ig (●) or rabbit anti-A10 antiserum (○).

Concrete proof awaits the successful separation and isolation of TsC populations in the reconstitution experiments.

Resistance to the P815 tumor can also be experimentally induced by an alternative method. Experiments with B16G had indicated that injection of this MAb several days after tumor challenge caused slower tumor growth and enhanced survival in the recipient mice than in controls. This occurred presumably in a manner analogous to the panning experiment described above, i.e., by the removal of a Ts_1 id^+ inducer population. With this result in mind, tumor therapy was attempted in which experimental animals were immunized subcutaneously with B16G-purified A10 in CFA. Animals immunized with PBS–CFA or BW 5147 ascites "purified" over B16G in CFA served as controls. When challenged with 3×10^3 P815 cells 4 weeks later, significant slowing of tumor growth, and enhancement of survival was seen in the A10–CFA group (Fig. 7). In addition, some 20–30% of A10-immunized mice are "cured" by this procedure and survive for the long term (100+ days). All control mice were dead by day 30–40. It is worth noting that the time at which tumor regression occurs (day 10–14 after tumor injection) coincides with the time that anti-P815 CTL occur naturally and can be detected in the spleens of tumor-bearing animals, an observation made in this laboratory several years ago.[18] Similar results were noted when irradiated A10 and BW 5147 or P815 cells in CFA were used as immunogens. Only mice immunized with A10 cells–CFA exhibited slowed tumor growth and enhanced survival (Table II).

We conclude that active immunization with a TsF leads to an antisuppressor response. The effect may be mediated by either antibody or an induced T-cell population; which has not yet been identified. The effect is general: animals immunized with A10 TsF–CFA exhibit greatly enhanced survival when challenged with L1210, another syngeneic DBA/2 tumor line. This argues against a completely anti-idiotypic response in generating these results; however, the induction of either a B16G analogue or a contrasuppressor T-cell population is possible. We have termed this form of therapy "suppressor vaccination therapy."

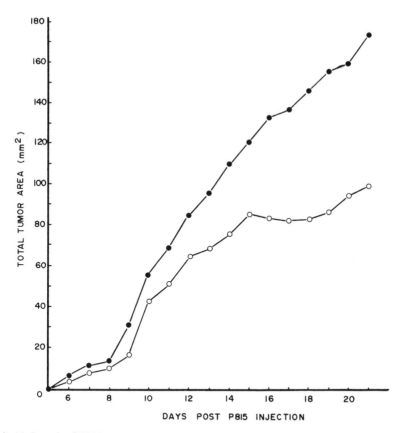

FIGURE 7. (a) Growth of P815 tumors in mice which had been actively immunized with affinity-purified A10 in 50% CFA (O) or had received 50% CFA–PBS (●). (b) Survival curve of the same mice. In this experiment 4/8 of the A10-immunized mice survived for 100 days after rejecting their tumors.

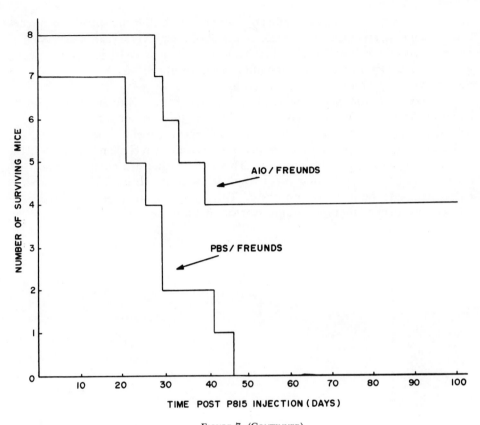

Figure 7. (Continued)

7. Summary

In the P815 system, we are dealing with a weakly immunogenic tumor which rarely undergoes spontaneous regression. It has been shown, however, that in animals bearing small tumors, a transient state of active immunity occurs and during that time CTL with specificity for P815 cells are detectable in their spleens. Whether or not these represent a type of autoreactive T cell has not been determined but it is certainly possible that they are reactive with cell surface constituents which are "self" in nature but are expressed preferentially on the tumor cells. Since CTL clones with specificity for P815 are now available, it should be possible to answer this question. Regardless of their specificity, these CTL are singularly ineffective in rejecting the P815 tumor, as, presumably, is the situation for CTL with specificities for autologous tumors in man. The ineffectiveness of the P815 CTL has been shown to be due, in part, to the development of TsC which override the CTL response. This sort of overwhelming suppressive response is not surprising when we consider that we may be dealing with a type of autoreactivity. In

TABLE II
Survival Time of Mice Actively Immunized with Irradiated
Tumor Cells in CFA 28 Days Prior to Challenge with 3×10^3
P815 Cells[a]

Treatment	Survival (days ± S.E.M.)	p_b
P815	21.1 ± 1.1	NS
EL-4	20.38 ± 2.8	NS
BW 5147	21.43 ± 2.9	NS
A10	30.1 ± 4.3	< 0.01
PBS–CFA	19.1 ± 0.6	—

[a]All animals received 10^7 irradiated (3000 rad) cells in 50% CFA. There were ten animals in each experimental group.
[b]Statistical analyses involved Student's t test of each value compared to the PBS–CFA control.

attempts to dissect out various elements of the suppressor cell network involved in the P815 system, we have succeeded in producing a T-cell hybridoma (A10) which, evidence suggests, may represent the fusion between a significant Ts_1 in this system and BW 5147 cells. In the preliminary experiments discussed here, we have shown that immunity to the P815 tumor in DBA/2 mice may be enhanced by manipulation of the immune response through administration of A10 or its soluble product. We suggest that this model for antisuppressor therapy may be useful in the application of such principles to the treatment of human neoplasias.

References

1. Forsman, L. M., Joupila, P. I., and Anderson, L. C., 1984, Sera from multiparous women contain antibodies mediating cytotoxicity against breast carcinoma cells, *Scand. J. Immunol.* **19:**135–139.
2. Ramey, W. G., Hasking, G. A., Munther, A. S., Swistel, A. J., Burrows, W. B., and Fitzpatrick, H. F., 1980, Detection of circulating lung tumor antigen sensitive T lymphocytes in the early stages of lung cancer, *Surgery* **88:**202–206.
3. Kelly, B. S., Stredulinsky, U., Vanden Hock, J., and Levy, J. G., 1981, Antibodies in the sera of patients with bronchogenic carcinoma that react with antigen from a tumor cell line, *Cancer Immunol. Immunother.* **12:**5–10.
4. Shoham, J., and Cohen, M., 1979, Antibody-dependent cellular cytotoxicity to human colon tumour cells, *Br. J. Cancer* **40:**234–243.
5. Shoham, J., and Cohen, M., 1979, Antibody-dependent cellular cytotoxicity to human colon tumour cells. II. Analysis of the antigens involved, *Br. J. Cancer* **40:**244–252.
6. Sethi, J., and Hirshant, Y., 1976, Complement-fixing antigen of human sarcomas, *J. Natl. Cancer Inst.* **57:**489–493.
7. Baldwin, R. W., Embleton, M. J., Jones, J. S. P., and Langman, M. J. S., 1973, Cell-mediated and humoral immune reactions of human tumours, *Br. J. Cancer* **12:**73–82.
8. Levy, N. L., 1978, Specificity of lymphocyte-mediated cytotoxicity in patients with primary intracranial tumors, *J. Immunol.* **121:**903–915.
9. Prehn, R. T., and Main, J. H., 1957, Immunity to methyl cholanthrene-induced sarcomas, *J. Nat. Cancer Inst.* **18:**789–796.

10. Klein, G., Sjögren, O., Klein, E., and Hellström, K. E., 1960, Demonstration of host resistance against MCA-induced sarcomas in the primary autochthonous host, *Cancer Res.* **20:**1561–1567.
11. Fisher, M. S., and Kripke, M. L., 1982, Suppressor T lymphocytes control the development of primary skin cancers in ultraviolet irradiated mice, *Science* **216:**1133–1134.
12. Greene, M. I., Sy, M.-S., Kripke, M. L., and Benacerraf, B., 1979, Impairment of antigen-presenting cell function by ultraviolet radiation, *Proc. Natl. Acad. Sci. USA* **76:**6591–6595.
13. Olsson, L., and Ebbesen, P., 1979, Immunoadjuvant treatment of primary grafted and spontaneous AKR-leukemia, *J. Immunol.* **122:**781–787.
14. Fujiwara, H., Hamaoka, T., Shearer, G. M., Yamamoto, H., and Terry, W. D., 1980, The augmentation of *in vitro* and *in vivo* tumor-specific T cell-mediated immunity by amplifier T lymphocytes, *J. Immunol.* **124:**863–869.
15. Fernandez-Cruz, E., Gilman, S. C., and Feldman, J. D., 1982, Immunotherapy of a chemically induced sarcoma in rats: Characterization of the effector T cell subset and nature of suppression, *J. Immunol.* **128:**1112–1121.
16. Mule, J..J., Rosenstein, M., Shu, S., and Rosenberg, S. A., 1985, Eradication of a disseminated syngeneic mouse lymphoma by systemic adoptive transfer of immune lymphocytes and its dependence on host components, *Cancer Res.* **45:**526–531.
17. Evans, R., and Duffy, T. M., 1985, Amplification of immune T lymphocyte function *in situ:* The identification of active components of the immunologic network during tumor rejection, *J. Immunol.* **135:**1498–1504.
18. Takei, F., Levy, J. G., and Kilburn, D. G., 1976, *In* vitro induction of cytotoxicity against syngeneic mastocytoma and its suppression by spleen and thymus cells from tumor bearing mice, *J. Immunol.* **116:**288–293.
19. Takei, F., Levy, J. G., and Kilburn, D. G., 1977, Characterization of suppressor cells in mice bearing syngeneic mastocytoma, *J. Immunol.* **118:**412–417.
20. Greene, M. I., Fujimoto, S., and Sehon, A. H., 1977, Regulation of the immune response to tumor antigens. III. Characterization of thymic suppressor factor(s) produced by tumor-bearing hosts, *J. Immunol.* **119:**757–764.
21. Fujimoto, S., Matsugawa, T., Nakagawa, K., and Tada, T., 1978, Cellular interaction between cytotoxic and suppressor T cells against syngeneic tumors in the mouse, *Cell. Immuno.* **38:**378–387.
22. Small, M., 1978, Opposing reactivities of subpopulations of T lymphocytes toward syngeneic tumor cells: Separation of early thymocytes, *J. Immunol.* **121:**1167–1172.
23. Bertschmann, M., and Lüscher, E. F., 1983, Stimulation of different pathways of T-cell functions by syngeneic tumor cells and soluble membrane proteins, *Cell. Immunol.* **78:**13–22.
24. Nagarkatti, M., and Kaplan, A. M., 1985, The role of suppressor T cells in BCNU-mediated rejection of a syngeneic tumor, *J. Immunol.* **135:**1510–1517.
25. Ninneman, J. L., 1978, Melanoma-associated immunosuppression through B cell activation of suppressor T cells, *J. Immunol.* **120:**1573–1579.
26. Chiorazzi, N., Fu, S. M., Montazeri, G., Kunkel, H. G., Lai, K., and Gee, T., 1979, T cell helper defect in patients with chronic lymphocytic leukemia, *J. Immunol* **122:**1087–1090.
27. Takei, F., Levy, J. G., and Kilburn, D. G., 1978, Characterization of a soluble factor which specifically suppresses the *in vitro* generation of cells cytotoxic for syngeneic tumor cells in mice, *J. Immunol.* **120:**1218–1224.
28. Maier, T., Levy, J. G., and Kilburn, D. G., 1980, The Lyt phenotype of cells involved in the cytotoxic response to syngeneic tumor and tumor specific suppressor cells, *Cell. Immunol* **56:**392–399.
29. Mills, C. D., and North, R. J., 1985, Lyl$^+$ suppressor T cells inhibit the expression of passively transferred anti-tumor immunity by suppressing the generation of cytolytic T cells, *Transplantation* **39:**202–208.
30. Maier, T., Kilburn, D. G., and Levy, J. G., 1981, Properties of a syngeneic and allogeneic antisera raised to tumor specific suppressor factor from DBA/2J mice, *Cancer Immunol. Immunother.* **12:**49–56.
31. Maier, T., and Levy, J. G., 1982, Anti-tumor effects of an anti-serum raised in syngeneic mice to a tumor specific T suppressor factor, *Cancer Immunol. Immunother.* **13:**134–139.
32. Maier, T., Stammers, A. T., and Levy, J. G., 1983, Characterization of a monoclonal antibody directed to a T cell suppressor factor, *J. Immunol.* **131:**1843–1848.

33. Perry, L. L., Benacerraf, B., and Greene, M. I., 1978, Regulation of the immune response to tumor antigen: Tumor antigen-specific suppressor factor(s) bear I-J determinants and induce suppressor T cells *in vivo, J. Immunol* **121:**2144–2147.
34. Drebin, J. A., Waltenbaugh, C., Schatten, S., Benacerraf, B., and Greene, M. I., 1983, Inhibition of tumor growth by monoclonal anti-I-J antibodies, *J. Immunol.* **130:**506–509.
35. Steele, J. K., Stammers, A. T., and Levy, J. G., 1985, Isolation and characterization of a tumor specific T suppressor factor from a T cell hybridoma, *J. Immunol.* **134:**2767–2778.
36. Ferguson, T. A., Beaman, K. D., and Iverson, C. M., 1985, Isolation and characterization of a T suppressor factor by using a monoclonal antibody, *J. Immunol.* **134:**3163–3171.
37. Takabayashi, K., Suzuki, N., Kanno, M., Imai, K., Tokuhisa, T., Tomioka, H., and Taniguchi, M., 1983, Cell cycle dependent expression of antigen-binding and I-J bearing molecules on suppressor cell hybridomas, *J. Immunol.* **130:**2552–2556.
38. Ptak, W., Gershon, R. K., Rosenstein, R. W., Murray, J. H., and Cone, R. E., 1983, Purification and characterization of TNP-specific immunoregulatory molecules produced by T cells sensitized by picrylchloride, *J. Immunol* **131:**2859–2863.
39. Healy, C. T., Kapp, J. A., and Webb, D. R., 1983, Purification and biochemcial analysis of antigen-specific suppressor factors obtained from the supernatant, membrane, or cytosol of a T cell hybridoma, *J. Immunol* **131:**2843–2847.
40. Benacerraf, B., 1982, Genetic control of the specificity of T lymphocytes and their regulatory products, *Prog. Immunol.* **4:**420–437.

19

The Generation and Down-Regulation of the Immune Response to Progressive Tumors

ROBERT J. NORTH, ANTONIO DiGIACOMO, AND EARL S. DYE

1. Introduction

Determining whether a given animal or human tumor possesses antigens capable of evoking a specific antitumor immune response is essential before a rational attempt can be made to treat the tumor by a therapeutic modality designed to augment specific antitumor immunity. The role of the experimental tumor immunologists is to obtain detailed information about the immunogenicity of animal tumors and about the antitumor immune response they evoke, with a view to supplying the clinical oncologist with knowledge about the type of immune response that needs to be boosted. Regardless of whether or not animal tumors are suitable models of the human disease, the fact is that animal and human tumors have a lot in common, including the capacity to grow progressively in their hosts. Therefore, even though an animal tumor may possess tumor-specific antigens, it is not destroyed by specific or nonspecific host defense mechanisms.

2. Tumor Immunogenicity as Revealed by Immunization

In spite of the fact that chemically induced, transplantable animal tumors grow progressively to kill their syngeneic hosts, it was shown by Foley in 1953[1] that they can be immunogenic. This means that they are capable of specifically immunizing their hosts against the growth of a subsequent implant of tumor cells.

ROBERT J. NORTH, ANTONIO DiGIACOMO, AND EARL S. DYE • Trudeau Institute, Inc., Saranac Lake, New York 12983.

The immunogenicity of chemically induced tumors was subsequently demonstrated on many occasions by a number of investigators who successfully immunized syngeneic mice, either by removing established primary tumors by surgery or ligation, or by giving mice a single or repeated injections of tumor cells made incapable of replicating by exposure to high doses of ionizing radiation (reviewed in Ref. 1–5).

It was later shown that the immunity acquired by immunized mice is cell-mediated, as opposed to antibody-mediated, in that it can be passively transferred from immunized donors to normal recipients with lymphoid cells, but not with serum (reviewed in Refs. 1, 5). More recently, the immunity has been shown to be mediated by T lymphocytes, as evidenced by the finding that the cells that passively transfer it are functionally eliminated by treatment with anti-Thy antibody and complement.[6] This is in keeping with the demonstration that lymphocytes from tumor-immune mice can give rise to tumor-specific cytolytic T cells when incubated with stimulator tumor cells *in vitro*.[7] It is too early to conclude, however, that cytolytic T cells are the ultimate effectors of antitumor immunity, because it has been demonstrated that the T cells that passivley transfer immunity display the $Ly1^+2^-$ surface phenotype of helper T cells.[8,9] This has led to the suggestion that tumor regression is achieved by a delayed-type hypersensitivity reaction, rather than by cytolytic T cells.[8] Similar results have come from an investigation of the T cells that passively transfer antiallograft immunity.[10] Even so, any interpretation of these results must involve a consideration of the recent demonstrations that $Ly1^+$, $L3T4^+$ T cells can be cytolytic.[11,12] This means that the surface phenotype of a T cell can no longer be used alone to determine whether the cell is a helper or a cytolytic effector. This is particularly so in the case of antitumor immunity where the meaning of "help" is not well defined. There remains a need, therefore, to functionally identify the type of T cells that mediates antitumor immunity *in vivo*, under conditions where the type and purity of donor T cells are known, and where a contribution by recipient T cells can be excluded.

3. Concomitant Immunity as the Unsuccessful Response to Progressive Tumor

Since the 1950s, tumor immunology has been involved, for the most part, with investigating the nature of immunity acquired by a host prophylactically immunized against growth of a tumor implant. However, from the point of view of immunotherapy, there is a need to know whether the host generates an immune response against a progressive tumor, because this is the situation with which the therapist is confronted.

In fact, there is ample evidence that progressively growing immunogenic tumors can evoke in their syngeneic hosts an underlying state of concomitant immunity that, although incapable of causing regression of the primary tumor, is capable of causing destruction of an implant of tumor cells. The subject has been reviewed recently by Gorelik.[13] In most cases, concomitant immunity to a challenge implant is specific. Moreover, it was shown some years ago[14] that it can be passively transferred from tumor bearers to normal recipients with T cells, according to the Winn assay. An important point to make about concomitant immunity, moreover, is that it is not evoked until the primary tumor grows to a certain size.[14] This means that, regardless of whether a smaller number of tumor cells is implanted, the tumor that eventually emerges must grow to a given size before the generation of concomitant immunity is triggered. In other words, it is probable that, in all cases, a tumor must be palpable and already progressively growing, before it can induce an immune response. Needless to say, the possible acquisition by a tumor bearer of an underlying mechanism of antitumor immunity gives rational reason for attempting therapy with immunoadjuvants capable of augmenting immune responses in general. Indeed, it is likely that augmentation of the level of concomitant immunity is the reason why intralesional injection of BCG or *C. parvum* results, under certain circumstances, in the regression of established tumors.[15] Even in these cases, however, the tumors need to be below a certain critical size to be susceptible.

This leads to next important point to be made about concomitant immunity, namely, that it can undergo progressive decay after the primary tumor grows beyond a certain size.[13] This subject has been intelligently dealt with by Vaage[16] who has reviewed the literature dealing with the generation and loss of concomitant immunity, as measured in terms of the acquisition and loss of immunity to growth of a challenge implant. The generation and decay of concomitant immunity also has been measured in terms of the generation and loss by the tumor-bearing host of T cells cytolytic for cells of the tumor *in vitro*. For example, it was shown by Takei *et al.*[17] and later by Tuttle *et al.*[18] and by North and Dye[19] that although early growth of the P815 mastocytoma induces the generation of cytolytic T cells capable of specifically lysing P815 tumor cells *in vitro*, these T cells are lost as the tumor grows larger in size.

The generation and subsequent decay of concomitant immunity has also been measured in terms of the acquisition and loss by the tumor-bearing host of T cells capable, on passive transfer, of causing regression of a relatively small tumor in irradiated recipients. It was shown[19–21] that passive transfer of spleen cells from mice bearing any one of three different immunogenic tumors can result in partial or complete regression of a small established tumor in γ-irradiated recipients, depending on the size of the donor tumor at the time the T cells are harvested. Based on the degree of regression of the recipient tumor, it was calculated[19–21] that tumor-sensitized T cells are generated progressively between days 6 and 9 of tumor growth, and that they are then progressively lost. The T cells that transfer concomitant immunity were shown to display the $Ly1^-2^+$ phenotype, in that they were functionally eliminated by treatment with anti-Ly2 antibody and complement, but not by treatment with anti-Ly1 antibody and complement.[19–21] Thus, they bear the phenotype of cytolytic T cells, and this is in keeping with evidence[19] showing that the generation and loss by a tumor bearer of T cells capable of causing regression of a recipient tumor occurs in unison with the generation and loss of cytolytic T cells.

Needless to say, the demonstration that a mouse bearing a progressive tumor can contain tumor-sensitized T cells in large enough numbers to cause the regression of a tumor in a recipient mouse seems highly paradoxical. However, the paradox is more apparent than real when one considers that the recipient tumor needs to be about half the size of the donor tumor at the time of passive transfer. This indicates that the donor would have been capable of rejecting its own tumor if its tumor had been the size of the recipient tumor at the time of passive transfer. In other words, it seems that the primary tumor is always above a size that is capable of being rejected by the level of concomitant immunity that exists at any one time.

This is not the case for secondary tumors. On the contrary, recent published results of a study by Dye[22] reveal that, even though a primary P815 mastocytoma grows progressively in its host, the concomitant immunity that it evokes enables the host to almost completely eliminate already seeded metastases in the draining lymph node and spleen. According to this study, tumor cells disseminate from a primary subcutaneous P815 tumor during the first 6 days of tumor growth. After day 6, however, there is a dramatic destruction of disseminated tumor cells which continues until about day 12. After day 12, there is a resumption of tumor dissemination and growth of metastases in the draining nodes and spleen, and this is associated temporally with the decay of concomitant immunity. Thus, concomitant immunity, when it is present, functions to retard the development of systemic disease. This explains why mice made T cell deficient by thymectomy and lethal irradiation, and protected with bone marrow cells (TXB mice) survive only half as long as immunocompetent mice after subcutaneous implantation of P815 tumor cells. TXB mice with already established systemic disease can be rescued, however, by intravenous infusion of tumor-sensitized T cells from concomitantly immune, tumor-bearing donors.[22]

Concomitant immunity, then, represents a significant, though eventually unsuccessful, immune response to tumor-specific transplantation antigens. Obviously, this form of immunity fails to reject and immunogenic tumor, because circumstances arise after a certain stage of tumor growth that cause the immunity to undergo decay.

4. The Decay of Concomitant Immunity Is Associated with the Generation of Suppressor T Cells

It goes without saying that the reason for postulating that tumor bearers possess suppressor T cells is to explain why immunogenic tumors, in spite of their possession of tumor-specific transplantation rejection antigens, nevertheless grow progressively to kill their syngeneic hosts. It seems reasonable to suggest that if suppressor T cells are in fact responsible for the failure of the host to reject its immunogenic tumor, then they must function either to suppress an immune response from being generated in the first place, or to down-regulate an already ongoing immune response, in the second. According to the discussion in Section 3, there is ample evidence to justify hypothesizing that, in the case of many immunogenic tumors, the latter possibility is likely to be the correct one. According to such a hypothesis, suppressor T cells should not be generated until concomitant immunity begins to undergo decay, and this means that, in the case of the tumors under study in this laboratory, suppressor T cells should not be generated until after 9 days of progressive tumor growth. This hypothesis would exclude the participation of those tumor-induced T suppressor cells that have been defined in terms of their capacity to suppress the expression, in immunized mice, of a tumor-specific delayed-type hypersensitivity reaction elicited by injection of nonreplicating tumor cells. According to those who have studied this model of suppression.[23,24] T cells capable of suppressing delayed sensitivity are generated very early in tumor growth (within 48 hr of implanting tumor cells) and are progressively lost after reaching peak numbers on about day 6 or tumor growth.[25] Therefore, an alternative type of suppressor cell must be invoked to explain the loss of immunity, and this leaves suppressor T cells that are generated later in tumor growth to be considered. Tumor-induced suppressor T cells that arise later in tumor growth have been described in this laboratory over the past few years.

The assay that was developed in this laboratory to measure the presence of tumor-induced suppressor cells was based on a consideration of the knowledge that, except for rare examples, it has proved next to impossible, until fairly recently, to cause the regression of an established tumor by the passive transfer of tumor-sensitized T cells from immunized donors. This contrasts with the ease of demonstrating that the same sensitized T cells after passive transfer can neutralize the growth of a tumor implant in a normal recipient. It was reasoned, therefore, that the expression of passively transferred immunity against an already established tumor is blocked by the presence in the tumor-bearing recipient of a population of tumor-induced suppressor T cells. It was reasoned, in turn, that if this is the case, it should be possible to demonstrate successful adoptive immunization against an established tumor by employing a tumor-bearing recipient that is incapable of generating suppressor T cells (or other T cells), because of having been made T cell deficient by thymectomy and lethal irradiation, and protected with syngeneic bone marrow (TXB mice). This prediction proved correct, as evidenced by the results of experiments,[26–28] which showed that intravenous infusion of splenic T cells from immunized donors caused regression of an established tumor growing in T-cell-deficient recipients, but failed to cause regression of the same-sized tumor growing in immunocompetent recipients. It was shown next that this mechanism of suppression of expression of adoptive immunity could be passively transferred with T cells. Thus, tumor regression that follows adoptive immunization of T-cell-deficient recipients with tumor-sensitized T cells from immune donors could be inhibited by infusing the recipients several hours later with spleen cells from mice bearing an established tumor, but not by infusion of spleen cells from normal mice.[26,27] The responsible suppressor cells were T cells, as evidenced by the finding that they could be functionally eliminated by treatment with anti-Thy-1.2 antibody and complement.

This physiological passive transfer assay of suppression provided the means to investigate the suppressor cells in terms of additional properties. For example, it was shown that suppressor cells capable of passively transferring suppression are destroyed by treating the tumor-bearing donor with cyclophosphamide.[29] This is in keeping with the additional finding that treatment of tumor-bearing mice with 100 mg/kg of cyclophosphamide facilitated the expression of adoptive immunity against their tumors by passively transferred sensitized T cells. Subjecting tumor-bearing mice to a sublethal dose of ionizing radiation also allows passively transferred tumor-sensitized T cells to cause regression of the tumors.[30] Again, this immunofacilitating action of irradiation could be inhibited by the passive transfer of radiosensitive T cells from tumor-bearing donors, but not from normal donors.

The ability to passively transfer suppression also allowed the surface phenotypes of the suppressor T cells to be determined. It was revealed[19,20] that the ability of suppressor T cells from donors with well-established tumors to inhibit, on passive transfer, the expression of adoptive immunity against an established tumor in T-cell-deficient recipients was eliminated by treatment with anti-Ly1.2 antibody and complement, but not by treatment with anti-Ly2.2 antibody and complement. More recently, it has been shown[39] that the function of the suppressor T cells can be ablated by treating them with anti-L3T4 antibody and complement. There can be no doubt, then, that the suppressor cells that function in the assay described in this laboratory display the $Ly1^+2^-$, $L3T4a^+$ surface phenotypes of helper T cells. They are in no way similar, therefore, to the T suppressors of tumor-specific delayed sensitivity which, in other laboratories[23,25] as well as in this one,[39] have been shown to display the $Ly1^-2^+$ phenotype.

Another important property of the suppressors of adoptive antitumor immunity is that they are specific for the tumor that evokes their production. This was demonstrated by reciprocal passive transfer experiments with T cells from donors bearing one of two different syngeneic DBA/2 tumors: the P815 mastocytoma and the P388 lymphoma.[31] It was shown first that adoptive immunity itself is specific, in that passive transfer of P815-sensitized T cells from a P815-immunized donor caused regression of a P815 tumor, but not a P388 tumor growing in T-cell-deficient recipients. Reciprocally, P388-sensitized T cells were capable of causing regression only of the P388 tumor. Suppression of adoptive immunity proved correspondingly specific, in that the expression of adoptive immunity against the P815 mastocytoma in T-cell-deficient recipients was inhibited by infusion of T cells from P815 tumor bearer, but not from a P388 tumor bearer. Conversely, T cells from a P388 tumor bearer could suppress the expression of adoptive immunity only against the P388 tumor. This type of evidence certainly is not in keeping with the idea that tumors cause a state of generalized immunosuppression. On the contrary, there is published evidence showing that tumor-bearing mice are capable of generating normal levels of antiallograft immunity[26] and heightened levels of antibacterial and antiviral immunity.[28]

In returning to the possible role of $Ly1^+2^-$, $L3T4^+$ suppressor T cells in the decay of concomitant immunity, the subject that needs to be dealt with next is the kinetics of production of these T cells. According to the discussion in Section 3, concomitant immunity begins undergoing decay after 9 days of progressive tumor growth, at least in the case of the murine tumors under study in the laboratory. Therefore, if suppressor T cells are responsible for this decay, their generation would be expected to begin after day 9 of tumor growth. The time course of production of suppressor T cells was investigated by determining changes, against time

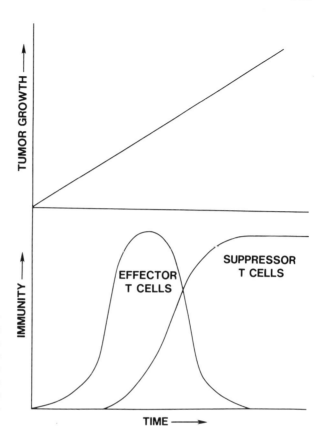

FIGURE 1. Diagrammatic representation of the generation and loss of immunity during growth of an immunogenic tumor. Evidence cited in the text shows that the premature loss of effector T cells is associated with progressive acquisition by the host of suppressor T cells.

of tumor growth, in the capacity of splenic T cells to inhibit, on passive transfer, the expression of adoptive immunity against an established tumor, according to the standard suppressor assay. It was demonstrated[19,20] that the T cells capable of suppressing the expression of adoptive immunity are acquired on about day 9 of tumor growth and increased in number thereafter. Therefore, the curve for the generation of suppressor T cells is approximately the reciprocal of the curve for the loss of effector T cells.[19,20] This is shown in Fig. 1.

5. Suppressor T Cells as the Down-Regulators of the Antitumor Immune Response

There can be no doubt, in light of the preceding discussion, that certain immunogenic tumors are capable of evoking a state a concomitant antitumor immunity in their hosts, and that this immunity undergoes eclipse after the tumor reaches a certain size. There also can be no doubt that, as concomitant immunity decays, the host generates a population of suppressor T cells, as defined by a given *in vivo* adoptive immunization assay. This does not represent causal evidence, however, that suppressor T cells down-regulate concomitant immunity. Causal evidence is supplied in the form of additional results[32] which show that intravenous infusion of suppressor T cells at an early stage of tumor growth prevents a recipient from generating concomitant immunity, as measured by failure to acquire the ability to inhibit growth of a tumor implant, and to generate T cells capable of passively transferring immunity to appropriate recipients.

Additional evidence that suppressor T cells down-regulate concomitant immunity is seen from the results of an ongoing study in this laboratory (R. J. North, to be published) of tumor regression caused by whole-body exposure to ionizing radiation. The therapeutic effect of ionizing irradiation was described originally by Hellstrom et al.[33] who demonstrated that sublethal X-irradiation of tumor-bearing mice can result in partial or complete tumor regression, depending on the immunogenicity of the tumor. The additional demonstrations that regression depended on the tumor being above a certain size, and that tumor regression could be inhibited by infusing the host with normal T cells immediately after irradiation, suggest that irradiation acts to prevent suppression of an ongoing immune response. Results recently obtained in this laboratory serve to confirm these published findings, and show, in addition, that tumor regression is associated with failure by the host to generate suppressor T cells, as measured by the basic suppressor assay described above. More to the point, irradiation-induced regression was found to be associated with the sustained presence in the host of T cells capable of passively transferring immunity to appropriate recipients.

According to those who have studied suppression of antitumor immunity *in vitro*, suppressor T cells function to inhibit the generation, rather than the function, of effector T cells.[17,34] Thus, it was shown[17,34] that addition of suppressor T cells from mice bearing a large tumor to a mixed lymphocyte–tumor cell reaction inhibited the production of cytolytic T cells. In contrast, the same suppressor T cells failed to inhibit the ability of already generated cytolytic T cells to lyse tumor targets.

These *in vitro* results are in keeping with the results of an investigation of the *in vivo* suppressor assay described in this laboratory. This investigation was based on a consideration of the repeated finding that passively transferred immunity against an established tumor in T-cell-deficient recipients is not expressed until after a delay of 6–8 days. This indicates that before tumor regression can commence, an effector mechanism needs to develop in the adoptively immunized recipients. The possiblity that passively transferred suppressor T cells function to prevent the recipient from generating cytolytic T cells was therefore investigated. It was shown[35] that immediately preceding the onset of regression of an established tumor in adoptively immunized TXB test recipients, a sizable cytolytic T-cell response was generated in the lymph node draining the tumor. In contrast, a greatly reduced cytolytic T-cell response was generated in the draining node of recipients infused with suppressor T cells immediately after being infused with immune T cells. Because there is convincing evidence[36] that the tumor-sensitized T cells routinely harvested from immunized donors are memory T cells, it was hypothesized that suppressor T cells function in the adoptive immunization assay described in this laboratory to suppress the ability of passively transferred memory

T cells to give rise to a secondary cytolytic T-cell response in the recipient. In this way, the generation of the effector mechanism needed to cause tumor regression is inhibited.

6. Conclusion

The purpose of the foregoing discussion is to present evidence for believing that most animal tumors that prove to be immunogenic, according to their ability to prophylactically immunize against a tumor implant, will also prove capable of evoking the generation of an underlying concomitant immune response during their progressive growth in syngeneic hosts. Available evidence strongly suggests that the concomitant immune response fails to cause tumor regression, because it is down-regulated by $Ly1^+2^-$, $L3T4^+$ suppressor T cells before the host has time to generate enough effector T cells. The finding that T cells with a helper phenotype can function as regulatory suppressor cells should not be considered surprising. There is now convincing evidence that T cells with this phenotype function as suppressors in other physiological models of unresponsiveness.[37,38]

The types of assays employed in this laboratory to investigate the antitumor immune response and its down-regulation by suppressor T cells in mice bearing a progressive tumor are depicted in Fig. 2.

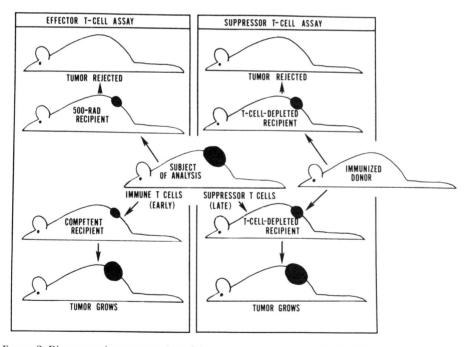

FIGURE 2. Diagrammatic representation of the *in vivo* assays employed in this laboratory for measuring the production of effector T cells and suppressor T cells during progressive growth of an immunogenic tumor.

ACKNOWLEDGMENTS. This work was supported by Grants CA-16642 and CA-27794 from the National Cancer Institute, Grant IM-431 from the American Cancer Society, and a grant from R. J. Reynolds Industries, Inc.

References

1. Foley, E. J., 1953, Antigenic properties of methylcholanthrene-induced tumors in mice at the strain of origin, *Cancer Res.* **13**:835–837.
2. Old, L. J., Boyse, E. A., Clarke, D. A., and Carswell, E. A., 1962, Antigenic properties of chemically-induced tumors, *Ann. N.Y. Acad. Sci.* **101**:80–106.
3. Sjogren, H. O., 1965, Transplantation methods as a tool for detection of tumor-specific antigens, *Prog. Exp. Tumor Res.* **6**:289–322.
4. Klein, G., 1966, Tumor antigens, *Annu Rev. Microbiol.* **20**:223–252.
5. Hellstrom, K. E., and Hellstrom, I., 1969, Cellular immunity against tumor antigens, *Adv. Cancer Res.* **12**:167–223.
6. Tuttle, R. L., and North, R. J., 1976, Mechanisms of antitumor action of *Corynebacterium parvum*: Replicating short-lived T cells as the mediators of potentiated tumor-specific immunity, *J. Reticuloendothel. Soc.* **20**:209–216.
7. Burton, R. C., Chism, S. E., and Warner, N. L., 1978, In vitro induction and expression of T cell immunity to tumor-associated antigens, *Contemp. Top. Immunobiol.* **8**:69–106.
8. Greenberg, P. D., Kern, D. E., and Cheever, M. A., 1985, Therapy of disseminated murine leukemia with cyclophosphamide and immune Ly-$1^+,2^-$ T cells: Tumor eradication does not require participation of cytotoxic T cells, *J. Exp. Med.* **161**:1122–1134.
9. Bhan, A. K., Perry, L. L., Cantor, H., McCluskey, R. T., Benacerraf, B., and Greene, M. I., 1981, The role of T cell sets in rejection of a methylcholanthrene-induced sarcoma (S1509a) in syngeneic mice, *Am. J. Pathol.* **102**:20–27.
10. Loveland, B. E., Hogarth, P. M., Ceredig, R., and McKenzie, I. F. C., 1981, Cells mediating skin graft rejection in the mouse. I. Lyt-1 cells mediate skin graft rejection, *J. Exp. Med.* **153**:1044–1057.
11. Lukacher, A. E., Morrison, L. A., Braciale, V. L., Malissen, B., and Braciale, T. J., 1985, Expression of specific cytolytic activity by H-2I region-restricted, influenza virus-specific T lymphocyte clones, *J. Exp. Med.* **162**:172–187.
12. Golding, H., Munitz, T. I., and Singer, A., 1985, Characterization of antigen-specific, Ia-restricted, L3T4$^+$ cytolytic T lymphocytes and assessment of thymic influence on their self specificity, *J. Exp. Med.* **162**:943–961.
13. Gorelik, E., 1983, Concomitant tumor immunity, *Adv. Cancer Res.* **39**:71–120.
14. North, R. J., and Kirstein, D. P., 1977, T cell-mediated concomitant immunity to syngeneic tumors: Activated macrophages as the expressors of nonspecific immunity to unrelated tumors and bacterial parasites, *J. Exp. Med.* **145**:275–292.
15. Bast, R. C., Bast, B. S., and Rapp, H. J., Critical review of previously reported animal studies of tumor immunotherapy with non-specific immunostimulants, 1976, *Ann. N.Y. Acad. Sci.* **277**:60–92.
16. Vaage, J., 1971, Concomitant immunity and specific depression of immunity by residual or reinjected syngeneic tumor tissue, *Cancer Res.* **31**:1655–1662.
17. Takai, F., Levy, J. G., and Kilburn, D. G., 1976, In vitro induction of cytotoxicity against syngeneic mastocytoma and its suppression by spleen and thymus cells from tumor bearing mice, *J. Immunol.* **116**:288–293.
18. Tuttle, R. L., Knick, V. C., Stopford, C. R., and Wolberg, G., 1983, In vivo and in vitro antitumor activity expressed by cells of concomitant immune mice, *Cancer Res.* **43**:2600–2605.
19. North, R. J., and Dye, E. S., 1985, Ly-1^+2^- suppressor T cells down-regulate the generation of Ly-1^+2^- effector T cells, *Immunology* **53**:47–56.
20. North, R. J., and Bursuker, I., 1984, The generation and decay of the immune response to a progressive fibrosarcoma: Ly-1^+2^- suppressor T cells down-regulate the generation of Ly-1^-2^+ effector T cells, *J. Exp. Med.* **159**:1295–1311.

21. North, R. J., 1984, The therapeutic significance of concomitant antitumor immunity: Ly-1$^-$2$^+$ T cells from mice with a progressive tumor cause regression of an established tumor in γ-irradiated recipients, *Cancer Immunol. Immunother.* **18**:69–74.
22. Dye, E. S., 1986, The antimetastatic function of concomitant immunity. II. Evidence that the genration of Ly-1$^+$2$^+$ effector T cells temporarily causes destruction of already disseminated tumor cells, *J. Immunol.* **136**:1510–1515.
23. Greene, M. I., 1980, The genetic and cellular basis of regulation of the immune response to tumor antigens, *Contemp. Top. Immunobiol.* **11**:81–116.
24. Hawrylko, E., 1982, Tumor bearer T cells suppress BCG-potentiated antitumor responses. 1. Requirements for their effectors, *Cell. Immunol.* **66**:121–138.
25. Hawrylko, E., Mele, C. A., and Stutman, O., 1982, Tumor bearer T cells suppress BCG-potentiated antitumor responses. II. Characteristics of the efferent phase suppressor, *Cell. Immunol.* **66**:139–151.
26. Berendt, M. J., and North, R. J., 1980, T cell-mediated immunosuppression of antitumor immunity: An explanation for progressive growth of an immunogenic tumor, *J. Exp. Med.* **151**:69–80.
27. Dye, E. S., and North, R. J., 1981, T cell mediated immunosuppression as an obstacle to adoptive immunotherapy of the P815 mastocytoma and its metastases, *J. Exp. Med.* **154**:1033–1042.
28. Bonventre, P. F., Nockol, A. D., Ball, E. J., Michael, J. G., and Bubel, H. C., 1982, Development of protective immunity against bacterial and viral infections in tumor-bearing mice is coincident with suppression, *J. Reticuloendothel. Soc.* **32**:25–34.
29. North, R. J., 1982, Cyclophosphamide-facilitated adoptive immunotherapy of an established tumor depends on the elimination of tumor-induced suppressor T cells, *J. Exp. Med.* **55**:1063–1074.
30. North, R. J., 1984, γ-Irradiation facilitates the expression of adoptive immunity against established tumors by eliminating suppressor T cells, *Cancer Immunol. Immunother.* **16**:175–181.
31. Dye, E. S., and North, R. J., 1984, Specificity of the T cells that mediate and suppress adoptive immunotherapy of established tumors, *J. Leukocyte Biol.* **36**:27–38.
32. Bursuker, I., and North, R. J., 1985, Suppression of generation of concomitant immunity by passively transferred suppressor T cells from tumor-bearing donors, *Cancer Immunol. Immunother.* **19**:215–218.
33. Hellstrom, K. E., Hellstrom, I., Kant, J. A., and Temerius, J. D., 1978, Regression and inhibition of sarcoma growth by interference with a radiosensitive T cell population, *J. Exp. Med.* **148**:799–804.
34. Frost, P., Prete, P., and Kerbel, R., 1982, Abrogation of the *in vitro* generation of the cytotoxic T cell response to a murine tumor: The role of suppressor cells, *Int. J. Cancer* **30**:211–217.
35. Mills, C. D., and North, R. J., 1983, Expression of passively transferred immunity against an established tumor depends on generation of cytolytic T cells in recipient: Inhibition by suppressor T cells, *J. Exp. Med.* **157**:1448–1460.
36. Dye, E. S., and North, R. J., 1984, Adoptive immunization against an established tumor with cytolytic versus memory T cells: Immediative versus delayed onset of regression, *Transplantation* **37**:600–605.
37. Hall, B. M., Jelbart, M. E., Gurley, K. E., and Dorsch, S. E., 1985, Specific unresponsiveness in rats with prolonged cardiac allograft survival after treatment with cyclosporine: Mediation of suppression by T helper/inducer cells, *J. Exp. Med.* **162**:1683–1694.
38. Mullen, C. A., Urban, J. L., VanWaes, C., Rowley, D. A., and Schreiber, H., 1985, Multiple cancers: Tumor burden permits the outgrowth of other cancers, *J. Exp. Med.* **162**:1665–1682.
39. DiGiacomo, A., and North, R. J., 1986, T cell suppressors of antitumor immunity. The production of Ly-1$^-$,2$^+$ suppressors of delayed sensitivity precedes the production of suppressors of protective immunity, *J. Exp. Med.* **164**:1179–1192.

20

Origin and Significance of Transplantation Antigens Induced on Cells Transformed by UV Radiation

LISA W. HOSTETLER AND MARGARET L. KRIPKE

1. Introduction

Many experimentally induced cancers express cell surface antigens not normally found on the tissue of origin. Some of these antigens induce complete or partial protection against a lethal challenge with the tumor cells. Because of their ability to induce resistance to tumor transplantation, they have been termed tumor-specific transplantation antigens (TSTA), and they are defined primarily on the basis of their ability to induce a protective immune response *in vivo*.

In addition to their practical significance, the existence of these TSTA raises some interesting philosophical questions about the origin and function of such structures. Because these antigens are potential targets for the immunologically mediated destruction of the cells, their rather common occurrence on tumors induced by certain chemical carcinogens seems paradoxical. Many investigators[1] have argued that TSTA must serve some essential physiological function in transformation or in the establishment or growth of a transformed clone of cells, otherwise they would be eliminated by means of immunoselection. Prehn,[2] for example, has proposed that a small amount of immune reactivity is actually beneficial for the survival of a newly developing tumor, and thus, TSTA confer a selective

---

LISA W. HOSTETLER AND MARGARET L. KRIPKE • Department of Immunology, University of Texas System Cancer Center, M. D. Anderson Hospital and Tumor Institute, Houston, Texas 77030.

advantage for the growth of tumor cells. Others have postulated that carcinogens must be immunosuppressive in order to induce tumors expressing TSTA.[3]

Studies on the antigenicity of tumors induced in inbred mice by polycyclic hydrocarbons have led to several key observations that are relevant to questions of the origin and possible significance of TSTA. First, antigens expressed on tumors induced by a single carcinogen are individually specific; that is, immunization with one tumor does not confer protection against challenge with a second tumor, even when the second tumor is induced in the same animal.[4] Second, the degree, or strength, of antigenicity of tumors induced by a single carcinogen varies considerably; however, each carcinogen tends to produce tumors with a discrete range of variability. For example, strong carcinogens, which induce tumors in a high proportion of animals in a short period of time, usually induce highly antigenic tumors, whereas weaker carcinogens tend to induce tumors that are less antigenic as a group.[3,4] Third, the tranformation by a single carcinogen of cells that are the progeny of a single cell also produces tumors that are individually specific and exhibit a broad range in their antigenic strength.[5] Fourth, TSTA are generally inherited by the progeny of a tumor cell.[4] Taken together, these findings have led to the idea that TSTA represent random alterations at the level of the cellular genome that are associated in some unknown manner with the transformation event. The fact that tumor cells sometimes lose their antigenic properties without losing their tumorigenic potential[6] indicates that although TSTA may be associated with the transforming event, they are not required for the maintenance of the transformed phenotype. In this chapter we reexamine questions of the origin and significance of TSTA in light of recent information gained from studies of skin cancers induced in mice by UV radiation.

2. The UV Tumor System

Skin cancers induced in inbred mice by repeated exposure to UV radiation in the UV-B (280–320 nm) range have unusual immunologic properties that make them particularly useful for studying tumor antigens. Under certain experimental conditions, a large proportion of such primary tumors exhibit a "regressor" phenotype. That is, following transplantation of small tumor fragments into syngeneic, normal recipients, the tumor tissue eventually disappears, whereas transplantation of similar fragments into immunosuppressed hosts results in the progressive and lethal growth of the tumor.[7,8] The ability of the tumors to grow in immunosuppressed hosts implied that the regression in normal animals resulted from immunologic rejection. Indeed, it has been shown using both *in vivo*[7] and *in vitro*[9–12] methods that tumor regression is accompanied by the development of immunity to the implanted tumor. The resulting transplantation immunity is specific for the immunizing tumor, and cross-protection generally cannot be induced against other UV-induced tumors using this procedure.[13] In this regard, UV-induced skin cancers resemble the tumors induced in mice and rats by chemical carcinogens. However, they exhibit a much higher degree of antigenicity as a group than chemically induced tumors.

The finding that a significant proportion of UV-induced skin cancers from several inbred mouse strains were immunologically rejected by normal, syngeneic mice raised the question of how such highly antigenic cancers could survive immunologic destruction in the primary host. Transplantation experiments addressing this question[14,15] led to the following conclusions: First, the inability of the primary host to reject these skin cancers was due to a systemic unresponsiveness of the host, rather than to inaccessibility of the tumor to immunologic attack. Second, the unresponsiveness of the primary host resulted from a systemic alteration brought about by exposing the animal to UV radiation and was unrelated to the presence of the primary tumor. Third, the UV-induced systemic alteration was immunologic because the unresponsiveness could be transferred with lymphoid cells.

Analysis of the cells responsible for the transfer of unresponsiveness revealed that in the lymphoid organs of UV-irradiated mice, T lymphocytes were present that could prevent tumor rejection in a secondary recipient.[16,17] Thus, the mechanism of unresponsiveness involves active suppression of immune rejection and is mediated by lymphocytes with the phenotypic markers Thy-1$^+$, Lyt-1$^-$, 2$^+$, I-J$^+$.[18,19] The UV-induced suppressor T lymphocytes (UV Ts) inhibit the rejection of all UV-induced skin cancers, even though these tumors exhibit individually-specific TSTA. However, they appear to recognize UV-induced tumors as a group, because they generally do not interfere with transplantation resistance to tumors induced by other carcinogens.[13,20] The inhibition by UV Ts of the immune response to a restricted set of non-cross-reacting TSTA, in particular those expressed on UV-induced skin cancers, provides an interesting model for investigating the specificity and mechanisms of immune regulation in a primary tumor system.

Another intriguing aspect of the UV-induced suppression of tumor rejection concerns the origin of the Ts in UV-irradiated mice. How these regulatory cells are activated following exposure of the skin to UV radiation is not clear, but two possibilities have been proposed. First, tumor antigens that induce the UV Ts may have physiochemical properties that enable them to stimulate Ts preferentially, perhaps by means of an affinity for cutaneous antigen-presenting cells that activate the Ts pathways.[21] Second, UV radiation can activate the Ts pathway to certain epicutaneously and subcutaneously administered antigens under defined conditions.[22-25] Perhaps the early immunologic alterations resulting from exposure of the skin to UV radiation cause Ts to be produced against antigens that otherwise would induce immune effector cells. Again, these hypotheses are not mutually exclusive, and both mechanisms may be operative.

Thus, UV-induced tumors have two unusual properties that make them particularly attractive for studies of tumor antigens. First, they are highly antigenic, and for this reason the detection of TSTA by transplantation tests and *in vitro* assays is much easier than in other systems. Also, both progressor and regressor tumors occur, enabling comparisons of the immunologic basis for their biologic behavior. Second, the tumors are recognized by UV Ts as a group, suggesting the presence of a common regulatory antigenic determinant. This means that the UV Ts can be used to identify UV-induced tumors, and they serve as a diagnostic marker for tumors with this etiology.

3. Antigens on UV Tumors

3.1. Transplantation-Defined Antigens

The first studies on the antigenicity of UV-induced skin tumors were performed by Graffi and co-workers in the early 1960s.[26] UV-induced fibrosarcomas were produced in strain XVII albino mice, and upon immunization of syngeneic mice they induced a high level of transplantation resistance. This resistance was specific for the individual tumor used for immunization,[27] which is also characteristic of tumors induced by chemical carcinogens.[28] The UV-induced fibrosarcomas were highly antigenic, even though the tumors arose after a long latency period.[29] This observation was surprising in light of the evidence from other systems that tumor antigenicity is inversely related to the latency period for tumor induction.[3,28,30] Because UV-induced tumors arise after a long latency period, one would expect them to be only weakly antigenic.

The earlier observations of Graffi and co-workers were confirmed by Kripke[7,8] who showed, in addition, that many tumors induced in mice by UV radiation failed to grow progressively in normal, syngeneic mice but grew in immunosuppressed mice. These results suggested that UV-induced tumors fail to grow in normal recipients because of a highly effective primary immune response against TSTA.

Immunity to progressor UV-induced tumors has been detected by comparing the susceptibility of immunized and unimmunized syngeneic mice to tumor challenge. With "regressor" tumors, the challenge dose of tumor cells is injected 1 day after sublethal X-irradiation. This procedure distinguishes between immune and nonimmune animals by eliminating the primary immune response, while the secondary response remains intact.[7] Animals immunized and challenged with the same tumor are highly resistant to challenge. The immunity generated is tumor-specific, in that mice challenged with a tumor different from the one used for immunization are susceptible to challenge as unimmunized mice.

Immunity to tumor antigens has also been demonstrated by the passive transfer of lymphocytes from immunized mice to immunosuppressed syngeneic recipients followed by challenge of the recipients with the immunizing tumor.[31] This experiment demonstrated that lymphoid cells from mice that have rejected a tumor are more effective in causing tumor rejection than lymphocytes from unimmunized animals. Such passive transfer of immunity depends upon the presence of T lymphocytes.[11]

The conclusion from these studies is that UV-induced murine skin cancers express very strong TSTA capable of inducing an immune response in normal, syngeneic mice. Furthermore, because the resulting transplantation immunity is specific for the immunizing tumor and cross-protection generally is not produced against other UV-induced tumors, these TSTA are considered to be individually specific. In addition to these individual TSTA, studies by Daynes and co-workers using hyperimmunized mice have suggested that common antigens might also be present on UV-induced tumors.[32,33] Recent experiments by Fortner and associates also demonstrated that hyperimmunization of mice with UV-induced tumor cells grown in tissue culture induced cross-protective immunity *in vivo* against other tumor cells that also were grown in medium containing fetal bovine serum. However, when mice were challenged with tumors grown in medium supplemented with mouse serum, only tumor-specific immunity was observed, implying that the cross-reactivity was due to adsorbed serum proteins (G. W. Fortner, personal communication). Thus, although immunologic cross-reactivity has been demonstrated by means of *in vivo* transplantation tests under certain conditions, the significance of this cross-protection and the relationship of these reactions to TSTA remain unclear.

The stability and uniformity of TSTA have also been examined in this system. Both progressor[34] and regressor[35] tumors have been cloned, and the clones have been tested for growth in normal and immunosuppresed mice and for cross-reactivity with the parent tumor. Regardless of the behavior of the parent tumor, both regressor and progressor clones were isolated from these tumors, although the majority of the clones reflected the behavior of the tumor from which they were derived. Thus, there was clear evidence of antigenic heterogeneity among cells derived from an individual tumor in terms of their antigenic strength as defined in this manner. Some clones were tested for their ability to immunize mice against challenge with the parent tumor. All clones induced protection,[35] but the amount of protection varied among the clones (M. L. Kripke, unpublished data). At least in these limited testings, there was no evidence of heterogeneity in the specificity of the TSTA, but only in the strength of the TSTA. This result is consistent with a recent study by Burnham *et al*,[36] demonstrating, by means of enzyme analysis, the clonal origin of these tumors.

The isolation of a regressor clone from a progressor tumor raises an interesting question as to how this clone managed to survive immune selection during tumor growth. It suggests that the progressor cells somehow protect the highly antigenic clone from immune attack, for example, by inducing suppressor cells, or by shedding antigens that divert the activity of immune cells. The existence of progressor clones within regressor tumors is easier to explain, particularly if they have the same antigenic specificity as the parent tumor. In this instance, the rejection of the regressor clones would bring about destruction of the progressor clones, either by means of antigen-specific effector cells or by nonspecific, "innocent bystander" killing.

Thus, individual cells within a UV-induced tumor appear to vary in their degree of antigenicity. In some instances, this could be due merely to differences in the rate of growth of the cells because this parameter influences the measurement of antigenicity. For example, a tumor with a rapid growth rate would be perceived as being less antigenic than a tumor of identical intrinsic antigenicity, but with a slower rate of growth. However, there are instances in which tumor cells clearly differ in intrinsic antigenicity, independent of their rate of division. Schreiber and colleagues[37] have produced progressor tumors from regressors by recovering the rare tumor that grows out in a normal animal. These appear to be true antigen-loss variants, and they probably represent the outgrowth of a progressor clone in response to immune selection. The existence of antigen-loss variants, which are phenotypically quite stable, points out the independence of TSTA and the transformed phenotype on tumorigenic cells in this system.

3.2. Cytotoxic Lymphocyte-Defined Antigens

Antigens on UV-induced tumors have also been demonstrated *in vitro* by the use of cytotoxic T lymphocyte (CTL) assays. In these tests, lymphocytes from tumor-immunized or normal mice are mixed with tumor cells in tissue culture. After a period of hours or days, the killing of tumor cells by CTL is quantitated by cell detachment or by the release of radioactivity from prelabeled tumor target cells. In some studies, the lymphocytes are restimulated by culturing them for several days with unlabeled tumor cells *in vitro* prior to assay with radiolabeled tumor target cells. By the use of such assays, CTL-defined tumor antigens can be detected on the surface of UV-induced tumors. Initial studies utilizing CTL assays were aimed at determining the relative antigenicity and the cross-reactivity of UV-induced tumors, and comparing the ability of UV-irradiated and unirradiated, tumor-immunized mice to generate tumor-specific CTL. A study carried out by Fortner and Kripke[9] revealed that nylon-wool-purified spleen cells from tumor-immunized, unirradiated, but not UV-irradiated, syngeneic mice exhibited a cytolytic response that was tumor-specific. Thorn[10] determined that only a secondary, not a primary, cytotoxic response to UV-induced tumors was impaired in UV-irradiated mice and confirmed that CTL activity generated by UV-induced tumors was tumor-specific. Subsequently, Daynes *et al.*[12] reported that the cytotoxic activity of cells from popliteal lymph nodes of mice immunized against syngeneic UV- or methylcholanthrene-induced tumors cross-reacted with a variety of tumor targets. They proposed that both unique and cross-reactive determinants were expressed on UV-induced tumors; however, it is possible that the nonspecific killing of tumor cells was mediated by cells other than CTL. More recently, Schreiber and colleagues[38,39] have isolated and cloned CTL lines from tumor-immunized mice. These cloned CTL are specific for the immunizing tumor and show no evidence of cross-reactivity with other UV-induced or non-UV-induced tumors.

Collectively, these studies demonstrate that UV-induced tumors can provoke a CTL response against individually specific antigens on UV-induced tumors. Although it is possible that there are also CTL directed against common determinants, the likelihood that NK cells or macrophages are responsible for the cross-reactions cannot be ruled out. In fact, experiments by Fortner and colleagues suggest that nonspecific killing[40] by lymphocytes from the spleens of mice immunized with UV-induced progressor tumors is indeed mediated by NK cells, rather than by CTL. In an interesting series of experiments, this group recently investigated differences in the response of mice to immunization with progressor and regressor UV-induced tumors. The studies indicated that tumor-specific CTL are induced by immunization with regressor tumors, but not progressor tumors. Only cytotoxic lymphocytes with the phenotypic characteristics of NK cell were produced following immunization with progressor tumors. These cytotoxic cells exhibited broad cross-reactivity with other types of tumors *in vitro* and could cause tumor rejection when mixed with tumor cells prior to inoculation. Nonetheless, the *in vivo* transplantation immunity induced by progressor tumors was highly tumor-specific. Based on these studies, Fortner postulated that progressor tumors activate a delayed hypersensitivity type of response in which recognition of the specific antigen leads to release of lymphokines. These lymphokines, in turn, activate nonspecific cytotoxic cells which do not require specific antigen recognition for target cell killing. Regressor tumors, on the other hand, stimulate the production of tumor-specific CTL, in addition to nonspecific cytotoxic cells (G. W. Fortner, personal communication). Thus, it appears that both specific and nonspecific mechanisms are involved in tumor cell killing, but as yet there is no convincing evidence that CTL recognize determinants shared by UV-induced tumors or by UV-induced and chemically induced tumors.

A question raised by the demonstration of tumor-specific CTL is whether these lymphocytes recognize the same antigen as the one responsible for tumor rejection *in vivo*. Evidence supporting the idea that CTL-defined antigens and TSTA are related, or even identical, was provided in an elegant series of experiments carried out by Schreiber and co-workers.[37] As mentioned above, they demonstrated that occasionally a regressor tumor will grow in a normal animal. Recovery and reinjection of such a tumor into normal mice resulted in progressive tumor growth. The failure of the tumor to be rejected in normal mice correlated with an inability of the tumor cells to be killed by cloned, tumor-specific CTL *in vitro*. In contrast, recovery of the tumors following growth in immunosuppressed mice did not alter the susceptibility of the tumor cells to *in vivo* rejection or CTL killing. The reciprocal experiment was also performed.[38,39] Antigen-loss variants were produced by *in vitro* selection using cloned CTL. The resulting variants were resistant to both CTL killing and tumor rejection *in vivo*, again indicating the CTL recognized the same antigen as that responsible for tumor immunity *in vivo*. Another important finding was made in these studies relating to the complexity of both TSTA and CTL-defined antigens. The antigen-loss variant produced by *in vitro* selection of a regressor UV-induced tumor (UV-1591) was used to immunize mice for CTL production, to see whether any other antigens remained after the loss of the selected antigen. Successive immunization and selection in this manner resulted in the identification of four independent, UV-1591-specific antigens, A, B, C, and D, that contribute to the antigenicity of this tumor. The significance of each antigen in tumor rejection was determined by injecting each tumor variant into normal mice. This analysis showed that the expression of the A and C, but not the B and D antigens was critical for tumor rejection. Thus, the TSTA is actually composed of at least two and possibly several more independent antigens on the UV-1591 regressor tumor. Although it is not known whether other tumors also express multiple determinants that contribute to their antigenic phenotype, this finding illustrates the potential complexity of what we call a TSTA.

3.3. Suppressor T Lymphocyte-Defined Antigens

As mentioned earlier, exposure of mice to UV radiation induces suppressor cells that play a role in the development of primary UV-induced skin cancers.[41] These UV Ts are specific for all UV-induced tumors, even though the tumors express individual TSTA.[7,27] However, the ability of the UV Ts to inhibit tumor rejection is specific for UV-induced tumors as a group, because they do not inhibit the rejection of tumors induced by other carcinogens.[20,22,42]

The apparent difference in the specificity of suppressor and effector cells in this system can be explained by two hypotheses. First, it is possible that during the course of UV irradiation a large multiplicity of antigenic changes occurs in the skin. These antigens would activate a corresponding multiplicity of suppressor clones, each of which is specific for a single antigen. A cancer cell bearing one of these antigenic changes would escape immune destruction because of the existence of a Ts clone with the appropriate specificity, and cancer cells lacking such antigens would not be detected because they would be subject to immunologic destruction. When examined as a whole population, however, the Ts would appear to have specificity for all UV-induced tumors.

The second hypothesis is that the UV Ts recognize a determinant that is common to all UV-induced tumors but is not present on tumors of a different etiology (Fig. 1). Inherent in this model is the idea that the suppressor cells recognize a different antigen or antigenic determinant than that recognized by the cells involved in tumor rejection. Thus, recognition of one structure would lead to regulation of the immune response to a different structure. It has been shown in several systems that the immune response is capable of such dual recognition.[43] Peptide fragments generated from digested proteins such as β-galactoside or hen egg lysozyme were capable of activating preferentially effector or regulatory components of the immune response.[44-47] Of note is the observation that certain portions of the protein molecule possessed antigenic determinants that activated regulatory (Th and Ts) cells. The activation of regulatory cells could, in turn, augment or suppress responses to any determinant on or attached to the same antigen. Although this phenomenon has been demonstrated using several well-defined protein molecules as antigens, the use of a common determinant to regulate the immune response against a set of non-cross-reacting TSTA has not been described

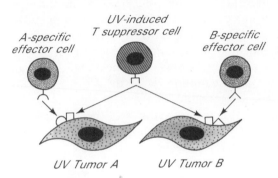

FIGURE 1. Model for the recognition of the unique and common UV-associated antigens on UV-induced tumors by tumor-specific immune effector cells and immunoregulatory suppressor cells.

previously with tumors or other etiologies. How Ts actually control the immune response to an antigenic determinant that they themselves do not seem to recognize is not clear.

The most direct test of the latter hypotheses is to clone T lymphocytes of the appropriate phenotype from UV-irradiated mice and determine their specificity. Such a study was recently carried out by Roberts[48] who isolated a Ts clone that inhibits the rejection of several UV-induced tumors, but does not interfere with immune response to a variety of exogenous antigens. It is not known whether cells from this clone affect transplantation immunity to other syngeneic tumors, but assuming that they do not, these results would indicate that UV Ts do indeed recognize a common determinant that is restricted to UV-induced tumors. However, this finding does not rule out the possible existence of tumor-specific suppressor cells, in addition to the Ts directed against a shared determinant. The approach of cloning lymphocytes from UV-irradiated mice in the absence of any antigenic stimulation or selection is not likely to permit the detection of a Ts clone that is specific for an individual UV-induced tumor. Nonetheless, the fact that cloned Ts can interfere with the rejection of several tumors expressing individually-specific TSTA indicates that a common determinant is present on these tumors. Because regulatory immune cells (Th and Ts) generally recognize class II major histocompatibiltiy antigens,[49] the UV-associated common antigen could be a class-II-like molecule that is recognized preferentially by UV Ts. Regardless of the identity of the antigen recognized by the UV Ts, these findings provide an interesting and unique model for investigating mechanisms of immune regulation in a tumor system.

3.4. Biochemically Defined Antigens

Characterization of the molecular structures of TSTA is crucial for understanding their origin and polymorphism and may provide insight into the process of neoplastic transformation. Attempts to purify TSTA from tumor cells have usually involved biochemical extraction or immunoprecipitation procedures. Isolating functionally active TSTA in sufficient quantity to permit structural characterization has proven to be quite difficult. Many TSTA are relatively weak to begin with, and dilution of soluble TSTA in irrelevant by-products of the extraction procedures further decreases the ability to detect and quantitate them. To assure that isolated TSTA have retained their antigenicity, they must be tested *in vivo* or *in vitro* for immunoprotection against the original tumor.

The extraction of cell surface antigens from whole cells or purified plasma membrane preparations is possible using detergents,[50-52] sonication,[53] and 3 M KCl.[54-57] The isolation of TSTA from UV-induced tumors has been attempted using a few of these methods.

The 3 M KCl extraction method has been used previously to solubilize tumor antigens from the plasma membranes of a variety of tumors.[54-57] Pasternak and co-workers[58,59] used this approach to solubilize tumor antigens from a skin tumor induced in strain XVII/Bln mice by UV radiation. Whole tumor cells[54] or isolated tumor cells membranes[58] were extracted, and the extracted proteins were tested for their ability to immunize *in vivo* against a challenge with the original tumor. One immunization with 1.0–1.2 mg of whole cell extract or 100 µg of extract from purified tumor membranes was able to induce immunity to lethal challenge with the UV tumor cells. Ransom *et al.*[60] also isolated a tumor rejection antigen from the membrane fraction of a UV-induced murine tumor. The tumor cells were ruptured using nitrogen cavitation, and the membrane fraction was collected by differential centrifugation and extracted in 3 M KCl. Both the membrane fraction and the extracted material induced significant immunity *in vivo*. The immunogenic material was purified further and characterized by gel filtration chromatography and gel electrophoresis. The resulting 76,000-dalton glycoprotein was immunogenic in an *in vivo* tumor rejection assay.

Recently, extraction of TSTA from tumor cells using the organic solvent 1-butanol has led to the isolation of TSTA from methylcholanthrene-induced tumors.[61–63] Extraction with butanol yields free membrane proteins without lysing or killing the tumor cells. Therefore, the butanol extracts are free of contaminating cytoplasmic proteins and contain sufficient amounts of TSTA for analysis and purification.[64] Butanol extraction of a C3H UV radiation-induced regressor tumor, UV-2240, resulted in a crude extract capable of inducing complete protection against tumor challenge with UV-2240 cells in syngeneic hosts.[65] The loss of TSTA from the butanol-extracted cells was verified in an *in vitro* CTL assay. UV-2240-specific CTL were primed *in vivo*, restimulated with UV-2240 *in vitro*, and assayed for their ability to lyse UV-2240 and butanol-extracted UV-2240 cells in a ^{51}Cr-release assay. Whereas UV-2240 cells were sensitive to cytolysis by UV-2240-specific CTL, butanol-extracted UV-2240 were resistant to cytolysis. Furthermore, incubation of an unrelated, chemically induced tumor with the crude butanol extract from UV-2240 cells rendered these cells sensitive to cytolysis by UV-2240-specific CTL. These results verify that at least certain UV-induced skin tumors exhibit cell surface TSTA that can be released by butanol extraction. However, recent attempts to extract TSTA from another regressor UV tumor, UV-1591, have been unsuccessful, suggesting that not all UV-induced TSTA are extractable with butanol (W. J. Simcik and S. J. LeGrue, unpublished data).

In spite of these attempts at purification of TSTA using biochemical approaches, no detailed structural information has resulted from these studies. The major difficulties that confound progress are the absence of a rapid, quantitative *in vitro* assay for the soluble antigen, the loss of biologic activity during purification and handling of the antigen, and the low yield of antigen after the initial purification steps. It would seem that a reasonable alternative approach to antigen purification would be to immunoprecipitate the TSTA using a tumor-specific monoclonal antibody. However, this also has proved extremely difficult. Numerous attempts to produce monoclonal antibodies or antisera specific for individual TSTA on UV-induced tumors have been largely unsuccessful. In spite of the high degree of antigenicity of these tumors *in vivo*, they are very inefficient in stimulating antibody production.[66,67] Recently, however, Schreiber and co-woerkers have produced monoclonal antibodies against several tumor-specific determinants on a single UV-induced tumor.[68,69] Studies utilizing these antibodies are described in the next section.

4. Origin of Antigenic Changes

The unusual antigenic properties of skin cancers induced by UV radiation raise many questions about the origin of these antigenic changes. What is UV radiation doing to cells to make them so antigenic? How does this differ from what chemical carcinogens do to cells? How unique is UV radiation in producing antigenic tumors? Although the answers to these questions are still not complete, some relevant information is available.

4.1. Comparison with Other Tumors

Over the years, attempts have been made to determine whether UV-induced tumors were truly unique, or whether other, closely related carcinogens produced skin tumors with similar antigenic properties. A few chemical carcinogens, such as acetyl-aminofluorine and 4-nitroquinoline oxide (4-NQO), are considered to be UV-mimetic because they activate the same biochemical pathways for repairing DNA damage as those induced by UV radiation. However, induction of murine sarcomas by s.c. injection of 4-NQO did not produce highly antigenic tumors (M. S. Fisher and M. L. Kripke, unpublished data). No regressor tumors were found, and no difference in the growth of the tumors in normal and immunosuppressed mice could be detected. This indicated that the 4-NQO-induced tumors were weakly antigenic, at best.

The chemical photosensitizer, 8-methoxypsoralen, in combination with longwave, or UV-A (320–400 nm), radiation also induces some of the same DNA repair mechanisms as UV radiation of shorter wavelengths. Skin tumors induced by psoralen and UV-A (PUVA) treatment also were only weakly antigenic, and no regressor tumors were found. In addition, these tumors were not recognized by UV Ts.[42]

It could be argued that PUVA and 4-NQO induce highly antigenic tumors that are immunologically rejected in the primary host before they can grow to a detectable size. However, in a recent study, PUVA tumors were induced in mice immunosuppressed by exposure to UV-B radiation at a different site. Even under these conditions of systemic immunosuppression, none of the resulting PUVA-induced tumors were highly antigenic (R. D. Granstein, W. L. Morison, and M. L. Kripke, unpublished data). This result makes it unlikely that PUVA treatment induced highly antigenic tumors that are eliminated because of immunoselection.

One difficulty with these experiments is that neither PUVA nor 4-NQO induces the same lesions in DNA as UV-B radiation. Thus, neither is truly UV-mimetic in this regard. Unfortunately, pyrimidine dimers, which are the main DNA alteration induced by short-wave UV radiation, are not produced in the DNA of living cells by any known chemical agent. UV radiation of wavelengths below 280 nm (UV-C) is very efficient in producing pyrimidine dimers and other lesions in the DNA of single cells. However, it is somewhat less effective than UV-B radiation in producing skin cancers in mice, presumably because it is strongly absorbed by epidermal proteins. Skin cancers induced in mice by 254-nm radiation from a germicidal lamp appear to have antigenic properties that are very similar to those of

UV-B-induced tumors: They are highly antigenic, and they are recognized by UV Ts from UV-B-irradiated mice.[70] Thus, only UV-B and UV-C radiation have been found to produce tumors with these antigenic characteristics.

A possible explanation for the unique antigenic properties of UV-induced tumors is that the antigens are present normally on cells of a particular lineage in the skin. Such cells would be highly susceptible to transformation by UV radiation but resistant to transformation by other carcinogens. This hypothesis was tested by examining the antigenic properties of cells transformed *in vitro* with UV-C radiation, X rays, or MCA.[71] The murine cell line used for the transformation studies was derived from a cloned, 10T 1/2 fibroblast line. Thus, all tumors were derived from the progeny of a single cell. The results indicated that cells transformed *in vitro* by UV radiation were highly antigenic and were recognized by the UV Ts, whereas those transformed by X rays or MCA were not. These studies showed that transformation of cells *in vitro* with UV radiation resulted in the expression of antigens similar to those on skin cancers induced *in vivo*. Furthermore, they demonstrated that the antigens were not a unique property of the particular cells transformed *in vivo* by UV radiation. This means that the antigens are related in some way to the action of UV radiation.

4.2. Molecular Analysis of TSTA on UV-1591 Tumor Cells

A possible clue to the indentity of these antigens has been provided recently in a series of very elegant studies on the regressor UV-induced tumor, UV-1591. Analysis of the TSTA using a monoclonal antibody specific for UV-1591 has produced provocative information about their possible molecular nature and genetic origin. In studies by Schreiber and colleagues,[68] a monoclonal antibody was used to precipitate a 45,000-dalton molecule that comigrated with a 12,000-dalton, β_2-microglobulin-like molecule. Although the UV-1591-specific monoclonal antibody failed to react with a wide variety of normal, embryonic and neoplastic syngeneic cells (*H-2^k* haplotype), allogeneic B10.D2 spleen cells of the *H-2^d* haplotype were lysed by the monoclonal antibody and complement. Further analysis using congenic mice indicated that the monoclonal antibody cross-reacts with H-2^d. This cross-reactivity was verified by transfecting mouse L-cells with allogeneic class I *H-2^d* genes and assaying the reactivity of the monoclonal antibody with the transfectants. The cross-reaction of the monoclonal antibody with allogeneic class I molecules suggests that UV-1591 expresses an inappropriate class I molecule, H-2^d. However, UV-1591 tumor cells reacted with three of four H-2L^d-specific antibodies and two of four H-2L^d-reactive CTL clones, but not with H-2D^d-specific antibodies. Thus, whereas the monoclonal antibody is specific for UV-1591 and cross-reacts with the H-2D^d molecule, UV-1591 cells appear to express H-2L^d-like epitopes.

This apparent paradox has been resolved by structural studies on the TSTA expressed by UV-1591. McMillan and co-workers[69] characterized the molecules expressed by UV-1591 that are recognized by the UV-1591-specific monoclonal antibody and by two H-2L^d-specific antibodies. Radioimmunoprecipitation of UV-1591 cell lysates with these antibodies, followed by two-dimensional gel electrophoresis, revealed that although UV-1591 expresses normal levels of H-2K^k and H-2D^k, it also expresses at least two additional, novel class I molecules. The first molecule is recognized by the UV-1591-specific monoclonal antibody and possesses properties of both H-2K^k and H-2D^k molecules by tryptic peptide mapping. This mosaic molecule also shares certain peptides with the H-2D^d molecule, which probably accounts for the cross-reactivity of the monoclonal antibody with H-2D^d molecules. The second molecule is recognized by anti-H-2L^d monoclonal antibodies, and although it is closely related to H-2L^d by tryptic peptide mapping, the molecule is distinct from the bona fide H-2L^d molecule expressed on BALB/c spleen cells.

Goodenow and co-workers[72] have cloned the genes encoding these novel class I molecules. These studies demonstrated that UV-1591 expresses at least three novel class I antigens not present on normal C3H tissue and that one of the antigens is derived from the *H-2K* gene. Furthermore, mouse fibroblast lines transfected with these novel class I genes are rendered susceptible to lysis by UV-1591-specific CTL. *In vitro* and *in vivo* analysis of the expression of the novel class I molecules on variants of UV-1591 reveals that the loss of the novel class I molecules from the tumor correlates with their immunogenicity.[68] For example, antigen-loss variants of UV-1591 cells selected for loss of the A antigen did not express the novel class I molecules but did express the normal H-2^k class I molecules. Therefore, at least one of these novel class I molecules appears to comprise the A antigen and is important in determining the rejection of UV-1591 *in vivo*. Therefore, in the case of UV-1591, tumor regression appears to correlate with the expression of at least one novel class I molecule, and the progressive growth of the variants correlates with the loss of this molecule.

The existence of a novel class I molecule that functions as a TSTA raises questions about the mechanism by which the antigen is generated. Comparison of the class I genes of UV-1591 and C3H tissue by Southern analysis and sequence analysis suggests that the novel class I antigens are generated by a series of recombination events among the endogenous class I genes of C3H mice.[68] This suggestion is consistent with the theory that recombination rather than mutation is responsible for the enormous polymorphism in the major histocompatibility complex of different strains of mice. Although it is possible that this recombination occurred in the germline of the mouse in which the UV-1591 tumor arose, this seems unlikely because of the high frequency of such antigenic tumors in several mouse strains. It is more likely that the novel class I antigen resulted from a somatic alteration at some point in the life history of the tumor.

These studies are extremely interesting and important because they provide the first detailed structural information on a TSTA that is not associated with a tumor virus. They also provide a possible explanation for the high degree of antigenicity of these tumors, in that they express molecules that resemble foreign histocompatibility antigens. An important question remaining unanswered is whether all UV-induced tumors resemble UV-1591 in expressing antigens derived from genes within the major histocompatibility complex, or whether this property is unique to UV-1591. The fact that monoclonal antibodies could be produced against UV-1591 but not against other regressor tumors implies that UV-1591 may be unusual among the UV-induced tumors. However, if all UV-induced, but not other skin cancers are found to express molecules resembling products of the *H-2* complex, many of which serve as immunologic recognition molecules, this could explain why the UV-induced tumors evoke such a strong immune response compared with other tumors.

4.3. Relationship between Antigenicity and Transformation

Although it seems certain that the TSTA on UV-induced tumors result from some specific action of the UV radiation, their relationship to neoplastic transformation by UV radiation is less clear. These antigens may be intimately involved in the process of transformation, or they may be a secondary and unrelated effect of the carcinogen treatment. Evidence addressing this point in other tumor systems is fragmentary and inconclusive. For example, previous attempts to induce tumor antigens in normal cells using chemical carcinogens were unsuccessful.[73] Conversely, the fact that tumors occasionally lose TSTA indicates that the antigens are not essential for maintenance of the tumorigenic phenotype.

Recent investigations indicate that tumor cells can acquire new surface antigens, as a result of exposure to certain carcinogenic agents *in vitro*.[74] The ability to increase the antigenicity of transformed cells to the level of regressor tumors suggests that antigenicity and neoplastic transformation may be completely independent events.

In the past, various methods have been used to augment the antigenicity of poorly immunogenic tumor cells. These have included infection with lytic virus,[75,76] infection with nonlytic Friend virus,[77-79] somatic hybridization of tumor cells with allogeneic cells,[80,81] and chemical coupling of haptens to the tumor cell surface.[82-84] In addition, various agents are able to increase the antigenicity of poorly antigenic tumors, resulting in their immunological rejection in syngeneic mice. An increase in the antigenicity of tumor cells was observed following treatment of murine tumors with chemotherapeutic drugs *in vivo* and *in vitro*,[85-89] and with mutagenic agents such as 4-NQO,[90] *N*-methyl-*N*-nitro-*N*-nitrosoguanidine (MNNG),[91-94] ethyl methane sulfonate (EMS),[95-99] and UV-C radiation.[99] Zbar and co-workers[100] also generated antigenic variants by treating cells of a guinea pig fibrosarcoma with MNNG. The observation that 5-azacytidine, a nonmutagenic, DNA hypomethylating agent, can augment antigenicity in various tumor lines[101] suggests that DNA hypomethylation, rather than mutation, may be responsible for the increased antigenicity.

The possibility is consistent with the fact that an extremely high frequency of antigenic variants was generated by treatment of tumor cells with MNNG,[91,93,102] EMS,[97] and UV-C radiation.[99] Many of the cloned lines derived from the carcinogen-treated cultures were immunologically rejected in normal syngeneic mice but grew in immunosuppressed mice. *In vivo* cross-reactivity and *in vitro* cytotoxicity assays indicated that the antigenic variant lines express one or more newly induced, unique antigens as well as antigens shared with the parent tumor. However, the relationship to the TSTA found on tumors induced by these agents *in vivo* cannot be assessed at present.

Recently, we have attempted to determine whether the unique and common antigens found on UV-B-induced tumors could be induced in cells that were already tumorigenic or whether their induction was associated only with a primary neoplastic transformation event. A nonantigenic tumor cell line, SF19, derived from a spontaneous C3H fibrosarcoma, was treated *in vitro* with UV-B radiation. Clonal cell lines from UV-irradiated and unirradiated SF19 cultures were isolated, expanded, and injected into normal and immunosuppressed mice to assess their antigenicity. Of the 39 cell lines established from UV-irradiated cultures of SF19 cells, 20 clones produced tumors only in immunosuppressed mice. In contrast, the parental SF19 tumor cell line and ten clones derived from it produced tumors in both types of recipients. This indicated that *in vitro* UV irradiation of a nonantigenic tumor cell line produces a high frequency of antigenic variant cells. The antigenic variants induced individually specific, non-cross-reactive immunity when mice were immunized with one variant and challenged with the same or another variant. As discussed earlier, this property is also characteristic of UV-B-induced skin tumors produced *in vivo*. However, this finding contrasts with earlier reports that immunogenic variants produced by MNNG or UV-C were at least weakly cross-reactive in immunoprotection experiments.[91,99,102]

To determine whether the antigenic variants expressed the common, UV-associated antigenic determinant that is recognized by UV Ts, cell transfer experiments were performed (Table I). Normal syngeneic mice that had been lethally X-irradiated (850 R) were reconstituted 24 hr later with spleen cells from UV-irradiated and untreated donors. The reconstituted mice were challenged 24 hr later with antigenic variant cells, a regressor UV-induced tumor (positive control), a chemically induced regressor tumor (negative control), or the parent tumor, SF19. Lymphoid cells from UV-irradiated donors did not mediate the rejection of the antigenic variants or the UV-induced tumor, and these lymphoid cells suppressed the ability of normal lymphoid cells to effect rejection. In contrast, the chemically induced regressor tumor was rejected, indicating that its rejection was unaffected by the presence of the UV Ts.

These results indicate that the antigenic variants exhibit a UV-associated common determinant that is recognized by the UV Ts. This determinant is otherwise found only on tumors induced with UV radiation.[15,17,20,42] The induction of the UV-associated antigen on a nonantigenic tumor cell by *in vitro* exposure to UV radiation implies that at least some UV-associated antigens arise as a consequence of exposing cells to UV radiation and that they occur independently of the initial neoplastic transformation event.

A second approach to investigating the relationship between tumor antigens and neoplastic transformation is to use DNA transfection. If antigenicity arises as

TABLE I
Effect of UV Radiation-Induced Suppressor Lymphocytes on the Growth of UV-Induced Antigenic Variants

Treatment of C3H mice	Number of mice with tumor/Number of mice injected with				
	SF19	Variant 1	Variant 2	UV-2240	MCA-113
850 R + 10^8 normal spleen cells	5/5	0/9	0/9	0/5	0/4
850 R + 10^8 UV-irradiated spleen cells	9/9	9/11	10/10	9/9	0/5
850 R + 5 × 10^7 normal plus 5 × 10^7 UV-irradiated spleen cells	9/10	6/10	9/9	10/10	0/4
Thymectomy + 450 R X-irradiation	10/10	8/10	10/10	10/10	5/5

a consequence of neoplastic transformation, then these two properties should be related genetically. Assuming this premise is true, then the genes encoding the two phenotypes, antigenicity and transformation, should cotransfer at high frequency following DNA transfection.[103–105]

Hopkins et al.[106] investigated the relationship between the gene(s) encoding TSTA of the MCA-induced Meth A sarcoma and the transformed phenoptype by transfecting DNA from the Meth A tumor into BALB/3T3 cells. After two successive rounds of DNA tranfection, 80% of the transformed foci also exhibited the Meth A TSTA as determined by *in vivo* transplantation immunity. These results suggested that the genes responsible for transformation and the unique TSTA of the Meth A sarcoma were closely associated. However, the subsequent finding that the Meth A TSTA was indistinguishable from antigens encoded by a transforming DNA virus makes this study difficult to interpret (N. Hopkins, personal communication). It is not clear whether the antigen on Meth A that cotransferred with the transformed phenotype truly represents a MCA-induced TSTA.

Similar experiments have been attempted using DNA from a UV-induced regressor tumor, UV-2240, and the syngeneic fibroblast line, 10T 1/2 (Ananthaswamy, unpublished data). UV-2240 DNA was cotransfected with the pSV40 plasmid, which caries the gene for resistance to the antibiotic G418. The transfected cells are selected for G418 resistance, as an indication that they have taken up foreign DNA, and a transformed phenotype. After expansion of such cells *in vitro*, they were tested for expression of UV-2240 antigens using UV-2240-specific CTL. None of the 12 transformants isolated were lysed by UV-2240-specific CTL. These preliminary results imply that UV-associated TSTA and the transformed phenotype are not linked at the genetic level, or at least that expression of the transformed phenotype does not necessarily result in expression of TSTA. Thus, this approach also points to the conclusion that the induction of transformed cells and the induction of UV-associated TSTA are two separate effects of UV irradiation.

5. Conclusions

It seems clear in the UV tumor system that at least some of the tumor antigens result from a specific, UV-induced alteration in the genome of the cell. The finding that normal cells transformed *in vitro* with UV radiation and tumorigenic cells exposed to UV radiation *in vitro* express a heritable determinant common to all UV-induced tumors, but not found on tumors induced by other carcinogens, strongly supports this conclusion. So far, this determinant is demonstrable only by virtue of its interaction with immunoregulatory cells, and its relationship with the individually specific TSTA is still obscure. That such a relationship exists is implied by the fact that suppressor cells directed against the common determinant inhibit the immune response to the individually specific TSTA. We favor the hypothesis that the common and individually specific determinants are present on a single molecule. As mentioned earlier, there are precedents for recognition of different epitopes on a single protein by regulatory and effector cells. Also, we know of no instance where the TSTA and the common antigen exist independently of each other. However, there is as yet no genetic or biochemical evidence to support this hypothesis.

We also conclude that whatever UV radiation is doing to cells to make them highly antigenic, it is different from what most other carcinogens do. Although this conclusion may seem surprising, it calls to mind studies concerning the selectivity of UV radiation in bacterial mutagenesis.[107] Although many chemical carcinogens and mutagens produce lesions at random sites in bacterial DNA, UV radiation produces lesions only in a few specific places. The basis for this selectivity is not known, but it represents an attractive model to explain the unusual antigenic properties of UV-induced tumors. If, as has been suggested by studies of UV-1591, the UV-induced TSTA represent alterations in MHC genes, it is quite possible that this region of the genome could contain such specific, UV-sensitive sites.

A third conclusion from these studies is that the induction of TSTA by UV radiation is probably independent of its transforming activity. There is no compelling evidence to suggest that these events are linked, aside from the fact that skin tumors produced by repeated UV irradiation exhibit both characteristics. We propose that the frequent association of these two effects of UV radiation is due to the following sequence of events: First, a high percentage of cells in the skin exposed to a few doses of UV radiation are induced to express UV-associated antigens. As indicated by our *in vitro* irradiation studies, expression of these antigens occurs with very high frequency after only a few exposures to UV-B radiation. Evidence from other studies is consistent with the hypothesis that the occurrence of these antigens on UV-irradiated skin precedes the appearance of detectable skin cancers. For example, grafts of preneoplastic, UV-irradiated skin may share certain antigenic properties with UV-induced tumors.[108,109] and some antisera have been shown to react with both UV-induced skin cancers and preneoplastic, UV-irradiated skin[110] Some time after induction of the antigenic changes, UV radiation induces, independently, a rare transformation event, resulting eventually in the outgrowth of a skin cancer. Because the antigenic alterations occur early and with high frequency, the subsequent transformation event is likely to occur in an antigenically altered cell.

Whether the studies on UV-induced tumors will serve as a model for the TSTA induced by other carcinogens is not really certain at present. Because of their unique properties, the UV-associated antigens may turn out to be completely different from the TSTA on chemically induced tumors, both in origin and in structure. Currently, there is no evidence for a common regulatory antigen among tumors induced by a chemical carcinogen. However, the identification of such an antigen would be extremely difficult in other systems in which the tumors are much less antigenic and carcinogen treatment does not lead to the induction of suppressor cells.

Regardless of whether UV-induced tumors will help our understanding of TSTA expressed on other types of tumors, it may be possible to use *in vitro* UV irradiation for therapeutic purposes to increase the antigenicity of tumor cells that are only weakly antigenic. In addition, understanding the antigenic makeup of UV-induced skin cancers and the immune responses against them may be helpful for certain human skin cancers. A small, but significant percentage of patients with sunlight-induced skin cancers develop recurrent or metastatic disease that is refractory to conventional treatment.[111] An immunologic approach to treatment of these cases may prove to be of considerable benefit. Immunologic studies on this tumor system have also demonstrated the importance of immunoregulatory cells in determining the outcome of carcinogenesis and have illustrated the potential complexity of tumor antigens and the influence of immunoselection. In addition, they have provided information on the induction and function of suppressor T lymphocytes. Finally, they have been instrumental in increasing our knowledge and awareness of the specialized immune surveillance system possessed by the skin.

ACKNOWLEDGMENTS. The authors gratefully acknowledge the support of PHS Grant CA40454-01 awarded by the National Cancer Institute, DHHS, and the Mary Kay Ash Foundation and the assistance of Ms. Alice Burnett in preparing the manuscript. L.W.H. is supported by a Rosalie B. Hite Predoctoral Fellowship.

References

1. Burnet, M. F., 1964, Immunological factors in the process of carcinogenesis, *Br. Med. Bull.* **20**:154–158.
2. Prehn, R. T., 1977, Immunostimulation of the lymphodependent phase of neoplastic growth, *J. Natl. Cancer Inst.* **59**:1043–1049.
3. Stjernsward, J., 1969, Immunosuppression by carcinogens, *Antibiot. Chemother. (Basel)* **15**:213–233.
4. Old, L. J., Boyse, E. A., Clarke, D. A., and Carswell, E. A., 1962, Antigenic properties of chemically induced tumors, *Ann. N.Y. Acad. Sci.* **101**:80–106.
5. Basombrio, M. A., and Prehn, R. T., 1972, Antigenic diversity of tumors chemically induced within the progeny of a single cell, *Int. J. Cancer* **10**:1–8.
6. Globerson, A., and Feldman, M., 1964, Antigenic specificity of benzo(a)pyrene-induced sarcomas, *J. Natl. Cancer Inst.* **34**:1229–1243.

7. Kripke, M. L., 1974, Antigenicity of murine skin tumors induced by ultraviolet light, *J. Natl. Cancer Inst.* **53:**1333–1336.
8. Kripke, M. L., 1977, Latency, histology, and antigenicity of tumors induced by ultraviolet light in three inbred mouse strains, *Cancer Res.* **37:**1395–1400.
9. Fortner, G. W., and Kripke, M. L., 1977, *In vitro* reactivity of splenic lymphocytes from normal and UV-irradiated mice against syngeneic UV-induced tumors, *J. Immunol.* **118:**1483–1487.
10. Thorn, R. T., 1978, Specific inhibition of cytotoxic memory cells produced against UV-induced tumors in UV-irradiated mice, J. Immunol. **121:**1920–1926.
11. Daynes, R. A., Schmitt, M. K., Roberts, L. K., and Spellman, C. W., 1979, Phenotypic and physical characteristics of the lymphoid cells involved in the immunity to syngeneic UV-induced tumors, *J. Immunol.* **122:**2458–2564.
12. Daynes, R. A., Fernandez, P. A., and Woodward, J. G., 1979, Cell-mediated immune response to syngeneic ultraviolet-induced tumors. II. The properties and antigenic specificities of cytotoxic T lymphocytes generated in vitro following removal from syngeneic tumor-immunized mice, *Cell. Immunol.* **45:**398–414.
13. Kripke, M. L., 1981, Immunologic mechanisms in UV radiation carcinogenesis, *Adv. Cancer Res.* **34:**69–106.
14. Kripke, M. L., and Fisher, M. S., 1976, Immunologic parameters of ultraviolet carcinogenesis, *J. Natl. Cancer Inst.* **57:**211–215.
15. Fisher, M.S., and Kripke, M. L., 1977, Systemic alteration induced in mice by ultraviolet light irradiation and its relationship to ultraviolet carcinogenesis, *Proc. Natl. Acad. Sci. USA* **74:**1688–1692.
16. Spellman, C. W., and Daynes, R. A., 1977, Modification of immunologic potential by ultraviolet radiation. II. Generation of suppressor cells in short-term UV-irradiated mice, *Transplantation* **24:**120–126.
17. Fisher, M. S., and Kripke, M. L., 1978, Further studies on the tumor-specific suppressor cells induced by ultraviolet radiation, *J. Immunol.* **121:**1139–1144.
18. Ullrich, S. E., and Kripke, M. L., 1984, Mechanisms in the suppression of tumor rejection produced in mice by repeated UV irradiation, *J. Immunol.* **133:**2786–2790.
19. Granstein, R. D., Parrish, J. A., McAuliffe, D. J., Waltenbaugh, C., and Greene, M. I., 1984, Immunologic inhibition of ultraviolet radiation-induced tumor suppressor cell activity, *Science* **224:**615–617.
20. Kripke, M. L., Thorn, R. M., Lill, P. H., Civin, C. I., Fisher, M. S., and Pazmino, N. H., 1979, Further characterization of immunologic unresponsiveness induced in mice by UV radiation: Growth and induction of non-UV-induced tumors in UV-irradiated mice, *Transplantation* **28:**212–217.
21. Granstein, R. D., Lowy, A., and Greene, M. I., 1984, Epidermal antigen presenting cells in activation of suppression: Identification of a new functional type of ultraviolet radiation-resistant epidermal cell, *J. Immunol.* **132:**563–565.
22. Kripke, M. L., 1984, Immunological unresponsiveness induced by ultraviolet radiation, *Immunol. Rev.* **80:**87–102.
23. Greene, M. I., Sy, M. S., Kripke, M. L., and Benacerraf, B., 1979, Impairment of antigen-presenting cell function by ultraviolet radiation, *Proc. Natl. Acad. Sci. USA* **76:**6592–6595.
24. Elmets, C. A., Bergstresser, P. R., Tigelaar, R. E., Wood, P. J., and Streilein, J. W., 1983, Analysis of the mechanism of unresponsiveness produced by haptens painted on skin exposed to low dose ultraviolet radiation, *J. Exp. Med.* **158:**781–794.
25. Noonan, F. P., DeFabo, E. C., and Kripke, M. L., 1981, Suppression of contact hypersensitivity by UV radiation and its relationship to UV-induced suppression of tumor immunity, *Photochem. Photobiol.* **34:**683–689.
26. Graffi, A., Pasternak, G., and Horn, K.-H., 1964, Die erzeugung von Resistenz gegen isolge transplantate UV-induzierter Sarkome der Maus, *Acta Biol. Med. Ger.* **12:**726–728.
27. Pasternak, G., Graffi, A., and Horn, K.-H., 1964, Der Nachweis individual spezifischer Antigenitat bei UV-induzierten Sarkomen der Maus, *Acta Biol. Med. Ger.* **13:**276–279.
28. Prehn, R. T., 1962, Specific isoantigenicities among chemically-induced tumors, *Ann. N.Y. Acad. Sci.* **101:**107–119.

29. Graffi, A., Horn, K.-H., and Pasternak, G., 1967, Antigenic properties of tumors induced by different chemical and physical agents, in: *Specific Tumor Antigens* (R. J. C. Harris, ed.), Munksgaard, Copenhagen, pp. 204–209.
30. Bartlett, G. L., 1972, Effect of host immunity on the antigenic strength of primary tumors, *J. Natl. Cancer Inst.* **49:**492–504.
31. Fisher, M. S., 1978, A systemic effect of UV-irradiation and its relationship to tumor immunity, *Natl. Cancer Inst. Monogr.* **50:**185–188.
32. Spellman, C. W., and Daynes, R. A., 1978b. Ultraviolet light induced murine suppressor lymphocytes dictate specificity of anti-ultraviolet tumor immune responses, *Cell. Immunol.* **38:**25–34.
33. Roberts, L. K., Lynch, D. H., and Daynes, R. A., 1982, Evidence for two functionally distinct cross-reactive tumor antigens associated with ultraviolet light and chemically induced tumors, *Transplantation* **33:**352–360.
34. Kripke, M. L., Gruys, E., and Fidler, I. J., 1978, Metastatic heterogeneity of cells from an ultraviolet light-induced fibrosarcoma of recent origin, *Cancer Res.* **38:**2962–2967.
35. Schmitt, M., and Daynes, R. A., 1981, Heterogeneity of tumorigenicity phenotype in murine tumors. I. Characterization of regressor and progressor clones isolated from a nonmutagenized ultraviolet regressor tumor, *J. Exp. Med.* **153:**1344–1359.
36. Burnham, D. K., Gahring, L. C., and Daynes, R. A., 1986, Clonal origin of tumors induced by ultraviolet radiation, *J. Natl. Cancer Inst.* **76:**151–158.
37. Urban, J. L., Burton, R. C., Holland, J. M., Kripke, M. L., and Schreiber, H., 1982, Mechanisms of syngeneic tumor rejection: Susceptibility of host-selected progressor variants to various immunological effector cells, *J. Exp. Med.* **155:**557–573.
38. Wortzel, R. D., Phillips, C., and Schreiber, H., 1983, Multiple tumor-specific antigens expressed on a single tumour cell, *Nature* **304:**165–167.
39. Wortzel, R. D., Urban, J. L., and Schreiber, H., 1984, Malignant growth in the normal host after variant selection in vitro with cytolytic T-cell lines, *Proc. Natl. Acad. Sci. USA* **81:**2186–2190.
40. Fortner, G. W., and Lill, P. H., 1985, Immune response to ultraviolet-induced tumors. I. Transplantation immunity developing in syngeneic mice in response to progressor ultraviolet-induced tumors, *Transplantation* **39:**44–49.
41. Fisher, M. S., and Kripke, M. L., 1982, Suppressor T lymphocytes control the development of primary skin cancers in ultraviolet-irradiated mice, *Science* **216:**1133–1134.
42. Kripke, M. L., Morison, W. L., and Parrish, J. A., 1982, Induction and transplantation of murine skin cancer induced by 8-methoxypsoralen plus UVA radiation, *J. Natl. Cancer Inst.* **68:**685–690.
43. Benjamin, D. C., Berzofsky, J. A., East, I. J., Gurd, F. R. N., Hannum, C., Leach, S. L., Margoliash, E., Michael, J. G., Miller, A., Prager, E. M., Reichlin, M., Sercarz, E. E., Smith-Gill, S. J., Todd, P. E., and Wilson, A. C., 1984, The antigenic structure of proteins: A reappraisal, *Annu. Rev. Immunol.* **2:**67–101.
44. Yowell, R. L., Araneo, B. A., Muller, A., and Sercarz, E. E., 1979, Amputation of a suppressor determinant on lysozyme reveals underlying T-cell reactivity to other detrreminants, *Nature* **279:**70–71.
45. Turkin, D., and Sercarz, E. E., 1977, Key antigenic determinants in regulation of the immune response, *Proc. Natl. Acad. Sci. USA* **74:**3984–3987.
46. Krzych, U., Fowler, A. V., and Sercarz, E. E., 1985, Repertoires of T cells directed against a large protein antigen, beta-galactoside. I. Helper cells have a more restricted specificity repertoire than proliferative cells, *J. Exp. Med.* **162:**311–323.
47. Oki, A., and Sercarz, E. E., 1985, T cell tolerance studied at the level of antigenic determinants. I. Latent reactivity to lysozyme peptides that lack suppressogenic epitopes can be revealed in lysozyme-treated mice, *J. Exp. Med.* **161:**879–911.
48. Roberts, L. K., 1986, Characterization of a cloned ultraviolet radiation (UV)-induced suppressor T cell line that is capable of inhibiting anti-UV tumor-immune responses, *J. Immunol.* **136:**1908–1916.
49. Klein, J., Juretic, A., Boxevanis, C. N., and Nagy, Z. A., 1981, The traditional and a new version of the mouse H-2 complex, *Nature* **291:**455–460.
50. Snary, D., Goodfellow, P., Hayman, M. J., Bodmer, W. F., and Crumpton, M. J., 1974, Subcellular separation and molecular nature of human histocompatibility antigens (HLA), *Nature* **247:**457–461.

51. Natori, T., Law, L. W., and Apella, E., 1977, Biological and biochemical properties of Nonidet-P40 solubilized and partially purified tumor-specific antigens of the transplantation type from membranes of a methylcholanthrene-induced sarcoma, *Cancer Res.* **37**:3406–3413.
52. Natori, T., Law, L. W., and Apella, E., 1978, Immunochemical evidence of a tumor-specific surface antigen obtained by detergent solubilization of the membranes of a chemically-induced sarcoma, *Cancer Res.* **38**:359–364.
53. Kahan, B. D., 1965, Isolation of a soluble transplantation antigen, *Proc. Natl. Acad. Sci. USA* **53**:153–161.
54. Reisfeld, R. A., Pellegrino, M. A., and Kahan, B. D., 1971, Salt extraction of soluble HL-A antigens, *Science* **172**:1134.
55. Meltzer, M. S., Leonard, E. J., Hardy, A. S., and Rapp, H. J., 1975, Protective tumor immunity induced by potassium chloride extracts of guinea pig hepatomas, *J. Natl. Cancer Inst.* **54**:1349–1354.
56. Pellis, N. R., Tom, B. H., and Kahan, B. D., 1984, Tumor-specific and allospecific immunogenicity of soluble extracts from chemically induced murine sarcomas, *J. Immunol.* **113**:708–711.
57. Pellis, N. R., and Kahan, B. D., 1975, Specific immunoprotection with 3M KCl solubilized tumor antigen, *J. Surg. Res.* **18**:263–269.
58. Pasternak, L., Pasternak, G., and Karsten, V., 1978, Immunogenicity of soluble extracts from a UV light-induced mouse sarcoma, *Cancer Immunol. Immunother.* **3**:272–275.
59. Pasternak, L., and Ristau, E., 1985, [Transplantation immunity to a UV-induced murine sarcoma by injection of a glycoprotein fraction from the tumor], *Arch. Geschwulstforsch.* **55**:17–21.
60. Ransom, J. H., Schengrund, C.-L., and Bartlett, G. L., 1981, Solubilization and partial characterization of a tumor-rejection antigen from an ultraviolet light-induced murine tumor, *Int. J. Cancer* **27**:545–554.
61. LeGrue, S. J., Kahan, B. D., and Pellis, N. R., 1980, Extraction of a murine tumor-specific transplantation antigen with 1-butanol. I. Partial purification by isoelectric focusing, *J. Natl. Cancer Inst.* **65**:191–196.
62. LeGrue, S. J., Allison, J., Macek, C., Pellis, N. R., and Kahan, B. D., 1981, Immunobiological properties of 1-butanol-extracted cell surface antigens, *Cancer Res.* **41**:3956–3960.
63. LeGrue, S. J., Pellis, N. R., Riley, L. B., and Kahan, B. D., 1985, Biochemical characterization of 1-butanol extracted murine tumor-specific transplantation antigens, *Cancer Res.* **45**:3164–3172.
64. LeGrue, S. J., 1985, Noncytolytic extraction of cell surface antigens using butanol, *Cancer Metastasis Rev.* **4**:209–219.
65. LeGrue, S. J., Simcik, W. J., Ananthaswamy, H. N., and Kripke, M. L., 1985, Extraction of tumor-associated antigens from UV-irradiation induced tumor using butanol, *Proc. Am. Assoc. Cancer Res.* **26**:A1229.
66. DeLuca, D., Kripke, M. L., and Marchalonis, J. J., 1979, Induction and specificity of antisera from mice immunized with syngeneic UV-induced tumors, *J. Immunol.* **123**:2696–2703.
67. DeWitt, C. W., 1981, Ultraviolet light induces tumors with both unique and host-associated antigenic specificities, *J. Immunol.* **127**:329–334.
68. Phillips, C., McMillan, M., Flood, P., Murphy, D. B., Forman, J., Lancki, D., Womack, J. E., Goodenow, R. S., and Schreiber, H. S., 1985, Identification of a unique tumor-specific antigen as a novel class I major histocompatibility molecule, *Proc. Natl. Acad. Sci. USA* **82**:5140–5144.
69. McMillan, M., Lewis, K. D., and Rovner, D. M., 1985, Molecular characterization of novel class I molecules expressed by a C3H-UV-induced fibrosarcoma, *Proc. Natl. Acad. Sci. USA* **82**:5485–5489.
70. Lill, P. H., 1983, Latent period and antigenicity of murine tumors induced in C3H mice by short wavelength ultraviolet radiation, *J. Invest. Dermatol.* **81**:342–346.
71. Fisher, M. S., Kripke, M. L., and Chan, G. L., 1984, Antigenic similarity between cells transformed by ultraviolet radiation in vitro and in vivo, *Science* **223**:593–594.
72. Goodenow, R. S., Vogel, J. M., and Linsk, R. L., 1985, Histocompatibility antigens on murine tumors, *Science* **230**:777–783.
73. Outzen, H. C., Andrews, E. J., Basombrio, M. A., Litwin, S., and Prehn, R. T., 1972, Attempted induction of tumor antigens in carcinogen-treated cells, *J. Natl. Cancer Inst.* **49**:1295–1302.
74. Boon, T., 1983, Antigenic tumor cell variants obtained with mutagens, *Adv. Cancer Res.* **39**:121–151.

75. Lindenmann, J., and Klein, P., 1967, Viral oncolysis: Increased immunogenicity of host cell antigen associated with influenza virus, *J. Exp. Med.* **126**:93–108.
76. Boone, C., and Blackman, K., 1972, Augmented immunogenicity of tumor cell homogenates infected with influenza virus, *Cancer Res.* **32**:1018–1022.
77. Kobayashi, H., Sendo, F., Shirai, T., Kaji, H., and Kodama, T., 1969, Modification in growth of transplantable rat tumors exposed to Friend virus, *J. Natl. Cancer Inst.* **42**:413–419.
78. Kobayashi, H., Shirai, T., Takeichi, N., Hosokawa, M., Saito, H., Sendo, F., and Kodama, T., 1970, Antigenic variant (WFT-2N) of a transplantable rat tumor induced by Friend virus, *Rev. Eur. Etud. Clin. Biol.* **15**:426–428.
79. Kobayashi, H., Gotohda, E., Hosokawa, M., and Kodama, T., 1975, Inhibition of metastasis in rats immunized with xenogenized autologous tumor cells after excision of the primary tumor, *J. Natl. Cancer Inst.* **54**:997–999.
80. Klein, G., and Klein, E., 1979, Induction of tumor cell rejection in the low responsive YAC–lymphoma strain A host combination by hybridization with somatic cell hybrids, *Eur. J. Cancer* **15**:551–557.
81. Toffaletti, D. L., Darrow, T. L., and Scott, D. W., 1983, Augmentation of syngeneic tumor-specific immunity by semiallogeneic cell hybrids, *J. Immunol.* **130**:2982–2986.
82. Martin, W. J., Wonderlich, J. R., Fletcher, F., and Inman, J. K., 1971, Enhanced immunogenicity of chemically-coated syngeneic tumor cells, *Proc. Natl. Acad. Sci. USA* **68**:469–472.
83. Galili, N., Naor, D., Asjo, B., and Klein, G., 1976, Induction of immune responsiveness in a genetically low-responsive tumor–host combination by chemical modification of the immunogen, *Eur. J. Immunol.* **6**:473–476.
84. Fujiwara, H., Hamaoka, I., Shearer, G., Yamamoto, H., and Terry, W., 1980, The augmentation of in vitro and in vivo tumor-specific T cell mediated immunity by amplifier T lymphocytes, *J. Immunol.* **124**:863–869.
85. Bonmasser, V. E., Bonmasser, A., Vadlamudi, S., and Goldin, A., 1970, Immunologic alteration of leukemic cells in vivo after treatment with an antitumor drug, *Proc. Natl. Acad. Sci. USA* **66**:1089–1095.
86. Tsukagoshi, S., and Hashimoto, Y., 1973, Increased immunosensitivity in nitrogen mustard-resistant Yoshida sarcoma, *Cancer Res.* **33**:1038–1042.
87. Schmid, F. A., and Hutchison, D. J., 1973, Decrease in oncogenic potential of L1210 leukemia by triazenes, *Cancer Res.* **33**:2161–2165.
88. Fuji, H., and Mihich, E., 1975, Selection for high immunogenicity in drug-resistant sublines of murine lymphomas demonstrated by plaque assay, *Cancer Res.* **35**:946–952.
89. Contessa, A. R., Bonmasser, A., Giampietri, A., Circolo, A., Goldin, A., and Fioretti, M. C., 1981, In vitro generation of a highly immunogenic subline of L1210 leukemia following exposure to 5-(3,3′-dimethyl-l-triazeno)imidazole-4-carboxamide, *Cancer Res.* **41**:2476–2482.
90. Koyama, K., and Ishii, K., 1969, Induction of non-transplantable mutant clones from an ascites tumor, *Gann* **60**:367–374.
91. Boon, T., and Kellermann, O., 1977, Rejection by syngeneic mice of cell variants obtained by mutagenesis of a malignant teratocarcinoma cell line, *Proc. Natl. Acad. Sci. USA* **74**:272–275.
92. Van Pel, A., Georlette, M., and Boon, T., 1979, Tumor cell variants obtained by mutagenesis of a Lewis lung carcinoma cell line: Immune rejection by syngeneic mice, *Proc. Natl. Acad. Sci. USA* **76**:5282–5285.
93. Van Pel, A., Vessiere, F., and Boon, T., 1983, Protection against two spontaneous mouse leukemias conferred by immunogenic variants obtained by mutagenesis, *J. Exp. Med.* **457**:1992–2001.
94. Uyttenhove, C., Van Snick, J., and Boon, T., 1980, Immunogenic variants obtained by mutagenesis of mouse matocytoma P815. I. Rejection by syngeneic mice, *J. Exp. Med.* **152**:1175–1183.
95. Kerbel, R. S., 1979, Immunologic studies of membrane mutants of a highly metastatic murine tumor, *Am. J. Pathol.* **97**:609–622.
96. Kerbel, R. S., Dennis, J. W., Largarde, A. E., and Frost, P., 1982, Tumor progression in metastasis: An experimental approach using lectin resistant tumor variants, *Cancer Metastasis Rev.* **1**:99–140.
97. Frost, P., Kerbel, R. S., Bauer, E., Tartamella-Biondo, R., and Cefalu, W., 1983, Mutagen treatment as a means for selecting immunogenic variants from otherwise poorly immunogenic malignant murine tumors, *Cancer Res.* **43**:125–132.

98. Carlow, D. A., Kerbel, R. S., Feltis, J. T., and Elliott, B. E., 1985, Enhanced expression of class I major histocompatibility complex gene (Dk) products on immunogenic variants of a spontaneous murine carcinoma, *J. Natl. Cancer Inst.* **75**:291–300.
99. Peppoloni, S., Herberman, R. B., and Gorelik, E., 1985, Induction of highly immunogenic variants of Lewis lung carcinoma tumor by ultraviolet irradiation, *Cancer Res.* **45**:2560–2566.
100. Zbar, B., Tanio, Y., Terata, N., and Hovis, J., 1984, Antigenic variants isolated from a mutagen-treated guinea pig fibrosarcoma, *Cancer Res.* **44**:5079–5085.
101. Kerbel, R. S., Frost, P., Liteplo, R., Carlow, D. A., and Elliott, B. E., 1984, Possible epigenetic mechanisms of tumor progression: Induction of high-frequency heritable but phenotypically unstable changes in tumorigenic and metastatic properties of tumor cell populations by 5-azacytidine treatment, *J. Cell. Physiol. Supp.* **3**:87–97.
102. Maryanski, J., Marchand, M., Uyttenhove, C., and Boon, T., 1983, Immunogenic variants obtained by mutagenesis of mouse mastocytoma P815. VI. Occasional escape from host rejection due to antigen-loss secondary variants, *Int. J. Cancer* **31**:119–123.
103. Graham, E. L., and Van der Eb, A. J., 1973, A new technique for the assay of infectivity of human adenovirus 5 DNA, *Virology* **52**:456–467.
104. Shih, C., Shilo, B.-Z., Goldfarb, M. P., Dannenburg, A., and Weinberg, R. A., 1979, Passage of phenotypes of chemically transformed cells via transfection of DNA and chromatin, *Proc. Natl. Acad. Sci. USA* **76**:5714–5718.
105. Cooper, G. M., and Neiman, P., 1980, Transforming genes of neoplasms induced by avian lymphoid leukosis viruses, *Nature* **287**:656–659.
106. Hopkins, N., Besmer, P., Deleo, A. B., and Law, L. W., 1981, High frequency cotransfer of the transformed phenotype and a tumor-specific transplantation antigen by DNA from the 3-methylcholanthrene-induced Meth A sarcoma of Balb/c Mice, *Proc. Natl. Acad. Sci. USA* **78**:7555–7559.
107. Miller, J. H., 1985, Mutagenic specificity of ultraviolet light, *J. Mol. Biol.* **182**:45–65.
108. Palaszynski, E. W., and Kripke, M. L., 1983, Transfer of immunological tolerance to UV radiation-induced skin tumors with grafts of UV-irradiated skin, *Transplantation* **36**:465–467.
109. Spellman, C. W., and Daynes, R. A., 1984, Cross-reactive transplantation antigens between UV-irradiated skin and UV-induced tumors, *Photodermatology* **1**:164–169.
110. Hong, S. R., and Roberts, L. K., 1987, Cross-reactive tumor antigens in the skin of mice exposed to subcarcinogenic doses of ultraviolet radiation, *J. Invest. Dermatol.* **88**:154–160.
111. Goepfert, H., Dichtel, W. J., Medina, J. E., Lindberg, R. D., and Luna, M. D., 1984, Perineural invasion in squamous cell carcinoma of the head and neck, *Am. J. Surg.* **148**:542–547.

21
Cellular and Molecular Mechanisms Involved in Tumor Eradication *in Vivo*

HIROMI FUJIWARA, TAKAYUKI YOSHIOKA, HIROTO NAKAJIMA, MASAHIRO FUKUZAWA, KOHICHI SAKAMOTO, MASATO OGATA, SHIGETOSHI SANO, JUN SHIMIZU, CHIHARU KIYOTAKI, AND TOSHIYUKI HAMAOKA

1. Introduction

Despite ample evidence of the role of T lymphocytes in rejecting tumor cells as well as allografts,[1] there are large gaps in our understanding of the effector mechanisms responsible for graft rejection. It is clearly important for both scientific and clinical reasons to define the effector mechanisms that take place within a rejecting tumor graft. The identification of an effector mechanism within a tumor graft does not necessarily imply a functional role for the mechanism in tumor destruction *in vivo*. Therefore, in considering the relevance of a particular mechanism, we must know whether the effector mechanism detected is subsidiary, necessary, or sufficient for tumor rejection *in vivo*.

Immune responses to tumor-associated transplantation antigens (TATA), similar to responses against nominal foreign antigens, consist of a complicated immune network formed by B cells, macrophages, and various subsets of T cells. This network includes TATA-recognition (afferent) and antitumor attacking (effer-

---

HIROMI FUJIWARA, TAKAYUKI YOSHIOKA, HIROTO NAKAJIMA, MASAHIRO FUKUZAWA, KOHICHI SAKAMOTO, MASATO OGATA, SHIGETOSHI SANO, JUN SHIMIZU, CHIHARU KIYOTAKI, AND TOSHIYUKI HAMAOKA • Department of Oncogenesis, Institute for Cancer Research, Osaka University Medical School, Osaka 553, Japan.

ent) phases. Therefore, when a particular cell type is essentially required for tumor rejection, it is also important to define the role(s) of such a cell type in each phase and to understand the whole pathway of cellular collaborations leading to the generation of the final effector mechanism.

Although the *in vitro* lysis of tumor target cells by cytotoxic T lymphocytes (CTL) has long been considered to be an appropriate model for studying tumor immunity,[2,3] it is now known that a T-cell subset which is distinct from the CTL population exerts its critical function in allograft rejection[4-7] and tumor cell eradication *in vivo*.[8-12] In this review we shall consider critically the evidence from our own and other studies that a T-cell subset distinct from the CTL population mediates *in vivo* tumor rejection and analyze the cellular and molecular mechanisms involved in pathways leading to ultimate tumor eradication *in vivo*.

2. The Role of Tumor-Specific Lyt-1^+2^- T-Cell Subset in Eradicating Tumor Cells *in Vivo*

It has long been accepted that CTL have a major role in tumor immunity and that the *in vitro* tumor target lysis by CTL primed *in vivo* to TATA is an appropriate model for studying tumor graft rejection.[2,3] In fact, CTL can be found in tumor masses undergoing rejection[2,13,14] and lymphoid tissues of TATA-primed mice.[3] Although many experiments have supported the concept that CTL can mediate tumor rejection *in vivo*, there is little direct evidence for this.

The *in vivo* activity of T-cell subsets against syngeneic tumors has been analyzed in several tumor models. Tumor neutralization of a syngeneic sarcoma in a Winn assay required Lyt-1^+2^+ immune cells.[15] Prevention of outgrowth of a Moloney lymphoma, in which immune cells were inoculated before the tumor challenges, required Lyt-1^-2^+ immune cells.[16] These immune cells were also responsible for the *in vitro* lysis of the corresponding tumor cells. However, recent analyses of cell types mediating tumor[8-12] as well as allograft rejection[4-7] have revealed that Lyt-1^+2^- T cells, which are distinct from T cells mediating cytotoxicity (Lyt-1^+2^+/1^-2^+), have a crucial role in rejecting grafts *in vivo*. For example, disseminated syngeneic leukemia cells could be eradicated by adoptive transfer of antitumor Lyt-1^+2^- T cells into recipients pretreated with noncurative, nonlethal chemotherapy with cyclophosphamide.[11]

We have also demonstrated the efficacy of Lyt-1^+2^- T cells in *in vivo* tumor rejection in several experimental systems. First, based on the T cell–T cell interaction concept, we have established experimental systems in which enhanced tumor-specific immunity could be generated by preinducing trinitrophenyl (TNP)-reactive,[17,18] vaccinia virus-reactive,[19] or haptenic muramyl dipeptide (MDP)-reactive helper T cells[20, and chapter 22, this volume] and by subsequently immunizing with TNP- or MDP-coupled or vaccinia-infected syngeneic tumor cells. The results demonstrated that mice immunized with these helper antigen (TNP, MDP, or vaccinia)-modified syngeneic tumor cells in the presence of the above helper T cells developed enhanced tumor-specific *in vivo* protective immunity in syngeneic plasmacytoma, leukemia, and hepatoma models.[17-20, and Chapter 22, this volume] Furthermore, the

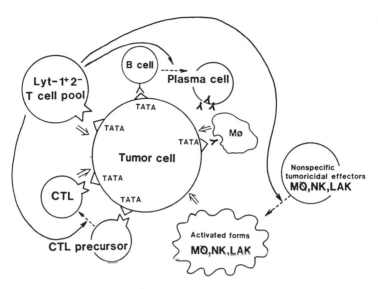

FIGURE 1. Role of Lyt-1^+2^- T cells in *in vivo* tumor cell eradication.

principle of T cell–T cell interaction mechanism for enhanced induction of tumor-specific immunity was successfully applied to the active tumor-specific immunotherapy model in which a growing tumor regresses by anti-TATA specific effector T cells augmented through collaboration with anti-TNP helper T cells.[22,23] In this model, the mice in which tumor regression occurred, retained a potent tumor-specific immunity. *In vivo* protective immunity that was generated in the above immunoprophylaxis as well as immunotherapy models was found to be mediated by the Lyt-1^+2^- T cells,[24] when Winn assays were performed in normal syngeneic recipient mice by using spleen cells from tumor-immunized or tumor-regressed mice.

Second, the effectiveness of Lyt-1^+2^- T cells was also observed in the effector cell analysis of lymphoid cells infiltrating into syngeneic plasmacytoma and hepatoma tumor masses (see Section 5 for details). Winn assays utilizing such tumor-infiltrating lymphoid cells have revealed that tumor neutralization was mediated only by tumor-specific Lyt-1^+2^- T cells which have no potential to generate *in vitro* cytotoxic activity as measured by the 4-hr ^{51}Cr release assay.[25]

Although this finding and results obtained by others[8-11] demonstrated the efficacy of the Lyt-1^+2^- T-cell subpopulation for the *in vivo* eradication of tumor cells, the mechanism of antitumor function by these Lyt-1^+2^- T cells was not determined. The potential mechanisms by which Lyt-1^+2^- T cells bring about tumor protection must reflect one or several of their functions. These include (1) anti-TATA helper T cells assisting CTL or antibody responses against TATA, (2) direct antitumor attackers exhibiting cytotoxicity as detected by a long-term, but not by a short-term (4 hr) cytotoxicity assay, or (3) T-cell-produced lymphokines able to activate other cell types such as macrophages which could act as the final tumoricidal effector cells (Fig. 1).

3. The Mechanisms Underlying Lyt-1$^+$2$^-$ T-Cell-Mediated Tumor Eradication in Vivo

3.1. Role of Lyt-1$^+$2$^-$ T Cells as Helper T Cells Assisting CTL or Antibody Responses to TATA

In most syngeneic tumor models, tumor-specific CTL activity can be detected in spleen or lymph node cells from tumor-immunized or tumor-regressed mice. Since the generation of CTL responses requires the participation of helper T cells expressing Lyt-1$^+$2$^-$ cell surface markers,[26–28] the anti-TATA Lyt-1$^+$2$^-$ T-cell subset might function as helper T cells assisting CTL responses against TATA.[29] In the above-mentioned studies,[9–11] the efficacy of the Lyt-1$^+$2$^-$ T-cell subset for *in vivo* tumor protection was examined using normal syngeneic mice as recipients. It is possible that CTL generated from the host's precursor T-cell pool are involved in the tumor eradication. The Lyt-1$^+$2$^-$ cells leading to CTL generation might still be relevant to tumor protection, but whether this pathway is necessary or sufficient for *in vivo* tumor protection remains unresolved. In order to know if an additional pathway mediated by Lyt-1$^+$2$^-$ T cells is functioning, it is of great importance to define an absolute requirement of CTL generation for *in vivo* protection.

This was approached by preparing T-cell-depleted B-cell mice and testing whether Lyt-1$^+$2$^-$ T cells, which had been demonstrated to generate no CTL response exert their protective function when adoptively transferred into B-cell mice depleted of CTL precursor potential.[12,30] For example, B-cell mice, prepared as shown in Fig. 2, were injected with various subpopulations of splenic T cells from tumor-immunized mice.[12] C3H/He B-cell mice that received Lyt-1$^-$2$^+$ T cells from syngeneic X5563 tumor-immunized mice exhibited an appreciable CTL response to X5563 tumor, whereas they failed to resist the intradermal challenge of X5563 tumor cells. In contrast, the adoptive transfer of Lyt-1$^+$2$^-$ anti-X5563 immune T cells into B-cell mice produced complete protection against subsequent tumor cell challenge (Fig. 2). In these models, no CTL response against X5563 tumor was detected in the above tumor-resistant B-cell mice. Thus, the possibility that immune Lyt-1$^+$2$^-$ T cells exert their activity by functioning as helper T cells to augment anti-X5563 CTL responses from CTL precursors recruited from the host is unlikely.

An absolute requirement of CTL generation for *in vivo* tumor rejection was also excluded in other tumor models. In the model of the CCl$_4$-induced MH134 hepatoma[19,31] and the MCA-11-A1 fibrosarcoma induced by methylcholanthrene (MCA), *in vivo* immune resistance was easily induced. Lymphoid cells from these tumor-immune mice failed to exhibit any significant antitumor CTL activity even after *in vitro* resensitization with tumor cells although these lymphoid cells contained Lyt-1$^+$2$^-$ T-cell activity mediating *in vivo* immunity.[19,31]

Antitumor antibody can be induced in some tumor systems,[19,32] and such antibody would function for tumor cell killing in collaboration with Fc receptor-bearing effector cells.[33] However, it is also unlikely that tumor-immune Lyt-1$^+$2$^-$ T cells must function as helper T cells assisting anti-TATA antibody responses for

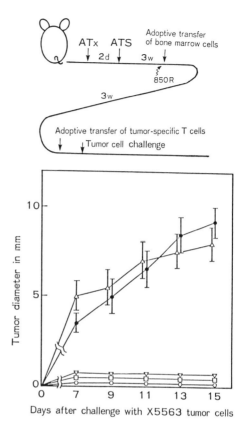

FIGURE 2. Immune protection in B-cell mice by adoptive transfer with Lyt-1^+2^- T-cell subset.[12] B-cell mice were adoptively transferred with 5×10^7 normal (●) or X5563-immunized spleen cells 1 day before challenge with 10^5 viable X5563 tumor cells. Immune spleen cells used for the adoptive transfer were untreated (○) or treated with either complement (C) alone (□), anti-Lyt-1.1 + C (△), or anti-Lyt-2.1 + C (▽).

exerting their activity, since mice immune to syngeneic plasmacytoma[19] and most of the MCA-induced fibrosarcomas[34-36] failed to exhibit any antibody activity in their sera as detected by indirect immunofluorescence antibody technique. Thus, these observations lead to the conclusion that tumor-specific Lyt-1^+2^- T cells do not necessarily require generation of antitumor CTL or antibody responses for the implementation of *in vivo* tumor-protective immunity and suggest the existence of additional Lyt-1^+2^- T-cell-mediated tumor rejection mechanism(s) other than the mechanisms leading to CTL and antibody generation.

3.2. Correlation of Tumor Protection and DTH Responses

An alternative mechanism by which Lyt-1^+2^- cells convey tumor resistance could be by the initiation of a delayed-type hypersensitivity (DTH) response. Sensitization of A/J mice with a syngeneic fibrosarcoma induces T cells that can adoptively transfer tumor-specific DTH responses to a secondary host.[9] There also appears to be a close association between graft rejection and DTH phenomena for H-2 differences[37] and H-Y differences.[38] Conversely, graft rejection and the detection of CTL are independent phenomena in the H-Y model. In the aforementioned tumor models, immunization of C3H/He mice with vaccinia virus-infected syngeneic X5563 plasmacytoma or MH134 hepatoma cells in the presence of vaccinia virus-reactive helper T-cell activity resulted in anti-tumor-specific CTL or antibody responses, respectively, along with potent *in vivo* immune resistances.[19] Although any significant CTL or antibody activity was not detected in the respective MH134- or X5563-immunized mice, C3H/He mice immunized to either type of the tumor exhibited strong antitumor DTH responses.[39] Such enhanced tumor-specific DTH responses were shown to be mediated by Lyt-1^+2^- T cells.[39] Moreover, B-cell mice which had been injected with tumor-immune Lyt-1^+2^- T cells and resisted the corresponding tumor challenge did not exhibit any significant anti-TATA CTL or antibody response, but they produced tumor-specific footpad swelling by injection of tumor cells as DTH-eliciting antigens.[12,31] These findings, taken together with others, indicate that graft rejection is associated with the detection of DTH but not always with CTL or antibody responses.

A DTH phenomenon consists of a complex immune response induced by Lyt-1^+2^- T cells which release a variety of lymphokines. These include lymphotoxin which might regulate direct cytolytic or cytostatic effects. Macrophage-activating factor (MAF), γ-interferon (γ-IFN), and interleukin-2 (IL-2) are released and are capable of activating other cell types.[40-42] Although the association of Lyt-1^+2^- T-cell-mediated DTH responses and tumor protection has been described, whether Lyt-1^+2^- T cells can produce these lymphokines and initiate a series of reactions that are involved in tumor rejection *in vivo* has not been determined. Our recent investigation has focused on this issue, which will be described in the following subsection.

3.3. Collaborative Antitumor Effect of Lyt-1^+2^- T Cells and Nonspecific Tumoricidal Effectors

While a Lyt-1^+2^- T-cell-mediated pathway distinct from that of CTL-mediated cytotoxicity functions *in vivo* for tumor growth inhibition, it remains to be tested whether these Lyt-1^+2^- T cells can exert their cytolytic or cytostatic effect directly on tumor cells. Although Lyt-1^+2^- T cells failed to exhibit any cytotoxic effect on tumor cells in a short-term cytotoxicity test such as the 4-hr ^{51}Cr release assay (our unpublished observation), it is still possible that they function as cytotoxic effectors in a long-term assay system. Since it is generally difficult to maintain the viability of tumor target cells *in vitro* for a long period, establishing an appropriate system in which the above Lyt-1^+2^- T-cell-mediated antitumor effect is assessed could lead to elucidation of the tumor-inhibiting mechanism mediated by this T-cell subset. In this respect, a diffusion chamber culture technique which has been utilized for various purposes[43,44] could be applied to investigating the above possibilities. This technique has enabled us to maintain viable tumor target cells for relatively long

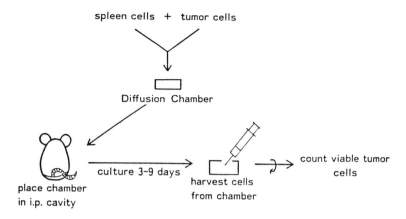

FIGURE 3. Experimental design using diffusion chamber.[45]

periods of time. The procedure for the construction and implantation of diffusion chambers is shown in Fig. 3. The Millipore diffusion chambers consisted of a Plexiglas ring covered at each end with a Millipore filter disk of 0.22-μm porosity. Suspensions of spleen cells and tumor cells were introduced in a 0.1-ml volume into the chambers and the latter were implanted in the intraperitoneal cavity of syngeneic mice. Viable tumor cells remaining in chambers after *in vivo* culture were counted on a morphological basis as previously described.[45] The results demonstrated that X5563 tumor cells admixed with normal syngeneic C3H/He spleen cells in the diffusion chamber resulted in continuous proliferation, whereas the X5563-immune Lyt-1^+2^- T cells exhibited an appreciable growth inhibition specific for X5563 tumor cells (Tables I–III). Most importantly, these tumor-specific Lyt-1^+2^- cells lost their antitumor activity by depletion of an adherent cell population contained in spleen cells, indicating the requirement of adherent cells for the Lyt-1^+2^- T-cell-mediated antitumor effect. This was substantiated by the fact

TABLE I
Time Course of Tumor Growth Inhibition in the Diffusion Chamber[a]

Culture in diffusion chamber[b]		Viable tumor cells ($\times 10^3$) after *in vivo* culture[c]		
Tumor cells	Spleen cells	Day 1	Day 3	Day 9
X5563	Normal	9.4 ± 0.6	19.9 ± 1.1	31.7 ± 3.3
X5563	X5563-immune	5.5 ± 0.1	4.5 ± 1.0	1.1 ± 0.3

[a]From Ref. 45.
[b]Spleen cells (10^7) from normal or X5563-immunized C3H/HeN mice were injected into a diffusion chamber together with 10^4 viable X5563 tumor cells, and the chamber was implanted in the peritoneal cavity of normal C3H/HeN mice.
[c]Various days after the implantation of the diffusion chambers, they were harvested and viable tumor cells were counted.

TABLE II
Specificity of Tumor Growth Inhibition in the Diffusion Chamber[a]

Culture in diffusion chamber		Viable tumor cells ($\times 10^3$) per chamber
Tumor cells	Spleen cells	
X5563	Normal	21.3 ± 4.4
	X5563-immune	7.9 ± 1.8
	MH134-immune	20.3 ± 3.8
MH134	Normal	36.4 ± 5.7
	X5563-immune	29.7 ± 5.5
	MH134-immune	9.1 ± 1.5

[a]From Ref. 45.

that immune Lyt-1^+2^- T cells depleted of adherent cells could restore their tumor-inhibiting effect when added back with the splenic or peritoneal resident adherent cell population (Table IV). Thus, the results obtained utilizing the diffusion chamber culture technique demonstrated that Lyt-1^+2^- T cells are unable to exhibit an antitumor effect by themselves and that their function depends on the existence of an adherent cell population. Although these observations illustrated the importance of the cellular interaction between Lyt-1^+2^- T cells and adherent cells, the mechanism underlying such an interaction remains to be determined.

Ample evidence has been presented for the requirement of adherent cells (macrophages) for the induction as well as implementation of immune responses. One is the requirement at an afferent limb of immune responses and adherent cells have a crucial role in presenting to T cells various types of antigens including protein, particulate, and cell surface antigens.[46–53] Another is that macrophages act as effector cells at the efferent limb of the cell-mediated immune response after being activated by lymphokines released from sensitized T cells.[54–56] The cellular interaction between tumor-immune Lyt-1^+2^- T cells and adherent cells as observed in the diffusion chamber system could also be considered in the context

TABLE III
Tumor Growth Inhibition in the Chamber by Tumor-Specific Lyt-1^+2^- T Cells[a]

Spleen cells	Treatment of spleen cells	Viable tumor cells ($\times 10^3$)	
		Expt 1	Expt 2
Normal	—	13.2 ± 0.2	24.3 ± 2.7
X5563-immune	—	5.6 ± 1.8	10.8 ± 1.1
	C	7.5 ± 1.2	4.3 ± 0.4
	Anti-Thy-1.2 + C	16.5 ± 4.4	ND
	Anti-Lyt-1.1 + C	16.8 ± 6.8	18.7 ± 0.7
	Anti-Lyt-2.1 + C	6.8 ± 1.6	6.9 ± 0.4

[a]From Ref. 45.

TABLE IV
Requirement of an Adherent Cell Population for Lyt-$1^{+}2^{-}$ T-Cell-Mediated
Tumor Growth Inhibition in the Diffusion Chamber[a]

Culture in diffusion chamber		Viable tumor cells ($\times 10^3$)	
Spleen cells (fraction)	Cells added back[b]	Expt 1	Expt 2
Normal (unfractionated)	—	15.9 ± 4.2	22.0 ± 1.0
Immune (unfractionated)	—	7.2 ± 2.7	10.5 ± 1.4
Immune (nylon-nonadherent)	—	20.1 ± 1.5	20.6 ± 4.5
Immune (nylon-nonadherent)	Adherent cells	6.6 ± 2.1	7.4 ± 0.6

[a]From Ref. 45.
[b]Cells added back were 10^6 splenic (Expt 1) or peritoneal resident adherent cells (Expt 2).

of the above mechanisms which consist of the requirement at the afferent, efferent, or both limbs. However, a single diffusion chamber system including both stages does not determine which mechanism(s) of requirement for adherent cells is relevant to the tumor growth inhibition. In this context, the double diffusion chamber model which we have developed more recently enabled us to investigate separately the involvement of adherent cells in both stages.[57]

The double diffusion chamber was constructed by separating both chambers from each other by a Millipore membrane (Fig. 4). When one chamber contained anti-MH134-specific Lyt-$1^{+}2^{-}$ T cells plus MH134 tumor cells and the other con-

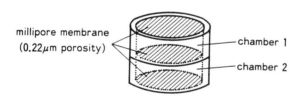

Culture in Double Diffusion Chamber				Viable Tumor Cells ($\times 10^3$) in	
Chamber 1		Chamber 2			
Tumor	Spleen Cells	Tumor	Spleen Cells	Chamber 1	Chamber 2
MH134	normal	X5563	normal	102.3±14.2	30.7±6.0
MH134	MH134-immune	X5563	normal	49.1±8.9	13.6±7.0

FIGURE 4. Double diffusion chamber culture system and tumor growth inhibition in both chambers. Viable MH134 tumor cells (1×10^4) and spleen cells (1×10^7) from normal or MH134-immunized C3H/HeN mice were injected into chamber 1. Viable X5563 tumor cells (2×10^4) and normal C3H/HeN spleen cells (1×10^7) were injected into chamber 2. Three days after the implantation, the double diffusion chamber was removed and viable tumor cells remaining in each chamber were individually counted and expressed as the mean cell number ±S.E.

tained normal spleen cells plus viable X5563 tumor cells of unrelated specificity, an appreciable growth inhibition of X5563 tumor cells in chamber 2 as well as MH134 tumor cells in chamber 1 was observed (Fig. 4).[57] The recognition of TATA by tumor-specific Lyt-1^+2^- T cells occurred only in chamber 1. Chamber 2 did not contain X5563 tumor-immune T cells but only normal spleen cells as an adherent cell source. Thus, chambers 1 and 2 in the double diffusion chamber system could represent the respective afferent and efferent stages of the antitumor effect. In this model the experiments of adherent cell depletion from each chamber have demonstrated the dual requirements of adherent cells.[57] First, Lyt-1^+2^- T cells require adherent cells at an afferent stage, since the depletion of adherent cells from chamber 1 fails to induce the tumor growth inhibition. This might include the failure to activate Lyt-1^+2^- T cells which act as the direct cytotoxic effector cells on tumor cells and/or to stimulate Lyt-1^+2^- T cells which release a lymphokine(s) capable of activating final tumoricidal effector cells such as macrophages. Although the former possibility remains to be determined, it is highly conceivable that Lyt-1^+2^- T cells are activated to produce a factor(s) which passes through cell-impermeable membranes and conveys tumoricidal potentials to adherent cells in other chambers and that such activation of tumor-specific Lyt-1^+2^- T cells depends on the existence of adherent cells.

Second, the growth inhibition of X5563 tumor cells in chamber 2 depends on the coexistence of adherent cells in chamber 2. When an adherent cell population was removed from cells in chamber 2 to be admixed with X5563 tumor cells, the growth inhibition of X5563 tumor cells was abolished. Thus, it is clear that adherent cells act as antitumor effector cells at the efferent phase. An earlier study from our laboratory has also demonstrated that the stimulation of tumor-immune Lyt-1^+2^- T cells with the corresponding tumor cells produced MAF able to induce nonspecific tumoricidal potentials in adherent cells.[58]

T-cell subsets or precursor T-cell subsets responsible for MAF production have been analyzed in only a few systems.[59-61] When spleen and lymph node T cells were stimulated with a polyclonal T-cell mitogen such as concanavalin A, both Lyt-2^+ and Lyt-2^- T-cell subpopulations produced MAF.[59] The MAF-producing capacity of alloreactive T-cell clones in long-term cultures was also Lyt phenotype-unrestricted.[62] Kelso and MacDonald[60] have demonstrated, however, that the precursor frequency of MAF-secreting cells in normal spleens upon allogeneic stimuli was three- to fourfold higher in the Lyt-2^- subpopulation than in the Lyt-2^+ subpopulation. Moreover, the average quantities of MAF produced by the Lyt-2^- subpopulation were about ten-fold higher than those produced by equivalent numbers of Lyt-2^+ responder cells. Although various T-cell subsets appear to have the potential to produce MAF, the surface phenotype of tumor-specific immune lymphoid cells responsible for MAF production has not been analyzed. We have demonstrated that the antitumor Lyt-1^+2^- T-cell subpopulation is predominantly responsible for producing MAF capable of inducing cytotoxic (cytostatic as well as cytolytic) potentials in macrophages. The results also indicate that the stimulation of such immune Lyt-1^+2^- T cells for MAF production was strictly TATA-specific, but once produced, MAF enabled macrophages to exhibit in a TATA-nonspecific way an appreciable cytotoxic effect on tumor cells which do not bear the Lyt-1^+2^- T-cell stimulating TATA.[58]

Such MAF preparation failed to exert any direct cytotoxic effect on tumor cells, whereas marked tumor growth inhibition was observed by injecting adherent cells into a single diffusion chamber along with MAF-containing culture supernatant, instead of connecting to the chamber possessing potentials to produce MAF (corresponding to chamber 1 in the double diffusion chamber system). These observations concerning MAF production from tumor-immune Lyt-1^+2^- T cells, together with the results obtained in the above-mentioned double diffusion chamber culture system, indicate that adherent cells in chamber 2 are activated by MAF which was produced by tumor-immune Lyt-1^+2^- T cells in chamber 1 and passed through a Millipore membrane to exert their tumor growth-inhibiting effect in chamber 2. This would enable us to construct a possible schema of Lyt-1^+2^- T-cell-mediated nonspecific tumor growth inhibition observed in the diffusion chamber as shown in Fig. 5.

There is also the possibility that tumor-specific Lyt-1^+2^- T cells capable of exhibiting a direct cytotoxic effect on the corresponding tumor cells can be activated by the presence of adherent cells. Although both MH134 and X5563 tumor cells used in the diffusion chamber culture system do not express class II major histocompatibility complex (MHC) antigens on their cell surfaces (unpublished observation), this does not exclude the potential expression of class II MHC (Ia) antigens on these tumor cells, since some Ia$^-$ normal[63–55] or tumor cells[66,67] have the capability of expressing Ia antigens under conditions in which γ-interferon (γ

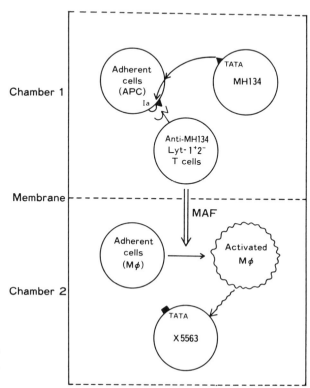

FIGURE 5. Possible mechanism of tumor growth inhibition in double diffusion chamber model.

-IFN) is produced. The postulation that Ia antigens on tumor cells in the chamber containing Lyt-1^+2^- T cells are induced by a factor(s) such as γ-IFN which could be produced by tumor-immune spleen cells may indicate that activated Lyt-1^+2^- T cells can induce a direct cytotoxic effect on tumor cells in a class II antigen-restricted way.

Although the *in vivo* cell culture system in conjunction with the double diffusion chamber has revealed the requirement of adherent cells (macrophages) as Lyt-1^+2^- T-cell-activated nonspecific tumoricidal effector cells, it is also possible that natural killer (NK), or lymphokine-activated killer (LAK) cells participate in nonspecific effector-mediated tumor eradication. Whether macrophages or LAK function predominantly as Lyt-1^+2^- T-cell-activated nonspecific effector cells might be determined by the following at least two factors. First, this could be influenced by the nature of TATA in the context of which type(s) of lymphokines(s) (IL-2, MAF, or γ-IFN) is predominantly produced by a given TATA-stimulated Lyt-1^+2^- T cells. Second, it also depends on the character of the target tumor cells which could determine the susceptibility to either macrophages, LAK, or both. Irrespective of these variables, it is highly likely that both types of nonspecific effector cells function at the tumor site in general.

4. Bystander Tumor Growth Inhibition

The expression of cell-mediated antitumor as well as antiallograft immunity *in vivo* is exquisitely specific. Several investigations have shown that the cellular immune responses can identify and destroy in an accurate way the histoincompatible allogeneic cells or corresponding tumor cells,[68] supporting the notion of target cell specificity.

Analysis of the tumor specificity of tumor rejection responses by using mixtures of the relevant and irrelevant tumor cells has produced discordant results. On the one hand, immune responses to a given syngeneic or allogeneic murine tumor failed to inhibit the local growth of admixed syngeneic tumor cells,[69-71] supporting the notion of target cell specificity at the effector phase as well. Other investigations have demonstrated that mixed inoculation of two different syngeneic tumors into recipients immunized to either one results in destruction of antigenically distinct bystander tumor cells along with the rejection of the corresponding tumor cells.[72-75]

Various factors or conditions influencing the expression of bystander tumor suppression have been reported. For example, the bystander effect is favored when specific and nonspecific tumors are mixed and inoculated together. This was accounted for by the histological studies of Galli *et al.*[75] Instead of models of tumor rejection requiring anatomic proximity between antigen-specific T cells and tumor cells, their results also suggested the importance of an alternative mechanism of tumor destruction involving leukocyte infiltration and microvascular damage. However, how tumor-specific immune T cells are involved in the above bystander phenomenon has not been investigated.

Studies demonstrating a pathway of tumor-specific Lyt-1^+2^- T-cell-initiated, nonspecific tumoricidal effector (macrophage)-mediated tumor growth inhibition raised the existence of an aspect of tumor growth-inhibiting mechanisms *in vivo* by a bystander effect, and permitted investigation of the possibility that tumor-specific Lyt-1^+2^- T cells have the potential to initiate growth inhibition of antigenically distinct bystander tumor cells. Our results demonstrated that when C3H/He mice immunized to syngeneic MH134 hepatoma were inoculated with antigenically distinct viable X5563 tumor cells (bystanders) together with the corresponding MH134 tumor cells at a single site, appreciable growth inhibition of X5563 tumor cells was observed. In this experiment, MH134 tumor-specific Lyt-1^+2^- T cells were responsible for such a bystander tumor inhibition.[76] These findings indicated that tumor-specific Lyt-1^+2^- T cells are capable of initiating *in vivo* growth inhibition of bystander as well as the corresponding tumor cells (Fig. 6A). The observations obtained in the above double diffusion chamber model support the existence of a bystander effect *in vivo* and account for the mechanism underlying *in vivo* bystander tumor growth inhibition. Moreover, our recent experiments which demonstrated that Lyt-1^+2^- T cells failed to exhibit this function in mice in which activities of NK cells or activated macrophages were suppressed by treatment with anti-asialo GM_1 antibody further confirmed the role of nonspecific tumoricidal effectors as activated by Lyt-1^+2^- T cells in eradicating tumor cells *in vivo*.[77] Thus, these results demonstrate the role of Lyt-1^+2^- T cells in mediating bystander tumor inhibition as well as provide supporting evidence for the existence of the mechanism of Lyt-1^+2^- T-cell-mediated tumor immunity *in vivo*.

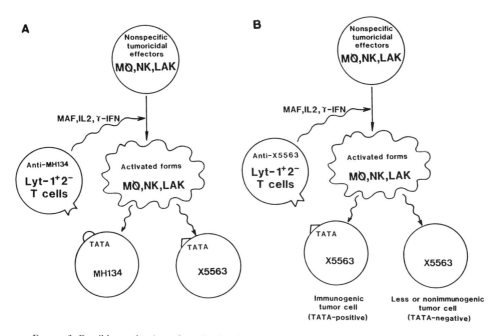

FIGURE 6. Possible mechanism of eradicating irrelevant (A) or nonimmunogenic (B) tumor cells.

5. Antitumor Immunity in the Tumor Mass *in Situ*

T-cell-mediated immune responses to TATA have been demonstrated in peripheral lymphoid organs of tumor-bearing,[77,78] tumor-immunized,[3] and tumor-regressed mice[2,23] by utilizing various cell-mediated immune assays. The fact that T cells as well as B cells, NK cells, and macrophages infiltrate into the tumor mass in various tumor models [2,79,80] has also supported the general concept that T cells have a major role in tumor-specific immunity. Although infiltration of lymphoid cells into animal and human tumor tissues is a well-observed phenomenon and a significant correlation has been reported between the prognosis of cancer patients and the degree of lymphoid cell infiltration in tumor masses,[81–85] there is little direct evidence demonstrating that T cells function to eradicate tumor cells in the tumor mass *in situ* and the observations described in the above literature are regarded as being indirect. Even in studies utilizing tumor-infiltrating lymphoid cells (TILC) isolated from the tumor mass, investigations focused on characterizing cell surface markers[79,80] or analyzing *in vitro* cytotoxic functions,[2,79] and *in vivo* functional analysis of TILC was investigated in a very limited model.[13,14] Therefore, isolating tumor-infiltrating lymphocytes (TIL) and testing their *in vivo* antitumor activity may lead to a better understanding of antitumor immune responses *in situ* for the tumor protection.

We have investigated the nature of TIL contained in TILC which could convey *in vivo* antitumor immune reactivity. The results demonstrated that Thy-1-positive cells (T cells) constitute approximately 30% of TILC (Table V) and that the Lyt-1^+2^- T-cell subset isolated from X5563 plasmacytoma and MH134 hepatoma tumor mass of syngeneic C3H/He recipient mice exhibited tumor-specific DTH responses as well as *in vivo* protective immunity in a Winn assay.[25] These results are consistent with the previous observation that the Lyt-1^+2^- T-cell subset in lymphoid organs has a crucial role in mediating DTH responses and tumor cell erad-

TABLE V
Cell Populations Constituting Tumor-Infiltrating Lymphoid Cells (TILC)[a]

Cells tested	Percent positivity			
	sIg-positive[b]	Thy-1-positive[b]	Mϕ[c]	LGL[d]
Normal spleen cells	46–52	23–27	4–7	4–6
Normal thymocytes	>99	<1	ND	ND
TILC[e]	29–35	16–29	10–20	15–25

[a]From Ref. 25.
[b]Cells were labeled with fluorescein isothiocyanate-conjugated monoclonal anti-Thy-1.2 (New England Nuclear) or anti-sIg (Cappel) antibody and fluorescence-staining cells were analyzed using a flow nicrofluorometer (Ortho Spectrum III). The scatter gates were set to exclude red blood cells and dead cells. The results represent the variation in three consecutive experiments.
[c]Mϕ (macrophages) were identified by esterase-staining.
[d]LGL (large granular lymphocytes) were identified morphologically by Wright–Giemsa staining.
[e]TILC were prepared from X5563 tumor masses of tumor-bearing C3H/HeN mice 2 weeks after the tumor cell inoculation.

ication in these tumor models[12,31] and support the observations described in the preceding section demonstrating the critical requirement of the Lyt-1^+2^- T-cell subset for *in vivo* tumor cell eradication.

We failed to detect any significant anti-X5563 or -MH134 specific CTL activity in TIL by utilizing the 4-hr ^{51}Cr release assay (unpublished observations). Although this does not necessarily exclude the possibility that cytotoxic Lyt-2^+ or Lyt-2^- T cells also function for tumor protection in the tumor site to some extent, it should be noted that tumor-specific DTH responses were observed to correlate closely with Lyt-1^+2^- T-cell-mediated tumor protection. A DTH phenomenon could be associated with the production of a variety of lymphokines including MAF, γ-IFN, and IL-2 by tumor-specific immune Lyt-1^+2^- T cells and the activation of nonspecific tumoricidal effector cells such as macrophages,[40] NK cells,[41] or LAK.[42] In addition, since TILC contained an appreciable proportion of esterase staining-positive and intracellular large granule-positive cells (presumably representing macrophages and NK cells, respectively) (Table V), it is conceivable that macrophages and/or NK cells as well as LAK contained in the Thy-1^+ T-cell population, which had been activated in the tumor mass *in situ* by tumor-specific Lty-1^+2^- T cells functioned to eradicate tumor cells. Such a postulate suggests that the existence of tumor-specific Lty-1^+2^- T cells in the tumor mass *in situ* results in the antitumor effect by continuously activating nonspecific tumoricidal effects through persistent restimulation with TATA.

6. The Recognition of TATA by Lyt-1^+2^- T cells

Murine and human T cells recognize antigen in the context of either class I or class II MHC molecules.[47,87,88] Expression of the Lyt-2^+ phenotype by murine T cells correlates primarily with class I MHC antigen reactivity, whereas Lyt-2^- T cells appear to interact with antigens in association with class II MHC determinants.[89-92] It has been confirmed that Lyt-1 antigen is expressed on almost all T cells without exhibiting any role in antigen recognition of T cells, whereas the expression of a new surface marker, L3T4 antigen that is relevant to Leu3/T4 antigen in the human system, correlates with class II MHC antigen reactivity.[93-96]

Although the proliferating cells in autoimmune MRL/1pr mice lack both L3T4 and Lyt-2 antigens,[97] normal mature (extrathymic) T cells express either L3T4 or Lyt-2, but not both antigens.[93,94] Thus, it has to be determined whether Lyt-1^+2^- T cells mediating tumor rejction in vivo express L3T4 antigen. Our recent studies have demonstrated that tumor-immune T cells depleted of Lyt-2^+ T cells express L3T4 antigens and that these L3T4$^+$ T cells mediate anti-tumor helper T cell activity assisting CTL response to the tumor[98] as well as tumor-neutralizing activity.[99]

L3T4 antigens have an important role in the recognition of antigens in association with class II MHC molecules.[94-96] However, most tumor cells express class I MHC, but not class II MHC. Although γ-IFN has been reported to have the potential to increase the expression of MHC antigens,[63-67] it is not easy to induce class II MHC on class II⁻ murine tumor cells including X5563 and MH134 tumor cells (our unpublished observations). The recognition of cell surface antigens on class II⁻ cells should be investigated in more detail. In studies analyzing the mechanism by which alloantigens are recognized by T cells, it has been demonstrated that MHC class I antigens on class I$^+$ class II$^-$ allogeneic cells can be recognized either by L3T4$^+$ Lyt-2$^-$ T cells after processing by and presenting on class II$^+$ self APC[100] or by L3T4$^-$ Lyt-2$^+$ T cells directly without involvement of self APC.[101] Although the results obtained in these totally *in vitro* systems are clear, these two categories of the recognition pathway for cell surface antigens should be further elucidated for its *in vivo* relevance. Nevertheless, the demonstration of the existence of the antigen-recognition pathway by L3T4$^-$ Lyt-2$^+$ T cells without help of self APC might contribute to constructing models for TATA recognition. Thus, whether Lyt-1$^+$2$^-$ L3T4$^+$ T cells can recognize directly TATA on tumor cells in the tumor mass *in situ* without involvement of class II MHC molecules or only TATA presented on class II$^+$ APC to produce lymphokines that activate ultimate effector cells (Fig. 5) remains to be further investigated.

7. The Role of Lyt-1$^+$2$^-$ T-Cell Subset in the Implementation of Allograft Rejections and Autoimmune Diseases

It is now recognized that there are several effector mechanisms with the potential to mediate allograft rejection. It has been demonstrated that both helper/inducer (Lyt-1$^+$2$^-$ or W3/25$^+$ Ox8$^-$) and cytotoxic/suppressor (Lyt-1$^+$2$^+$, Lyt-1$^-$2$^+$, or W3/25$^-$ Ox8$^+$) subsets of T cells restore the capacity to effect the rejection.[4–7, 102,103] However, the central role of helper/inducer T cells in mediating rejection is highlighted by adoptive transfer experiments in which these cells are the only cells required to restore the capacity of either heavily irradiated or adult thymectomized, irradiated, and bone marrow-reconstituted (B cell) rodents to reject grafts.[4–7] Whatever the ultimate mediator is, these observations parallel the results obtained in studies of syngeneic tumor graft rejection. This could also apply for the mechanism of induction of some types of tissue-specific autoimmune diseases. Antithryoglobulin Lyt-1$^+$2$^-$ T-cell lines caused a most severe thyroiditis within several days of i.v. injection in normal recipient mice, in irradiated mice, or in athymic nude mice.[104] Other examples were provided by experimental allergic encephalitis[105] or spontaneous murine nonobese diabetes[21] models, in which the Lyt-1$^+$2$^-$ T-cell subset was required for the initiation of disease.

Irrespective of the demonstration of the critical role of Lyt-1$^+$2$^-$ T cells in inducing allograft rejection and tissue-specific autoimmune diseases as well as in eradicating tumor cells, this does not exclude the involvement of CTL or antibody responses in initiating allograft rejection and autoimmune disease. We believe that the special importance of helper/inducer T cells (Lyt-1$^+$2$^-$ or W3/25$^+$ Ox8$^-$) derives from their capacity to produce factors that are required to activate CTL, macrophages, LAK, and B cells, which could contribute as elements of the total immune system.

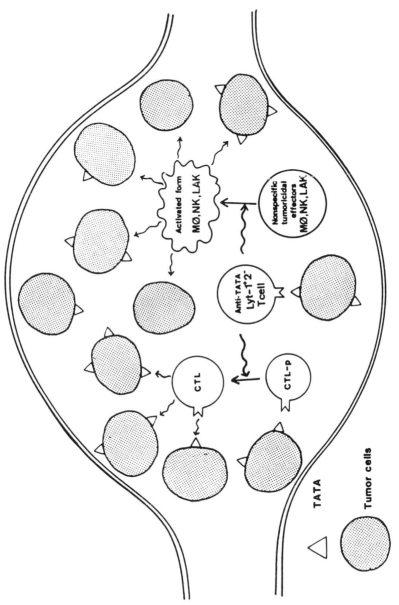

FIGURE 7. Possible tumor destruction mechanism by tumor-infiltrating lymphoid cells.

8. Summary and Conclusions

This chapter has reviewed recent findings concerning the mechanism(s) of tumor-specific Lyt-1^+2^- T-cell-mediated tumor eradication *in vivo*. The tumor-immune Lyt-1^+2^- T-cell subset, which is distinct from T cells mediating *in vitro* cytotoxicity (Lyt-$1^+2^+/1^-2^+$), had a crucial role in rejecting tumor cells when adoptively transferred into T-cell-deprived B-cell mice. Lyt-1^+2^- T cells do not necessarily require recruitment of the host's cytotoxic T-cell precursors for implementation of *in vivo* immunity. It could be that CTL are involved in the lysis of virus-infected target cells in virus-induced tumor models or autoimmune diseases and that they also function as antitumor mediators in general. Nevertheless, the recent studies illustrate the efficacy of the pathway of tumor-specific Lyt-1^+2^- T-cell-mediated tumor eradication *in vivo*. It appears that Lyt-1^+2^- T cells exert their antitumor effect by releasing MAF, γ-IFN, and IL-2 and by activating ultimate nonspecific tumoricidal effector cells such as macrophages, NK, or LAK.

Although TATA have been demonstrated in a vast majority of experimentally induced or spontaneously occurring tumors, it could also be that each tumor mass consists of tumor cells expressing quantitatively and/or qualitatively different TATA (Fig. 7). If tumor-specific T cells such as CTL are allowed to generate and directly attack tumor cells as the sole antitumor effector, less or nonimmunogenic tumor cells existing in each tumor mass might escape from the host's specific immune attack. In this context, it should be noted that a pathway of Lyt-1^+2^- T-cell-mediated nonspecific tumoricidal effector cell activation could eradicate less immunogeneic tumor cells as well (Fig. 6B). Thus, illustrating the existence of such a nonspecific aspect involved in tumor-specific immunity could provide a theoretical basis for the establishment of tumor-specific immunotherapy attempting to eradicate the tumor mass putatively containing less immunogenic tumor cells.

In conclusion, tumor-specific Lyt-1^+2^- T cells could exert their antitumor effect by activating ultimate nonspecific tumoricidal effector cells as well as tumor-specific CTL precursor cells through a variety of lymphokines. Such a pathway leading to the activation of nonspecific tumoricidal effectors contributes to the eradication of less or nonimmunogenic tumor cells in addition to immunogenic tumor cells, since their tumor cell killing does not require the recognition of TATA.

References

1. Cerottini, J. C., and Brunner, K. T., 1974, Cell-mediated cytotoxicity, allograft rejection and tumor immunity, *Adv. Immunol.* **18**:67–132.
2. Herberman, R. B., Holden, H. T., Varesio, L., Taniyama, T., Pucetti, P., Kirchner, H., Gerson, J., White, S., Keisari, Y., and Haskill, J. L., 1980, Immunologic reactivity of lymphoid cells in tumors, *Contemp. Top. Immunobiol.* **10**:61–78.
3. Levy, J. P., and Leclerc, J. C., 1977, The murine sarcoma virus-induced tumor: Exception or general model in tumor immunology? *Adv. Cancer Res.* **24**:1–66.

4. Loveland, B. E., Hogarth, P. M., Ceredig, R. H., and Mckenzie, I. F. C., 1981, Cells mediating graft rejection in the mouse. I. Lyt-1 cells mediate skin graft rejection, *J. Exp. Med.* **153**:1044–1057.
5. Dallman, M. J., and Mason, D. W., 1983, Cellular mechanisms of skin allograft rejection in the rat, *Transplant Proc.* **15**:335–338.
6. Lowry R. P., Gurley, K. E., Blackbrun, J., and Forbes, R. D. C., 1983, Delayed type hypersensitivity and lymphocytotoxicity in cardiac allograft rejection, *Transplant. Proc.* **15**:343–346.
7. Tilney, N. L., Kupiec-Weglinski, J. N., Heidecke, C. D., Lear, P. A., and Strom, T. B., 1984, Mechanisms of rejection and prolongation of vascularized organ allografts, *Immunol. Rev.* **77**:185–216.
8. Nelson, M., Nelson, D. S., Mckenzie, I.F.C., and Blanden, R. V., 1981, Thy and Ly markers on lymphocytes initiating tumor regression, *Cell. Immunol.* **60**:34–42.
9. Bhan, A. K., Perry, L. L., Cantor, H., McCluskey, R. T., Benacerraf, B., and Greene, M. I., 1981, The role of T cell sets in the rejection of a methylcholanthrene-induced sarcoma (S1509a) in syngeneic mice, *Am. J. Pathol.* **102**:20–27.
10. Fernandez-Cruz, E., Woda, B. A., and Feldman, J. D., 1980, Elimination of syngeneic sarcomas in rats by a subset of T lymphocytes, *J. Exp. Med.* **152**:823–841.
11. Greenberg, P. D., Cheever, M. A., and Fefer, A., 1981, Eradication of disseminated murine leukemia by chemoimmunotherapy with cyclophosphamide and adoptively transferred immune syngeneic Lyt-1$^+$2$^-$ lymphocytes, *J. Exp. Med.* **154**:952–963.
12. Fujiwara, H., Fukuzawa, M., Yoshioka, T., Nakajima, H., and Hamaoka, T., 1984, The role of tumor-specific Lyt-1$^+$2$^-$ T cells in eradicating tumor cells in vivo. I. Lyt-1$^+$2$^-$ T cells do not necessarily require recruitment of host's cytotoxic T cell precursors for implementation of in vivo immunity, *J. Immunol.* **133**:1671–1676.
13. Ibayashi, Y., Uede, T., Uede, T., and Kikuchi, K., 1985, Functional analysis of mononuclear cells infiltrating into tumors: Differential cytotoxicity of mononuclear cells from tumors of immune and nonimmune rats, *J. Immunol.* **134**:648–653.
14. Kreider, J. W., Howell, L. E., and Burtlet, G. L., 1984, T cells required for the expression of tumor protection immunity (TPI) and delayed hypersensitivity (DH) to 13762A rat mammary carcinoma, *Proc. Am. Assoc. Cancer Res.* **25**:281.
15. Shimizu, K., and Shen, F. W., 1979, Role of different T cell sets in the rejection of syngeneic chemically induced tumors, *J. Immunol.* **122**:1162–1165.
16. Leclerc, J. C., and Cantor, H., 1980, T cell-mediated immunity to oncornavirus-induced tumor. II. Ability of different T cell sets to prevent tumor growth in vivo, *J. Immunol.* **124**:851–854.
17. Hamaoka, T., Fujiwara, H., Teshima, K., Aoki, H., Yamamoto, H., and Kitagawa, M., 1979, Regulatory functions of hapten-reactive helper and suppressor T lymphocytes. III. Amplification of a generation of tumor-specific killer T lymphocyte activities by suppressor T-cell-depleted hapten-reactive T lymphocytes, *J. Exp. Med.* **149**:185–199.
18. Fujiwara, H., Hamaoka, T., Shearer, G. M., Yamamoto, H., and Terry, W. D., 1980, The augmentation of in vitro and in vivo tumor-specific T cell-mediated immunity by amplifier T lymphocytes, *J. Immunol.* **124**:863–869.
19. Shimizu, Y., Fujiwara, H., Ueda, S., Wakamiya, N., Kato, S., and Hamaoka, T., 1984, The augmentation of tumor-specific immunity by virus-help. II. Enhanced induction of cytotoxic T lymphocyte and antibody responses to tumor antigens by vaccinia-reactive helper T cells, *Eur. J. Immunol.* **14**:839–843.
20. Hamaoka, T., Takai, Y., Kosugi, A., Mizushima, Y., Shima, J., Kusama, T., and Fujiwara, H., 1985, The augmentation of tumor-specific immunity using haptenic muramyl dipeptide (MDP) derivatives. I. Synthesis of a novel haptenic MDP derivative cross-reactive with bacillus Calmette guerin and its application to enhanced induction of tumor immunity, *Cancer Immunol. Immunother.* **20**:183–188.
21. Hanafusa, T., Sugihara, S., Fujino-Kurihara, H., Miyagawa, J., Miyazaki, A., Yoshioka, T., Yamada, K., Nakajima, H., Asakawa, H., Kono, N., Fujiwara, H., Hamaoka, T., and Tarui, S., 1986, The induction of insulitis by adoptive transfer with L3T4$^+$ Lyt-2$^-$ T cells in T cell-depleted NOD mice, *Diabetes* (in press).
22. Fujiwara, H., Moriyama, Y., Suda, T., Tsuchida, T., Shearer, G. M., and Hamaoka, T., 1984,

Enhanced TNP-reactive helper T cell activity and its utilization in the induction of amplified tumor immunity which results in tumor regression, *J. Immunol.* **132:**1571–1577.

23. Fujiwara, H., Aoki, H., Yoshioka, T., Tomita, S., Ikegami, R., and Hamaoka, T., 1984, Establishment of tumor-specific immunotherapy model utilizing TNP-reactive helper T cell activity and its application to autochthonous tumor system, *J. Immunol.* **133:**509–514.

24. Yoshioka, T., Fukuzawa, M., Takai, Y., Wakamiya, N., Ueda, S., Kato, S., Fujiwara, H., and Hamaoka, T., 1986, The augmentation of tumor-specific immunity by virus-help. III. Enhanced generation of tumor-specific Lyt-1^+2^- T cells is responsible for augmented tumor immunity in vivo, *Cancer Immunol. Immunother.* **21:**193–198.

25. Tomita, S., Fujiwara, H., Yamane, Y., Sano, H., Nakajima, H., Izumi, Y., Arai, H., Kawanishi, Y., Tsuchida, T., and Hamaoka, T., 1986, Demonstration for intratumoral infiltration of tumor-specific Lyt-1^+2^- T cells mediating delayed type hypersensitivity response and in vivo protective immunity, *Gann* **77:**182–189.

26. Cantor, H., and Boyse, E. A., 1975, Functional subclasses of T lymphocytes bearing different Ly antigens. II. Cooperation between subclasses of Ly$^+$ cells in the generation of killer activity, *J. Exp. Med.* **141:**1390–1399.

27. Pilarski, L. M., 1977, A requirement for antigen-specific helper T cells in the generation of cytotoxic T cells from thymocyte precursors, *J. Exp. Med.* **145:**709–725.

28. Wagner, H., Rollinghoff, M., Pfizenmaier, K., Hardt, C., and Johnscher, G., 1980, T–T cell interactions during in vitro cytotoxic T lymphocyte (CTL) responses. II. Helper factor from activated Lyt-1^+ T cells is rate limiting, i) in T cell responses to nonimmunogenic alloantigens, ii) in thymocyte responses to allogeneic stimulator cells, and iii) recruits allo- or H-2 restricted CTL precursors from the Lyt 123^+ T subset, *J. Immunol.* **124:**1058–1067.

29. Fujiwara, H., Yoshioka, T., Shima, J., Kosugi, A., Itoh, K., and Hamaoka, T., 1986, Helper T cells against tumor-associated antigens (TAA): Preferential induction of helper T cell activities involved in anti-TAA cytotoxic and antibody responses, *J. Immunol.* **136:**2715–2719.

30. Greenberg, P. H., Kern, D. E., and Cheever, M. A., 1985, Therapy of disseminated murine leukemia with cyclophosphamide and immune Lyt-1^+2^- T cells: Tumor eradication does not require participation of cytotoxic T cells, *J. Exp. Med.* **161:**1122–1134.

31. Fukuzawa, M., Fujiwara, H., Yoshioka, T., Itoh, K., and Hamaoka, T., 1985, Tumor-specific Lyt-1^+2^- T cells can reject tumor cells in vivo without inducing cytotoxic T lymphocyte responses, *Transplant Proc.* **17:**599–605.

32. Old, L. J., 1981, Cancer immunology: The search for specificity, *Cancer Res.* **41:**361–373.

33. Kawase, I., Komuta, K., Ogura, T., Fujiwara, H., Hamaoka, T., and Kishimoto, S., 1985, Murine tumor cell lysis by antibody-dependent macrophage-mediated cytotoxicity using syngeneic monoclonal antibodies, *Cancer Res.* **45:**1663–1668.

34. Klein, G., Sjogren, H. O., Klein, E., and Hellstrom, K. E., 1960, Demonstration of resistance against methylcholanthrene induced sarcomas in the primary autochthonous host, *Cancer Res.* **20:**1561–1572.

35. Old, L. J., Boyse, E. A., Clarke, D. A., and Garwell, E., 1962, Antigenic properties of chemically induced tumors, *Ann. N.Y. Acad. Sci.* **101:**80–106.

36. Law, L. W., Rogers, M. J., and Appella, E., 1980, Tumor antigens on neoplasms induced by chemical carcinogens and by DNA- and RNA-containing viruses: Properties of the solubilized antigens, *Adv. Cancer Res.* **32:**201–235.

37. Loveland, B. E., and Mckenzie, I. F. C., 1982, Delayed type hypersensitivity and allograft rejection in the mouse: Correlation of effector cell phenotype, *Immunology* **46:**313–320.

38. Liew, F. Y., and Simpson, E., 1980, Delayed-type hypersensitivity responses to H-Y: Characterization and mapping of Ir genes, *Immunogenetics* **11:**255–266.

39. Takai, Y., Kosugi, A., Yoshioka, T., Tomita, S., Fujiwara, H., and Hamaoka, T., 1985, T–T cell interaction in the induction of delayed type hypersensitivity (DTH) responses: Vaccinia virus-reactive helper T cell activity involved in enhanced in vivo induction of DTH responses and its application to augmentation of tumor-specific DTH responses, *J. Immunol.* **134:**108–113.

40. Cohen, S., 1980, Lymphokines in delayed hypersensitivity, in: *Progress in Immunology, IV* (M. Fougereau and J. Dausset, eds.), Academic Press, New York, pp. 860–879.

41. Kuribayashi, K., Gillis, S., Kern, E., and Henney, C. S., 1981, Murine NK cell cultures: Effect of interleukin 2 and interferon on cell growth and cytotoxic activity, *J. Immunol.* **126:**2321–2327.
42. Grimm, E. A., Mazumder, A., Zhang, H. Z., and Rosenberg, S. A., 1982, Lymphokine-activated killer cell phenomenon: Lysis of natural killer-resistant fresh solid tumor cells by interleukin 2-activated autologous human peripheral blood lymphocytes, *J. Exp. Med.* **155:**1823–1824.
43. Algire, G. H., Weaver, J. M., and Prehn, R. T., 1954, Growth of cells in vivo in diffusion chambers. I. Survival of homografts in immunized mice, *J. Natl. Cancer Inst.* **15:**493–507.
44. Capalko, E. E., Albright, J. F., and Bennett, W. E., 1964, Evaluation of the diffusion chamber culture technique for study of the morphological and functional characteristics of lymphoid cells during antibody production, *J. Immunol.* **92:**243–251.
45. Fujiwara, H., Takai, Y., Sakamoto, K., and Hamaoka, T., 1985, The mechanism of tumor growth inhibition by tumor-specific Lyt-1^+2^- T cells. I. Anti-tumor effect of Lyt-1^+2^- T cells depends on the existence of adherent cells, *J. Immunol.* **135:**2187–2191.
46. Unanue, E. R., 1981, The regulatory role of macrophages in antigenic stimulation. II. Symbiotic relationship between lymphocytes and macrophages, *Adv. Immunol.* **31:**1–136.
47. Rosenthal, A. S., and Shevach, E. M., 1983, Function of macrophages in antigen recognition by guinea pig T lymphocytes. I. Requirement for histocompatible macrophages and lymphocytes, *J. Exp. Med.* **138:**1194–1212.
48. Hedrick, S. M., Matis, L. A., Hecht, T. T., Samelson, L. E., Longo, D. L., Heber-Katz, E., and Schwartz, R. H., 1982, The fine specificity of antigen and Ia determinant recognition by T cell hybridoma clones specific for pigeon cytochrome c, *Cell* **30:**141–152.
49. Pettinelli, C. B., Ahman, G. B., and Shearer, G. M., 1980, Expression of both I-A and I-E/C subregion antigens on accessory cells required for in vitro generation of cytotoxic T lymphocytes against alloantigens or TNBS-modified syngeneic cells, *J. Immunol.* **124:**1911–1916.
50. Wagner, H., Feldman, M., Boyle, W., and Schrader, J. W., 1972, Cell mediated immune responses in vitro. III. The requirement for macrophages in cytotoxic reactions against cell bound and subcellular alloantigens, *J. Exp. Med.* **136:**331–343.
51. Golding H., and Singer, A., 1984, Role of accessory cell processing and presentation of shed H-2 alloantigens in allospecific cytotoxic T lymphocyte responses, *J. Immunol.* **133:**597–605.
52. Weinberger, O., Germain, R. N., Springer, T., and Burakoff, S. J., 1982, Role of syngeneic Ia accessory cells in the generation of allospecific CTL responses, *J. Immunol.* **129:**694–697.
53. Taniyama, T., and Holden, H. T., 1979, Requirement of histocompatibility macrophages for the induction of a secondary cytotoxic response to syngeneic tumor cells in vitro, *J. Immunol.* **123:**43–49.
54. Nathan, C. F., Brukner, L. H., Silverstein, S. C., and Cohn, Z. A., 1979, Extracellular cytolysis by activated macrophages and granulocytes. I. Pharmacologic triggering of effector cells and the release of hydrogen peroxide, *J. Exp. Med.* **149:**84–99.
55. Mantovani, A., 1983, Origin and function of tumor-associated macrophages in murine and human neophasms, in: *Progress in Immunology. V* (Y. Yamamura and T. Tada, eds.), Academic Press, New York, pp. 1001–1008.
56. Adams, D. O., Lewis, J. G., and Johnson, W. J., 1983, Multiple modes of cellular injury by macrophages: Requirement for different forms of effector activation, in: *Progress in Immunology V* (Y. Yamamura and T. Tada, eds.), Academic Press, New York, pp. 1009–1018.
57. Sakamoto, K., Fujiwara, H., Nakajima, H., Yoshioka, T., Takai, Y., and Hamaoka, T., 1986, The mechanism of tumor growth inhibition of tumor-specific Lyt-1^+2^- T cells. II. Requirements of adherent cells for activating Lyt-1^+2^- T cells as well as for functioning as anti-tumor effectors activated by factor(s) from Lyt-1^+2^- T cells, *Gann* **77:**1142–1152.
58. Nakajima, H., Fujiwara, H., Takai, Y., Izumi, Y., Sano, S., Tsuchida, T., and Hamaoka, T., 1985, Studies on macrophage-activating factor (MAF) in anti-tumor immune responses. I. Tumor specific Lyt-1^+2^- T cells are required for producing MAF able to generate cytolytic as well as cytostatic macrophages, *J. Immunol.* **135:**2199–2205.
59. Guerne, P. -A., Piguet, P. -F., and Vassalli, P., 1983, Positively selected Lyt-2^+ and Lyt-2^-ms mouse T lymphocytes are comparable, after Con A stimulation, in release of IL 2 and of lymphokines

acting on B cells, macrophages, and mast cells, but differ in interferon production, *J. Immunol.* **130**:2225–2230.

60. Kelso, A., and MacDonald, H. R., 1982, Precursor frequency analysis of lymphokine-secreting alloreactive T lymphocytes: Dissociation of subsets producing interleukin 2, macrophage-activating factor, and granulocyte–macrophage colony-stimulating factor on the basis of Lyt-2 phenotype, *J. Exp. Med.* **156**:1366–1379.

61. Biondi, A., Roach, J. A., Schlossman, S. F., and Todd, R. F., 1984, Phenotypic characterization of human T lymphocyte populations producing macrophage-activating factor (MAF) lymphokines, *J. Immunol.* **133**:281–285.

62. Kelso, A., and Glasebrook, A. L., 1984, Secretion of interleukin 2, macrophage-activating factor, interferon, and colony-stimulating factor by alloreactive T lymphocyte clones, *J. Immunol.* **132**:2924–2931.

63. de Waal, R. M. W., Bogman, M. J. J., Maass, C. N., Cornelissen, L. M. H., Tex, W. J. M., and Koene, R. A. P., 1983, Variable expression of Ia antigens on the vascular endothelium of mouse skin allografts, *Nature* **303**:426–429.

64. Hall, B. M., Bishop, G. A., Duggin, G. G., Horvath, J. S., Philips, J., and Tiller, D. J., 1984, Increased expression of HLA-DR antigens on renal tubular cells in renal transplants: Relevance to the rejection response, *Lancet* **2**:247–251.

65. Pober, J. S., Gimbrone, M. A., Cotran, R. S., Jr., Reiss, C. S., Burakoff, S. J., Fiers, W., and Ault, K. A., 1983, Ia expression by vascular endothelium is inducible by activated T cells and by human γ interferon, *J. Exp. Med.* **157**:1339–1353.

66. Callahan, G. N., 1984, Soluble factors produced during an immune response regulate Ia antigen expression by murine adenocarcinoma and fibrosarcoma cells, *J. Immunol.* **132**:2649–2657.

67. Nickloff, B. J., Basham, T. Y., Merigan, T. C., and Morhenn, V. B., 1985, Immunomodulatory and antiproliferative effect of recombinant alpha, beta, and gamma interferons on cultured human malignant squamous cell lines, *J. Invest. Dermatol.* **84**:487–490.

68. Hearberman, R. B., 1974, Cell-mediated immunity to tumor cells, *Adv. Cancer Res.* **19**:207–263.

69. Klein, G., and Klein, E., 1956, Genetic studies of the relationship of tumor host cells: Detection of an allelic difference at a single gene locus in a small fraction of a large tumor-cell population, *Nature* **178**:1389–1391.

70. Ringertz, N., Klein, E., and Revesz, L., 1959, Growth of small compatible tumor implants in presence of admixed radiation-killed or incompatible tumor cells, *Cancer* **12**:697–707.

71. Weissman, I. L., 1973, Tumor immunity in vivo: Evidence that immune destruction of tumor leaves "bystander" cells intact, *J. Natl. Cancer Inst.* **51**:443–448.

72. Zbar, B., Wepsic, H. T., Borsos, T., and Rapp, H. J., 1970, Tumor-graft rejection in syngeneic guinea pigs: Evidence for a two-step mechanism, *J. Natl. Cancer Inst.* **44**:473–481.

73. Jessup, J. M., and Madden, R. E., 1971, Further evidence of a non-specific rejection in hosts immunized with non-cross-reacting tumors, *Proc. Am. Assoc. Cancer Res.* **12**:55.

74. Bast, R. C., Jr., Zbar, B., and Rapp, H. J., 1975, Local antitumor activity of a primary and an anamnestic response to a syngeneic guinea pig hepatoma, *J. Natl. Cancer Inst.* **55**:989–994.

75. Galli, S. J., Bast, R. C., Jr., Bast, B. S., Isomura, T., Zbar, B., Rapp, H. J., and Dvorak, H. F., 1982, Bystander suppression of tumor growth: Evidence that specific targets and bystanders are damaged by injury to a common microvasculature, *J. Immunol.* **129**:880–899.

76. Yoshioka, T., Fujiwara, H., Takai, Y., Ogata, M., Shimizu, J., and Hamaoka, T., 1986, The role of tumor-specific Lyt-1^+2^- T cells in eradicating tumor cells in vivo. II. Lyt-1^+2^- T cells have potentials to reject antigenically irrelevant (bystander) tumor cells on the activation with the specific tumor cells, *Cancer Immunol. Immunother.* (in press).

77. Yoshioka, T., Sato, S., Fujiwara, H., Hamaoka, T., 1986, The role of anti-asialo GM1 antibody-sensitive cells in the implementation of tumor-specific T cell-mediated immunity in vivo, *Gann* **77**:825–832.

78. Jones, J. A., Robinson, G., Rees, R. C., and Baldwin, R. W., 1978, Immune response of the draining and distal lymph nodes during the progressive growth of a chemically induced transplantable rat hepatoma, *Int. J. Cancer* **21**:171–178.

79. Glaser, M., Herberman, R. B., Kirchner, H., and Djeu, J. Y., 1974, Study of cellular immune

response to Gross virus induced lymphoma by the mixed lymphocyte–tumor cell interaction, *Cancer Res.* **34:**2165–2171.
80. Buessow, S. C., Paul, R. D., and Lopez, D. M., 1984, Influence of mammary tumor progression on phenotype and function of spleen and in situ lymphocytes in mice, *J. Natl. Cancer Inst.* **73:**249–255.
81. Ishii, Y., Matsuura, A., Takami, T., Ueda, T., Ibayashi, Y., Ueda, T., Imamura, M., Kikuchi, K., and Kikuchi, Y., 1984, Lymphoid cell subpopulations infiltrating into autologous rat tumors undergoing rejection, *Cancer Res.* **44:**4053–4058.
82. Martin, R. F., and Bechwith, J. B., 1968, Lymphoid infiltrates in neuroblastomas: Their occurrence and prognostic significance, *J. Pediatr. Surg.* **3:**161–164.
83. Kikuchi, K., Ishii, Y., Ueno, H., and Koshiba, H., 1976, Cell-mediated immunity involved in autochthonous tumor rejection in rats, *Ann. N.Y. Acad. Sci.* **276:**188–206.
84. Bennet, S. S., Futrell, J. W., Roth, J. A., Hoye, R. C., and Kechem, A. S., 1971, Prognostic significance of histologic host response in cancer of the larynx of hypolarynx, *Cancer* **28:**1255–1265.
85. Black, M. M., Freeman, C., Mork, T., Harvei, S., and Cutler, S. J., 1971, Prognostic significance of microscopic structure of gastric carcinomas and their regional lymph node, *Cancer* **27:**703–711.
86. Shimokawara, I., Imamura, M., Yamanaka, N., Ishii, Y., and Kikuchi, K., 1982, Identification of lymphocyte subpopulations in human breast cancer tissue and its significance: An immunoperoxidase study with anti-human T- and B-cell sera, *Cancer* **49:**1456–1464.
87. Shearer, G. M., Rehn, T. G., and Garbarino, C. A., 1975, Cell mediated lympholysis of trinitrophenyl-modified autologous lymphocytes: Effector cell specificity of modified cell surface components controlled by the H-2K and H-2D serological region of the murine major histocompatibility complex, *J. Exp. Med.* **141:**1348–1364.
88. Zinkernagel, R. M., Althage, A., Cooper, S., Kreel, G., Klein, P. A., Seftoh, B., Flahery, L., Stimpfling, J., Schreffler, D., and Klein, J., 1978, Ir genes in H-2 regulate generation of antiviral cytotoxic T cell: Mapping to K or D and dominance of unresponsiveness, *J. Exp. Med.* **148:**592–606.
89. Okada, M., and Henney, C. S., 1980, The differentiation of cytotoxic T cells in vitro. II. Amplifying factor(s) produced in primary mixed lymphocyte cultures against K/D stimuli require the presence of Lyt-2$^+$ cells but not Lyt-1$^+$ cells, *J. Immunol.* **125:**300–307.
90. Pierres, A., Schmitt-Verhulst, A. -M., Buferne, M., Golstein, P., and Pierres, M., 1982, Characterization of an Lyt-1$^+$ cytolytic T-cell clone specific for a polymorphic domain of the I-A^K molecule, *Scand. J. Immunol.* **15:**619–625.
91. Swain, S. L., 1981, Significance of Lyt phenotypes: Lyt-2 antibodies block activities of T cells that recognize class I major histocompatibility complex antigens regardless of their function, *Proc. Natl. Acad. Sci. USA* **78:**7101–7105.
92. Swain, S. L., Dennert, G., Wormsley, S., and Dutton, R. W., 1981, The Lyt phenotype of a long-term allospecific T cell line: Both helper and killer activities to IA are mediated by Ly-1 cells, *Eur. J. Immunol.* **11:**175–180.
93. Dialynas, D. P., Quan, Z. S., Wall, K. A., Pierres, A., Quintans, J., Loken, M. R., Pierres, M., and Fitch, F. W., 1983, Characterization of the murine T cell surface molecule, designated L3T4, identified by monoclonal antibody GK-1.5: Similarity of L3T4 to the human Leu-3/T4 molecule, *J. Immunol.* **131:**2445–2451.
94. Dialynas, D. P., Wilde, D. B., Marrack, P., Pierres, A., Wall, K. A., Havran, W., Otten, G., Loken, M. R., Pierres, M., Kappler J., and Fitch, F. W., 1983, Characterization of the murine antigenic determinant, designated L3T4a, recognized by monoclonal antibody GK1.5: Expression of L3T4a by functional T cell clones appears to correlate primarily with class II MHC antigen-reactivity, *Immunol. Rev* **74:**29–56.
95. Wilde, D. B., Marrack, P., Kappler, J., Dialynas, D. P., and Fitch, F. W., 1983, Evidence implicating L3T4 in class II MHC antigen and reactivity: Monoclonal antibody GK-1.5 (anti-L3T4a) blocks class II MHC antigen-specific proliferation, release of lymphokines, and binding by cloned murine helper T lymphocyte lines, *J. Immunol.* **131:**2178–2183.
96. Swain, S. L., Dialynas, D. P., Fitch, F. W., and English, M., 1984, Monoclonal antibody to L3T4

blocks the function of T cells specific for class 2 major histocompatibility complex antigens, *J. Immunol.* **132:**1118–1123.

97. Wofsy, D., Hardy, R. R., and Seaman, W. E., 1984, The proliferating cells in autoimmune MRL/lpr mice lack L3T4, an antigen on "helper" T cells that is involved in the response to class II major histocompatibility antigens, *J. Immunol.* **132:**2686–2689.

98. Kosugi, A., Yoshioka, T., Suda, T., Sano, H., Takahama, Y., Fujiwara, H., and Hamaoka, T., The activation of L3T4$^+$ helper T cells assisting the generation of anti-tumor Lyt-2$^+$ cytotoxic T lymphocytes: Requirement of Ia-positive antigen-presenting cells for processing and presentation of tumor antigens, *J. Leukocyte Biol.* (in press).

99. Yoshioka, T., Sato, S., Ogato, M., Sano, H., Shino, J., Yamamoto, H., Fujiwara, H., Hamaska, T., 1987, Role of tumor-specific Lyt-2$^+$ T cells in tumor growth inhibition in vivo. I. Mediation of in vivo tumor-neutralizing activity by Lyt-2$^+$ as well as L3T4$^+$ T cell subsets, *Jpn. J. Cancer Res. (Gann),* (in press).

100. Singer, A., Kruisbeek, A. M., and Andrysiak, P. M., 1984, T cell–accessory cell interactions that initiate allospecific cytotoxic T lymphocyte responses: Existence of both Ia-restricted and Ia-unrestricted cellular interaction pathways, *J. Immunol.* **132:**2199–2209.

101. Mizuochi, T., Golding, H., Rosenberg, A. S., Glimcher, L. H., Malek, T. R., and Singer, A., 1985, Both L3T4$^+$ and Lyt2$^+$ helper T cells initiate cytotoxic T lymphocyte responses against allogeneic major histocompatibility antigens but not against trinitrophenyl-modified self, *J. Exp. Med.* **162:**427–443.

102. Hall, B. M., de Saxe, I., and Dorsch, S. E., 1983, The cellular basis of allograft rejection in vivo: Restoration of first set rejection of heart grafts by T helper cells in irradiated rats, *Transplantation* **36:**700–705.

103. Le Francois, L., and Bevan, M. J., 1984, A reexamination of the role of Lyt-2$^+$ T cells in murine skin graft rejection, *J. Exp. Med.* **159:**56–67.

104. Maron, R., Zerubavel, R., Friedman, A., and Cohen, I. R., 1983, T lymphocyte line specific for thyroglobulin produces or vaccinates against autoimmune thyroiditis in mice, *J. Immunol.* **131:**2316–2322.

105. Hanser, S. L., Weiner, H. L., Bhan, A. K., Shapiro, M. E., Che, M., Aldrich, W. R., and Letvin, N. L., 1984, Lyt-1 cells mediate acute murine experimental allergic encephalomyelitis, *J. Immunol.* **133:**2288–2290.

22
Application of T Cell–T Cell Interaction to Enhanced Tumor-Specific Immunity Capable of Eradicating Tumor Cells *in Vivo*

Toshiyuki Hamaoka, Yasuyuki Takai, Atsushi Kosugi, Junko Shima, Takashi Suda, Yumiko Mizushima, Soichiro Sato, and Hiromi Fujiwara

1. Introduction

Investigations have attempted to delineate the consequences of malignant transformation of cells by the appearance of new cell surface structures [tumor-associated antigens (TAA) or tumor-associated transplantation antigens (TATA)] that could be identified by specific antiserum or by their ability to induce a specific cellular immune response. Considerable efforts have been undertaken to establish the significance of these tumor cell surface structures by correlating their cell surface expression with changes that take place during the course of neoplastic disease. The most compelling evidence for the existence of TATA comes from the study of chemically induced tumors of inbred rodents. These tumors express neoantigens capable of immunizing syngeneic or autochthonous hosts against subsequent challenge with the same tumor.[1–4]

Despite such evidence for the expression of TATA on neoplastic cell surfaces, autochthonous tumor- or syngeneic TATA-positive tumor-bearing hosts fail to reject these malignant cells. This could result from the weak immunogenicity of TATA and the consequences of a variety of suppressive mechanisms.[5–7] For this reason, various approaches to increasing a host's ability to generate immune responses to tumors as well as to the elimination of factors causing tumor-specific or -nonspecific immunosuppression would have to be attempted.

In order to enchance antitumor immune responses, we must know which types of immune responses are generated against TATA and which effector mechanism(s) is crucially responsible for eradicating tumor cells *in vivo*. Intensive studies concerning these issues are reviewed elsewhere in this volume. In this chapter, we shall consider critically our own and other studies seeking to enhance antitumor immune responses and to define conditions in which enhanced tumor-specific immunity leading to the eradication of a growing tumor can be obtained.

2. Modification of Tumor Cells with Various Chemical or Biological Substances for Enhanced Induction of Antitumor Effector T Cells

2.1. Theoretical Basis for Increasing the Size of Antitumor Effector T Cell Population

The limited immunogenicity of TATA might be attributed to a restricted number of effective T-cell clones against TATA. Therefore, an attempt to expand such a small number of anti-TATA effector clones is a plausible approach for augmenting a host's capability of eliciting immune resistance to tumors. There has been considerable emphasis on the modification of immunizing tumor cells with various chemical or biological substances because of its potential for enhanced induction of tumor immunity. The theoretical framework for the role of cell cooperation in tumor immunity, and speculations on manipulations that might augment tumor rejection have been presented by Mitchison.[8] He postulated that the immunogenicity of tumor antigens might be augmented by coupling new antigens such as haptens, proteins, new transplantation antigens, viral coat antigens, or xenogeneic cell antigens to tumor cell surfaces. Subsequently, a variety of experimental observations suggested that such a mechanism may indeed result in increased resistance to *in vivo* tumor growth.[9–16] Although low responsiveness of syngeneic hosts to TATA was overcome to some extent, the augmenting activity was not consistent. Improvement of this approach was required for its application to immunotherapy.

The requirement for a T-cell subset that augments the reactivities of other functionally distinct subsets of lymphocytes has been demonstrated for helper T cells in humoral[17] and cell-mediated immune responses.[18–25] It has also been established that although the activation of helper T cells is strictly antigen-specific, their ability to augment T-cell responses is antigen-nonspecific,[26,27] indicating that helper T cells against a given cell surface antigen are capable of assisting T-cell responses to other cell surface antigens. In the above studies of modified tumor cells, additional antigens introduced onto the tumor cell surface could contribute

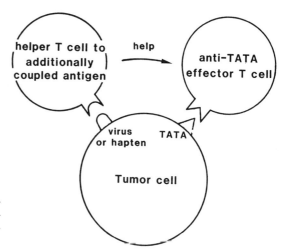

FIGURE 1. A model for T cell–T cell interaction between helper T cell to additionally coupled (helper) antigen and anti-TATA effector T cell.

to the induction of potent helper T cells. If these antigens act as helper determinants to enhance cellular as well as antibody responses to TATA coexisting on tumor cells, preinducing helper T cells against helper antigens and subsequently immunizing helper antigen-modified tumor cells in the presence of helper T cells might result in more effective anti-TATA immune responses. Thus, the T cell–T cell interaction model is based on the hypothesis that the most efficient physiological cellular mechanism responsible for the generation of anti-TATA effector T-cell responses can be induced by close linkage of helper T cells and anti-TATA effector precursor T cells in the microenvironment of antigen-coupled tumor cells (Fig. 1).

2.2. Induction of Helper T Cells Against Antigens Used for Tumor Cell Modification: Selection of an Appropriate Helper Antigen System

Induction of anti-TATA effector T-cell responses via T cell–T cell interactions to helper antigen-modified tumor cells is a prerequisite in the system. We have used haptenic determinants as modifying antigens in initial studies because of the advantages that the haptenic structure is known and cell surfaces can be modified without adversely influencing cell viability.

In contrast to antihapten B-cell (antibody) responses by hapten-conjugated heterologous carrier proteins, antihapten T-cell responses can be generated by immunizing mice with hapten-conjugated self-components. For example, administration of hapten-conjugated mouse isologous γ-globulin (MGG)[28] or hapten-modified syngeneic spleen cells,[29,30] or sensitization of skin with hapten dissolved in an appropriate organic solvent[31] induces various types of antihapten T-cell responses. These include the induction of (1) cytotoxic T lymphocytes (CTL), (2) T cells for delayed-type hypersensitivity (DTH), and (3) helper and/or suppressor T cells. Hapten-reactive helper T-cell activity can be assessed by measuring the augmented induction of antihapten CTL responses from normal spleen cells or thymocytes when cocultured with 850-R X-irradiated hapten-primed spleen cells upon stimulation with hapten-modified syngeneic cells.[31]

Although haptens are useful and convenient reagents for modifying cell surfaces, their utilization should be considered in the context of the known genetic control involved in the induction of hapten-reactive helper T cells. The magnitude of major histocompatibility complex (MHC)-restricted antihapten T-cell responses has been shown to be controlled by the genes mapping inside[32] as well as outside[33,34] the MHC. The immune response (Ir) gene effect was shown to be expressed at helper as well as CTL precursor levels.[35] In attempting to augment T-cell responses by hapten-reactive helper T cells, therefore, the choice of hapten suitable for an individual bearing a given MHC specificity is of great importance. For example, C3H/He *(H-2^k)* and C57BL/6 *(H-2^b)* strains are high and low responders, respectively, to the trinitrophenyl (TNP) hapten.[32] However, C57BL/6 mice exhibit high responses against another hapten, N-iodoacetyl-N'-(5-sulfonic-1-naphthyl) ethylene diamine (AED). In fact, these AED-reactive, but not TNP-reactive, helper T cells were able to enchance generation of CTL responses *in vitro* against syngeneic tumor cells.[36] Thus, the genetic defect involving helper T-cell generation against TNP plus H-2^b self can be circumvented by utilizing helper T cells against an appropriate hapten. This emphasizes the existence of Ir gene control in the induction of hapten-reactive T cells to be utilized for T cell–T cell interactions.

Various helper antigens suitable for clinical tumor systems should be scrutinized in considering any potential clinical application of this T cell–T cell interaction to tumor immunotherapy. These include utilization of tuberculin (purified protein derivative, PPD[37]) and vaccinia virus[38] as tumor cell surface modifying antigens, since helper T cells could be easily primed to these antigens in humans. A new attempt was also made to synthesize a haptenic compound which is cross-reactive to *Mycobacterium tuberculosis* or BCG[39] and is easily coupled onto tumor cells (described in Section 4). Which types of modifying antigens are the most appropriate as helper antigens could be determined by various factors including its immunobiological nature such as immunogenicity and Ir gene control as well as the feasibility of quality control of the modifying agent and standardization of the coupling condition for clinical application.

3. Augmented Induction of Tumor-Specific Immunity by T Cell–T Cell Interaction Between Helper T Cells and Anti-TATA Effector Cells

3.1. General Protocol

The hapten-reactive helper (TNP-reactive helper) model enabled us to construct a general protocol based on the aforementioned theory (Fig. 2). Thus, the conditions have been defined under which enhanced T-cell-mediated immunity

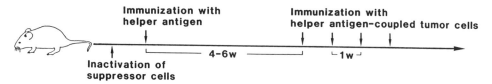

FIGURE 2. A protocol for enhanced induction of tumor-specific *in vivo* protective immunity.

against TATA could be generated by preinducing hapten (TNP)-reactive helper T cells and by subsequently immunizing with TNP-modified syngeneic tumor cells in TNP high-responder mice. It should also be noted that preinactivation of suppressor cells specific to TNP by using TNP-conjugated nonimmunogenic synthetic copolymer of D-glutamic acid and D-lysine (TNP-D-GL)[28] or nonspecific suppressor cells in general by 150-R whole-body X-irradiation[24] or cyclophosphamide inoculation[40] before TNP priming resulted in more potent TNP-reactive helper T-cell activity. The studies demonstrated that C3H/He and BALB/c mice immunized with TNP-modified respective syngeneic plasmacytoma[28] and leukemia cells[41] in the presence of potent TNP-reactive helper T cells exhibited augmented generation of tumor-specific cytotoxic T-cell activity against the corresponding tumor cells. The *in vivo* antitumor-protective activity was also augmented in the mice in which TNP-reactive helper T cells had been generated prior to immunization with TNP syngeneic tumor cells. Spleen cells from these mice exhibited complete tumor neutralization in a Winn assay, and these mice also exhibited an appreciable degree of growth inhibition of challenged tumor cells. In contrast, mice in which potent TNP helper activity had not been generated failed to generate such an *in vivo* effective antitumor immunity after immunization with TNP tumor cells. The magnitude of tumor-neutralizing activity developed in the presence of the amplified TNP helper T cells was as much as 20-fold greater than for control when titrated by Winn assay at various effector-to-target ratios, and this augmented tumor-neutralizing activity was also tumor specific and T cell mediated. Thus, the above experimental system provided an effective manipulation for eliciting enhanced tumor-specific immunity.

3.2. Enhanced Induction of Various Types of Antitumor Immune Responses

While TATA have been demonstrated to various extents to be expressed on the vast majority of experimentally induced or spontaneously occurring tumor cells, there is also evidence that the expression of TATA is hetergeneous qualitatively as well as quantitatively. In methylcholanthrene (MCA)-induced tumor models, TATA exhibit extreme polymorphism.[3,4,42,43] In addition, there appears to be another aspect of heterogeneity as recognized by the generation of selective immune responses against a given type of TATA. For example, MCA-induced tumors generate individually specific T-cell-mediated immunity, but they do not elicit demonstrable humoral immunity.[42,43] In contrast, melanomas demonstrate individually distinct TATA capable of inducing humoral responses.[43]

Earlier studies from our laboratory utilizing three syngeneic tumor models[38,44,45] have provided typical examples of preferential immune responses to TATA. In these models, immune responses were analyzed using tumor-immunized mice obtained by supplementing vaccinia virus-reactive helper T cells. Vaccinia virus-reactive helper T-cell activity was generated in C3H/HeN mice by intraperitoneal inoculation of live virus.[24] Immunization of these mice with vaccinia virus-infected syngeneic X5563 plasmacytoma or MH134 hapatoma cells led to the augmentation of immune resistance against challenge with viable tumor cells as compared to the level of resistance observed in control mice not primed to vaccinia virus.[38] *In vitro* cytotoxicity tests utilizing spleen cells and serum from mice which resulted in augmented tumor resistance by virus help have revealed that spleen cells from C3H/HeN mice immune to the X5563 plasmacytoma exhibited appreciable anti-X5563 CTL activity, whereas serum from these mice failed to display any antibody response.[38] In contrast, MH134-immune mice exhibited potent anti-MH134 antibody, but not CTL responses.[38]

Although the failure of X5563-TATA and MH134-TATA to elicit the respective anti-TATA antibody and CTL responses could be attributed to the defects of (1) responding B or CTL precursors, (2) helper T cells, or (3) both of them, recent studies utilizing assay systems for anti-TATA helper T-cell activity provide direct demonstration of the defects at the helper T cell level in anti-X5563 antibody and anti-MH134 CTL responses.[46] In these assay systems, spleen cells from these mice were tested for anti-TATA helper T-cell activity capable of augmenting (1) anti-TNP CTL and (b) anti-TNP antibody responses from anti-TNP CTL and B-cell precursors (responding cells) by the stimulation with TNP-modified X5563 or MH134 tumor cells. The results demonstrated that cultures of responding cells plus 850-R X-irradiated tumor-immunized spleen cells (helper cells) failed to enhance anti-TNP CTL or antibody responses when *in vitro* stimulation was provided by either unmodified tumor cells or TNP-modified syngeneic spleen cells (TNP-self). In contrast, these cultures resulted in appreciable augmentation of anti-TNP CTL or antibody response when stimulated by TNP-modified tumor cells. Interestingly, immunization with X5563 tumor cells resulted in anti-TATA helper T-cell activity involved in CTL, but not in antibody responses. Conversely, TATA of MH134 tumor cells induced selective generation of anti-TATA helper T-cell activity responsible for antibody response.

In addition, there also appears to exist defects at responding B cell or CTL precursor level. Immunization with vaccinia virus can produce vaccinia-specific helper T-cell activities responsible for CTL and antibody responses against other antigens coexisting with vaccinia-related antigens on stimulating cells.[24] Therefore, the findings that supplementation of the vaccinia helper T-cell system failed to induce anti-X5563 antibody or anti-MH134 CTL responses suggest that existence of defects of the responding B cells or CTL precursors in these anti-TATA responses. The parallelism in the defects of responding B cells plus B cell-helpers and CTL precursors plus CTL-helpers might be valid. Although the immunoregulatory mechanism(s) generating such a parallelism is obscure, these results indi-

TABLE I
Functional Heterogeneity of TATA

Tumor model	Immune responses[a] to TATA					
	CTL	CTL–Th[b]	B cell	B cell–Th[b]	DTH	In vivo protection
X5563 plasmacytoma	+	+	−	−	+	+
MH134 hepatoma	−	−	+	+	+	+

[a] These results are summarized from studies of Refs. 38, 46, and 47.
[b] These represent the respective anti-TATA helper T cells capable of assisting CTL or B-cell (antibody) responses against TATA.

cate that there exists qualitative TATA heterogeneity as shown by the preferential induction of CTL and antibody as well as CTL-helper and B cell-helper (These are summarized in Table I).

It should be noted that immunization of vaccinia virus-primed mice with virus-infected X5563 or MH134 tumor cells resulted in tumor-specific DTH responses[47] along with the induction of *in vivo* protective immunity in each tumor system. This indicates that utilization of helper T cells specific for modifying antigens has the potential to augment all three major immune responses to TATA and could contribute to a better understanding concerning the correlation between the type(s) of enhanced anti-TATA immune responses and augmented *in vivo* protection.

3.3. Analysis of a Cell Population Responsible for Enhanced *in vivo* Immune Protection

The mechanism by which syngeneic tumors can be rejected *in vivo* is still unclear. Although the role of T lymphocytes in the rejection of syngeneic tumors as well as allograft has been convincingly demonstrated,[48–50] it remains to be determined which type(s) of T cells is responsible for tumor rejection *in vivo*. It has been assumed that CTL, which bear the surface markers of Lyt-$1^+2^+3^+$ or Lyt-$1^-2^+3^+$, are capable of directly lysing syngeneic tumor cells as well as allogeneic cells and responsible for the graft rejection. Recent observations concerning analysis of the *in vivo* functional T-cell subset(s), however, have revealed the crucial role of Lyt-1^+2^- T cells in eradicating tumor cells *in vivo*.[51–54] We have also shown that Lyt-1^+2^- T cells depleted of the Lyt-2^+ T-cell subpopulation containing CTL or CTL precursors provide effective *in vivo* protective immunity.[55] The efficacy of tumor-immune Lyt-1^+2^- T cells was confirmed by preparing T-cell-depleted B-cell mice and by testing whether the Lyt-1^+2^- T-cell subpopulation causes the rejection of tumor cells when adoptively transferred into B-cell mice.[56,57] The results demonstrated that adoptive transfer of Lyt-1^+2^- but not Lyt-1^-2^+ immune T cells into B-cell mice produced complete protection against subsequent tumor cell challenge. In X5563 and MH134 tumor models, no CTL or antibody response against the tumor was detected in the above tumor-resistant B-cell mice into which had been transferred Lyt-1$^+$2$^-$ T cells.

In the preceding subsection, although the supplementation of T cell–T cell interaction mechanism using virus-help resulted in enhanced induction of antitumor CTL or antibody response along with *in vivo* resistance, whether such an enhanced response contributed to the augmented induction of *in vivo* resistance was not clear. Conversely, the above studies using B-cell recipient mice imply that *in vivo* tumor eradication does not necessarily require the induction of antitumor CTL or antibody response. In fact, when the cellular mechanism responsible for enhanced induction of *in vivo* resistance by virus-reactive helper T cells was analyzed, the generation of tumor-specific Lyt-1$^+$2$^-$ T cells was found to be augmented.[58] These results indicate that the augmented induction of tumor-specific Lyt-1$^+$2$^-$ T cells was essential for enhanced *in vivo* tumor immunity via virus-help (Fig. 3). The cellular and molecular mechanisms by which Lyt-1$^+$2$^-$ T cells eradicate tumor cells *in vivo* are described in Chapter 21 of this volume.

4. Application of T Cell–T Cell Interaction Mechanism to Establishing Tumor-Specific Immunotherapy Models

4.1. Solid Tumor System

Based on the results in the preceding section, we have designed an experimental protocol of a tumor-specific immunotherapy model.[40] This model includes (1) the induction of hapten-reactive helper T-cell activity, (2) implantation with a lethal dose of viable syngeneic tumor cells, and (3) subsequent injection of hapten into the tumor mass for haptenation of tumor cells *in situ* (Fig. 4). In this protocol, potent TNP-reactive helper T-cell activity was induced by the combined treatment with a single inoculation of cyclophosphamide (Cy) and skin painting with trinitrochlorobenzene (TNCB). C3H/He mice in which TNP-reactive helper T-cell activity had been generated were inoculated intradermally with 10^6 syngeneic X5563 tumor cells. Six days after tumor-cell inoculation, 0.15 ml TNCB (0.5%) in olive oil was injected into the tumor mass (7- to 10-mm diameter). At this stage, there was no significant difference in the mean diameter in any group of tumor-bearing mice. The tumor continued to grow for 2–3 days after administration of TNP into the tumor mass. As shown in Table II, *in situ* TNP modification of tumor cells in tumor-bearing mice in which potent TNP-reactive helper T-cell activity had been generated led to an appreciably high incidence of complete regression of growing tumors (group C). This contrasted with the fact that tumor regression was not evident when the injection of TNP into the tumor mass was made in the absence of TNP-reactive T cells (group B).

The successful regression of syngeneic tumors by TNP immunotherapy permitted us to test the applicability of this immunotherapy protocol to a 3-methylcholanthrene (MCA)-induced autochthonous tumor system.[59] Autochthonous tumors were induced in C3H/HeN mice by subcutaneous injection of 0.5 mg MCA

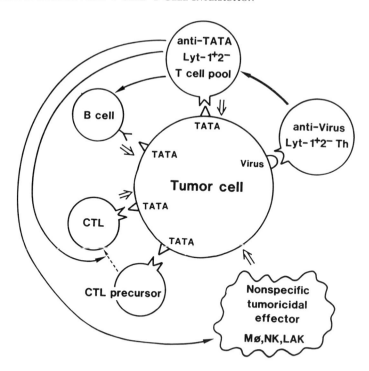

FIGURE 3. T cell–T cell interaction between Lyt-1^+2^- helper T cells to helper (virus) antigens and anti-TATA Lyt-1^+2^- T cells.

in 0.1 ml olive oil. Four weeks after the MCA injection, the experimental group received the combined treatment of cyclophosphamide injection and TNCB painting in order to induce amplified TNP-reactive helper T-cell activity (TNH-Th$^+$). The control group did not receive the regimen aimed to induce TNP helper T cells in (TNP-TH$^-$). The mice were seen to develop fibrosarcomas around 8 weeks after MCA treatment. At 9 weeks after MCA injection, 20–30% had tumors of 6–9 mm in diameter. These mice were collected and treated with intratumoral injection of 0.15 ml of 1% TNCB. The other two control groups included in this experiment in addition to the above groups were (1) MCA injection only, and (2) MCA injection plus combined treatment of cyclophosphamide and TNCB painting. Although all of the tumors in the three control groups except that for one animal in the TNP-

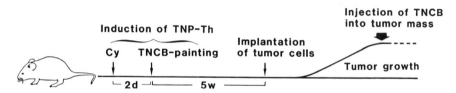

FIGURE 4. A protocol for TNP immunotherapy of solid tumor model.

TABLE II
Regression of Solid Tumor in TNP Immunotherapy Model[a]

Group	Induction of TNP-helper	Intratumoral TNP injection	Incidence of tumor regression
A	−	−	0/25
B	−	+	1/25
C	+	+	19/36

[a]From Ref. 40.

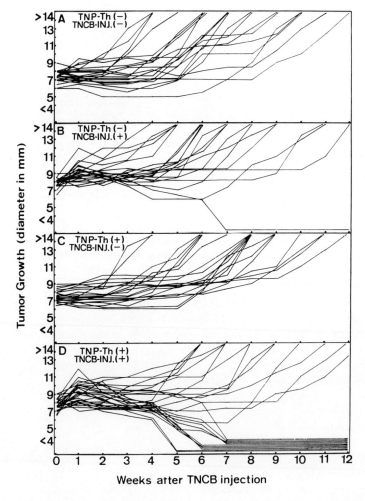

FIGURE 5. TNP immunotherapy in MCA-induced autochthonous fibrosarcoma model. Experiments employed four groups: group A, MCA injection only; group B, MCA injection and then intratumoral TNCB injection; group C, MCA injection and then induction of TNP helpers alone (but no intratumoral TNCB injection); group D, MCA injection, induction of TNP helpers, and then intratumoral TNCB injection. From Ref. 59.

Th⁻ group continued to grow and to kill animals, an appreciable number (11 of 25) of tumors in the TNP-Th⁺ group regressed. Tumor growth in individual animals of four groups is shown in Fig. 5. These results demonstrate that the efficacy of the TNP immunotherapy protocol is not limited to transplantable tumor systems but is also applicable to autochthonous tumors.

At least two major mechanisms are postulated to be responsible for the above tumor-regression process observed in the group generating amplified TNP helper T-cell activity. An application of TNP to the skin (TNCB painting) induces various immune responses to TNP including DTH, cytotoxic responses, and antibody formation in addition to the generation of TNP-reactive helper T-cell activity. Injection of TNCB into the tumor mass of TNP-primed mice would activate anti-TNP CTL, B cell, and DTH responses in the tumor mass by stimulation with TNP-modified tumor cells, and eventually induce a tumoricidal effect on TNP-modified tumor cells as well as unmodified tumor cells by the release of myriad nonspecific immunoreactive molecules including lymphokines (bystander effect).

However, it could also be that anti-TNP helper T-cell activity was involved in the augmented induction of antitumor T-cell responses by T cell–T cell interaction as discussed in the above sections, and that this enhanced antitumor T-cell immunity played a role in rejecting the growing tumor. The intradermal challenge of 10^5 viable X5563 tumor cells into the above tumor-regressor mice revealed that most of the mice could reject the tumor-cell challenge, indicating that tumor regression is accompanied by the concurrent induction of anti-X5563 immune resistance.[59] The development of much more potent anti-X5563 immune resistance in such tumor-regressor mice was confirmed by comparing tumor-neutralizing activity of spleen cells from these mice to that from mice whose tumor was surgically resected 7 days after implantation. Winn assays performed utilizing these two groups of spleen cells at various spleen-to-tumor ratios revealed tht appreciably stronger tumor-neutralizing activity was generated in the regressor mice by TNP immunotherapy than in the surgically treated mice.[59] Although these results only indirectly demonstrate the involvement of enhanced tumor-specific immunity in the rejection process, a critical role of tumor-specific immunity augmented by helper T cells against antigens used for tumor cell modification would be clearly demonstrated in the immunotherapy of the leukemia model described in the next subsection.

4.2. Leukemia System

Another immunotherapeutic protocol was designed for the model of murine chronic leukemia, BCL_1 (Fig. 6). In this model, BALB/c mice possessing TNP-reactive helper T cells were rendered leukemic by intravenous injection of 10^4 viable BCL_1 leukemia cells. One week after the intravenous dissemination of leukemia cells, the mice were immunized with 10,000-R X-irradiated BCL_1 cells modified with TNP. The results shown in Table III demonstrate that the immunization of leukemic mice with TNP leukemia cells in the presence of TNP helper T-cell activity results in an appreciable therapeutic effect (group F). This also reinforces the theoretical validity of the tumor-specific active immunotherapy model based on the T cell–T cell interaction mechanism.

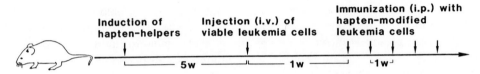

FIGURE 6. A protocol for hapten immunotherapy of leukemia model.

With regard to applying this model to tumor-bearing patients, the success of such an approach will depend on what type of haptenic reagents can be used for *in situ* modification of human tumor cells or *in vitro* modification of immunizing tumor cells. With reference to this issue, TNP may serve as one of the potential candidates for immediate clinical use. In addition, a new haptenic determinant recently synthesized by us for modification of tumor cells, i.e., a series of carbodimide derivatives of muramyl dipeptide (MDP), may represent another group of substances convenient for clinical use. MDP is known to be a common structural unit of the cell wall component of some bacteria such as mycobacterium to which humans have already been sensitized by spontaneous infection or by vaccination with BCG. Moreover, MDP is known to be the minimum structural requirement for adjuvant activity of the mycobacterial cell wall. Therefore, if any MDP derivative capable of easily binding to tumor cells is cross-reactive with mycobacterial cell wall components, such a derivative may provide an appropriate reagent useful for human tumor cells in conjunction with its potential adjuvant activity. We have found that 6-O-butyryl-N-acetylmuramyl-L-alanyl-D-isoglutamine-N-hydroxy-5-norbornene-endo-dicarboximide ester (herein designated as L4-MDP and shown in Fig. 7) may represent one such hapten and that priming mice with BCG induced significant L$_4$-MDP-reactive helper T-cell activity. Subsequent immunization of these BCG-primed mice with L4-MDP-coupled syngeneic tumor cells resulted in enhanced generation of tumor-specific immunity.[39] Moreover, when the TNP-

TABLE III
Active Immunotherapy Utilizing TNP Helpers in BCL$_1$ Leukemia Model[a]

Group	TNP-Th	Immunization with	Percent leukemic at 8 weeks	Percent survival at 16 weeks
A	−	−	100	0
B	+	−	100	0
C	−	BCL$_1$	100	0
D	+	BCL$_1$	100	0
E	−	TNP-BCL$_1$	100	14
F	+	TNP-BCL$_1$	57	43

[a]BALB/c mice in which TNP-helpers (TNP-Th) had (groups B, D, F) or had not been generated (groups A, C, E) were inoculated i.v. with 10^4 viable BCL$_1$ leukemia cells. These mice either were not immunized (groups A, B) or were immunized i.p. with unmodified or TNP-BCL$_1$ cells (10,000 R) five times at 1-week intervals. Percent incidence of leukemia and percent surviving at 8 and 16 weeks, respectively, after i.v. inoculation of viable leukemia cells are shown.

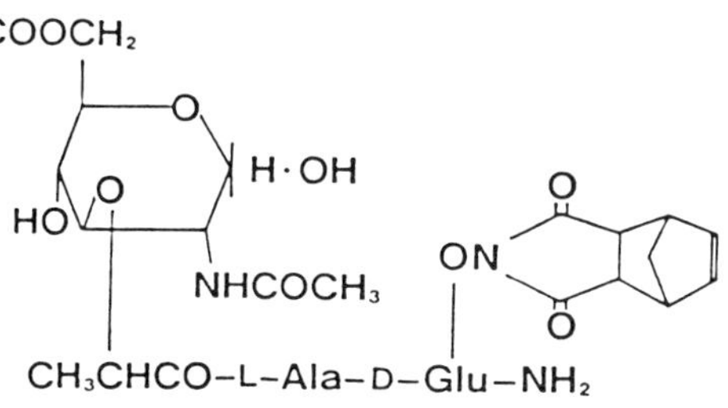

FIGURE 7. Structure of haptenic MDP, 6-o-Butyryl-N-acetylmuramyl- L-alanyl-D-isoglutamine-N-hydroxy-5-norbornene-2,3-dicarboxyimidyl ester (L4-MDP-ONB).

helper system in the immunotherapy model of leukemia was replaced by this BCG-MDP-hapten system, more potent therapeutic effect was observed and more than 90% of leukemic mice were completely cured (Table IV).[60]

Although various types of tumor cell modification using chemicals,[10-12,28,41] virus,[9,13,38] and proteins[37] have been devised, the use of L4-MDP hapten is of great advantage for the following reasons. First, coupling of L4-MDP to cell surfaces can be easily achieved, since the active ester group linked to the MDP provides a high degree of reactivity to the free amino residues on cell surfaces. The binding efficiency of L4-MDP and TNBS was compared on syngeneic cells modified with limiting concentration of TNBS or L4-MDP-ONB by immunofluorescence antibody technique utilizing anti-TNP or L4-MDP antiserum (unpublished results), and was found to be equivalent. While a large number of haptens have been demonstrated to generate potent T-cell responses when coupled to syngeneic cells, the magnitude of such T-cell responses are influenced by genes inside as well as outside the MHC as described in Section 2.2. In contrast, our recent observation has revealed that L4-MDP-reactive T-cell responses can be induced to the same extent in BALB/c ($H-2^d$) and C57BL/6 ($H-2^b$) as well as C3H/HeN ($H-2^k$) strains (manuscript in preparation). Thus, the L4-MDP hapten could provide additional advantages particu-

TABLE IV
Active Immunotherapy Utilizing MDP-Helpers in BCL_1 Leukemia Model[a,b]

BCG sensitization	Immunization with	Incidence of leukemia (%) at weeks			
		5	10	15	20
−	−	100	100	D[c]	D
+	−	60	100	D	D
+	BCL_1	30	90	D	D
+	MDP-BCL_1	0	0	6.7	6.7

[a] From Ref. 60.
[b] Protocol was almost the same as that used in Table III for TNP system except for BCG sensitization for MDP-helper induction instead of TNP-helper induction.
[c] D = death of all mice in group.

larly when attempts are made to augment effector T-cell responses to TATA via T cell–T cell interactions in individuals bearing various MHC specificities.

The third and most advantageous aspect of the L4-MDP system is concerned with its potential future clinical applicability. The present approach utilizing L4-MDP-reactive helper T cells in BCG-primed hosts is reminiscent of our previous work,[37] which demonstrated that immunization of PPD-coupled syngeneic tumor cells to mycobacterium-primed mice resulted in potent immunotherapeutic as well as immunoprophylactic effects against tumor growth. However, PPD has suffered from difficulties in quality control and standardization of its coupling capacity onto the tumor cell surface. In this context, it would be of great value that a BCG cross-reactive chemical compound (L4-MDP) has been successfully synthesized and that such a new haptenic MDP compound is capable of substituting for PPD as a cell surface modifying antigen. It should also be noted that since MDP per se is an immunopotentiating molecule which can activate and augment T-cell activity in general,[61-63] L4-MDP hapten could have the dual potential of exerting its function as a haptenic determinant in T cell–T cell interaction mechanism, and as an immunopotentiating molecule when the MDP moiety in the MDP–tumor cell complex is cleaved off.

5. Summary

The theoretical basis and practical manipulation of T cell–T cell interaction responsible for enhanced generation of antitumor effector T cells have been reviewed in this chapter.

Recent studies concerning molecular mechanisms of T cell–T cell interactions have revealed the involvement of various types of lymphokines.[91] These include interleukin-1 (IL-1), IL-2, killer helper factor, and others. Moreover, the DNA sequence of genes coding for some of the lymphokines has been determined, and such gene-translated lymphokine products have now become available. Along this line of the molecular mechanisms of T cell–T cell interaction, direct infusion into the tumor or systemic administration of lymphokines may be one plausible regimen to eradicate the tumor by inducing intralesional effector-cell activity against TATA. However, more precise information for possible lymphokine-mediated cascade reactions in the effector T-cell generation against TATA is required for the establishment of such a regimen. Moreover, it remains undetermined which combinations of the exogenous lymphokines actually function *in vivo* to augment the tumor-specific immunity.

Irrespective of which types of lymphokines and T cell–T cell interactions among various T-cell subsets are involved in the implementation of host immune responses against tumor, the more feasible and practical approach is to use both molecular and cellular mechanisms for the augmented induction of effector T-cell activity against TATA. We have defined conditions under which enhanced tumor-specific immunity can be obtained by preinducing potent helper T-cell activity reactive to a certain antigenic determinant such as hapten, followed by immunization with tumor cells coupled with the corresponding antigenic determinant.

This system induces the most efficient and physiological cellular and molecular mechanisms responsible for the generation of tumor-specific effector T-cell activity *in vivo* by virtue of the close linkage of hapten-reactive helper T cells and TATA-specific effector-precursor T cells in the microenvironment of hapten-coupled tumor cells.

We have demonstrated that this protocol resulted in enhanced *in vivo* tumor protective immunity but also was applicable to the immunotherapy model and that a growing solid tumor as well as disseminated leukemia cells could be effectively eradicated. Thus, these results emphasize the role of T cell–T cell interactions in augmenting tumor-specific immunity. This protocol utilizing hapten-reactive helper T cells provides an effective maneuver for tumor-specific immunotherapy.

References

1. Foley, F. J., 1953, Antigenic properties of methylcholanthrene-induced tumors in mice of the strain of origin, *Cancer Res.* **13**:835–837.
2. Prehn, R. T., and Main, J. M., 1957, Immunity to methylcholanthrene-induced sarcomas, *J. Natl. Cancer Inst.* **18**:769–778.
3. Klein, G., Sjögren, H. D., Klein, E., and Hellström, K. E., 1960, Demonstration of host resistance against methylcholanthrene-induced sarcomas in the primary autochthonous host, *Cancer Res.* **20**:1561–1572.
4. Old, L. J., Boyse, E. A., Clarke, D. A., and Garswell, E., 1962, Antigenic properties of chemically induced tumors, *Ann. N.Y. Acad. Sci.* **101**:80–120.
5. Hellström, K. E., and Hellström, I., 1974, Lymphocyte-mediated cytotoxicity and blocking serum activity to tumor antigens, *Adv. Immunol.* **18**:209–277.
6. Kamo, I., and Friedman, H., 1977, Immunosuppression and the role of suppressive factors in cancer, *Adv. Cancer Res.* **25**:271–321.
7. Fujimoto, S., Greene, M. I., and Sehon, A. A., 1976, Regulation of the immune response to tumor antigens. I. Immunosuppressor cells in tumor-bearing mice, *J. Immunol.* **116**:791–799.
8. Mitchison, N., 1970, Immunologic approach to cancer, *Transplant. Proc.* **2**:92–103.
9. Lindenmann, J., and Klein, P., 1967, Viral oncolysis: Increased immunogenicity of host cell antigen associated with influenza virus, *J. Exp. Med.* **126**:93–125.
10. Martin, W. J., Wunderlich, J. R., and Fletcher, F., 1971, Enhanced immunogenicity of chemically coated syngeneic tumor cells, *Proc. Natl. Acad. Sci. USA* **68**:469–472.
11. Hashimoto, Y., and Yamanoha, B., 1976, Induction of transplantation immunity by dansylated tumor cells, *Gann* **67**:315–319.
12. Galli, N. Naor, D., Asjö, B., and Klein, G., 1976, Induction of immune responsiveness in a genetically low responsive tumor host combination by chemical modification of the immunogen, *Eur. J. Immunol.* **6**:473–476.
13. Kobayashi, H., Kodama, T., and Gotohda, E., 1977, Xenogenization of tumor cells, *Hokkaido Univ. Med. Libr. Ser.* **9**:1–110.
14. Klein, G., and Klein, E., 1978, Induction of tumor cell rejection in the low responsive YAC-lymphoma strain A host combination by immunization with somatic cell hybrids, *Eur. J. Cancer* **15**:551–557.
15. Prager, M. D., and Gordon, W. C., 1979, Immunoprophylaxis and therapy with lipid conjugated lymphoma cells, *Gann Monogr. Cancer Res.* **23**:143–150.
16. Wallack, M. K., Meyer, M., Bourgoin, A., Dere, J.-F., Leftheriotis, E., Carcagne, J., and Koprowski, H., 1983, A preliminary trial of vaccinia oncolysates in the treatment of recurrent melanoma with serologic responses to the treatment, *J. Biol. Res. Modif.* **2**:586–596.
17. Katz, D. H., and Benacerraf, B., 1972, The regulatory influence of activated T cells on B cell responses to antigens, *Adv. Immunol.* **15**:1–94.

18. Cantor, H., and Asofsky, R., 1970, Synergy among lymphoid cells mediating the graft-versus-host response. II. Synergy in graft-versus-host reactions produced by BALB/c lymphoid cells of different anatomic origin, *J. Exp. Med.* **13**:223–234.
19. Cantor, H., and Boyse, E. A., 1975, Functional subclasses of T lymphocytes bearing different Ly antigens. II. Cooperation between subclasses of Ly^+ cells in the generation of killer activity, *J. Exp. Med.* **141**:1390–1399.
20. Baum, L. L., and Pilarski, L. M., 1978, In vitro generation of antigen-specific helper T cells that collaborate with cytotoxic T cell precursors, *J. Exp. Med.* **148**:1579–1591.
21. Wagner, H., Hardt, C., Heeg, K., Pfizenmaier, K., Solback, W., Bartlett, R., Stockinger, H., and Rollinghoff, M., 1980, T–T cell interactions during cytotoxic T lymphocyte (CTL) responses: T cell derived helper factor (interleukin 2) as a probe to analyze CTL responsiveness and thymic maturation of CTL progenitors, *Immunol. Rev.* **51**:215–255.
22. Leung, K. N., and Ada, G. L., 1981, Effect of helper T cells on the primary in vitro production of delayed-type hypersensitivity to influenza virus, *J. Exp. Med.* **153**:1029–1043.
23. Tucker, M. J., and Bretscher, P. A., 1982, T cells cooperating in the induction of delayed-type hypersensitivity act via the linked recognition of antigenic determinants, *J. Exp. Med.* **155**:1037–1049.
24. Fujiwara, H., Shimizu, Y., Takai, Y., Wakamiya, N., Ueda, S., Kato, S., and Hamaoka, T., 1984, The augmentation of tumor-specific immunity by virus-help. I. Demonstration of vaccinia virus-reactive helper T cell activity involved in enhanced induction of cytotoxic T lymphocyte and antibody responses, *Eur. J. Immunol.* **14**:171–175.
25. Kosugi, A., Takai, Y., Ogata, M., Fujiwara, H., and Hamaoka, T., 1985, T–T cell interaction in the in vitro induction of delayed type-hypersensitivity (DTH) responses: Demonstration of vaccinia virus-reactive helper T cell activity involved in enhanced induction of DTH responses, *J. Leukocyte Biol.* **37**:629–639.
26. Pfizenmaier, K., Delzeit, R., Rollinghoff, M., and Wagner, H., 1980, T–T cell interaction during in vitro cytotoxic T lymphocyte responses. III. Antigen-specific T helper cells release nonspecific mediator(s) able to help induction of H-2 restricted cytotoxic T lymphocyte responses across cell-impermeable membranes, *Eur. J. Immunol.* **10**:577–582.
27. Fujiwara, H., Levy, R. B., and Shearer G. M., 1981, Analysis of Ir gene control of cytotoxic response to hapten-modified self: Helper T cells specific for a sulfhydryl hapten can substitute for an anti-INP-H-2^b self helper cell defect, *J. Immunol.* **127**:940–945.
28. Hamaoka, T., Fujiwara, H., Teshima, K., Aoki, H., Yamamoto, H., and Kitagawa, M., 1979, Regulatory function of hapten-reactive helper and suppressor T lymphocytes. III. Amplification of generation of tumor-specific cytotoxic effector T lymphocytes by suppressor T cell-depleted hapten-reactive T lymphocytes, *J. Exp. Med.* **149**:185–199.
29. Greene, M. I., Sugimoto, M., and Benacerraf, B., 1978, Mechanisms of regulation of cell-mediated immune responses. I. Effect of the route of immunization with TNP-coupled syngeneic cells on the induction and suppression of contact sensitivity to picryl chloride, *J. Immunol.* **120**:1604–1611.
30. Finberg, R., Greene, M. I., Benacerraf, B., and Burakoff, S. J., 1979, The cytolytic T lymphocyte response to trinitrophenyl-modified syngeneic cells. I. Evidence for antigen-specific helper T cells, *J. Immunol.* **123**:1205–1209.
31. Fujiwara, H., and Shearer, G. M., 1980, Studies on in vivo priming of TNP-reactive cytotoxic effector cell system. II. Augemented secondary cytotoxic response by radioresistant T cells and demonstration of differential helper activity as a function of the stimulating dose of TNP-self, *J. Immunol.* **124**:1271–1276.
32. Shearer, G. M., Schmitt-Verhulst, A. -M., Pettinelli, C. B., Miller, M. W., and Gilheany, P. E., 1979, H-2 linked genetic control of murine T cell-mediated lympholysis to autologous cells modified with low concentrations of trinitrobenzene sulfonate, *J. Exp. Med.* **149**:1407–1423.
33. Fujiwara, H., and Shearer, G. M., 1981, Non-H-2 associated genetic regulation of cytotoxic responses to hapten-modified syngeneic cells: Effect on the magnitude of secondary response and helper T cell regulation after in vivo priming, *Eur. J. Immunol.* **11**:700–704.
34. Ogata, M., Shimizu, J., Tsuchida, T., Takai, Y., Fujiwara, H., and Hamaoka, T., 1986, Non-H-2-linked genetic regulation of cytotoxic responses to hapten-modified syngeneic cells. I. Non-H-2-

linked Ir gene defect expressed on T cells is not predetermined at the stage of bone marrow cells, *J. Immunol.* **136**:1178–1185.
35. Fujiwara, H., and Shearer, G. M., 1981, Genetic control of cell-mediated lympholysis to TNP-modified murine syngeneic cells. I. Expression of Ir gene function at the cytotoxic precursor and helper cell level in the response to TNP-H-2^b-self, *J. Immunol.* **126**:1047–1051.
36. Mizushima, Y., Fujiwara, H., Takai, Y., Shearer, G. M., and Hamaoka, T., 1985, Genetic control of hapten-reactive helper T cell responses and its implications for the generation of augmented antitumor cytotoxic responses, *J. Natl. Cancer Inst.* **74**:1269–1273.
37. Takatsu, K., Hamaoka, T., Tominaga, A., and Kanamasa, Y., 1980, Augmented induction of tumor-specific resistance by priming with PPD-coupled syngeneic tumor cells, *J. Immunol.* **125**:2367–2373.
38. Shimizu, Y., Fujiwara, H., Ueda, S., Wakamiya, N., Kato, S., and Hamaoka, T., 1984, The augmentation of tumor-specific immunity by virus-help. II. Enhanced induction of cytotoxic T lymphocyte and antibody responses to tumor antigens by vaccinia-reactive helper T cells, *Eur. J. Immunol.* **14**:839–843.
39. Hamaoka, T., Takai, Y., Kosugi, A., Mizushima, Y., Shima, J., Kusama, T., and Fujiwara, H., 1985, The augmentation of tumor-specific immunity by utilizing haptenic muramyl dipeptide (MDP) derivatives. I. Synthesis of a novel haptenic MDP derivative cross-reactive with bacillus Calmette Guerin and its application to enhanced induction of tumor immunity, *Cancer Immunol. Immunother.* **20**:183–188.
40. Fujiwara, H., Moriyama, Y., Suda, T., Tsuchida, T., Shearer, G. M., and Hamaoka, T., 1984, Enhanced TNP-reactive helper T cell activity and its utilization in the induction of amplified tumor immunity which results in tumor regression, *J. Immunol.* **132**:1571–1577.
41. Fujiwara, H., Hamaoka, T., Shearer, G. M., Yamamoto, H., and Terry, W. D., 1980, The augmentation of in vitro and in vivo tumor-specific T cell-mediated immunity by amplifier T lymphocytes, *J. Immunol.* **124**:863–869.
42. Law, L. W., Rogers, M. J., and Appella, E., 1980, Tumor antigens on neoplasms induced by chemical carcinogens and by DNA- and RNA-containing viruses: Properties of the solubilized antigens, *Adv. Cancer Res.* **32**:201–235.
43. Old, L. J., 1981, Cancer immunology: The search for specificity, *Cancer Res.* **41**:361–375.
44. Fujiwara, H., Hamaoka, T., Teshima, K., Aoki, H., and Kitagawa, M., 1976, Preferential generation of killer or helper T-lymphocyte activity directed to the tumor-associated transplantation antigens, *Immunology* **31**:239–248.
45. Fujiwara, H., Hamaoka, T., and Kitagawa, M., 1978, Suppressive effect of pretreatment with X-irradiated tumor cells on the induction of syngeneic murine tumor system. I. Suppressive versus sensitizing effect on the same pretreatment in 2 tumor systems with different immune responses, *Gann* **69**:793–803.
46. Fujiwara, H., Yoshioka, T., Shima, J., Kosugi, A., Itoh, K., and Hamaoka, T., 1986, Helper T cells against tumor-associated antigens (TAA): Preferential induction of helper T cell activities involved in anti-TAA cytotoxic T lymphocyte and antibody responses, *J. Immunol.* **136**:2715–2719.
47. Takai, Y., Kosugi, A., Yoshioka, T., Tomita, S., Fujiwara, H., and Hamaoka, T., 1985, T–T cell interaction in the induction of delayed type hypersensitivity (DTH) responses: Vaccinia virus-reactive helper T cell activity involved in enhanced in vivo induction of DTH responses and its application to augmentation of tumor-specific DTH responses, *J. Immunol.* **134**:108–113.
48. Cerottini, J. C., and Brunner, K. T., 1974, Cell-mediated cytotoxicity, allograft rejection and tumor immunity, *Adv. Immunol.* **18**:67–132.
49. Herberman, R. B., Holden, H. T., Varesio, L., Taniyama, T., Pucetti, P., Kirchner, H., Gerson, J., White, S., Keisari, Y., and Haskill, J. L., 1980, Immunologic reactivity of lymphoid cells in tumors, *Contemp. Top. Immunobiol.* **10**:61–78.
50. Levy, J. P., and Leclerc, J. C., 1977, The murine sarcoma virus-induced tumor: Exception or general model in tumor immunology? *Adv. Cancer Res.* **24**:1–66.
51. Nelson, M., Nelson, D. S., Mckenzie, I. F. C., and Blanden, R. V., 1981, Thy and Ly markers on lymphocytes initiating tumor regression, *Cell. Immunol.* **60**:34–42.
52. Bhan, A. K., Perry, L. L., Cantor, H., McCluskey, R. T., Benacerraf, B., and Greene, M. I., 1981, The role of T cell sets in the rejection of a methylcholanthrene-induced sarcoma (S1509A) in syngeneic mice, *Am. J. Pathol.* **102**:20–27.

53. Fernandez-Cruz, E., Woda, B. A., and Feldman, J. D., 1980, Elimination of syngeneic sarcomas in rats by a subset of T lymphocytes, *J. Exp. Med.* **152**:823–841.
54. Greenberg, P. D., Cheever, M. A., and Fefer, A., 1981, Eradication of disseminated murine leukemia by chemoimmunotherapy with cyclophosphamide and adoptively transferred immune syngeneic Lyt-1^{+}2^{-} lymphocytes, *J. Exp. Med.* **154**:952–963.
55. Fukuzawa, M., Fujiwara, H., Yoshioka, T., Itoh, K., and Hamaoka, T., 1984, Effector cell analysis of tumor cell-rejection in vivo in two syngeneic tumor systems exhibiting distinct in vitro cytotoxic mechanisms, *Gann* **75**:912–919.
56. Fujiwara, H., Fukuzawa, M., Yoshioka, T., Nakajima, H., and Hamaoka, T., 1984, The role of tumor-specific Lyt-1^{+}2^{-} T cells in eradicating tumor cells in vivo. I. Lyt-1^{+}2^{-} T cells do not necessarily require recruitment of host's cytotoxic T cell precursors for implementation of in vivo immunity, *J. Immunol.* **133**:1671–1676.
57. Fukuzawa, M., Fujiwara, H., Yoshioka, T., Itoh, K., and Hamaoka, T., 1985, Tumor-specific Lyt-1^{+}2^{-} T cells can reject tumor cells in vivo without inducing cytotoxic T lymphocyte responses, *Transplant. Proc.* **17**:599–605.
58. Yoshioka, T., Fukuzawa, M., Takai, Y., Wakamiya, N., Ueda, S., Kato, S., Fujiwara, H., and Hamaoka, T., 1985, The augmentation of tumor-specific immunity by virus-help. III. Enhanced generation of tumor-specific Lyt-1^{+}2^{-} T cells is responsible for augmented tumor immunity in vivo, *Cancer Immunol. Immunother.* **21**:193–198.
59. Fujiwara, H., Aoki, H., Yoshioka, T., Tomita, S., Ikegami, R., and Hamaoka, T., 1984, Establishment of tumor-specific immunotherapy model utilizing TNP-reactive helper T cell activity and its application to autochthonous tumor system, *J. Immunol.* **133**:509–514.
60. Fujiwara, H., Shima, J., Sano, H., Kosugi, A., Nakajima, H., and Hamaoka, T.: Tumor-specific immunotherapy by active immunization with haptenic muramyl dipeptide derivative-coupled tumor cells, in: *Cellular Immunotherapy of Cancer,* (R. L. Truitt, R. P. Gale, and M. M. Bortin, eds.), Alan R. Liss Inc., Yew York (in press).
61. Igarashi, T., Okada, M., Azuma, I., and Yamamura, Y., 1977, Adjuvant activity of synthetic N-acetylmuramyl-L-alanyl-D-isoglutamine and related compounds on cell-mediated cytotoxicity in syngeneic mice, *Cell. Immunol.* **34**:270–278.
62. Iribe, H., and Koga, T., 1984, Augmentation of the proliferative response of thymocytes to phytohemagglutinin by the muramyl dipeptide, *Cell. Immunol.* **88**:9–15.
63. Lowy, I., Bona, C., and Chedid, L., 1977, Target cells for the activity of a synthetic adjuvant: Muramyl dipeptide, *Cell. Immunol.* **29**:195–199.

23

Antigens Expressed by Melanoma and Melanocytes

Studies of the Immunology, Biology, and Genetics of Melanoma

ALAN N. HOUGHTON, LAURA J. DAVIS,
NICOLAS C. DRACOPOLI, AND ANTHONY P. ALBINO

1. Introduction

Serological studies with human sera and monoclonal antibodies have identified a large number of distinct antigen systems on melanoma cells. A major focus of the mapping of these antigens has been the identification of determinants that can serve as targets for recognition by the host. In the course of these studies, more has been revealed about the antigenic profile of normal cells than about specific antigenic changes that occur during the process of malignant transformation. The cell components that have been best characterized are differentiation antigens. The investigation of these antigens on melanoma cells has provided considerable information about antigenic changes occurring during normal differentiation of the normal cell counterpart, the melanocyte. The search continues for molecules that are expressed specifically during events of malignant transformation and tumor progression. Three categories of antigens on melanoma cells have been identified: (1) differentiation antigens; (2) antigens expressed during transformation; and (3) antigens that serve as targets for the immune system in the host with cancer. In this chapter, we discuss our recent studies, rather than review the substantial amount of information that has been published about melanoma antigens.

ALAN N. HOUGHTON, LAURA J. DAVIS, NICOLAS C. DRACOPOLI, AND ANTHONY P. ALBINO • Department of Medicine and Division of Immunology, Memorial Sloan–Kettering Cancer Center, New York, New York 10021.

2. Differentiation Antigens

The process of differentiation is accompanied by changes in the expression of certain antigens.[1] Differentiation antigens have been identified in a wide range of cell types in both mouse and human tissues, and a large number of antigens have been described that distinguish one cell type from another. Melanocytes are cells located at the base of the epidermis and are distinctive because they are pigmented (synthesizing the pigment melanin in specialized organelles, melanosomes, using the enzyme tyrosinase). We have described 65 antigen systems on melanocytes and melanoma cells—30 of these antigens are broadly distributed on most cell types, 22 have been shown to have an intermediate distribution, and 13 are restricted to melanoma cells and closely related cell types of neuroectodermal origin. Two antigen systems have been particularly well characterized.

In an initial analysis of monoclonal antibodies (MAbs) raised against human melanoma cells, Dippold et al. described an IgG3 MAb, designated R24, that has a high degree of serological specificity for cultured melanoma cells.[2] R24 antibody was shown to react with cultured melanomas and a proportion of astrocytomas but not with cell lines of epithelial origin. Subsequent studies showed that R24 reacted with skin melanocytes.[3] When tested against frozen sections of normal and malignant tissues, R24 reacted with 58 of 60 melanomas, with four astrocytomas, and occasionally with epithelial cancers.[4] In normal tissues, R24 reacted with melanocytes, adrenal medulla, islet cells in the pancreas, fetal thymus, a small population of peripheral blood lymphocytes, epithelial cells of the parotid gland, and cells in the central nervous system. No reactivity was found with normal lung, liver, spleen, heart, intestine, kidney, or bladder. It was clear from this study that most tissues reacting with R24 arose from the neuroectoderm. A detailed study of the human brain revealed that R24 reacted with ependymal cells and subependymal glia, dorsal root ganglia, and sympathetic neurons, and with multiple neuronal groups in the spinal cord, brain stem, diencephalon, and hippocampus.[5]

Initial characterization of the antigen detected by R24 demonstrated a heat-stable, protease-resistant, and neuraminidase-sensitive component, features consistent with a glycolipid structure.[2] A more detailed study showed that the antigen was GD_3 ganglioside, based on carbohydrate composition, partial structural analysis, and comigration of antigen reactivity with GD_3 by thin-layer chromatography.[6] Melanoma cells have relatively high ganglioside composition compared to other cell types, with GD_3 and GM_3 being the two most prominent gangliosides in cultured melanoma cells. GD_3 can be detected, albeit at lower levels, in normal melanocytes.[7] Recent data have suggested that the increased expression of GD_3 in melanomas is related to events occurring during transformation[8] (see Section 3). The accumulation of GD_3 in melanomas may have several explanations, including increased levels of sialyltransferases resulting in increased synthesis, depressed levels of N-acetylgalactosylaminyl transferase(s) resulting in decreased degradation, or high levels of β-N-acetylgalactosaminidase producing increased degradation of other gangliosides (GD_3 is a precursor for the biosynthesis of other gangliosides). An interesting finding has been that GD_3 is a normal component of a large variety of cells and tissues.[6] The apparent discrepancy between the restricted specificity of R24 when tested against intact cells and tissue sections and the ubiquitous presence of GD_3 in other cell and tissue types can be accounted for by the facts that: (1) melanomas have much higher levels of gangliosides, and GD_3 in particular, than do most cells, and (2) GD_3 is localized to the cell surface in melanomas whereas in most other cell types it is likely to be an intracellular biosynthetic precursor to other gangliosides.

R24 has several interesting biological features and properties. First, reactivity of R24 is generally restricted to melanomas and melanocytes and closely related cell types of neuroectodermal origin. In this respect, cell surface expression of GD_3 is a marker for cells of certain neuroectodermal lineages, including the melanocyte pathway. Second, it is a potent mediator of complement-mediated cytotoxicity and antibody-dependent cellular cytotoxicity, and is effectively able to kill melanoma target cells by immune mechanisms.[9,10] R24 and other anti-GD_3 MAbs also cause profound aggregation and growth inhibition of target cells.[11,12,13] Finally, R24 has been found to induce proliferation and activation of T lymphocytes (K. Welte and A. N. Houghton, unpublished observation). These properties of R24 may be important in the regression of melanoma tumors observed in a small number of patients treated in a clinical phase 1 trial of R24 antibody.[14]

Several groups have reported MAbs that react with a set of high-molecular-weight glycoproteins.[4,15–17] The molecular weights of the antigens detected by this group of antibodies include a component of approximately 250,000 (250k) and a very-high-molecular-weight component (> 500k). The most detailed studies of this antigen system have been performed with the 9.2.27 MAb.[18,19] 9.2.27 immunoprecipitates a 250k N-linked sialoglycoprotein and a sulfate-containing, high-molecular-weight component that is sensitive to alkaline β-elimination and to chondroitinase. These characteristics are all consistent with a chondroitin sulfate proteoglycan. The proteoglycans are major components of extracellular and pericellular matrices, and, in keeping with these characteristics, the proteglycan antigen on melanoma cells does not behave as in integral membrane protein since it is readily extracted by mild dissociating conditions and is easily shed into chemically defined culture medium. 9.2.27 recognizes the 250k glycoprotein that serves as the core glycoprotein for the high-molecular-weight proteoglycan.

Both GD_3 and the high-molecular-weight proteoglycan are restricted antigens that can be detected on almost all melanoma cell lines and melanoma tumor tissues. In our experience, most restricted antigens have been found to identify subsets of melanomas. What distinguishes GD_3 and the proteoglycan from other restricted antigens is their expression by nearly all melanoma tumors. On the other hand, most antigens that we have analyzed are expressed on only a proportion of cell lines. It seemed likely that this diversity in melanoma phenotypes reflects a corresponding diversity in the phenotype of normal cells undergoing melanocyte differentiation. Just as one specific pathway of differentiation can be distinguished from other pathways by antigen phenotype, it is probable that patterns of antigen expression change as cells progress through different phases in a particular lineage.

To pursue this idea, we analyzed the antigens of melanocytes, using a recently described method for growing normal melanocytes in tissue culture,[20] and compared their antigen expression to a panel of melanoma cell lines. Antibodies directed against antigens expressed on malignant melanoma cells were tested for reactivity with fetal, newborn, and adult melanocytes.

Melanoma antigens could be grouped into four categories on the basis of their expression on melanocytes: (1) those not detected on fetal, newborn, or adult melanocytes; (2) those expressed on fetal and newborn but not generally on adult; (3) those expressed on adult but not generally on fetal or newborn; and (4) those detected equally on fetal, newborn, and adult melanocytes.[3] Eight antigens were selected because they defined subsets of melanomas and had distinct patterns of expression of fetal/newborn and adult melanocytes. Three antigens (M-2, M-3, and HLA-DR) were presumed to be early markers of melanocyte differentiation because they were expressed on melanomas but not on cultured melanocytes. Antigens M-4, M-5, and M-6 appeared on fetal and newborn melanocytes but not adult melanocytes and therefore marked an intermediate stage of melanocyte differentiation. Antigens M-9 and M-10 appeared to be late markers in the melanocyte lineage because they were strongly expressed on adult melanocytes as compared to fetal or newborn melanocytes.

The surface phenotypes of a panel of melanoma cell lines corresponded to early, intermediate, or late stages of melanocyte differentiation. Evidence for the basis of these subsets of melanomas came from comparison of the pattern of cell surface antigens with other differentiation characteristics, including pigmentation, morphology, and tyrosinase levels. Most melanomas expressing early markers were epithelioid, lacked pigmentation, and expressed no tyrosinase. The intermediate group of melanomas resembled fetal and newborn melanocytes with a bipolar shape. Melanocytes expressing late markers were frequently polydendritic, similar to adult melanocytes, with heavy pigmentation and high levels of tyrosinase activity. From these studies of differentiation, three stages of melanocyte differentiation were defined—precursor, intermediate, and mature. The features of the melanocyte precursor were inferred from the characteristics of melanomas expressing early differentiation.

Recently, we have analyzed in more detail three antigen systems that are late markers and are expressed only by pigmented cells: (1) MAb TA99 reacts with a 75k glycoprotein (gp75) present in mature melanosomes, (2) MAb CF21 detects a heat-labile, protease-sensitive antigen also located in mature melanosomes, and (3) MAb C350 recognizes a 180k cell surface glycoprotein.[21] TA99 and CF21 have been shown to recognize distinct antigenic determinants by antibody competition experiments. Reactivity of all three MAbs was absolutely restricted to melanocytic cells and related neuroectoderm-derived, pigment-producing cells. In cultured cells, all three MAbs reacted only with normal melanocytes and pigmented melanomas but not with nonpigmented melanomas or nonmelanocytic cell types.

In tests of a large panel of normal tissues by immunohistochemical methods, TA99 stained only cutaneous and choroidal melanocytes and pigmented retinal epithelial cells, and CF21 reacted only with cutaneous melanocytes. Although TA99 reacted strongly with melanocytes in both fetal and adult skin, CF21 reacted strongly with fetal melanocytes but only weakly with adult melanocytes. C350 did not stain any normal tissues. All nevi and primary melanoma specimens and 93% of metastatic melanomas expressed at least one of these three antigens. There was no reactivity against 62 nonmelanoma tumors. TA99 (gp75) expression decreased with the stage of tumor progression; 100% of nevus specimens and 90% of primary melanomas were TA99$^+$, but only 40% of metastatic melanomas expressed the TA99 antigen. These markers will prove useful for studies of melanocyte differentiation and malignant transformation, subsetting melanocytic lesions and distinguishing tissues and tumors of melanocytic origin.

3. Ia Antigens on Melanoma and Melanocytes

Although Ia or class II major histocompatibility antigens have generally been associated with cells of the immune system, recent studies have shown the presence of Ia on cells with no immune function, including melanomas. While Ia can be demonstrated on a proportion of both cultured and noncultured melanomas, no Ia antigens have been detected on normal melanocytes, either *in vitro* or *in vivo*.[3,22,23] A possible explanation for this finding is that Ia antigens are expressed on an early, as yet unidentified cell in the normal melanocyte lineage and that Ia$^+$ melanomas arise from this Ia$^+$ progenitor. According to this view, Ia antigens would be classified as differentiation antigens in the melanocyte pathway. The finding that melanomas corresponding to early or intermediate stages of melanocyte differentiation have generally the highest expression of Ia antigens, while well-differentiated melanomas are Ia$^-$, supports this view.[3] In addition, we have induced clonally-derived early, Ia$^+$ melanomas to differentiate into late, Ia$^-$ phenotypes, again suggesting that Ia expression is linked to the differentiation program of the melanoma cell. Since normal melanocytes from fetal, newborn, or adult skin do not express Ia antigens, they would represent later stages in melanocyte differentiation. This view depends of course on the existence of an Ia$^+$ melanocyte precursor, but the identification and isolation of such a cell is still under way.

An alternative explanation is that Ia expression is a consequence of events occurring during malignant transformation. To investigate the relationship of Ia expression in melanomas to malignant transformation, we infected melanocytes with transforming amphotropic pseudotypes of Harvey (Ha-MSV) and Kirsten (Ki-MSV) murine sarcoma viruses.[8] The Ha-MSV and Ki-MSV retroviruses contain oncogenes of the *ras* family, and were chosen because previous studies indicated that 10% of cultured melanomas have an activated *ras* gene allele (either Ha-*ras* or N-*ras*), and no other conclusive perturbation (i.e., rearrangement or amplification) in 15 other known oncogenes.[8,24] Two weeks postinfection, islands of Ki-MSV- or Ha-MSV-infected melanocytes appeared that had a distinct morphology. Uninfected melanocytes grow as bipolar spindle-shaped cells without much intercellular contact. Cells infected with the amphotropic MuLV helper virus had the same morphology as normal melanocytes. Melanocytes infected with either Ki-MSV or Ha-MSV were more polygonal and grew as clusters of cells with a tendency to pile up. Northern blot analysis confirmed the presence of Ha-*ras*- or Ki-*ras*-specific viral mRNAs.

Proliferation of melanocytes *in vitro* depends upon the addition of exogenous growth factors to the medium. Factors that support the growth of melanocytes include 12-*O*-tetradecanoyl phorbol 13-acetate (TPA), and growth factors derived from fetal fibroblasts, astrocytoma and melanoma cells.[20,25] In the absence of growth factors, melanocytes rapidly senesce and die. In contrast, cultured melanomas grow vigorously in the absence of TPA, suggesting that malignant transformation of melanocytes is associated with the acquisition of autonomy from exogenous growth factors. In the presence of TPA, the growth rate of melanocytes infected with Ki-MSV or Ha-MSV was comparable to unifected control cells and to a companion culture infected with amphotropic helper virus. In the absence of exogenous growth factors, uninfected melanocytes and melanocytes infected with helper virus died rapidly (within 6–12 days). Ki-MSV- and Ha-MSV-infected cultures also senesced but at a slower rate (within 2–4 weeks). Thus, expression of *ras* oncogenes could not induce stable actively proliferating, growth-factor-independent melanocyte cell lines.

Infection with Ha-MSV or Ki-MSV induced the strong expression of class II antigens within 3–5 days in all of four infected melanocyte cultures. Immunological analysis by dual-labeling fluorescence indicated that the same cells which expressed Ia antigens also expressed viral p21 protein. Little or no change was detected in the expression of class I histocompatibility antigens or the expression of melanocyte stage-specific differentiation antigens. Melanocytes infected with a temperature-sensitive mutant of Ki-MSV were also studied. In addition to viral p21 protein, Ia antigens were expressed in equivalent amounts at both permissive and nonpermissive temperatures, suggesting that once Ia antigen expression has been established, continued expression is not dependent on functional p21. To determine whether Ia expression can be induced by Ki-MSV in other cell types, we infected early passage human fetal foreskin fibroblasts and an early passage, Ia-negative melanoma. No Ia antigen induction was seen in either case even though approximately 90% of the cells were producing Ki-MSV p21 and infectious transforming virus.

There was essentially no change in differentiation-related characteristics upon infection with *ras*-oncogene-containing viruses. Infected melanocytes (1) remained pigmented, indicating the presence of melanin, (2) had tyrosinase levels that were comparable to uninfected cells, and (3) expressed melanosome markers TA99 and CF21, and the cell surface glycoprotein detected by C350. One well-characterized melanoma marker whose expression did change was GD_3. Melanocytes infected with Ki-MSV showed a substantial increase in the expression of GD_3 as measured by immune-rosetting assays. Direct quantitation of GD_3 expression using ^{125}I-labeled antibody indicated that the level of GD_3 in Ki-MSV-infected melanocytes increased by greater than five- to tenfold over uninfected melanocytes or melanocytes infected with helper virus. Amplified GD_3 expression has been found in other transformation systems. Astrocytoma cells have an increased expression of this ganglioside compared to normal glial cells, and rat fibroblasts transfected with adenovirus E1A transforming gene have augmented amounts of GD_3.[26,27]

4. Heterogeneity of Melanoma Cells

An important aspect of the regulation of expression of melanoma antigens is the identification and characterization of genes that control antigen expression.

The chromosomal mapping of genes controlling expression of these antigens has proceeded through the analysis of rodent–human somatic cell hybrids. In this way, several cell surface antigens of melanoma cells and melanocytes have been mapped, including gp130 (chromosome 11), gp240/>500 (chromosome 15), and gp140/30 (chromosome 12).[28,29] Other strategies involve contransfection of DNA from melanoma cells with a cloned selectable gene vector into rodent cells. For instance, gp130 had been transfected and expressed in recipient mouse cells.[30] Transfection of these genes into mouse cells should make it possible to isolate genes and clone them.

Genetic studies of melanoma cells have provided a vantage for evaluating tumor heterogeneity. Phenotypic heterogeneity of cells within a single tumor is a feature of human cancers, including melanoma.[31,32] Several mechanisms may lead to tumor heterogeneity, including epigenetic and genetic events. The remarkable heterogeneity that can occur in the tumors of patients with malignant melanoma has been emphasized by studies of melanoma cells from patient DX. A panel of six melanoma cell lines was established from metastatic sites in patient DX over the course of 2 years.[32] It is presumed that these six cell lines are clonal in origin since five of the six express a common class 1 unique (clonotypic) antigen. Substantial phenotypic heterogeneity was observed in all six cell lines, including morphology, pigmentation, and antigen expression, such that each cell has a distinct phenotype. It is evident that part of these phenotypic differences reflect the corresponding diversity occurring during normal melanocyte differentiation. However, two studies have implied that additional factors, particularly genetic, are involved in the heterogeneity of the tumors from patient DX.[24,33]

Loss of genetic material appears to occur randomly in melanoma cells.[33] When DNA probes recognizing restriction-fragment-length polymorphisms (RFLP) were used to screen melanoma cells and autologous B lymphocytes or fibroblasts from 24 patients, loss of genetic material was observed at approximately 27% of informative loci (heterozygote loci in constitutive cells became hemizygous or homozygous in tumor cells). These losses occurred at loci on eight different chromosomes and the frequency of losses at individual loci varied between 8 and 67%. A variety of abnormalities including unequal sister chromatid exchange, nondisjunction, aberrant mitotic recombination, and interstitial deletion may be involved in the loss of RFLP. One possible explanation to consider is that genetic loss occurs continually in somatic cells, and that the RFLP loss simply reflects the genotype of the normal precursor cell. Alternatively, the timing of RFLP loss could coincide or follow events associated with malignant transformation. The results of studies of melanomas from patient DX suggest that the second explanation is more likely. Three of four melanoma cell lines from patient DX have lost different allelic fragments at heterozygous loci on chromosomes 3, 13, 15, and 17. In the search for transforming genes in melanoma cells by Albino et al., 4 of 30 melanoma cell lines were found to contain transforming *ras* genes in the NIH/3T3 assay.[24] It is of particular interest that only one of five melanomas cell lines from patient DX contained an activated *ras* gene (N-*ras*). This provocative finding suggests that activated *ras* genes may not be involved in the origin or maintenance of melanoma tumorigenicity, but arise as a consequence of genetic instability. A plausible idea is that widespread allelic loss provides the basis for the extensive phenotypic heterogeneity of melanoma cells. In this view, genetic loss would permit the expression of recessive N-*ras* mutations (or other oncogene loci), perhaps permitting these genes to play a role in the pathogenesis of melanoma.

5. Mouse MAb Detecting GD_3: Initial Application to the Treatment of Patients with Melanoma

We have evaluated the IgG3 mouse MAb R24 in the treatment of 12 patients with melanoma.[14] Seven women and five men were treated with R24. The median age was 40 and the median performance status was 70 on a Karnofsky scale. All patients had skin and soft tissue metastases and half of the patients had visceral metastases (lung, liver, brain). All patients had been treated for metastatic disease previously—5 had received chemotherapy, 2 radiation therapy, and 7 interferon, intralesional, or vaccine treatments.

R24 was administered by intravenous infusion in 100–200 ml of 0.9% normal saline with 5% human serum albumin. Skin tests with 0.1 µg R24 were performed before the first treatment to test for immediate hypersensitivity to mouse immunoglobulin. The doses and schedule of treatment were 1 or 10 mg/m^2 every other day for eight treatments or 30 mg/m^2 per day by continuous infusion on days 1–5 and 8–12. Three patients were treated with 1 and 30 mg/m^2 and six patients with 10 mg/m^2.

No side effects were observed at the 1 mg/m^2 dose level. Patients receiving doses of 10 mg/m^2 or higher developed urticaria and pruritus, usually developing 2–4 hr after starting treatment. The intensity of side effects was related to the dose and rate of infusion. Skin reactions characteristically appeared over tumor-bearing areas. Six patients went on to develop more generalized urticarial reactions over the face, trunk, and/or limbs. One patient developed mild, self-limiting wheezing after rapid infusion (10 mg/hr). Diphenhydramine was effective in controlling side effects but was used only for uncomfortable reactions. At the dose level of 30 mg/m^2, R24 was given by continuous infusion to maintain a rate of less than 5 mg/hr. Two patients treated at this dose developed mild nausea and vomiting between 4 and 8 hr after the start of therapy. Temperature elevation was seen in two patients, up to 37.8°C. Side effects were generally most marked after initial treatments, and were minimal or absent by the end of therapy. No renal, hepatic, or hematopoietic toxicity was observed, and no change in vision, skin pigmentation, or neurological status was noted after therapy.

Three patients had partial responses, with greater than 50% reduction in the size of all measurable lesions. Two patients had mixed responses with greater than 50% reduction in some tumors but not others. Responses lasted 10 weeks to 44 weeks.

Serological analyses of serum from patients showed that peak R24 levels were related to the doses of antibody administered. Peak levels were 0.8 µg/ml (1 mg/m^2), 7 µg/ml (10 mg/m^2), and 58 µg/ml (30 mg/m^2). Human IgG antibodies directed against mouse immunoglobulin were detected between 8 and 40 days after the start of therapy. Studies of tumor biopsies taken before the start of treatment demonstrated heterogeneity of the expression of GD_3. There was no detection of antigenic modulation during therapy. Progression of tumors after treatment was not due to outgrowth of GD_3-negative cells. Biopsies of tumors persisting or growing after treatment with R24 showed GD_3-positive melanoma cells. The amount of R24 reaching tumors was related to the dose of antibody administered, and R24 was detected most strongly after treatment with 30 mg/m^2. Biopsies of lesions taken during treatment confirmed inflammatory reactions. There were increased numbers of mast cells with evidence of degranulation. Complement deposition was observed, including C3, C5, and C9. Infiltration with T3/Ia/T8-positive lymphocytes was seen after treatment.

ACKNOWLEDGMENTS. This work was supported by NCI Grants PO1 CA-33049, CA-37907, and CA-32152, The Alcoa Foundation, and the Louis and Anne Abrons Foundation, Inc. A.N.H. is the recipient of a Cancer Research Institute Investigatorship. We thank Jeanie Melson for assistance in preparing the manuscript.

References

1. Boyse, T. A., and Old, L. J., 1969, Some aspects of normal and abnormal cell surface genetics, *Annu. Rev. Genet.* **3**:269–291.
2. Dippold, W. G., Lloyd, K. O., Li, L. T. C., Ikeda, H., Oettgen, H. F., and Old, L. J., 1980, Cell surface antigens of human malignant melanoma: Definition of six new antigenic systems with mouse monoclonal antibodies, *Proc. Natl. Acad. Sci. USA* **77**:6114–6118.
3. Houghton, A. N., Eisinger, M., Albino, A. P., Craincross, J. G., and Old, L. J., 1982, Surface antigens of melanocytes and melanomas: Markers of melanocyte differentiation and melanoma subsets, *J. Exp. Med.* **156**:1755–1766.
4. Real, F. X., Houghton, A. N., Albino, A. P., Cordon-Cardo, C., Melamed, M. R., Oettgen, H. F., and Old, L. J., 1985, Surface antigens of melanomas and melanocytes defined by mouse monoclonal antibodies: Specificity analysis and comparison of antigen expression in cultured cells and tissues, *Cancer Res.* **45**:4401–4411.
5. Graus, F., Cordon-Cardo, C., Houghton, A. N., Melamed, M., and Old, L. J., 1984, Distribution of the ganglioside GD_3 in the human nervous system detected by R24 mouse monoclonal antibody, *Brain Res.* **324**:190–194.
6. Pukel, C. S., Lloyd, K. O., Travassos, L. R., Dippold, W. G., Oettgen, H. F., and Old, L. J., 1982, GD_3, a prominent ganglioside of human melanoma: Detection and characterization by mouse monoclonal antibody, *J. Exp. Med.* **155**:1133–1147.
7. Carubia, J. M., Yu, R. K., Macala, L. J., Kirkwood, J. M., and Varga, J. M., 1984, Gangliosides of normal and neoplastic human melanocytes, *Biochem. Biophys. Res. Commun.* **120**:500.
8. Albino, A. P., Houghton, A. N., Eisinger, M., Lee, J. S., Kantor, R. R. S., Oliff, A. I., and Old, L. J., 1986, Class II histocompatibility antigen expression in human melanocytes transformed by Hs-MSV and Ki-MSV retroviruses, *J. Exp. Med.* **164**:1710–1722.
9. Vogel, C.-W., Welt, S., Carswell, E., Old, L. J., and Müller-Eberhard, H. J., 1983, A murine IgG3 monoclonal antibody to a melanoma antigen that activates complement in vitro and in vivo, *Immunobiology* **164**:309–313.
10. Knuth, A., Dippold, W. G., Houghton, A. N., Meyer zum Buschenfelde, K.-H., Oettgen, H. F., and Old, L. J., 1984, ADCC reactivity of human melanoma cells with mouse monoclonal antibodies, *Proc. Am. Assoc. Cancer. Res.* **25**:1005.
11. Dippold, W. G., Knuth, A., and Meyer zum Buschenfelde, K.-H., 1983, Inhibition of human melanoma cell growth in vitro by monoclonal anti-GD_3 ganglioside antibody, *Cancer Res.* **44**:806–810.

12. Cheresh, D. A., Harper, J. R., Schulz, G., and Reisfeld, R. A., 1984, Location of gangliosides GD_2 and GD_3 in adhesion plaques and on the surface of human melanoma cells, *Proc. Natl. Acad. Sci. USA* **81**:5767–5771.
13. Cheresh, D., Pierschbacher, M. D., Herzig, M. A., and Mujoo, K., 1986, Disialogangliosides GD_2 and GD_3 are involved in the attachment of human melanoma and neuroblastoma cells to extracellular matrix proteins, *J. Cell Biol.* **102**:688–696.
14. Houghton, A. N., Mintzer, D., Cordon-Cardo, C., Welt, S., Fliegel, B., Vadhan, S., Carswell, E., Melamed, M., Oettgen, H. F., and Old, L. J., 1985, Mouse monoclonal IgG3 antibody detecting GD_3 ganglioside: A phase I trial in patients with malignant melanoma, *Proc. Natl. Acad. Sci. USA* **82**:1242–1246.
15. Morgan, A. C., Galloway, D. R., and Resfeld, R. A., 1981, Production and characterization of monoclonal antibodies to a melanoma-specific glycoprotein, *Hybridoma* **1**:27–36.
16. Imai, K., Ng, A. K., and Ferrone, S., 1981, Characterization of monoclonal antibodies to human melanoma-associated antigens, *J. Natl. Cancer Inst.* **66**:489–496.
17. Carrel, S., Accolla, R. S., Carmagnola, R. L., and Mach, J.-P.,1980, Common human melanoma-associated antigen(s) detected by monoclonal antibodies, *Cancer Res.* **40**:2523–2528.
18. Bumol, T., and Reisfeld, R., 1982, Unique proteoglycan complex defined by monoclonal antibody on human melanoma cells, *Proc. Natl. Acad. Sci. USA* **79**:1245–1249.
19. Bumol, T., Walker, L. E., and Reisfeld, R. A., 1984, Biosynthetic studies of proteoglycans in human melanoma cells with a monoclonal antibody to a core glycoprotein of chondroitin sulfate proteoglycans, *J. Biol. Chem.* **259**:12733–12741.
20. Eisinger, M., and Marko, O., 1982, Selective proliferation of normal human melanocytes in vitro in the presence of phorbol ester and cholera toxin, *Proc. Natl. Acad. Sci. USA* **79**:2018–2022.
21. Thomson, T., Real, F. X., Murakami, S., Cordon-Cardo, C., Old, L. J., and Houghton, A. N., 1986, Differentiation antigens of melanocytes and melanoma: Analysis of melanosome and cell surface markers of human pigmented cells with monoclonal antibodies (submitted for publication).
22. Winchester, R. J., Wang, C.-Y., Gibofsky, A., Kunkel, H. G., Lloyd, K. O., and Old, L. J., 1978, Expression of Ia-like antigens on cultured human malignant melanoma cell lines, *Proc. Natl. Acad. Sci. USA* **75**:6235–6239.
23. Wilson, B. S., Indiveri, F., Pellegrino, M. A., and Ferrone, S., 1979, DR (Ia-like) antigens on human melanoma cells, *J. Exp. Med.* **149**:658–668.
24. Albino, A. P., Le Strange, R., Oliff, A. I., Furth, M. E., and Old, L. J., 1984, Transforming ras genes from human melanoma: A manifestation of tumor heterogeneity? *Nature* **308**:69–71.
25. Eisinger, M., Marko, O., Ogata, S., and Old, L. J., 1985, Growth regulation of human melanocytes: Mitogenic factors in extracts of melanoma, astrocytoma, and fibroblast cell lines, *Science* **229**:984–986.
26. Traylor, D. T., and Hogan, E. L., 1980, Gangliosides of human cerebral astrocytoma, *J. Neurochem.* **34**:126.
27. Nakakuma, H., Sanai, Y., Shiroki, K., and Nagi, Y., 1984, Gene regulated expression of glycolipids: Appearance of GD_3 ganglioside in rat cells on transfection with transforming gene E1 of human adenovirus type 12 DNA and its transcriptional units, *J. Biochem.* **96**:1471.
28. Rettig, W., Dracopoli, N. C., Goetzger, T. A., Spengler, B. A., Biedler, J. L., Oettgen, H. F., and Old, L. J., 1984, Somatic cell genetic analysis of human cell surface antigens: Chromosomal assignments and regulation of expression in rodent–human hybrid cells, *Proc. Natl. Acad. Sci. USA* **81**:6437–6441.
29. Dracopoli, N. C., Rettig, W. J., Goetzger, T. A., Houghton, A. N., Spengler, B. A., Oettgen, H. F., Biedler, J. L., and Old, L. J., 1984, Three human cell surface antigen systems determined by genes on chromosome 12, *Somat. Cell Mol. Genet.* **10**:475–481.
30. Albino, A. P., Graf, L. H., Kantor, R. R. S., Mclean, W., Silagi, S., and Old, L. J., 1985, DNA-mediated transfer of human melanoma cell surface glycoprotein, gp130: Identification to transfectants by erythrocyte rosetting, *Mol. Cell. Biol.* **5**:692–698.
31. Heppner, G. H., and Miller, B. E., 1983, Tumor heterogeneity: Biological implications and therapeutic consequences, *Cancer Metastasis Rev.* **2**:5–23.

32. Albino, A. P., Lloyd, K. O., Houghton, A. N., Oettgen, H. F., and Old, L. J., 1981, Heterogeneity in surface antigen expression and glycoprotein expression of cell lines derived from different metastases of the same patient: Implications for the study of tumor antigens, *J. Exp. Med.* **154**:1764–1778.
33. Dracopoli, N. C., Houghton, A. N., and Old, L. J., 1985, Loss of polymorphic restriction fragments in malignant melanoma: Implications for tumor heterogeneity, *Proc. Natl. Acad. Sci. USA* **82**:1470–1474.

24
Malignant Transformational Changes of the Sugar Chains of Glycoproteins and Their Clinical Application

AKIRA KOBATA

1. Introduction

Tumor cell products, which are not produced or are produced only in small amounts in normal cells, are called *tumor markers*. These materials are considered to serve effectively for the diagnosis and prognosis of tumors, and are studied for their application to clinical use. Although many of the tumor markers so far reported are glycoproteins, most such studies have focused on their polypeptide moieties and very few on their carbohydrate moieties. This was mainly because the sugar chains of tumor markers were mostly asparagine-linked sugar chains which were difficult to analyze. However, by the establishment of a series of sensitive techniques to study the structures of asparagine-linked sugar chains, the structural characteristics of the sugar chains of several tumor markers have been elucidated. In this chapter, I will introduce the results of studies on the sugar chains of two typical tumor markers and elaborate on the importance of this new field of cancer research.

AKIRA KOBATA • Department of Biochemistry, Institute of Medical Science, University of Tokyo, Shirokanedai, Minato-ku, Tokyo 108, Japan.

2. Structural Rules of the Sugar Chains of Glycoproteins*

At the outset, I will present an outline of the structural characteristics of the sugar chains found in glycoproteins. The sugar chains of glycoproteins can be classified into two groups by the structures of their linkage regions to the polypeptide backbones. Sugar chains attached to the polypeptide by an *O*-glycosidic linkage from *N*-acetylgalactosamine to serine or threonine have been called *mucin-type* sugar chains because these structures were widely found in mucin secreted from epithelial cells constructing mucous tissues. The second major class of sugar chains are called *asparagine-linked*. These oligosaccharides are linked *N*-glycosidically from *N*-acetylglucosamine (GlcNAc) to the amide nitrogen of asparagine. A variety of structural differences have been found between the two sugar chain groups. The most prominent difference is the presence or absence of mannose. No mannose is included in the mucin-type sugar chains so far reported. In contrast, asparagine-linked sugar chains always contain mannose. Therefore, one can determine the type of sugar chains present in a glycoprotein sample by investigating its monosaccharide composition.

Asparagine-linked sugar chains have much higher structural rules than mucin-type sugar chains. Structurally, they can be classified into three subgroups (Fig. 1). Oligosaccharides that contain only mannose in addition to the *N,N'*-diacetyl chitobiose structure located at the reducing termini are classified as *high mannose type*. A heptasaccharide—Manα1→6(Manα1→3)Manα1→6(Manα1→3)Manβ1→4GlcNAcβ1→4GlcNAc—is included as a common core of these oligosaccharides, and variation is formed by the number and location of the Manα1→2 residues linked to the three nonreducing terminal α-mannosyl residues of the core portion. *Complex-type* sugar chains contain a core pentasaccharide—Manα1→6(Manα1→3)Manβ1→4GlcNAcβ1→4GlcNAc—and structural variation arises by the structure and number of the outer chain moieties linked to the two α-mannosyl residues of the core. In addition, an α-fucosyl residue linked at C-6 of the proximal *N*-acetylglucosamine residue of the core and a β-*N*-acetylglucosaminyl residue linked at C-4 of the β-mannosyl residue of the core (this residue is called *bisecting GlcNAc* to discriminate it from other *N*-acetylglucosamine residues) occur as common variants. The outer chains are often a trisaccharide—NeuAcα2→6Galβ1→4GlcNAc—but many other outer chains as will be described later have been found.

The last subgroup was named *hybrid type* because they were found to have structural features characteristic of both high-mannose and complex-type sugar chains. They have a common pentasaccharide core—Manα1→6(Manα1→3)Manβ1→4GlcNAcβ1→4GlcNAc—and outer chains composed of *N*-acetylglucosamine, galactose, and sialic acids are linked to one α-mannosyl residue as in the case of complex-type sugar chains, while one or two α-mannosyl residues are linked to another α-mannosyl residue of the core as in the case of high-mannose-type sugar chains. Presence or absence of bisecting GlcNAc and of an α-fucosyl residue

---

* All sugars mentioned in this chapter have a D configuration except fucose, which has an L configuration.

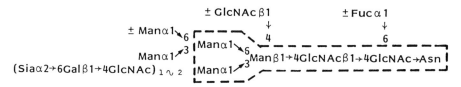

FIGURE 1. General structures of the three types of asparagine-linked sugar chains. Oligosaccharide portions enclosed by dashed lines are common core portions.

linked to the proximal N-acetylglucosamine residue of the core also causes the variation of oligosaccharides in this subgroup. High-mannose-type sugar chains have been determined to be precursors of the biosynthesis of complex-type and hybrid-type sugar chains.

The complex-type sugar chains occur as a variety of different antennary sugar chains as summarized in Table I. In tri- and tetraantennary sugar chains, the GlcNAcβ1→4(GlcNAcβ1→2)Man branch is mostly located on the Manα1→3 side, and the GlcNAcβ1→6(GlcNAcβ1→2)Man branch on the Manα1→6 side. However, the branchings are reversed in the case of membrane glycoproteins of calf thymocytes. This evidence may reflect the species- or organ-specific difference of β-N-acetylglucosaminyl transferases. Various kinds of outer chains starting from the N-acetylglucosamine residue have been found to occur in complex-type sugar chains. In Table II, structures of some representative outer chains are summarized. By combination of the different antennaries and different outer chains, a great number of different complex-type sugar chains can be formed. This large variation may be the molecular basis of the signals of cell-to-cell and cell-to-glycoprotein recognition played in multicellular organisms.[12]

TABLE I
Complex-Type Sugar Chains with Different Numbers of Outer Chains

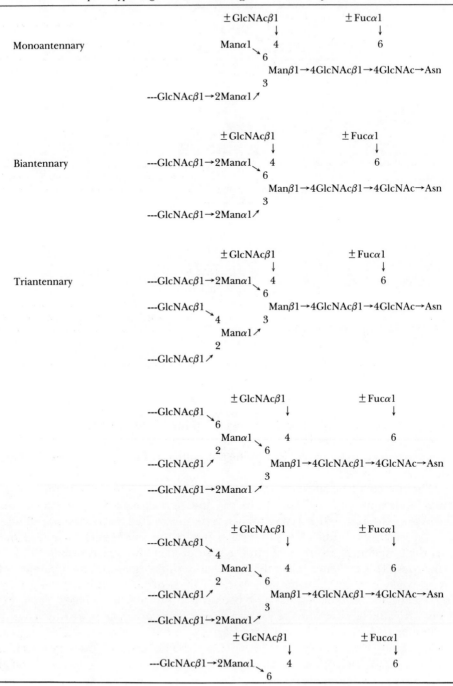

TABLE I. (CONTINUED)

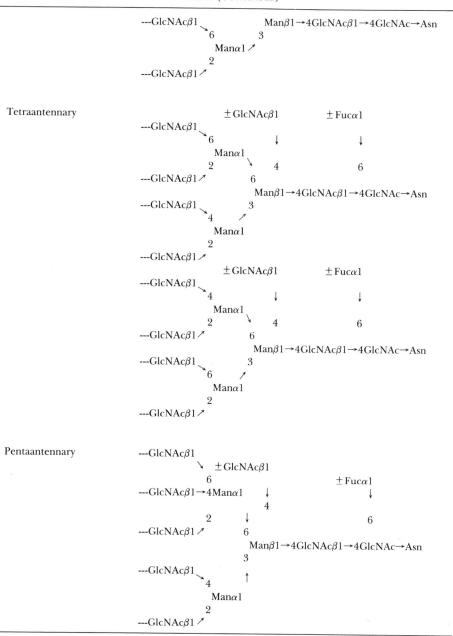

TABLE II.
Structures of Various Outer Chains Found in Complex-Type Asparagine-Linked Sugar Chains

---

$\pm \text{Sia}\alpha 2 \rightarrow 6\text{Gal}\beta 1 \rightarrow 4\text{GlcNAc}\beta 1$

$\text{Sia}\alpha 2 \rightarrow 3\text{Gal}\beta 1 \rightarrow 4\text{GlcNAc}\beta 1$

$$\begin{array}{c} \pm \text{Sia}\alpha 2 \\ \downarrow \\ 6 \end{array}$$
$\text{Sia}\alpha 2 \rightarrow 3\text{Gal}\beta 1 \rightarrow 3\text{GlcNAc}\beta 1$

$$\begin{array}{c} \text{Sia}\alpha 2 \\ \downarrow \\ 6 \end{array}$$
$\text{Sia}\alpha 2 \rightarrow 4\text{Gal}\beta 1 \rightarrow 3\text{GlcNAc}\beta 1$

$$\begin{array}{c} \text{Fuc}\alpha 1 \\ \downarrow \\ 3 \end{array}$$
$\text{Gal}\beta 1 \rightarrow 4\text{GlcNAc}\beta 1$

$$\begin{array}{c} \text{Fuc}\alpha 1 \\ \downarrow \\ 3 \end{array}$$
$\text{Gal}\beta 1 \rightarrow 4\text{GlcNAc}\beta 1 \rightarrow 3\text{Gal}\beta 1 \rightarrow 4\text{GlcNAc}\beta 1$

$$\begin{array}{c} \text{Fuc}\alpha 1 \\ \downarrow \\ 3 \end{array}$$
$\text{Gal}\beta 1 \rightarrow 4\text{GlcNAc}\beta 1 \rightarrow 3\text{Gal}\beta 1 \rightarrow 4\text{GlcNAc}\beta 1$

$\text{NeuAc}\alpha 2 \rightarrow 6\text{Gal}\beta 1 \rightarrow 4\text{GlcNAc}\beta 1 \rightarrow 3\text{Gal}\beta 1 \rightarrow 4\text{GlcNAc}\beta 1$

$(\text{Gal}\beta 1 \rightarrow 4\text{GlcNAc}\beta 1 \rightarrow 3)_n \text{Gal}\beta 1 \rightarrow 4\text{GlcNAc}\beta 1$

$\text{SO}_4 - (3 \text{ or } 4)\text{GalNAc}\beta 1 \rightarrow 4\text{GlcNAc}\beta 1$

---

3. γ-Glutamyl Transpeptidase (γ-GT) and Hepatoma

γ-GT is one of the glycoenzymes which bind intrinsically to the plasma membrane. The enzyme, reported initially by Hanes *et al.* in 1950,[2] catalyzes the hydrolytic cleavage of γ-glutamyl residue of glutathione as well as its transfer to amino acids and peptides. It is exclusively located at the apical side of the plasma membrane of epithelial cells in various mammalian organs, and is composed of two subunits (heavy and light) both of which contain sugars. The heavy subunit is associated with the plasma membrane by its hydrophobic amino acid cluster located near its *N*-terminal region. The light subunit contains the catalytic site and is associated with the heavy subunit by noncovalent bonds. One of the characteristic features of γ-GT is its molecular multiformity. For example, the rat kidney enzyme could be separated into at least 12 isozymic forms by isoelectric focusing.[3] No structural difference has been found in the polypeptide moieties of these isozymic γ-GTs. However, these enzymes contain 2 mol of neutral sugar chains and different numbers of acidic sugar chains in one molecule.[4] Since the total number of sialic acid residues included in the acidic sugar chains showed a reciprocal relationship to the isoelectric point of each isozymic form,[4] difference in the sugar moiety was concluded to be responsible for the multiformity of γ-GT. It is well established that in N-glycosidically linked glycoproteins, like γ-GT, the asparagine residue bearing the sugar chain is invariably part of the tripeptide sequence -Asn-X-Ser- or -Asn-X-Thr-, where X represents one of 20 amino acids except proline.[5] However, asparagine in this sequence is not always glycosylated,[6] and one possible explanation for this is that posttranslational folding of the polypeptide chain may inhibit transfer of the oligosaccharide from the lipid intermediate to the asparagine residue.[7] Possibly, the competition between folding of the polypeptide moiety and glycosylation at the potential asparagine site is responsible for γ-GT molecules with different numbers of sugar chains. Two of these sugar chains might be prohibited from sialylation by the steric effect of the polypeptide moiety.

Although γ-GTs purified from various organs are immunologically identical, their heterogeneity varied according to the tissues.[8] The physicochemical characteristics of γ-GT may also change according to the physiological state of the cells, especially in malignant transformation.[9-11] Fiala et al. reported that γ-GT activity is tremendously elevated in rat hepatoma induced by 3′-methyl-4-dimethylaminoazobenzene.[12] Since this elevation was observed in the preneoplastic nodule of liver[13] which is considered as a precancerous state, and the γ-GT level in serum reflects the enzyme level in liver, the enzyme was expected to be a useful marker for the diagnosis of hepatoma. However, clinical studies performed in succession revealed that the level of γ-GT in serum is also elevated in many noncancerous hepatic diseases such as alcoholic hepatitis and biliary obstruction. This high incidence of false-positives is one of the biggest problems in using the enzyme level as a diagnostic marker of hepatoma.

γ-GTs produced in azo dye-induced rat hepatoma have more acidic pI values than those of γ-GTs in normal liver.[9] The difference mostly disappeared during sialidase digestion, indicating that transformational changes are induced in the carbohydrate moieties of γ-GTs. Comparative study of the structures of oligosaccharides released by hydrazinolysis of γ-GTs purified from rat AH-6 hepatoma and from rat liver revealed the following three prominent changes.[14]

First, the average number of sugar chains in one molecule of γ-GT increased from 3.7 in normal liver to 14.0 in hepatoma. It was improbable that the increment of sugar chains was induced by the structural change of the polypeptide moiety of γ-GT, because the enzyme from various tumor and normal tissues showed the same antigenicity and similar amino acid compositions. In view of the occurrence of γ-GT molecules with different number of sugar chains, the results may indicate that the enzymes with higher sugar contents are more favorably synthesized in hepatoma cells.

Comparative study of the oligosaccharides released from the two enzymes revealed the whole structures shown in Fig. 2. More prominent structural difference among the sugar chains of the two enzymes have become evident by this study. The structures of sugar chains of liver γ-GT indicated that they are bi-, tri-, and tetraantennary complex-type sugar chains with completed outer chains such as Siaα2→Galβ1→4GlcNAc and Siaα2→Galβ1→4GlcNAcβ1→Galβ1→4GlcNAc. In contrast, many of the complex-type sugar chains of hepatoma γ-GT have incomplete outer chains and no N-acetyllactosamine repeating structure. Approximately 16% of the total sugar chains of this enzyme are a series of high-mannose-type sugar chains, which are well known as the processing intermediates in the biosynthesis of complex-type sugar chains. Therefore, attenuation of the sugar chains as widely reported in glycolipids[15] is also induced by malignant transformation in the case of this membrane glycoprotein.

Another important finding is that bisecting GlcNAc were detected in more than half of the total sugar chains of hepatoma γ-GT. Comparative study of the sugar chains of γ-GT purified from liver and kidney of many mammals revealed that extensive organ-specific difference as well as species-specific difference occur in the sugar chains of this enzyme.[16] In Fig. 3, structures of the major sugar chains of each γ-GT sample are summarized. Although the structures of sugar chains show variety according to animal species, one of the common features is that bisecting GlcNAc is not found in the sugar chains of liver enzymes while it is always present in the sugar chains of kidney enzymes. Not only the sugar chains of liver γ-GTs, but those of all other glycoproteins of liver origin, so far studied, lack the bisecting GlcNAc. Therefore, it is most probable that N-acetylglucosaminyl transferase III,[17] which forms the bisecting GlcNAc, is not manifested in the liver of mammalian species including man. If so, detection of the bisecting GlcNAc residue in the sugar chains of hepatoma γ-GT is important, because the expression of N-acetylglucosaminyl transferase III[17] in hepatoma can be considered as one of the dedifferentiation phenomena characteristic of malignant transformation. By making use of the transformational change of liver γ-GT, we have recently developed a new method to distinguish human serum γ-GT associated with primary hepatoma from that of nonhepatoma patients.[18]

4. Appearance of Unusual Sugar Chains in Human Chorionic Gonadotropin (hCG) Produced in Choriocarcinoma

hCG is a glycoprotein hormone produced in trophoblastic cells. It is composed of two subunits with molecular weights of 16,000 (α) and 30,000 (β). These two subunits are strongly bound by electrostatic and hydrophobic interactions, and cannot be dissociated in neutral urea solution; however, it can be in acidic urea solution.[19] Although its dissociated subunits do not show any biological activity, they restore hormonal activity upon reassociating in a sodium bicarbonate solution.[20] The complete amino acid sequences of α and β subunits were determined by Canfield's group[21] at Columbia University and Bahl's group[22,23] at the

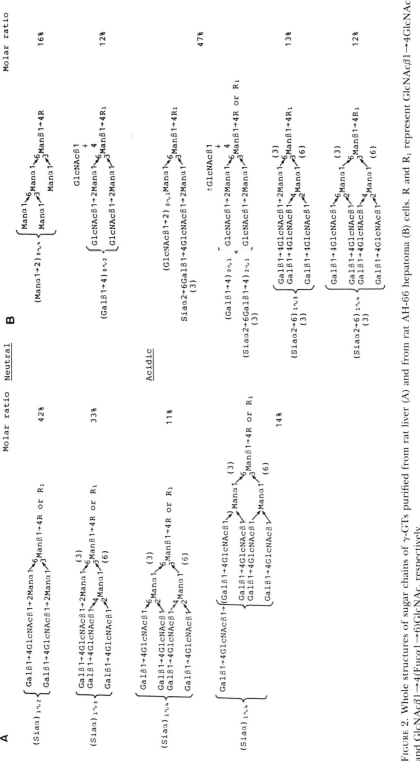

FIGURE 2. Whole structures of sugar chains of γ-GTs purified from rat liver (A) and from rat AH-66 hepatoma (B) cells. R and R_1 represent GlcNAcβ1→4GlcNAc and GlcNAcβ1→4(Fucα1→6)GlcNAc, respectively.

Kidney

Rat and Cattle

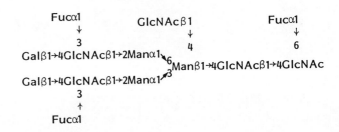

Mouse

```
                  Fucα1         GlcNAcβ1        Fucα1
                    ↓              ↓              ↓
                    3              4              6
   Galβ1→4GlcNAcβ1→2Manα1
                             ⁶\
                               Manβ1→4GlcNAcβ1→4GlcNAc
                             ³/
   Galβ1→4GlcNAcβ1→2Manα1
                    3
                    ↑
                  Fucα1
```

Liver

Rat

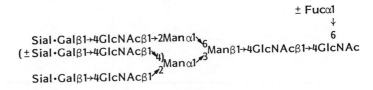

Cattle and Mouse

```
               NeuAcα2
                  ↓
                  6
   NeuAcα2→3Galβ1→3GlcNAcβ1→2Manα1
                                  ⁶\
                                    Manβ1→4GlcNAcβ1→4GlcNAc
                                  ³/
   NeuAcα2→3Galβ1→3GlcNAcβ1→2Manα1
                  6
                  ↑
               NeuAcα2
```

FIGURE 3. Structures of the major sugar chains of γ-GTs purified from kidney and liver of various mammals.

State University of New York (Buffalo) as shown in Fig. 4. As shown, both subunits of hCG contain two asparagine-linked sugar chains. The β subunit contains four mucin-type sugar chains in addition. In 1975, Moyle et al.[24] discovered that the

α-SUBUNIT

1
ALA – PRO – ASP – VAL – GLN – ASP – CYS – PRO – GLU – CYS – THR – LEU –
                                                    10
GLN – GLU – ASP – PRO – PHE – SER – GLN – PRO – GLY – ALA – PRO –
                                    20
ILE – LEU – GLX – CYS – MET – GLY – CYS – CYS – PHE – SER – ARG – ALA –
                            30
TYR – PRO – THR – PRO – LEU – ARG – SER – LYS – THR – MET – LEU –
            40
VAL – GLN – LYS – ASN – VAL – THR – SER – GLU – SER – THR – CYS – CYS –
              CHO   60
VAL – ALA – LYS – SER – TYR – ASN – ARG – VAL – THR – VAL – MET – GLY –
                                            70
GLY – PHE – LYS – VAL – GLU – ASN – HIS – THR – ALA – CYS – HIS – CYS –
                      CHO   80
SER – THR – CYS – TYR – TYR – HIS – LYS – SER
                  90

β-SUBUNIT

1
SER – LYS – GLU – PRO – LEU – ARG – PRO – ARG – CYS – ARG – PRO – ILE –
CHO   10
ASN – ALA – THR – LEU – ALA – VAL – GLU – LYS – GLU – GLY – CYS – PRO –
                                    20
VAL – CYS – ILE – THR – VAL – ASN – THR – THR – ILE – CYS – ALA – GLY –
                              CHO
                              30
TYR – CYS – PRO – THR – MET – THR – ARG – VAL – LEU – GLN – GLY – VAL –
                  40   60
LEU – PRO – ALA – LEU – PRO – GLN – VAL – VAL – CYS – ASN – TYR – ARG –
      50   70
ASP – VAL – ARG – PHE – GLU – SER – ILE – ARG – LEU – PRO – GLY – CYS –
                                          80
PRO – ARG – GLY – VAL – ASN – PRO – VAL – VAL – SER – TYR – ALA – VAL –
                                90
ALA – LEU – SER – CYS – GLN – CYS – ALA – LEU – CYS – ARG – ARG – SER –
                        100
THR – THR – ASP – CYS – GLY – GLY – PRO – LYS – ASP – HIS – PRO – LEU –
                                                                    120
THR – CYS – ASP – ASP – PRO – ARG – PHE – GLN – ASP – SER – SER – SER –
      110   CHO
CHO   130   CHO
SER – LYS – ALA – PRO – PRO – PRO – SER – LEU – PRO – SER – PRO – SER –
                                                140
ARG – LEU – PRO – GLY – PRO – SER – ASP – THR – PRO – ILE – LEU – PRO –
                        CHO
145
GLN

FIGURE 4. Amino acid sequences of the hCG subunits. CHOs represent the sugar chains.

sugar chains of hCG play key roles in the expression of biological activity of the hormone. Since then, several research groups[25,26] have reported supporting data for this notion by demonstrating that the hormonal activity of hCG is lost by elimination of its sugar moiety by enzymatic or chemical methods. Structures of all asparagine-linked sugar chains in the hCG molecule were elucidated by Endo et al.[27] and that of mucin-type sugar chains by Kessler et al.[28] as shown in Fig. 5. Structure A-1 was reported later by Kessler et al.[29]

Sugar chains of glycoproteins are synthesized by the concerted action of glycosyl transferases. Unlike protein biosynthesis, template is not included in the biosynthetic mechanism. Therefore, structures of sugar chains produced are determined by the specificity of each glycosyl transferase for a particular nucleotide sugar and for a particular glycose acceptor, and by its ability to synthesize a particular type of sugar linkage. Accordingly, microheterogeneity is considered an inherent characteristic of the sugar moieties of glycoproteins. Structurally, A-2 to A-5

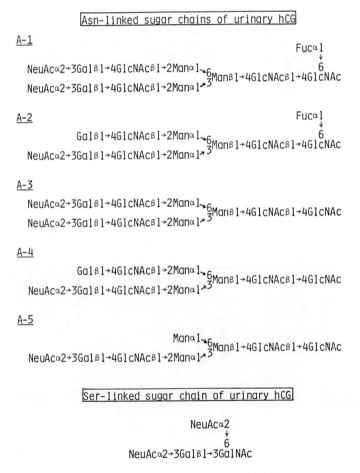

FIGURE 5. Structures of sugar chains present in the hCG molecule.

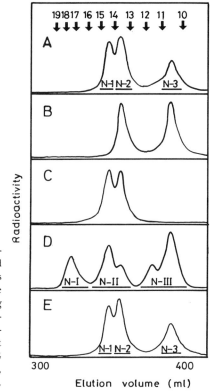

FIGURE 6. Gel permeation chromatographies of desialylated oligosaccharide fractions obtained from hCG and its subunits. Radioactive oligosaccharide fractions obtained from the following glycoprotein samples were subjected to Bio-Gel P-4 column chromatography using water as irrigant. The arrows indicate the eluting positions of glucose oligomers and numbers indicate the glucose units. A, hCG purified from placenta; B, α subunit of placental hCG; C, β subunit of placental hCG; D, hCG purified from urine of a choriocarcinoma patient; E, hCG purified from urine of a hydatidiform mole patient.

in Fig. 5 can be considered as a series of incomplete sugar chains of A-1. Therefore, the multiple sugar chains of hCG can be regarded as a reflection of the microheterogeneity. However, comparative study of the oligosaccharide patterns of α and β subunits as described below indicated that they are not simple, incomplete biosynthetic products.[30]

The radioactive oligosaccharide mixtures obtained from placental hCG and from its two subunits were desialylated by exhaustive sialidase digestion and the neutral oligosaccharide mixtures, thus obtained, were analyzed by Bio-Gel P-4 column chromatography. The fractionation pattern obtained from palcental hCG is shown in Fig. 6A. This pattern was identical to that obtained from urinary hCG (data not shown), indicating that the two hCG samples have N-1, N-2, and N-3 in a molar ratio of 1:2:1. Structures of these three oligosaccharides were elucidated as shown in Fig. 7. Further important and interesting evidence was revealed by the study of two subunits of hCG. Bio-Gel P-4 column chromatography of the desialylated radioactive oligosaccharide fractions of the two hCG subunits indicated that the α subunit contains N-2 and N-3 in approximately equal amounts (Fig. 6B), while the β subunit contains N-1 and N-2 also in equal amounts (Fig. 6C). Because both subunits contain two glycosylated asparagine residues in their polypeptide chains, the results strongly suggested that N-1, N-2, and N-3 are linked specifically

N-1

$$\begin{array}{l}\text{Gal}\beta1\rightarrow4\text{GlcNAc}\beta1\rightarrow2\text{Man}\alpha1\searrow_6\\ \text{Gal}\beta1\rightarrow4\text{GlcNAc}\beta1\rightarrow2\text{Man}\alpha1\nearrow^3\end{array}\text{Man}\beta1\rightarrow4\text{GlcNAc}\beta1\rightarrow4\text{GlcNAc}\overset{\text{Fuc}\alpha1}{\underset{6}{\downarrow}}$$

N-2

$$\begin{array}{l}\text{Gal}\beta1\rightarrow4\text{GlcNAc}\beta1\rightarrow2\text{Man}\alpha1\searrow_6\\ \text{Gal}\beta1\rightarrow4\text{GlcNAc}\beta1\rightarrow2\text{Man}\alpha1\nearrow^3\end{array}\text{Man}\beta1\rightarrow4\text{GlcNAc}\beta1\rightarrow4\text{GlcNAc}$$

N-3

$$\begin{array}{l}\text{Man}\alpha1\searrow_6\\ \text{Gal}\beta1\rightarrow4\text{GlcNAc}\beta1\rightarrow2\text{Man}\alpha1\nearrow^3\end{array}\text{Man}\beta1\rightarrow4\text{GlcNAc}\beta1\rightarrow4\text{GlcNAc}$$

FIGURE 7. Structures of oligosaccharides in fractions N-1, N-2, and N-3 in Fig. 6A.

at the four asparagine loci of the hCG molecule. Therefore, N-2 and N-3 should not be considered as incomplete biosynthetic products of N-1: the sugar chain of one asparagine locus of the β subunit is never fucosylated like in the case of that of another asparagine locus of this subunit. The sugar chain of one asparagine locus of the α subunit remains as nonfucosylated monoantennary oligosaccharide, while the other is converted to a nonfucosylated biantennary sugar chain. In the trophoblast, where hCG is produced, these four neutral sugar chains must be sialylated to various extents. However, hCG molecules with nonsialylated sugar chains must be trapped by liver parenchymal cells while the hormone circulates in the blood, and only sialylated hCG is excreted through renal glomeruli into the urine.

The specific distribution of the three different sugar chains at the four asparagine loci of the hCG molecule cannot be explained by our current knowledge of the biosynthetic mechanism of the sugar chains of glycoproteins. An unknown control mechanism involving steric effects from the polypeptide moiety may also play a role in the formation of asparagine-linked sugar chains of hCG. However, from the viewpoint of the functional role of the sugar chains, the strict distribution of different types of asparagine-linked sugar chains in hCG is of particular interest. If the sugar chains in hCG act as functional signals for target cells as suggested recently by a variety of evidence, the four asparagine-linked sugar chains may each contain a different message.

hCG is detected not only in the blood and urine of normal pregnant women, but also in the body fluid of patients with trophoblastic diseases classified histopathologically as hydatidiform mole and choriocarcinoma. Because a number of differences have been found in the carbohydrate moieties of glycoproteins produced in normal and malignant cells, it was interesting to investigate whether the sugar chains of hCG, the formation of which is strictly regulated in normal trophoblasts, reflect malignant transformational change. Nishimura et al.[31] have purified urinary hCG from a patient with choriocarcinoma and have characterized its biochemical properties. Choriocarcinoma hCG also consists of α and β subunits. Although its amino acid composition and immunoreactivities against anti-hCG, anti-αhCG, and anti-βhCG are the same as normal urinary hCG, its carbohydrate composition is different. It shows extremely low biological activity *in vivo* but about three times higher receptor binding activity *in vitro* than normal urinary hCG. Comparative study of the oligosaccharides released from choriocarcinoma hCG by hydrazinolysis and those from normal hCG revealed a variety of structural differences.[32] By paper electrophoresis, it was found that the sugar chains of choriocarcinoma hCG were not sialylated. When fractionated by Bio-Gel P-4 column chromatography, the oligosaccharide fraction from choriocarcinoma hCG had an elution profile (Fig. 6D) quite different from those of desialylated oligosaccharide mixtures obtained from urinary hCG and placental hCG. Structural study of each radioactive component revealed that fractions N-I, N-II, and N-III in Fig. 6D were mixtures respectively of two, four, and two oligosaccharides shown in Fig. 8. Remember that only *E, F,* and *H* were found in normal hCGs.

Further studies of the sugar chains of hCGs purified from urine of three additional choriocarcinoma patients and three hydatidiform mole revealed many new aspects of the transformational change of hCG.[33] The deletion of sialic acid residues in the sugar chains is not a common phenomenon detected in hCGs produced by choriocarcinoma. However, abnormality of the neutral portion of the asparagine-linked sugar chains found in the first case of choriocarcinoma hCG was also found to occur in all three cases of choriocarcinoma: urinary hCGs from three choriocarcinoma patients contained eight oligosaccharides shown in Fig. 8, although the molar ratio of each oligosaccharide was different in the four tumor glycohormones. In strict contrast, the neutral portion of the asparagine-linked sugar chains of the three mole hCGs gave exactly the same fractionation pattern as normal hCGs upon Bio-Gel P-4 column chromatography (Fig. 6E). Studies of each oligosaccharide in fractions N-1, N-2, and N-3 revealed that their structures were the same as shown in Fig. 7. Therefore, the appearance in hCG of the five oligosaccharides A, B, C, D, and G shown in Fig. 8 could well be a specific characteristic of malignant transformation of the trophoblast.

The apparently complicated structural change found in the sugar chains of choriocarcinoma hCG can be explained by the increase or induction of two particular glycosyl transferases in the trophoblast during carcinogenesis. Total amounts of fucosylated sugar chains in choriocarcinoma hCGs were approximately 50%, which were double those found in normal and mole hCGs. Therefore, increase of the fucosyl transferase responsible for the formation of the Fucα1→6GlcNAc group appears to be one of the specific characteristics of malignant transformation of the trophoblast. This enzyme may have wider substrate specificity than that in normal trophoblasts, because oligosaccharide G in Fig. 8 was not found in nor-

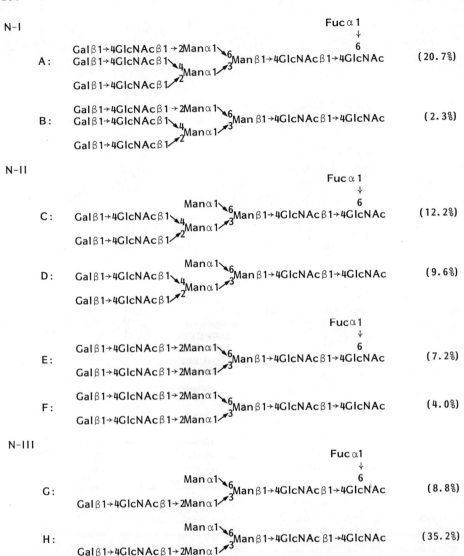

FIGURE 8. Structures of oligosaccharides in fractions N-I, N-II, and N-III in Fig. 6D. Numbers in parentheses are the percent molar ratio of the oligosaccharides.

mal and mole hCGs. Note that oligosaccharides A, B, C, and D in Fig. 8 can be derived respectively from oligosaccharides E, F, G, and H by adding the Galβ1→4GlcNAcβ1→4 group. This structural relationship suggested that the N-acetylglucosaminyl transferase responsible for the formation of the GlcNAcβ1→4Manα1→ group is newly expressed by malignant transformation of trophoblasts.

From a clinical point of view, the structural changes found in the sugar chains of choriocarcinoma hCGs are expected to be useful for the diagnosis and prognosis of choriocarcinoma. Furthermore, the change may provide a new direction for the immunotherapy of human choriocarcinoma, because oligosaccharides C and D have never been found in normal human glycoproteins so far studied.

5. Concluding Remarks

Since the finding of Meezan et al.[34] that the plasma membrane glycoproteins of virus-transformed cells contain larger fucosylated sugar chains than those of normal counterparts, a number of differences have been discerned in the carbohydrate moieties of glycoproteins produced in normal and malignant cells.[35–42] Most of these differences, however, were found by using rather indirect methods such as lectin binding and comparative study of the sizes of glycopeptides obtained by exhaustive Pronase digestion of membrane glycoprotein mixture. The two examples introduced in this chapter indicate that elucidation of whole sugar chain structures is essential for the effective development and clinical use of this line of study. As exemplified by the study of γ-GT, transformational change useful for clinical application can be disclosed through such study.

As revealed through study of the sugar chains of hCG, abnormal sugar chains produced in cancer cells can include those which have never been detected in normal human glycoproteins. As already discussed, these tumor-specific sugar chain structures might be produced because tumor glycosyl transferases acquired wider substrate specificities. Such phenomenon was also found in the case of glycosyl transferases that form the sugar chains of glycolipids.[43] Therefore, elucidation of the biochemical basis of the glycosyl transferase modification in tumor cells is one of the important problems to be solved in the future.

In addition to the changes induced in glycosyl transferases, another possible mechanism forming tumor-specific sugar antigen can be considered. I shall explain it by way of an illustration from the biosynthesis of blood group determinants. The antigens which belong to the ABO blood group system are sugars linked to either proteins or lipids.[44] For example, blood group A antigenic determinant is composed of four monosaccharides—N-acetylgalactosamine, fucose, galactose, and N-acetylglucosamine—and formed by the pathway shown in Fig. 9. Because A enzyme, an α-N-acetylgalactosaminyl transferase, can transfer an N-acetylgalactosamine residue only to the Fucα1→2Gal group,[45] action of the α-1,2-fucosyl transferase is a prerequisite for the formation of blood group A determinant. In humans of blood type A, blood type A antigen is found not only on the surface of erythrocytes but in a variety of glycoproteins in saliva, semen, and mucins secreted from the gastrointestinal tract. However, approximately 15% of blood type A individuals cannot produce blood type A antigen in their secretory glycoproteins although the antigen can be detected on their erythrocytes. Individuals with such a phenotype are called nonsecretors to discriminate them from secretors who can produce blood group A substances in secretion. The biochemical basis of the nonsecretor phenotype is as follows.[46] The α-1,2-fucosyl transferase in Fig. 9 is coded by a structural gene *H*. This *H* gene can be transcribed in hematopoietic

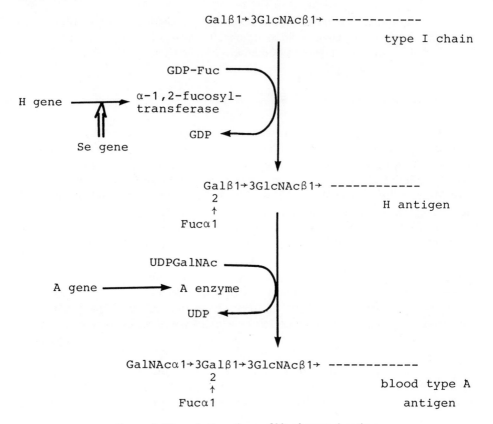

FIGURE 9. Biosynthetic pathway of blood group A antigen.

tissues without the help of any other gene. However, transcription of the *H* gene in secretory tissues required another regulatory gene *Se*. Nonsecretor individuals, whose genotypes are *se/se*, cannot produce α-1,2-fucosyl transferase in secretory tissues. Therefore, the biosynthetic scheme in Fig. 9 stops at the first step and the precursor sugar chains can never be converted to blood type A antigenic determinants in the secretory tissues although A enzyme does exist in these tissues.

Many of the glycosyl transferases in the human body might occur in a relationship similar to α-1,2-fucosyl transferase and A enzyme shown in Fig. 9. Suppose that an enzyme in the situation of α-1,2-fucosyl transferase is translated only at the fetal stage and another enzyme in the situation of A enzyme only at the adult stage. Under such a circumstance, antigen corresponding to the A antigen can never be formed throughout human life. If the first enzyme in the situation of α-1,2-fucosyl transferase is translated in cancer, the final antigenic sugar chain will be formed in this cancer tissue. Such an antigen can be regarded as a true tumor-specific antigen and is expected to be a useful target for cancer immunotherapy as well as a tumor marker for diagnosis. Actually, many carbohydrate epitopes have recently been shown to be useful markers for the diagnosis of tumors by introduction of monoclonal antibody technique in the field of tumor antigen study.[47-49]

References

1. Kobata, A., 1979, Structures and functions of the sugar chains of cell surface glycoproteins, *Cell Struct. Funct.* **4:**169–181.
2. Hanes, C. S., Hird, F. J. R., and Isherwood, F. A., 1950, Synthesis of peptides in enzymic reactions involving glutathione, *Nature* **166:**288–292.
3. Tate, S. S., and Meister, A., 1976, Subunit structure and isozymic forms of γ-glutamyl transpeptidase, *Proc. Natl. Acad. Sci. USA* **73:**2599–2603.
4. Yamashita, K., Tachibana, Y., Hitoi, A., Matsuda, Y., Tsuji, A., Katunuma, N., and Kobata, A., 1983, Differences in the sugar chains of two subunits and of isozymic forms of rat kidney γ-glutamyltranspeptidase, *Arch. Biochem. Biophys.* **227:**225–233.
5. Marshall, R. D., 1972, Glycoproteins, *Annu. Rev. Biochem.* **41:**673–702.
6. Struck, D. K., and Lennarz, W. J., 1980, The function of saccharide-lipids in synthesis of glycoproteins, in: *The Biochemistry of Glycoproteins and Proteoglycans* (W. J. Lennarz, ed.), Plenum Press, New York, pp. 35–83.
7. Pless, D. D., and Lennarz, W. J., 1977, Enzymatic conversion of proteins to glycoproteins, *Proc. Natl. Acad. Sci. USA* **74:**134–138.
8. Miura, T., Matsuda, Y., Tsuji, A., and Katunuma, N., 1980, Immunological cross-reactivity of γ-glutamyltranspeptidases from human and rat kidney, liver, and bile, *J. Biochem.* **89:**217–222.
9. Taniguchi, N., 1974, Purification and some properties of γ-glutamyltranspeptidase from azo-dye induced hepatoma, *J. Biochem.* **75:**473–480.
10. Novogrodsky, A., Tate, S. S., and Meister, A., 1976, γ-Glutamyl transpeptidase, a lymphoid cell-surface marker: Relation to blastogenesis, differentiation and neoplasia, *Proc. Natl. Acad. Sci. USA* **73:**2414–2418.
11. Jaken, S., and Mason, M., 1978, Differences in the isoelectric focusing patterns of γ-glutamyl transpeptidase from normal and cancerous rat mammary tissue, *Proc. Natl. Acad. Sci. USA* **75:**1750–1753.
12. Fiala, S., Fiala, A. E., and Dixon, B., 1972, γ-Glutamyltranspeptidase in transplantable, chemically induced rat hepatomas and "spontaneous" mouse hepatomas, *J. Natl. Cancer Inst.* **48:**1393–1401.
13. Fiala, S., and Fiala, E. S., 1973, Activation by chemical carcinogens of γ-glutamyltranspeptidase in rat and mouse liver, *J. Natl. Cancer Inst.* **51:**151–158.
14. Yamashita, K., Hitoi, A., Taniguchi, N., Yokosawa, N., Tsukada, Y., and Kobata, A., 1983, Comparative study of the sugar chains of γ-glutamyltranspeptidases purified from rat liver and rat AH-66 hepatoma cells, *Cancer Res.* **43:**5059–5063.
15. Hakomori, S., 1975, Structures and organization of cell surface glycolipids dependency on cell growth and malignant transformation, *Biochim. Biophys. Acta* **417:**55–89.
16. Yamashita, K., Hitoi, A., Tateishi, N., Higashi, T., Sakamoto, Y. and Kobata, A., 1983, Organ-specific difference in the sugar chains of γ-glutamyltranspeptidase, *Arch. Biochem. Biophys.* **225:**993–996.
17. Narasimhan, S., 1982, Control of glycoprotein synthesis: UDP-GlcNAc: glycopeptide β4-N-acetylglucosaminyltransferase III, an enzyme in hen oviduct which adds GlcNAc in β1→4 linkage to the β-linked mannose of the trimannosyl core of N-glycosyl oligosaccharides, *J. Biol. Chem.* **257:**10235–10242.
18. Hitoi, A., Yamashita, K., Ohkawa, J., and Kobata, A., 1984, Application of a *Phaseolus vulgaris* erythroagglutinating lectin agarose column for the specific detection of human hepatoma γ-glutamyltranspeptidase in serum, *Gann* **75:**301–304.

19. Canfield, R. E., and Morgan, F. J., 1973, Human chorionic gonadotropin (hCG): Purification and biochemical characterization, in: *Methods in Investigative and Diagnostic Endocrinology* (S. A. Berson and R. S. Yallow, eds.), North-Holland, Amsterdam, pp. 727–733.
20. Aloi, S. M., Edelboch, H., Ingham, K. C., Morgan, F. J., Canfield, R. E., and Ross, G. T., 1973, The rate of dissociation and reassociation of the subunits of human chorionic gonadotropin, *Arch. Biochem. Biophys.* **159**:497–504.
21. Morgan, F. J., Birken, S., and Canfield, R. E., 1973, Human chorionic gonadotropin: A proposal for the amino acid sequence, *Mol. Cell. Biochem.* **2**:97–106.
22. Bellisario, R., Carlsen, R. B., and Bahl, O. P., 1973, Human chorionic gonadotropin: Linear amino acid sequence of the alpha subunit, *J. Biol. Chem.* **248**:6797–6807.
23. Carlsen, R. B., Bahl, O. P., and Swaminathan, N., 1973, Human chorionic gonadotropin: Linear amino acid sequence of the beta subunit, *J. Biol. Chem.* **248**:6810–6827.
24. Moyle, W. R., Bahl, O. P., and Merz, L., 1975, Role of the carbohydrate of human chorionic gonadotropin in the mechanism of hormone action, *J. Biol. Chem.* **250**:9163–9169.
25. Kalyan, N. K., Lippes, H. A., and Bahl, O. P., 1982, Role of carbohydrate in human chorionic gonadotropin, *J. Biol. Chem.* **257**:12624–12631.
26. Goverman, J. M., Parson, T. F., and Pierce, J. G., 1982, Enzymatic deglycosylation of the subunits of chorionic gonadotropin, *J. Biol. Chem.* **257**:15059–15064.
27. Endo, Y., Yamashita, K., Tachibana, Y., Tojo, S., and Kobata, A., 1979, Structures of the asparagine-linked sugar chains of human chorionic gonadotropin, *J. Biochem.* **85**:669–679.
28. Kessler, M. J., Mise, T., Ghai, R. D., and Bahl, O. P., 1979, Structure and location of the O-glycosidic carbohydrate units of human chorionic gonadotropin, *J. Biol. Chem.* **254**:7909–7914.
29. Kessler, M. J., Reddy, M. S., Shah, R. H., and Bahl, O. P., 1979, Structures of N-glycosidic carbohydrate units of human chorionic gonadotropin, *J. Biol. Chem.* **254**:7901–7908.
30. Mizuochi, T., and Kobata, A., 1980, Different asparagine-linked sugar chains on the two polypeptide chains of human chorionic gonadotropin, *Biochem. Biophys. Res. Commun.* **97**:772–778.
31. Nishimura, R., Endo, Y., Tanabe, K., Ashitaka, Y., and Tojo, S., 1981, The biochemical properties of urinary human chorionic gonadotropin from the patients with trophoblastic diseases, *J. Endocrinol. Invest.* **4**:349–358.
32. Mizuochi, T., Nishimura, R., Derappe, C., Taniguchi, T., Hamamoto, T., Mochizuki, M., and Kobata, A., 1983, Structures of the asparagine-linked sugar chains of human chorionic gonadotropin produced in choriocarcinoma, *J. Biol. Chem.* **258**:14126–14129.
33. Mizuochi, T., Nishimura, R., Taniguchi, T., Utsnomiya, T., Mochizuki, M., Derappe, C., and Kobata, A., 1985, Comparison of carbohydrate structure between human chorionic gonadotropin present in urine of patients with trophoblastic diseases and healthy individuals, *Jpn. J. Cancer Res. (Gann)* **76**:752–759.
34. Meezan, E., Wu, H., Black, P., and Robbins, P. W., 1969, Comparative studies on the carbohydrate-containing membrane components of normal and virus-transformed mouse fibroblasts, *Biochemistry* **8**:2518–2524.
35. Buck, C. A., Glick, M. C., and Warren, L. A., 1970, A comparative study of glycoproteins from the surface of control and Rous sarcoma virus transformed hamster cells, *Biochemistry* **9**:4567–4576.
36. Buck, C. A., Glick, M. C., and Warren, L. A., 1971, Glycopeptides from the surface of control and virus-transformed cells, *Science* **172**:169–171.
37. Smets, L. A., Van Beek, W. P., and Van Rooij, H., 1976, Surface glycoproteins and concanavalin-A-mediated agglutinability of clonal variants and tumor cells derived from SV40-virus-transformed mouse 3T3 cells, *Int. J. Cancer* **18**:462–468.
38. Von Nest, G., and Grimes, W. J., 1977, A comparison of membrane components of normal and transformed BALB/C cells, *Biochemistry* **16**:2902–2908.
39. Warren, L., Fuhrer, J. P., and Buck, C. A., 1972, Surface glycoproteins of normal and transformed cells: A difference determined by sialic acid and a growth-dependent sialyl transferase, *Proc. Natl. Acad. Sci. USA* **69**:1838–1842.
40. Van Beek, W. P., Smets, L. A., and Emmelot, P., 1973, Increased sialic acid density in surface glycoprotein of transformed and malignant cells—A general phenomenon? *Cancer Res.* **33**:2913–2922.

41. Van Beek, W. P., Smets, L. A., and Emmelot, P., 1975, Changed surface glycoprotein as a marker of malignancy in human leukaemic cells, *Nature* **253**:457–460.
42. Glick, M. C., Schlesinger, H., and Hummeler, K., 1976, Glycopeptides from the surface of human neuroblastoma cells, *Cancer Res.* **36**:4520–4524.
43. Hakomori, S., 1985, Aberrant glycosylation in cancer membrane as focused on glycolipids: Overview and perspectives, *Cancer Res.* **45**:2405–2414.
44. Hakomori, S., and Kobata, A., 1974, Blood group antigens, in: *The Antigens*, Vol. II (M. Sela, ed.), Academic Press, New York, pp. 79–140.
45. Kobata, A., and Ginsburg, V., 1970, UDP-N-acetyl-D-galactosaminyltransferase, a product of gene that determines blood type A in man, *J. Biol. Chem.* **245**:1484–1490.
46. Shen, L., Grollman, E. F., and Ginsburg, V., 1968, An enzymatic basis for secretor status and blood group substance specificity in humans, *Proc. Natl. Acad. Sci. USA* **59**:224–230.
47. Magnani, J. L., Nilsson, B., Brockhaus, M., Zopf, D., Steplewski, Z., Koprowski, H., and Ginsburg, V., 1982, A monoclonal antibody defined antigen associated with gastrointestinal cancer is a ganglioside containing sialylated lacto-N-fucopentaose II, *J. Biol. Chem.* **257**:14365–14369.
48. Fukushima, K., Hirota, M., Terasaki, P. I., Wakisaka, A., Togashi, H., Chida, D., Suyama, N., Fukushi, Y., Nudelman, E., and Hakomori, S., 1984, Characterization of sialylated Lewis X as a new tumor-associated antigen, *Cancer Res.* **44**:5279–5285.
49. Nilsson, O., Mansson, J. -E., Lindholm, L., Halmgren, J., and Svennerholm, L., 1985, Sialyllactotetraosylceramide, a novel ganglioside antigen detected in human carcinomas by a monoclonal antibody, *FEBS Lett.* **182**:398–402.

25
Ganglioside Involvement in Tumor Cell–Substratum Interactions

DAVID A. CHERESH

1. Introduction

The process of human tumor proliferation and metastasis undoubtedly involves surface structures that interface with the local environment, i.e., extracellular matrix and host tissue. Therefore, it is imperative to understand the molecular and cell biological events at the tumor cell surface that are associated with proliferation and metastasis. To this end, a major research effort has been directed toward understanding the molecular events associated with tumor cell–substratum interactions under the assumption that to metastasize, a tumor cell must be able to attach and invade the basement membrane. A number of investigators have developed monoclonal antibodies (MAbs) directed to a variety of surface antigens that have allowed for the initial characterization of some of the surface structures involved in cell–substratum interactions. Several of these MAbs define carbohydrate antigens present on glycoproteins, proteoglycans, or glycolipids, which comprise the tumor cell glycocalyx, the portion of the tumor cell surface that makes initial contact with its environment. Recent technological advances have enabled investigators to completely characterize the structure of a number of rather complex tumor-associated antigens defined by MAbs recognizing carbohydrate epitopes.[1-12] These reagents have provided useful structural and functional information regarding a number of antigenic determinants on oligosaccharides. Of particular interest are the recent reports of MAbs directed to ganglioside (sialic acid-bearing glycolipids) antigens.[4-12] In this regard, a number of MAbs directed to these determinants were shown to recognize highly restricted markers for a variety of tumor types.[1-12] The use of such reagents may help strengthen and extend

---

DAVID A. CHERESH • Department of Immunology, Scripps Clinic and Research Foundation, La Jolla, California 92037.

observations that implicate these molecules as putative cell surface receptors for hormones,[13] growth factors,[14,15] viruses,[16] and toxins,[17] and provide further evidence for their possible role in cell–substratum interactions.[5,8,18–26] The surface expression of gangliosides has also been linked to particular stages of differentiation or development,[2,4] and more knowledge in this area may thus help to define relevant oncofetal antigens on the tumor cell.

Although gangliosides are preferentially expressed on tissues derived from the neural crest, they are present on the surface of all eukaryotic cells. Evidence concerning the functional properties ascribed to these molecules has come primarily from studies involving their exogenous addition to cells in culture, under the assumption that the added gangliosides could appropriately embed in the lipid bilayer of the cells being examined. Additional information regarding the functional properties of gangliosides has been derived from their large-scale biochemical extraction from specific tissues, organs, or cultured cell lines. Most recently, a number of investigators have detected new ganglioside structures using MAbs and have attempted to define their functional properties.[5–12] In this regard, our laboratory has produced and characterized several MAbs to the disialogangliosides GD_2 and GD_3, which are present on the surface of human melanoma and neuroblastoma cells.[8] Using these MAbs as molecular and functional probes, we examined the function of these gangliosides on the surface of human melanoma and neuroblastoma cells.[8,9,21,22] This chapter will review a number of studies which have implicated gangliosides in cell–substratum interactions and summarize in detail studies performed in our laboratory using MAbs as functional probes to determine the role of gangliosides in the attachment of neuroectodermally derived tumor cells to extracellular matrix components. This report is not intended to be a comprehensive review of all the extensive work in ganglioside research but rather will address various key studies implicating the oligosaccharide portion of gangliosides as playing an instrumental role in the attachment of cells to components within the extracellular matrix.

2. Evidence for the Involvement of Gangliosides as Potential Cell Surface Receptors for Fibronectin

Initial reports indicating that gangliosides are involved in cell attachment to the extracellular matrix came from experiments involving their exogenous addition to cultured cells. Kleinman et al.[18] demonstrated that the addition of 0.3–0.6 μmole/ml of certain gangliosides to media effectively prevented the attachment of Chinese hamster ovary cells to collagen-coated culture dishes containing serum as a fibronectin source. In addition, the larger or more heavily sialylated gangliosides were more potent inhibitors since $GT_1 > GD_{1a} > GM_1 > GM_2$, and GM_3 had no measurable effect. These effects were shown not to be due to toxicity since removal of the gangliosides allowed the cells to reattach normally and addition of gangliosides to cells after their attachment had no effect on their growth rate. The active portion of the gangliosides apparently involved the oligosaccharide moiety since removal of the lipid (ceramide) component had no effect on the inhibitory activity.

This activity, however, could not be demonstrated by a variety of simple sugars or various polysaccharides including: heparin, heparan sulfate, keratan sulfate, dermatan sulfate, and chondroitin sulfate, even at relatively high concentrations. The functional domain of the gangliosides appeared to involve terminal sialic acid since mild periodate oxidation (which specifically clips the exocyclic arm on terminal unsubstituted sialic acid) greatly reduced the ability of gangliosides to interfere with cell adhesion. The investigators conducting this study ultimately concluded that sialic acid-containing glycoconjugates on the cell surface may act as receptors for fibronectin. This hypothesis was supported by observations of Yamada et al.[19] who demonstrated that purified soluble gangliosides inhibited both cell spreading and fibronectin-mediated hemagglutination. In addition, the fibronectin-mediated restoration of normal parallel cell alignment and fibroblastlike morphology to SV1 tumor cells was found to be inhibited by the presence of mixed gangliosides. For example, the addition of brain gangliosides (at 5×10^{-4} M) to these cells was capable of blocking the effects of 20 µg/ml fibronectin. This inhibition was found to be dose-dependent and competitive since it could be overcome by additional fibronectin. Moreover, the effectiveness of ganglioside inhibitors increased with increasing sialic acid content and the entire inhibitory activity was retained in the isolated oligosaccharide moiety. A series of other lipids were also tested for their ability to inhibit cell–substratum interactions. The most active of these were the more negatively charged phospholipids, such as phosphatidylserine and phosphatidylinositol. The quantity required for inhibitory activity, however, was an order of magnitude higher than that required for gangliosides. Based on these data, Yamada et al.[19] concluded that the mechanism by which fibronectin binds to the cell surface involves a receptor that contains a negatively charged lipid, e.g., a ganglioside. This hypothesis was strengthened by Perkins et al.[20] who demonstrated that cell attachment and spreading could be inhibited by di- and trisialogangliosides. The concentration required to achieve 50% inhibition was approximately tenfold less than that reported by Kleinman et al.[18] These results suggested that fibronectin could actually bind to gangliosides through a low-affinity interaction. It was therefore proposed that one potential function for gangliosides at the cell surface may be to aid in the organization of fibronectin within the extracellular matrix. In fact, a potential role for gangliosides in binding fibronectin at the cell surface is supported by the observation that fibronectin is synthesized but not retained at the surface of a cell line (NCTC 2071) that is unable to synthesize the more complex gangliosides.[27] Growth of this cell line in medium supplemented with gangliosides led to a dramatic increase in the amount of fibronectin localized at the cell surface.

These observations were supported and extended by the findings of Spiegel and co-workers[24,25] who demonstrated that rhodamine- and fluorescein-labeled gangliosides added to cultured fibroblasts could be taken up by the cells in a time- and temperature-dependent manner. In dense cultures of these fibroblasts, a large fraction of fluorescent gangliosides were organized in a fibrillar network and became immobilized, as determined by photobleaching recovery measurements. Using antifibronectin antibodies and indirect immunofluorescence, these gangliosides were observed to codistribute the fibrillar fibronectin.[25] The fact that some cell lines that contain little di- or trisialogangliosides were capable of attaching and spreading on a fibronectin substrate[25] suggests that gangliosides probably cannot act alone as a receptor for fibronectin. Taken together, these studies implicate cell surface gangliosides as playing a cooperative role in the very complex interactions between cells and their extracellular matrix.

3. Gangliosides GD_2 and GD_3 in Focal Adhesion Plaques of Human Melanoma Cells

In our attempt to define antigenic markers associated with human melanoma and neuroblastoma cells, we produced a number of MAbs that react with surface antigens on these cells. Several of these antibodies were shown to be specifically directed to the oligosaccharide portion of GD_2 (MAbs 14.18 and 126) or GD_3 (MAbs MB3.6 and 11C64)[6-9,21,22] and did not cross-react with any of the other ganglioside structures in Fig. 1 or with carbohydrate determinants on glycoproteins.[22] To determine the cell surface distribution of GD_2 and GD_3, indirect immunofluorescence microscopy was performed on human melanoma cells using these antiganglioside MAbs. As shown in Fig. 2, M21 human melanoma cells that are attached to a fibronectin-coated glass coverslip displayed intense staining with an anti-GD_2 (MAb 126,IgM) localizing this antigen in a number of microprocesses that emanate from the cell surface and frequently make direct contact with the fibronectin substrate. This is in contrast to the staining pattern obtained with MAb 9.2.27, which recognizes a melanoma-associated glycoprotein antigen, i.e., a chondroitin sulfate proteoglycan core protein, also expressed on the surface of these cells. In this case, MAb 9.2.27 bound preferentially to the apical surface of these cells and failed to localize near the focal points of attachment.

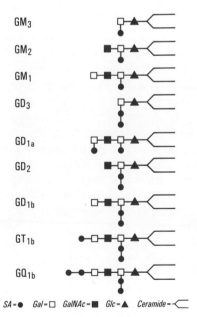

FIGURE 1. Composition of GD_2, GD_3, and various other gangliosides. The sugar composition of a number of gangliosides including GD_2 and GD_3 that are recognized by specific monoclonal antibodies. Gangliosides are termed according to the nomenclature of Svennerholm.[52]

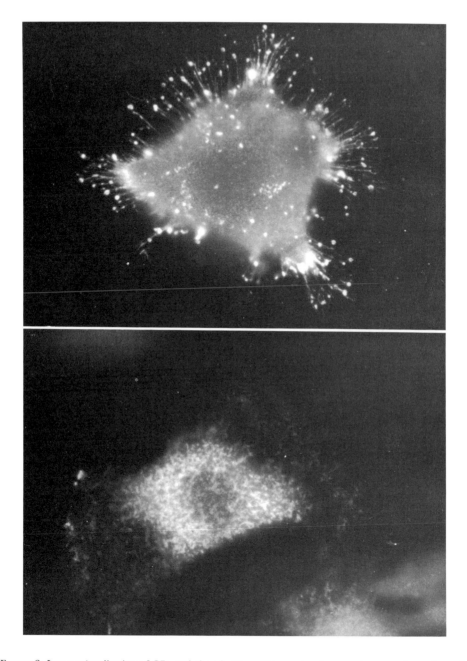

FIGURE 2. Immunolocalization of GD_2 and chondroitin sulfate proteoglycan antigens on the surface of M21 human melanoma cells. M21 cells were allowed to attach to glass coverslips fixed in 3% paraformaldehyde and stained by indirect immunofluorescence with anti-GD_2 MAb 126 (top) or anti-chondroitin sulfate MAb 9.2.27 (bottom) as previously described.[8,22] The cells were photographed through a Zeiss microscope equipped with epifluorescence at 63×.

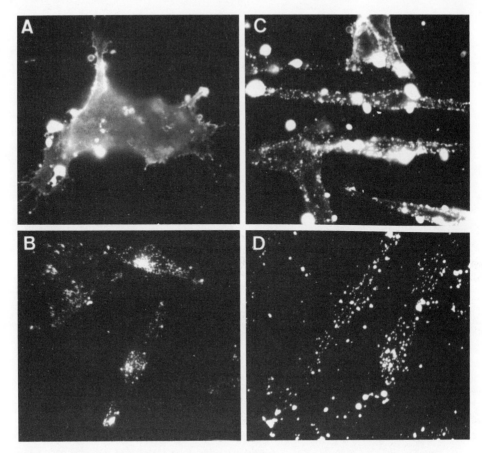

FIGURE 3. Detection of GD_2 and GD_3 on human melanoma cells by indirect immunofluorescence. Melur (A–D) or M21 (E–H) melanoma cells were grown on glass coverslips, fixed, incubated with 10 µg/ml of either MAb 126 (anti-GD_2) or MB3.6 (anti-GD_3) and then stained by indirect immunofluorescence. Melur or M21 cells were stained with MAb 126 (A, B, E, and F) or MAb MB3.6 (C, D, G, and H). Adhesion plaques were prepared by treating coverslips with EDTA as previously described[8] and appear in B, D, F, and H.

Additional immunofluorescence microscopy experiments have been performed on M21 human melanoma cells that were grown on glass coverslips and removed with EDTA, a process known to generate substrate-attached focal adhesion plaques.[8] As shown in Fig. 3, both GD_2 and GD_3 could be localized within these structures using MAbs directed to these gangliosides. Other membrane-associated antigens, however, were not detected within these structures, indicating that focal adhesion plaques are probably not composed of indiscriminant membrane fragments (Fig. 4). Moreover, a biochemical analysis of adhesion plaques generated by human melanoma cells that were labeled metabolically with [^{3}H]glucosamine demonstrated that both GD_2 and GD_3 are enriched at these points of melanoma cell contact.[8] These observations suggested that surface expression of GD_2 and

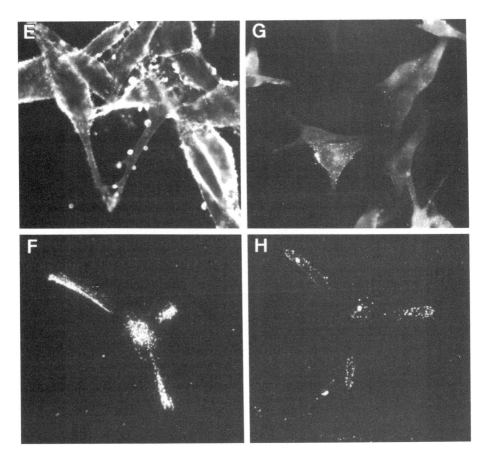

FIGURE 3. (CONTINUED)

GD$_3$ may be involved in the attachment process of human melanoma cells. In support of this hypothesis, Okada et al.[26] demonstrated that the detergent-insoluble substrate attachment matrix (DISAM), i.e., adhesion plaque, produced by glass-attached BHK cells was relatively rich in ganglioside as compared to whole cells. Moreover, the ratio of ganglioside (GM$_3$) to neutral glycolipids was higher in DISAM than in whole cells and transformed BHKpy cells showed an even greater enrichment of ganglioside in DISAM. Therefore, Okada et al.[26] suggested that glycolipids at cell attachment sites have some functional role in regulating BHK cell attachment. More recently, Mugnai et al.[23] demonstrated that adhesion plaques isolated from both normal and virus transformed BALB/c 3T3 cells contain an enrichment of the more complex gangliosides such as GD$_{1a}$ relative to the simpler gangliosides. This result could not be attributed to the presence of plasma membranes in the substrate-attached material, supporting the reuslts from our laboratory.[8] Thus, three independent studies using different cell systems have shown that gangliosides were heavily localized at the interface between membrane

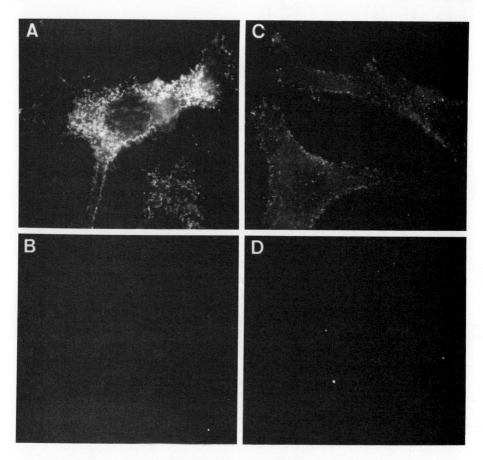

FIGURE 4. Detection of proteoglycan and histocompatibility antigens on human melanoma cells by indirect immunofluorescence. Melur melanoma cells were grown on glass coverslips and stained as described in Fig. 3, using MAb 9.2.27 (A) or W6/32 (B) as primary antibodies. The coverslips were stained with these same antibodies after the cells were removed with EDTA as described in Fig. 3. (C) 9.2.27; (D) W6/32.

and substrate upon which the cells were attached. Other reports have indicated that in addition to gangliosides, cell-associated adhesion plaques also contain cytoskeletal proteins[28-30] and proteoglycans.[30-32] In fact, chemical analysis of such adhesion plaques has demonstrated the presence of approximately 1% of the cell protein and phospholipid but as much as 5–15% of their carbohydrate,[33] suggesting that complex carbohydrates, whether on a protein or lipid backbone, may play a major role in cell attachment processes. At this time it is unclear whether such glycoconjugates act independently or in a synergistic manner with various other surface components to promote the attachment and spreading of cells. It is for this reason that MAbs directed to defined oligosaccharide moieties will be invaluable probes to dissect the specific role of these molecules in the cell attachment process.

4. MAbs Directed to GD_2 and GD_3 Inhibit Human Melanoma and Neuroblastoma Cell Attachment to Extracellular Matrix Components

Based on the surface distribution of GD_3 and GD_2 on human melanoma cells, the question arose as to whether MAbs directed to the oligosaccharide portion of these molecules could perturb cell attachment. To address this issue, M21 cells

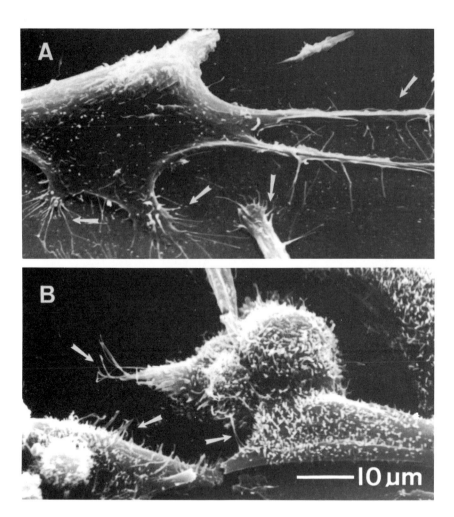

FIGURE 5. Detection of M21 human melanoma cells from a fibronectin substrate by anti-GD_2 MAb. M21 cells were allowed to attach and spread on a fibronectin-coated coverslip for 1 hr. The cells were overlaid with 100 µg/ml irrelevant MAb (A) or anti-GD_2 MAb 14.18 (B) for 3 hr at 37°C. The cells were fixed and prepared for scanning electron microscopy as previously described.[22] Arrows correspond to attachment-promoting microprocesses.

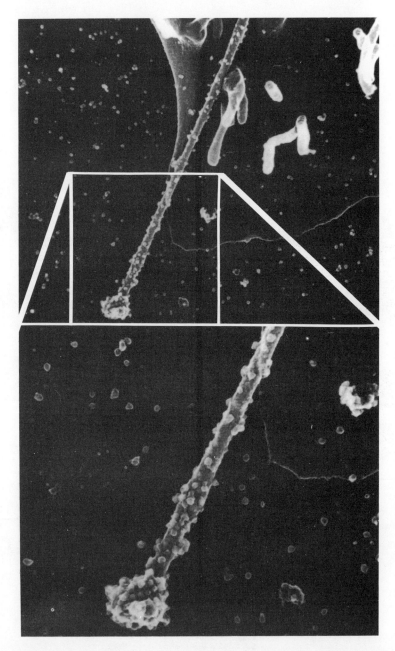

FIGURE 6. Immunolocalization of GD_2 on M21 microprocesses by scanning immunoelectron microscopy. M21 cells were allowed to attach and spread on fibronectin-coated coverslips, fixed with 0.5% glutaraldehyde, and stained using MAb 14.18 as primary antibody followed by anti-mouse Ig conjugated to colloidal gold (20–30 nm) as previously described.[22] Insets on the left and right are magnified 2 and 10×, respectively.

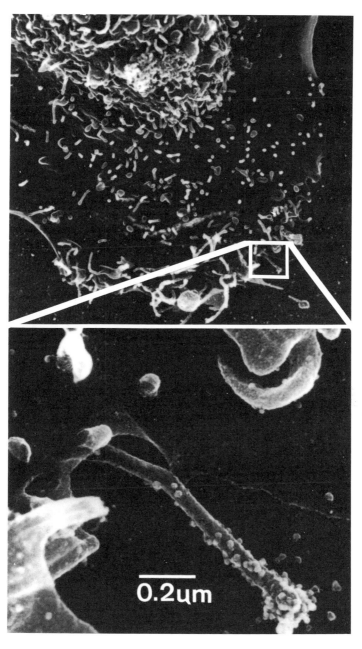

FIGURE 6. (CONTINUED)

were grown on fibronectin-coated plastic dishes, allowed to attach, spread, and reach confluency. The monolayer was then overlaid with growth media containing purified anti-GD_2, anti-GD_3, anti-HLA, or a nonbinding control MAb. Only cells treated anti-GD_2 or GD_3 in this manner showed a dose-dependent detachment and cell rounding within 2–3 hr of antibody exposure. The percentage of cells in the population that were affected was essentially identical to the relative percentage of GD_2 (99%)- and GD_3 (62%)-positive cells in the population as determined by flow cytometric analysis.[21] In contrast, neither anti-HLA antibody, which binds to M21 cells, nor the nonbinding control MAb had any demonstrable effect on the cell monolayer. Furthermore, MAbs directed to either GD_2 or GD_3 had no effect on cells that did not express these antigens. These results are consistent with those reported by Dippold et al.[34] who demonstrated that MAb R24, also directed to GD_3, could cause a specific melanoma cell detachment. However, in this case human melanoma cells were grown on a plastic surface.

Ultrastructural analyses of antiganglioside MAb-induced inhibition of melanoma cell attachment indicated that once the cells rounded up from the substrate, their attachment-promoting microprocesses became dislodged from the fibronectin-coated surface (Fig. 5). Indirect immunolocalization wtih the anti-GD_2 MAb 14.18 as primary antibody and goat anti-mouse Ig conjugated to 30-nm gold particules in conjunction with scanning electron microscopy demonstrated that GD_2 is heavily expressed on these M21 melanoma cell attachment-promoting microprocesses as they near the fibronectin substrate (Fig. 6). In addition, immunolocalization by transmission electron microscopy on cross sections and longitudinal sections of melanoma cell microprocesses demonstrates that GD_2 is enriched on these structures as compared to the cell body (Fig. 7A). This technique has also allowed for the detection of this ganglioside directly at the points of cell contact (arrows) with the fibronectin substrate (Fig. 7B,C). In fact, when M21 cells previously grown in suspension were allowed to adhere to immobilized fibronectin, GD_2 and GD_3 redistributed on the cell surface from a random or uniform to a punctated distribution into these microprocesses within 20 min at 37°C.[22] When these same cells were removed from the substrate with the divalent cation chelator EDTA, the gangliosides rapidly redistributed back to a uniform distribution on the cell surface.[22] Moreover, as M21 cells round up during mitosis, the GD_2 and GD_3 distribution on the cell surface becomes uniform,[22] further suggesting that the surface expression of these molecules correlates with their state of adhesion.

Pretreatment of human melanoma cells with MAbs directed to gangliosides could also prevent their attachment to a variety of extracellular matrix components.[20] A kinetic analysis of this effect demonstrated that the antiganglioside MAbs primarily inhibited the early stages (within 5 min) of cell attachment to either fibronectin or laminin (Fig. 8). In addition to the observed effects on cell attachment, pretreatment of M21 cells with antiganglioside MAbs could also inhibit their attachment to vitronectin, collagen, and a heptapeptide (glycyl-L-arginyl-glycyl-L-aspartyl-L-seryl-L-prolyl-L-cysteine) that constitutes the cell attachment site of fibronectin.[21] Moreover, when melanoma cells containing GD_2 and GD_3 were incubated with MAbs to both of these molecules, an additive inhibition of attachment was observed.[21] The specificity of this inhibition was demonstrated

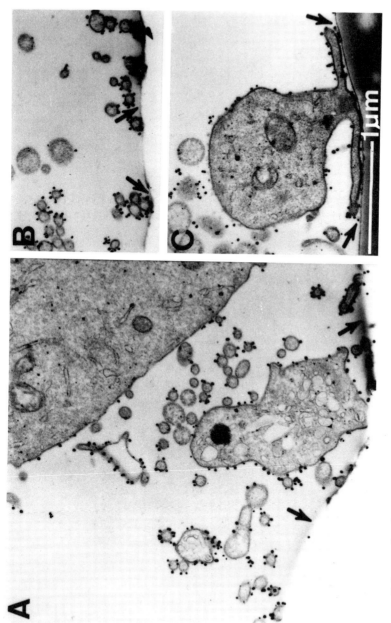

FIGURE 7. Immunolocalization of GD_2 on the cell surface and microprocesses of M21 cells by transmission electron microscopy. M21 cells were allowed to attach to fibronectin-coated coverslips and stained with MAb 14.18 and anti-mouse Ig conjugated to gold (7–9 nm) as previously described.[22] Arrows correspond to the fibronectin substrate. A includes a cross section of the cell body (upper right) as well as microprocesses. B demonstrates cross sections of microprocesses making direct contact with the substrate. C demonstrates longitudinal sections of two microprocesses making direct contact with the substrate.

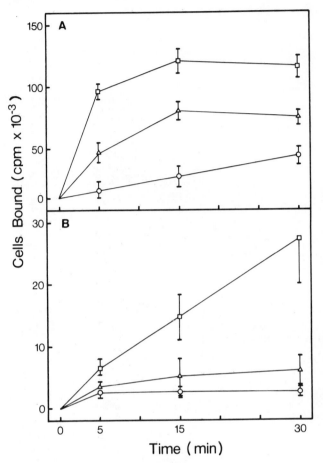

FIGURE 8. Effects of monoclonal antiganglioside antibodies on the kinetics of M21 human melanoma cell attachment to fibronectin or laminin. Metabolically labeled M21 human melanoma cells (5×10^3) were allowed to attach for various times to microtiter wells coated with either fibronectin (A) or laminin (B). Prior to addition to substrate-coated wells, the cells were allowed to react with MAbs W6/32 (anti-HLA, □), 11C64 (anti-GD$_3$, △), or 3F8 (anti-GD$_2$, ○) for 1 hr at 4°C and washed free of excess antibody. The data are expressed as the total number of cells bound (cpm bound) at the designated times as previously described.[21] Each point represents the mean ± S.D. of five replicates.

since MAbs of various isotypes directed to either protein or carbohydrate epitopes expressed on a number of major melanoma or neuroblastoma cell surface antigens had no effect on cell attachment.[21]

The role of gangliosides in melanoma cell attachment to a more physiological substrate has also been examined. As shown in Table I, pretreatment of M21 human melanoma cells with anti-GD$_2$ antibodies resulted in a 92% inhibition of attachment to a matrix laid down on tissue culture plastic by bovine endothelial cells. The fact that MAbs to both GD$_2$ and GD$_3$ can inhibit cell attachment to a number of different substrates suggests that gangliosides may play a general role in cell–substratum interactions rather than acting specifically as a surface receptor for a given substrate. In support of this hypothesis, a strong interaction could not be observed between radiolabeled fibronectin and gangliosides separated on a thin-layer chromatogram.[21] Alternatively, gangliosides, being strongly anionic, may play a role in the electrostatic requirements for optimal cell–substratum interactions. In this regard, controlled periodate oxidation of the terminal, unsubstituted sialic acid residues on the cell surface not only specifically destroyed the antigenic epitopes on GD$_2$ and GD$_3$, but also inhibited melanoma and neuroblastoma cell

TABLE I
Anti-GD$_2$ MAb-Induced Inhibition of M21 Human Melanoma Cell Attachment to Bovine Endothelial Extracellular Matrix[a,b]

MAb	Time of binding (min)		
	5[c]	15[c]	30[c]
Media (control)	498 ± 88 (100)	1295 ± 154 (100)	2349 ± 95 (100)
9.2.27 (antiproteoglycan)	733 ± 27 (147)	1163 ± 210 (90)	2641 ± 457 (108)
14.18 (anti-GD$_2$)	144 ± 62 (29)	154 ± 31 (12)	188 ± 73 (8)

[a]Bovine endothelial extracellular matrix was prepared as previously described.[(53)]
[b]M21 cells metabolically labeled with [^{3}H]leucine were allowed to attach to the substrate for various times at 37°C as previously described.[(21)] The data are expressed as the mean cpm ± S.D. of four replicates and as percent of control binding in absence of MAb.
[c]Cpm bound ± S.D. (percent of control).

attachment.[(21)] In fact, as shown in Fig. 9, the periodate-induced ganglioside oxidation and the inhibition of the attachment of two human melanoma and one neuroblastoma cell line were equally dose-dependent. As expected, treatment of cells with neuraminidase, which results in the removal of most of the surface sialic acid, also inhibited cell attachment.[(21)] These findings suggest that cell–substratum interactions may depend in part on the electrostatic environment provided by terminal sialic acid residues of cell surface gangliosides and possibly other anionic glycoconjugates.

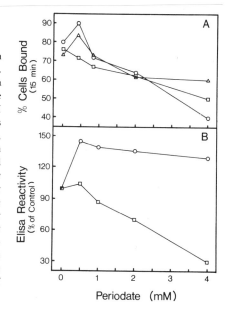

FIGURE 9. Effects of cell surface periodate oxidation on cell attachment and ganglioside antigen expression. Metabolically labeled M21 and 983b human melanoma cells or SK-NAS human neuroblastoma cells were treated with growth media containing 0.5–4.0 mM sodium-meta-periodate on ice in the dark for 30 min as previously described.[(21)] (A) After washing, M21 (○), 983b (△), or SK-NAS (□) cells were allowed to attach to fibronectin-coated microtiter wells for 15 min and the percent cells bound determined as previously described.[(21)] Each point represents the mean of four replicates. (B) M21 cells were treated with periodate and washed as above and then dried onto microtiter wells for antigen detection by ELISA as previously described[(21)] where MAbs 9.2.27 (○) or MB3.6 (anti-GD$_3$, □) were used as primary antibody. Each point represents the percent of control binding, i.e., in the absence of periodate, and is expressed as the mean of duplicate values.

The role played by cell surface gangliosides acting as specific receptors for various adhesive proteins may also be in doubt because of the recent reports indicating the presence of specific glycoprotein cell surface receptors for the extracellular matrix components, i.e., fibronectin,[35-37] vitronectin,[38] and laminin.[39] These receptors were shown to recognize the peptide RGDS (arginyl-glycyl-L-aspartyl-L-seryl), known to constitute the cell attachment site for both fibronectin[35] and vitronectin[38] and thus may represent a class of cell surface-associated extracellular matrix receptors. This class of cell surface receptors is also known to require the divalent cation Ca^{2+}, for activity since EDTA treatment of cells eliminates their attachment activity.[40,41] Moreover, the fibronection receptor on the cell surface was shown to be resistant to proteolytic cleavage by trypsin in the presence of physiological levels of Ca^{2+},[40] indicating that this cation binds tightly with the receptor and may possibly help to orient its interaction with fibronectin. The role of gangliosides or other membrane components in cell attachment may therefore involve their capacity to form a Ca^{2+}-dependent complex with one or more of these glycoprotein receptors in the membrane. The formation of such a complex may result in the activation of a receptor for a given substrate. It is conceivable that Ca^{2+}, known to be required for fibronectin-mediated cell attachment and also capable of binding very tightly to gangliosides,[42-44] is within a complex containing the negatively charged ganglioside on the cell surface, thus allowing gangliosides to distribute preferentially into certain domains on the cell surface. This contention is consistent with the results of Sharom and Grant[45] who demonstrated that physiological levels of Ca^{2+} lead to cross-linking and condensing of ganglioside headgroups in membranes by complexing carboxyl residues of sialic acid. Therefore, in the presence of divalent cations, it is possible that laterally mobile carbohydrate-bearing components such as gangliosides show a tendency to cluster about complex glycoproteins containing one or several carboxyl groups. Alternatively, gangliosides may act as a secondary receptor of low-affinity ionic interaction where the sum of many such interactions may lead to a higher avidity of a cell for its substrate. Our data and the results of others[8,18-27] would be consistent with either hypothesis.

5. Four Possible Models for the Role of Gangliosides in Cell Attachment

As shown in Fig. 10, four hypotheses (I–IV) are proposed to account for the potential role of gangliosides in cell attachment. In the first model (I), cell surface gangliosides and a separate fibronectin RGDS-dependent receptor are proposed to be capable of recognizing two distinct domains on the fibronectin molecule. In this case, the sum of these interactions would lead to an overall increased avidity of the cell surface for the immobilized fibronectin molecule. In partial support of this hypothesis, McCarthy et al.[46] demonstrated that the 33,000-dalton tryptic/catheptic carboxy-terminal heparin-binding fragment of fibronectin could stimulate murine melanoma cell attachment and spreading. This fragment, which is known not to contain the RGDS peptide, must therefore interact with an alterna-

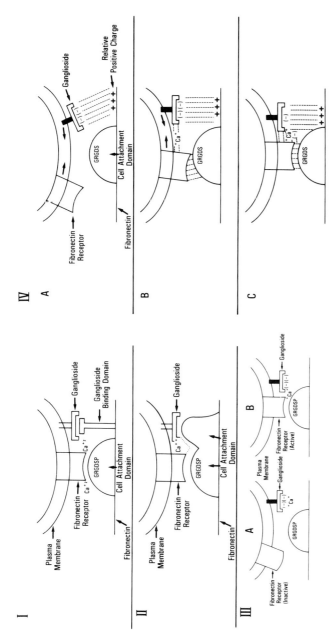

FIGURE 10. Schematic speculative diagrams comparing the modes of possible ganglioside involvement in cell attachment to an RGDS-containing substrate. *I* proposes the presence of two independent binding domains on the substrate, one of which is recognized by a cell surface RGDS-dependent receptor and the other by a cell surface ganglioside. *II* indicates the presence of a Ca^{2+}-dependent receptor/ganglioside complex on the cell surface that recognizes two closely associated domains on the substrate. *III* suggests that an inactive form of the receptor in A, after forming a Ca^{2+}-dependent complex in B, produces an active receptor that can now bind the RGDS domain of the substrate. *IV* involves a time-dependent mechanism where the ganglioside provides the initial electrostatic attraction between the cell and the substrate as in A. In B, a Ca^{2+}-dependent receptor complex forms that allows for optimal cell–substratum interaction as in C.

tive cell surface receptor. In this regard, it has been shown that cell surface proteoglycans are capable of interacting with this domain of fibronectin.[47,48] It is not unlikely that other oligosaccharides such as those found on gangliosides may also be capable of interacting with this portion of the fibronectin molecule. The second model (II) is somewhat similar to the first; however, in this case the assumption is made that the ganglioside-binding domain is closely associated with the actual RGDS-dependent fibronectin receptor. This receptor/ganglioside complex may be held together by a Ca^{2+} bridge and as in I, the sum of the two interactions may be expected to increase the avidity of the cell–fibronectin interaction. However, observations that antiganglioside MAbs could inhibit melanoma cell attachment to the immobilized peptide (RGDS) argue against models I and II since they assume that gangliosides only interact with an independent portion of the fibronectin molecule. At present, our favored hypothesis involves a combination of hypotheses III and IV. In each of these cases, the assumption is made that in order to achieve optimal cell attachment, a complex must form that in hypothesis III would lead to the activation of the glycoprotein receptor in the membrane for optimal interaction with its specific receptor. Hypothesis IV, on the other hand, assumes that the charged headgroup of the ganglioside establishes the initial attractive forces in a nonspecific manner until the specific receptor can become oriented in the complex which in time would lead to formation of a synergistic interaction culminating in optimal cell attachment. In this latter hypothesis, it is possible that the formation of the receptor complex would also serve to activate the specific interaction of the receptor with its substrate. In preliminary experiments we found evidence for such a complex existing on the surface of human melanoma cells. Specifically, both GD_2 and GD_3 copurify with the fibronectin or vitronectin receptors when purified by affinity chromatography from a detergent lysate of M21 human melanoma cells. Moreover, immunolocalization by either indirect immunofluorescence microscopy or transmission electron microscopy demonstrated that these gangliosides codistribute with the fibronectin and vitronectin receptors on M21 cells when attached and spread on a solid substrate containing either of these adhesive proteins.

Depending on the particular oligosaccharide moiety, a ganglioside may serve to induce a preferred orientation or create a suitable electrostatic environment for optimal receptor–ligand interaction on the cell surface. In this regard, gangliosides on a number of cell types have been shown to interact with and potentially modulate the function of cell surface receptors for thyroid-stimulating hormone,[13] platelet-derived growth factor (PDGF),[15] and the serotonin receptor.[49] The fact that GD_2 and GD_3 respectively represent the major gangliosides on the surface of human neuroblastoma and melanoma cells and apparently play a role in their attachment to the extracellular matrix suggests the possibility that structurally similar gangliosides on other cell types may have an analogous function. MAbs directed to other gangliosides on such cells will be helpful in addressing this issue.

6. Perspectives

It will ultimately be important to determine whether MAbs directed to ganglioside antigens on the tumor cell surface that are capable of preventing cell–

substratum interactions *in vitro* can also do so *in vivo*. In fact, recent evidence suggests that anti-GD_3 therapy in man[50] and in mice[51] may be effective in treating the proliferation of metastatic melanoma. Thus, in a recently reported experiment,[51] 12 athymic nude mice were injected subcutaneously with 2.5×10^6 human melanoma cells into the right and left flank, 6 of which were treated with eight intraperitoneal injections of 75 μg of purified MAb MB3.6 (anti-GD_3) on days 2, 4, 9, 11, 16, 21, and 23; 6 controls received no antibody. The mean tumor volume of each animal was measured at various times after tumor inoculation for up to 43 days. All of the control animals showed progressively growing tumors that ranged in volume from 35 to 350 mm³, with a mean volume of 194 mm³. The mean volume in the experimental group on day 43 was 11.2 mm³, which represents a 94% decrease compared to controls. Two representative examples of animals of the control (right) and of the experimental group (left) are shown in Fig. 11. The

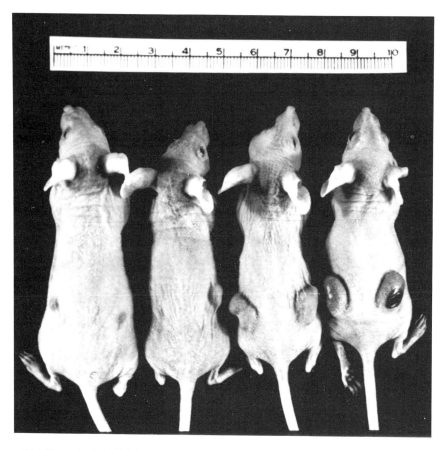

FIGURE 11. Effects of MAb MB3.6 (anti-GD_3) on human melanoma tumor formation in the athymic nude mouse. Twelve athymic nude mice were injected with 2.5×10^6 human melanoma cells subcutaneously in the right and left flank. The two representative animals on the left received six intraperitoneal injections of 75 μg of MAb MB3.6 at various times after tumor inoculation, and the two representative animals on the right received phosphate-buffered saline. These photographs were taken after tumors were allowed to develop for 43 days.

possibility that such an effect is due to the capacity of anti-GD_3 MAb to interfere with melanoma cell–substratum interactions remains to be determined. It is possible that by preventing or inhibiting attachment to host tissues, a tumor cell becomes much more susceptible to immunological effector mechanisms and is thus more easily cleared. Alternatively, inhibition of tumor cell attachment *in vivo* may reduce their capacity to obtain blood supply and nutrients and thereby inhibit their ability to establish a metastatic foci. The answer to these questions may help to elucidate the complex process of tumor cell metastasis and perhaps aid in its control.

ACKNOWLEDGMENTS. The author wishes to acknowledge Dr. R. A. Reisfeld for helpful comments and Ms. Bonnie Pratt Filiault for preparing the manuscript. Figures 5, 7, and 8 are reproduced from Refs. 21 and 22 in *The Journal of Cell Biology* and Figs. 3 and 4 are reproduced from Ref. 8.

The author's research was supported by Grant CA 28420 from the National Institutes of Health as well as a J. Ernest Ayre Fellowship from the National Cancer Cytology Center.

This is Scripps Publication 4305-Imm.

References

1. Hakomori, S., 1984, Tumor-associated carbohydrate antigens, *Annu. Rev. Immunol.* **2**:103–126.
2. Hakomori, S., and Kannagi, R., 1983, Glycosphingolipids as tumor-associated and differentiation markers, *J. Natl. Cancer Inst.* **71**:231–251.
3. Feizi, T., 1985, Carbohydrate antigens in human cancer, *Cancer Surv.* **4**:245–269.
4. Feizi, T., 1985, Demonstration by monoclonal antibodies that carbohydrate structures of glycoproteins and glycolipids are onco-developmental antigens, *Nature* **314**:53–57.
5. Cheresh, D. A., 1985, Structural and functional properties of ganglioside antigens on human tumors of neuroectodermal origin, *Surv. Synth. Path. Res.* **4**:97–109.
6. Cheresh, D. A., Varki, A. P., Varki, N. M., Stallcup, W. B., Levine, J., and Reisfeld, R. A., 1984, A monoclonal antibody recognizes an O-acylated sialic acid in a human melanoma-associated ganglioside, *J. Biol. Chem.* **259**:7453–7459.
7. Cheresh, D. A., Reisfeld, R. A., and Varki, A. P., 1984, O-acetylation of disialoganglioside GD3 by human melanoma cells creates a unique antigenic determinant, *Science* **225**:844–846.
8. Cheresh, D. A., Harper, J. R., Schulz, G., and Reisfeld, R. A., 1984, Localization of gangliosides GD2 and GD3 in adhesion plaques and on the surface of human melanoma cells, *Proc. Natl. Acad. Sci. USA* **81**:5767–5771.
9. Schulz, G., Cheresh, D. A., Varki, N. M., Yu, A., Staffileno, L. K., and Reisfeld, R. A., 1984, Detection of ganglioside GD2 in tumor tissues and in sera of neuroblastoma patients, *Cancer Res.* **44**:5914–5920.
10. Koprowski, H., Herlyn, M., Steplewski, Z., and Sears, H. F., 1981, Specific antigen in serum of patients with colon carcinoma, *Science* **212**:53–55.
11. Pukel, C. S., Lloyd, K. O., Trabassos, L. R., Dippold, W. G., Oettgen, H. F., and Old, L. J., 1982, GD3, a prominent ganglioside antigen of human melanoma, detection, and characterization by mouse monoclonal antibody, *J. Exp. Med.* **155**:1133–1147.
12. Nudelman, E., Hakomori, S., Kannagi, R., Levery, S., Yeh, M.-Y., Hellstrom, K. E., and Hellstrom, I., 1982, Characterization of a human melanoma-associated ganglioside antigen defined by a monoclonal antibody, 4.2, *J. Biol. Chem.* **257**:12752–12756.
13. Mullin, B. R., Fishman, P. H., Lee, G., Aloj, S. M., Ledley, F. D., Winand, R. J., Kohn, L. D., and

Brady, R. O., 1976, Thyrotropin–ganglioside interactions and their relationship to the structure and function of thyrotropin receptors, *Proc. Natl. Acad. Sci. USA* **73**:842–846.
14. Robb, R. J., 1986, The suppressive effect of gangliosides upon IL2-dependent proliferation as a function of inhibition of IL2 receptor association, *J. Immunol.* **136**:971–976.
15. Bremer, E. G., Hakomori, S., Bowen-Pope, D. F., Raines, E., and Ross, R., 1984, Ganglioside-mediated modulation of cell growth, growth factor binding, and receptor phosphorylation, *J. Biol. Chem.* **259**:6818–6825.
16. Suzuki, Y., Matsunaga, M., and Matsumoto, M., 1985, GM3-NeuAc: A new influenza virus receptor which mediates the adsorption–fusion process of viral infection, *J. Biol. Chem.* **260**:1362–1365.
17. van Heyningen, W. E., 1974, Gangliosides as membrane receptors for tetanus toxin, cholera toxin, and serotonin, *Nature* **249**:415–417.
18. Kleinman, H. K., Martin, G. R., and Fishman, P. H., 1979, Ganglioside inhibition of fibronectin-mediated cell adhesion to collagen, *Proc. Natl. Acad. Sci. USA* **76**:3367–3371.
19. Yamada, K. M., Kennedy, D. W., Grotendorst, G. R., and Momoi, T., 1981, Glycolipids: Receptors for fibronectin? *J. Cell. Physiol.* **109**:343–351.
20. Perkins, R. M., Kellie, S., Patel, B., and Critchley, D. R., 1982, Gangliosides as receptors for fibronectin, *Exp. Cell Res.* **141**:231–243.
21. Cheresh, D. A., Pierschbacher, M. D., Herzig, M. A., and Mujoo, K., 1986, Disialogangliosides GD2 and GD3 are involved in the attachment of human melanoma and neuroblastoma cells to extracellular matrix proteins, *J. Cell Biol.* **102**:688–696.
22. Cheresh, D. A., and Klier, F. G., 1986, Disialoganglioside GD2 distributes preferentially into substrate-associated microprocesses on human melanoma cells during their attachment to fibronectin, *J. Cell Biol.* **102**:1887–1897.
23. Mugnai, G., Tombaccini, D., and Ruggieri, S., 1984, Ganglioside composition of substrate-adhesion sites of normal and virally-transformed BALB/c 3T3 cells, *Biochem. Biophys. Res. Commun.* **125**:142–148.
24. Spiegel, S., Schlessinger, J., and Fishman, P. H., 1984, Incorporation of fluorescent gangliosides into human fibroblasts: Mobility, fate, and interaction with fibronectin, *J. Cell Biol.* **99**:699–704.
25. Spiegel, S., Yamada, K. M., Horn, B. E., Moss, J., and Fishman, P. H., 1985, Fluorescent gangliosides as probes for retention and organization of fibronectin by ganglioside-deficient mouse cells, *J. Cell Biol.* **100**:721–726.
26. Okada, Y., Mugnai, G., Bremer, E. G., and Hakomori, S., 1984, Glycosphingolipids in detergent-insoluble substrate attachment matrix (DISAM) prepared from substate attachment material (SAM): Their possible role in regulating cell adhesion, *Exp. Cell Res.* **155**:448–456.
27. Yamada, K. M., Critchley, D. R., Fishman, P. H., and Moss, J., 1983, Exogenous gangliosides enhance the interaction of fibronectin with ganglioside-deficient cells, *Exp. Cell Res.* **143**:295–302.
28. Revel, J. P., Hoch, P., and Ho, D., 1974, Adhesion of culture cells to their substratum, *Exp. Cell Res.* **84**:207–218.
29. Rosen, J. J., and Culp, L. A., 1977, Morphology and cellular origins of substrate-attached material from mouse fibroblasts, *Exp. Cell Res.* **107**:139–147.
30. Lark, M. W., Laterra, J., and Culp, L. A., 1985, Close and focal contact adhesion of fibroblasts to a fibronectin-containing matrix, *Fed. Proc.* **44**:394–403.
31. Garner, J. A., and Culp, L. A., 1981, Aggregation competence of proteoglycans from substratum adhesion sites of murine fibroblasts, *Biochemistry* **20**:7350–7359.
32. Lark, M. W., and Culp, L. A., 1981, Turnover of heparan sulfate proteoglycans from substratum adhesion sites of murine fibroblasts, *J. Biol. Chem.* **256**:6773–6782.
33. Culp, L. A., 1974, Substrate-attached glycoproteins mediating adhesion of normal and virus-transformed mouse fibroblasts, *J. Cell Biol.* **63**:71–83.
34. Dippold, W. G., Knuth, A., and zum Bushenfelde, K. H. M., 1984, Inhibition of human melanoma cell growth *in vitro* by monoclonal anti-GD3 ganglioside antibody, *Cancer Res.* **44**:806–810.
35. Pytela, R., Pierschbacher, M. D., and Ruoslahti, E., 1985, Identification and isolation of a 140 Kd cell surface glycoprotein with properties expected of a fibronectin receptor, *Cell* **40**:191–198.
36. Brown, P. J., and Juliano, R. L., 1985, Selective inhibition of fibronectin-mediated cell adhesion by monoclonal antibodies to a cell-surface glycoprotein, *Science* **228**:1448–1450.
37. Giancotti, F. G., Tarone, G., Knudsen, K., Damsky, C., and Comoglio, P. M., 1985, Cleavage of a

135 Kd cell surface glycoprotein correlates with loss of fibroblast adhesion to fibronectin, *Exp. Cell Res.* **156:**182–190.
38. Pytela, R., Pierschbacher, M. D., and Ruoslahti, E., 1985, A 125/115 kDa cell surface receptor specific for vitronectin interacts with the arginine-glycine-aspartic acid adhesion sequence derived from fibronectin, *Proc. Natl. Acad. Sci. USA* **82:**5766–5770.
39. Rao, C. N., Barsky, S. H., Tervanova, B. P., and Liotta, L., 1983, Isolation of a tumor cell laminin receptor, *Biochem. Biophys. Res. Commun.* **11:**804–808.
40. Akiyama, S. K., and Yamada, K. M., 1985, The interaction of plasma fibronectin with fibroblastic cells in suspension, *J. Biol. Chem.* **260:**4492–4500.
41. Takeichi, M., 1977, Functional correlation between adhesive properties and some cell surface proteins, *J. Cell Biol.* **75:**464–474.
42. Abramson, M. B., Yu, R. K., and Zaby, V., 1972, Ionic properties of beef brain gangliosides, *Biochim. Biophys. Acta* **280:**365–372.
43. Goldenring, J. R., Otis, L. C., Yu, R. K., and DeLorenzo, R. J., 1985, Calcium/ganglioside-dependent protein kinase activity in rat brain membrane, *J. Neurochem.* **44:**1229–1234.
44. Jaques, L. W., Brown, E. B., Barrett, J. M., Brey, W. S., and Weltner, W., 1977, Sialic acid, a calcium-binding carbohydrate, *J. Biol. Chem.* **252:**4523–4538.
45. Sharom, F. J., and Grant, C. W. M., 1978, A model for ganglioside behavior in cell membranes, *Biochim. Biophys. Acta* **507:**280–293.
46. McCarthy, J. B., Hagen, S. T., and Furcht, L. T., 1986, Human fibronectin contains distinct adhesion- and motility-promoting domains for metastatic melanoma cells, *J. Cell Biol.* **102:**179–188.
47. Perkins, R. M., Ji, H. T., and Hynes, R. O., 1979, Cross-linking of fibronectin to proteoglycans at the cell surface, *Cell* **16:**944–952.
48. Woods, A., Hook, M., Kjellen, L., Smith, C. G., and Rees, D. A., 1984, Relationship of heparin sulfate proteoglycans to the cytoskeleton and extracellular matrix of cultured fibroblasts, *J. Cell Biol.* **99:**1743–1753.
49. Berry-Kravis, E., and Dawson, G., 1985, Possible role of gangliosides in regulating an adenylate cyclase-linked 5-hydroxytryptamine (5-HT$_1$) receptor, *J. Neurochem.* **45:**1739–1747.
50. Houghton, A. N., Mintzer, D., Cordon-Cardo, L., Welt, S., Fliegel, B., Vadhan, S., Carswell, E., Melamed, M. R., Oettgen, H. F., and Old, L. J., 1985, Mouse monoclonal IgG3 antibody detecting GD3 ganglioside. A phase I trial in patients with malignant melanoma, *Proc. Natl. Acad. Sci. USA* **82:**1242–1246.
51. Cheresh, D. A., Honsik, C. J., Staffileno, L. K., Jung, G., and Reisfeld, R. A., 1985, Disialoganglioside GD3 on human melanoma serves as a relevant target antigen for monoclonal antibody-mediated tumor cytolysis, *Proc. Natl. Acad. Sci. USA* **82:**5155–5159.
52. Svennerholm, L., 1963, Chromatographic separation of human brain gangliosides, *Neurochemistry* **10:**613–623.
53. Friedman, R., Fuks, Z., Ovadia, H., and Vlodavsky, I., 1985, Differential structural requirements for the induction of cell attachment, proliferation, and differentiation by the extracellular matrix, *Exp. Cell Res.* **157:**181–194.

26
Specific Adoptive Therapy of Disseminated Tumors

Requirements for Therapeutic Efficacy and Mechanisms by Which T Cells Mediate Tumor Eradication

PHILIP D. GREENBERG, DONALD E. KERN, JAY P. KLARNET, MICHAEL C. V. JENSEN, KENNETH H. GRABSTEIN, AND MARTIN A. CHEEVER

1. Introduction

There is little difficulty demonstrating under experimental conditions that T cells can specifically recognize and kill autochthonous or MHC-compatible tumor cells. However, the all too common observation that neoplastic cells can develop and grow progressively in apparently immunocompetent hosts suggests that the cellular immune response is an inadequate defense mechanism for protection from fatal tumors. There are many reasons why the host immune system may be ineffective for controlling tumor growth. The most obvious, and most problematic with regard to immunologic intervention, is that the tumor may not express any determinants immunogenic to the host. Although some tumors may lack such determinants, a substantial amount of evidence is accumulating to suggest that many tumors do express potentially immunogenic surface antigens. The presence of these antigens

---

PHILIP D. GREENBERG • Departments of Medicine and Microbiology/Immunology, University of Washington, and Division of Oncology, Fred Hutchinson Cancer Research Center, Seattle, Washington 98195. DONALD E. KERN AND MICHAEL C. V. JENSEN • Department of Microbiology/Immunology, University of Washington, Seattle, Washington 98195. JAY P. KLARNET AND MARTIN A. CHEEVER • Department of Medicine, University of Washington, and Division of Oncology, Fred Hutchinson Cancer Research Center, Seattle, Washington 98195. KENNETH H. GRABSTEIN • Immunex Corporation, Seattle, Washington, 98101.

on tumor cells implies that during progressive growth the tumor may have elicited a quantitatively inadequate response, or may have induced suppressor cells which down-regulated the immune response. Recent technologic and biologic advances have made it possible to purify potential effector T cells and to identify weak antigen-specific T-cell responses by expanding the number of specifically reactive T cells during *in vitro* culture. Application of these *in vitro* techniques for analysis of the immune response to tumors has permitted detection of apparent tumor-specific T-cell immunity in cancer patients with a variety of solid tumors and hematologic malignancies.[1-6] Thus, methods to augment such tumor immunity are being explored as potential cancer therapies.

Several distinct approaches for enhancing host responses to tumors are being examined, including selective *in vivo* depletion of potential suppressor cells, sensitization to tumor antigens presented in a matrix more immunogenic than intact tumor cells, and administration of immunoadjuvants such as purified lymphokines. An additional method, which is being studied in our laboratory, involves the adoptive transfer into a tumor-bearing host of specifically immune T cells. This approach is based on the supposition that one way to modify an ineffective immune response would be to infuse additional effector cells, such as could be accomplished by expanding host tumor-reactive T cells *in vitro*. Animal models using a wide variety of tumors have been developed, and are being used to identify the potential and to elucidate the principles of such adoptive immunotherapy. In these models, the transfer of tumor-specific T cells can be shown to mediate the eradication of locally established and/or disseminated tumors. The extent to which the insights gained from these studies can be generalized to man is uncertain, but the demonstration of tumor-specific T cells in some cancer patients,[1-6] and the identification of clearly immunogenic membrane determinants on human tumors such as HTLV-positive leukemias,[5] suggests that there should be circumstances in which this approach will be useful therapeutically.

In this review, we will describe studies from our laboratory examining the adoptive therapy of disseminated murine leukemias and lymphomas. For most studies, mice bearing a disseminated syngeneic tumor have been treated with donor lymphocytes obtained from syngeneic mice previously immunized to the tumor. The use of syngeneic donor cells has served as a prototype of what might be accomplished with appropriate modulation of autologous lymphocytes derived from a tumor-bearing host. Several basic principles underlying the requirements for recognition and destruction of established tumors by transferred cells have been identified in these models, and it has become apparent that straightforward infusion of large numbers of potential effector cells into tumor-bearing hosts is unlikely to result in rapid and complete eradication of tumor cells. Therefore, studies evaluating the problems that necessitate treatment of the tumor-bearing host prior to adoptive therapy in order to modify the underlying host–tumor relationship so that following cell transfer tumor immunity can be expressed will be summarized, and the results of studies of adoptive therapy from our laboratory will be presented, with particular attention to:

1. The requirement for T cells to mediate a prolonged *in vivo* antitumor effect in order to completely eradicate the tumor cells,

2. The T cells responsible for recognition and eradication of disseminated tumor,
3. The potential role of macrophages in tumor eradication, and
4. The generation and use of tumor-specific T cells that have been expanded to large numbers *in vitro* before transfer

2. Large Tumor Burdens as an Obstacle to Adoptive Therapy

The biologic consequences of a large-growing tumor present considerable obstacles to effective cellular therapy. Thus, even under optimal conditions, in which the tumor is known to be immunogenic and sensitive to immune-mediated lysis, and in which immune T cells transferred into the host at the time of tumor transplantation result in tumor rejection, the infusion of immune T cells into hosts after the tumor has grown and become established only rarely has a detectable antitumor effect. This does not reflect a simple arithmetic relationship in which a greater number of tumor cells requires infusion of a greater number of effector cells, since in most instances the administration of even very large numbers of specifically immune T cells to increase the effector to tumor target ratio at the time of cell transfer has little antitumor activity. Several reasons have been identified that may explain this observed lack of efficacy. One inherent problem is that the tumor cell mass is proliferating, so that even if all the tumor cells are initially accessible to the effector population, the number of targets may increase at a rate that exceeds the capacities of the transferred effector cells. Large solid tumors can also behave in some respects as immunologically privileged sites, with the poorly vascularized areas inside the tumor mass being relatively inaccessible to immune effector cells. Moreover, tumor cells can produce factors that act locally to nonspecifically interfere with immunologic effector functions at the site of the tumor nodule,[7] as well as secrete factors that induce nonspecific suppression of host immune function, thus potentially creating local and systemic environments that are not conducive to immune-mediated tumor rejection.

The major obstacle to adoptive therapy of large tumors, however, does not appear to be the actual tumor size, but rather that large growing immunogenic tumors induce suppressor T cells (Ts) that prevent the expression of adoptively transferred tumor immunity. This has been convincingly demonstrated in adoptive therapy studies comparing treatment of tumor-bearing hosts that are immunologically intact with treatment of hosts that previously have been rendered T deficient by thymectomy followed by lethal irradiation and reconstitution with T-depleted bone marrow. The transfer of specifically immune T cells into T-deficient tumor-bearing hosts resulted in complete rejection of a large established syngeneic fibrosarcoma or a disseminated mastocytoma, whereas the same immune T cells were totally ineffective when transferred into tumor-bearing normal hosts.[8,9] This lack of antitumor activity in the normal hosts was due to Ts, since the therapeutic efficacy of donor T cells in the T-deficient hosts was completely inhibited if T cells from tumor-bearing normal hosts were administered concurrently with the immune T cells.[8,9] The importance of tumor-induced Ts has been substantiated by studies in other tumor therapy models in which the efficacy of donor T cells was enhanced if the host was immunologically suppressed prior to the transfer of immune cells,[10,11] and in which the inoculation of antibodies that specifically deplete Ts enhanced cell-mediated rejection of an established sarcoma.[12] In addition to tumor-induced Ts, normal immunocompetent hosts appear to have immunoregulatory Ts circuits that interfere with the adoptive transfer of immunity.[13–15] Thus, treatment of the host to diminish Ts activity induced by a growing immunogenic tumor and/or participating in host regulation of immune reactivity appears to be necessary for effective adoptive therapy.

The method used to reduce the tumor burden and/or modify host immunoregulatory circuits has varied depending on the model being studied and the therapeutic effect needed. In some settings, surgical resection of the large tumor mass responsible for inducing Ts permits the expression of adoptively transferred tumor immunity. For example, in a lung carcinoma model, transferred immune cells were ineffective in hosts with a large growing primary tumor, but could prevent the outgrowth of micrometastatic pulmonary lesions if the primary tumor mass was removed surgically.[16] In other settings, depletion of tumor-induced Ts can be used to permit expression of antitumor activity against a large tumor. For example, in mice bearing an advanced locally growing fibrosarcoma, transferred immune T cells which had no apparent effect against the progressing primary mass could eradicate the large tumor if the host was depleted of Ts prior to cell transfer by treatment with cyclophosphamide at doses with no significant direct antitumor activity.[11]

Several models have been developed in our laboratory for the treatment of disseminated syngeneic leukemias and lymphomas by the adoptive transfer of immune cells. These models have been termed adoptive chemoimmunotherapy (ACIT), since the mice bearing advanced tumors are first treated with cyclophosphamide and then receive an infusion of immune T cells. Therapy of these disseminated tumors with large doses of immune cells in the absence of cyclophosphamide has no detectable *in vivo* antitumor effect. The use of cyclophosphamide in these models surmounts several major obstacles to the efficacy of immunotherapy. Cyclophosphamide has been shown in adoptive therapy models to ablate both tumor-induced Ts and host immunoregulatory Ts which interfere with the adoptive transfer of immunity.[11,14] Moreover, in the ACIT model, this drug has been shown to have a direct tumoricidal effect, producing a several log reduction in the number of viable tumor cells in the host and thereby removing the stimulus to further induction of Ts.[17] Thus, in ACIT as well as most adoptive immunotherapy models, modification of the host–tumor relationship is necessary to facilitate expression of tumor immunity.

3. Requirement for Specifically Immune T Cells to Mediate a Prolonged Antitumor Effect to Completely Eradicate Tumor by ACIT

The ACIT model most extensively studied in our laboratory is the treatment of disseminated FBL-3 erythroleukemia in C57BL/6 (B6) mice. Briefly, B6 host mice, inoculated intraperitoneally with 5×10^6 FBL-3 leukemia cells on day 0, are left untreated until day 5, at which time disseminated tumor is detectable in the peripheral blood and lymphoid organs.[17] Mice receiving no therapy on day 5 die 1 to 2 weeks later with progressive ascites, splenomegaly, and lymphadenopathy. Chemotherapy alone on day 5 with 180 mg/kg cyclophosphamide cures no mice but does prolong survival for 2 to 3 weeks, whereas immunotherapy alone by transfer of immune T cells on day 5 has no apparent therapeutic effect. However, combined treatment on day 5 with cyclophosphamide followed in 6 hr (i.e., to allow for drug clearance) by transfer of immune T cells can eradicate the tumor and cure mice.

Efficacy in ACIT requires transfer of specifically immune T cells, which exhibit dose-dependent activity. Transfer of low cell doses results in prolonged survival but very few mice are cured, whereas transfer of higher cell doses (i.e., $> 10^7$ cells) can cure nearly 100% of mice. The therapeutic effect is dependent only on the T cells present in the transferred cells, since immune donor spleen cells depleted of T cells have no therapeutic activity, and purified T cells retain the therapeutic efficacy present in unfractionated spleen cells.[18] Specificity for tumor antigens has been demonstrated in studies examining ACIT of two antigenically distinct tumors in B6 mice,[19] and MHC restriction of T-cell specificity has been demonstrated in studies with F_1 hybrid mice examining ACIT of two parental strain tumors sharing cross-reactive tumor antigens but disparate for MHC antigens.[20]

The kinetics of tumor elimination following ACIT of FBL leukemia with immune T cells has been examined by sacrificing mice after potentially curative treatment and examining the peripheral blood and spleen for remaining viable tumor cells.[17] Comparisons of the amount of tumor present in mice treated on day 5 with cyclophosphamide alone to the tumor present in mice treated with cyclophosphamide and immune cells demonstrated that the immune cells had an immediate effect on reducing the tumor cells persisting following chemotherapy. However, eradication of all the remaining tumor cells did not occur rapidly, since viable tumor cells could still be detected for up to 30 days following curative therapy. Thus, complete elimination of disseminated leukemia following ACIT requires a prolonged time period. Similar requirements for a prolonged *in vivo* effect have been observed in adoptive therapy models for the treatment of advanced locally growing solid tumors.[8,10]

The prolonged time period required for complete elimination of disseminated tumors presumably reflects the need for T cells to traffic to all sites of residual tumor and express antitumor activity. To determine if persistent T-cell immunity is required during the entire period of tumor eradication, we have depleted mice of T cells *in vivo* at different time points after ACIT and examined the effect on tumor elimination. T-cell depletion was achieved by inoculating mice for 3 consecutive days with monoclonal IgG2a anti-Thy antibody in doses previously shown to eliminate for more than 10 days peripheral and splenic T cells. Mice receiving only cyclophosphamide and 2×10^7 immune cells were cured of leukemia, whereas mice depleted of T cells by infusion of anti-Thy antibody beginning on day 20 (i.e., 15 days after cell transfer) or day 30 died of recurrent tumor (Fig. 1). If T depletion was delayed until day 40, 50% of mice died of disseminated leukemia. No

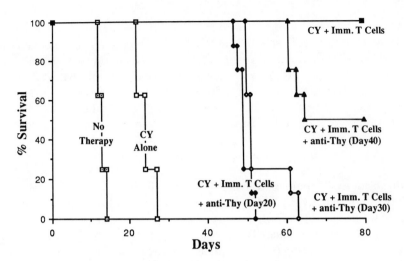

FIGURE 1. Requirement for persistent T-cell immunity to eradicate tumors. B6/Thy-1.1 mice were inoculated with 5×10^6 FBL on day 0, and received no therapy or were treated on day 5 with either 180 mg/kg cyclophosphamide (CY) alone or CY and 2×10^7 spleen cells from B6/Thy-1.1 donor mice primed *in vivo* to FBL. Beginning on day 20, 30, or 40, the designated groups of mice were depleted of T cells *in vivo* by i.p. inoculation for 3 consecutive days of 0.5 ml of a 1:10 dilution of ascites containing IgG2a monoclonal antibody to Thy-1.1.

deaths from tumor were observed if T depletion was delayed until day 50. These results emphasize the importance of persistent T-cell antitumor activity for complete tumor eradication, and suggest that immunosuppression of hosts after adoptive immunotherapy might significantly interfere with therapeutic efficacy and result in a potentially lethal recurrence of tumor.

The prolonged T-cell response necessary for tumor eradication could reflect a persistent contribution by transferred donor tumor-reactive T cells surviving in the host or a requirement for a host T-cell response to complete tumor elimination. Host T-cell responsiveness would be expected to be diminished by the cyclophosphamide therapy on day 5, but should recover long before tumor eradication is complete.[21,22] To permit analysis of the origin of T cells contributing to tumor immunity, host and donor mice congenic at the Thy locus were used in adoptive therapy studies. The two strains of mice utilized, B6 and congenic B6.PL(74NS), denoted B6/Thy-1.1, are genetically identical except for the Thy locus. Thus, with the use of appropriate monoclonal antibodies, T cells from B6 mice, which express Thy-1.2 can be distinguished from T cells from B6/Thy-1.1 mice, which express Thy-1.1. The B6 strain is a genetic low responder to Thy antigens,[23] and no immune response to either Thy allele was detectable following cell transfer in these adoptive therapy studies.[24] B6 host mice bearing disseminated FBL were treated on day 5 with cyclophosphamide and transfer of 2×10^7 immune B6/Thy-1.1 donor cells, and then sacrificed at day 60 to permit assessment of host and donor T cells after tumor elimination was complete. Analysis of the Thy phenotype of spleen cells at day 60 by fluorescence analysis with conjugated monoclonal antibodies revealed that a small number of donor T cells were still present, comprising 1.3% of the cells present in the spleen.[24] The reactivity of host and residual donor

TABLE I
Spleen Cells Present at Day 60 in Hosts Cured of Tumor by ACIT: Contribution of Host and Donor T Cells to Tumor Reactivity

Source of responder lymphocytes[a]	Cytotoxic T-cell generation test culture			Antibody-mediated depletion after sensitization culture[d]	Percent specific lysis of target[e]	
	Antibody-mediated depletion before culture[b]	Irradiated stimulator[c]			FBL	BALB/c
B6/Thy-1.1$_{\alpha FBL}$→B6	—	FBL		—	41	2
		BALB/c			0	44
B6/Thy-1.1$_{\alpha FBL}$→B6	αThy-1.2	FBL		—	35	1
		BALB/c			0	0
B6/Thy-1.1$_{\alpha FBL}$→B6	αThy-1.1	FBL		—	3	0
		BALB/c			0	36
B6/Thy-1.1$_{\alpha FBL}$→B6	—	FBL		αThy-1.2	34	0
		BALB/c			0	8
B6/Thy-1.1$_{\alpha FBL}$→B6	—	FBL		αThy-1.1	2	0
		BALB/c			0	48

[a] Responder spleen cells were obtained on day 60 from B6 hosts cured of disseminated FBL by treatment on day 5 with cyclophosphamide plus 2×10^7 spleen cells from B6/Thy-1.1 donors immune to FBL (B6/Thy-1.1$_{\alpha FBL}$→B6).
[b] Before culture, responder cells were incubated with either monoclonal αThy-1.2 plus complement or monoclonal α Thy-1.1 plus complement.
[c] Responder cells (6×10^6) were cultured for 5 days with either 0.3×10^6 irradiated syngeneic FBL tumor stimulator cells or 1.5×10^6 irradiated allogeneic BALB/c spleen cells. These culture conditions have been shown to detect primary allogeneic responses, but only secondary and no primary anti-FBL responses.
[d] After 5-day sensitization culture, effector cells were incubated with either monoclonal αThy-1.2 plus complement or monoclonal αThy-1.1 plus complement.
[e] Effector cells obtained after 5-day sensitization culture were tested in a 4-hr chromium release assay for cytotoxicity to labeled syngeneic FBL tumor or allogeneic BALB/c spleen cell blasts at an E/T of 20:1.

T cells was determined by examining *in vitro* the phenotype of T cells responding to FBL and to an allogeneic control stimulator (Table I). Depletion of host T cells with anti-Thy-1.2 ablated the response to BALB/c H-2^b alloantigens but had little effect on tumor reactivity, whereas depletion of donor T cells with anti-Thy-1.1 ablated tumor reactivity but had little effect on the alloresponse. Thus, although the majority of immunocompetent T cells, as reflected by alloreactivity, were of host origin, host T cells did not make a substantial contribution to the antitumor response. These results suggest that, for optimal adoptive therapy, it will be necessary to transfer immune T cells capable of persisting for the entire time period required for complete tumor elimination.

4. Nature of T Cells Responsible for Recognition and Eradication of Disseminated Leukemia

T cells can be divided into two major subpopulations, an L3T4$^+$ Lyt-2$^-$ T-cell subset that is restricted to recognizing antigens in the context of class II MHC antigens and contains most of the helper T cells (Th), and an L3T4$^-$ Lyt-2$^+$ T-cell subset that is restricted to recognizing antigens in the context of class I MHC antigens and contains most of the cytotoxic T cells (Tc). To examine the mechanisms by which transferred T cells mediate their antitumor effect, the ability of T-cell subsets to recognize and lyse FBL targets and to treat disseminated FBL was evaluated. FBL is an erythroleukemia which expresses Friend retrovirus-associated tumor antigens and class I MHC antigens, but does not express class II MHC antigens either constitutively or after incubation with recombinant γ-interferon (γ-IFN).[25]

The contribution of each T-cell subset to the generation and expression of the *in vitro* cytolytic response to FBL was assessed. Unfractionated FBL-primed spleen cells cultured with FBL generated a strong specific cytotoxic response to FBL. Consistent with the tumor phenotype, all T cells capable of lysing FBL were class I-restricted Lyt-2^+ CTL.[26] However, the generation of such Tc required the presence of either IL-2-producing class II-restricted L3T4$^+$ Th or addition of exogenous IL-2 during the *in vitro* sensitization. The L3T4$^+$ subset could not recognize FBL directly, and stimulation of these Th required Ia$^+$ macrophages to present tumor antigens in the context of class II determinants. Pretreatment of macrophages with lysozomotropic agents such as chloroquine prevented activation of the L3T4$^+$ cells, implying that biochemical processing was necessary for tumor antigen recognition. Thus, the T-cell response to FBL includes both class II-restricted, IL-2-producing, L3T4$^+$ Th that recognize processed tumor antigen and class I-restricted, Lyt-2^+ Tc that directly recognize and lyse FBL tumor cells.

The efficacy of each of these purified T-cell subsets was examined in adoptive therapy of FBL. Spleen cells depleted of Th had minimal activity in ACIT of disseminated FBL, whereas spleen cells depleted of cytolytic Lyt-2^+ T cells were nearly as effective on the basis of cell number as unfractionated immune spleen cells.[27] The observed efficacy of the purified L3T4$^+$ population was surprising, since this subset cannot directly recognize or lyse FBL tumor cells. However, these Th can secrete lymphokines which activate other effector cells. Although the above studies demonstrated that it was not necessary to transfer donor CTL, it remained possible that CTL were still required for tumor eradication since the donor Th could potentially secrete IL-2 and induce a host CTL response. Therefore, to exclude any possible contribution of host or donor CTL to tumor lysis, purified donor L3T4$^+$ T cells were infused into T-deficient ATXBM (adult, thymectomized, irradiated, and reconstituted with T-depleted bone marrow) hosts bearing disseminated FBL leukemia.[25] In this setting, treatment with cyclophosphamide and 10^7 immune L3T4$^+$ T cells was still found to cure 100% of mice (Table II). To confirm that CTL did not participate in tumor therapy, mice were sacrificed during and after the period of tumor elimination by ACIT and the T-cell compartment analyzed. Neither phenotypic Lyt-2^+ T cells nor functional CTL could be detected, but specifically immune L3T4$^+$ T cells capable of secreting lymphokines in response to FBL were present.[25] Thus, under circumstances in which no CTL could contribute to tumor eradication, transferred L3T4$^+$ T cells promoted complete elimination of disseminated leukemia. These studies emphasize the importance of T-cell-dependent mechanisms other than direct lysis by CTL for eliminating tumor cells.

TABLE II
Therapy with T-Cell Subsets of T-Deficient B6 Hosts Bearing Disseminated FBL

Treatment regimen[a]			Therapeutic outcome[b]	
Drug	Cells	IL-2	Median survival (days)	Percent surviving
—	—	—	10	0
—	Unfractionated immune T cells	—	10	0
—	—	+	10	0
CY	—	—	25	0
CY	—	+	25	0
CY	Unfractionated immune T cells	—	>80	92
CY	Immune L3T4$^+$ T cells	—	>80	100
CY	Immune Lyt-2$^+$ T cells	—	32	0
CY	Unfractionated cultured immune T cells	—	>80	56
CY	Cultured immune cytolytic Lyt-2$^+$ T cells	—	35	0
CY	Cultured immune cytolytic Lyt-2$^+$ T cells	+	>80	100

[a] T-deficient ATXBM hosts were inoculated with 5×10^6 FBL on day 0, and then left untreated or treated on day 5 with 180 mg/kg CY, CY and 10^7 immune donor cells of the designated type, or CY and 10^7 immune cells plus 2.4×10^3 U IL-2 daily on days 5–10. Immune cells were obtained from donors sensitized *in vivo* to FBL. Cultured immune cells were obtained by 5-day *in vitro* stimulation with FBL of these *in vivo* primed cells to induce cytolytic activity in the Lyt-2$^+$ population. T-cell subsets were obtained either by treatment with anti-Lyt-2 and complement to remove CTL, or anti-Lyt-1 and complement to remove helper cells.

[b] Mice were observed for mortality for 80 days. Each treatment group contained 9 to 12 mice.

The limited efficacy observed when purified CTL, that can directly recognize and kill FBL tumor cells, were used in ACIT to treat disseminated leukemia was unexpected. CTL have been shown to be effective in other therapy models,[28–32] and would be predicted to provide benefit in ACIT under properly defined conditions. Since previous studies have shown that donor T cells must proliferate and persist to be effective in ACIT,[17,24] we examined whether the minimal therapeutic activity observed following transfer of CTL might reflect a requirement by purified CTL for factors derived from Th such as IL-2 to survive and expand *in vivo* and to mediate a significant antitumor effect. At the time these studies were performed, anti-L3T4 was not available, and Th were depleted by treatment with anti-Lyt-1 and complement, which was shown to effectively remove helper T cells and leave residual Lyt-2$^+$ CTL.[27,33] Therefore, tumor-bearing mice were treated with cyclophosphamide and Lyt-2$^+$ CTL followed by administration of exogenous recombinant IL-2 for 6 consecutive days in doses previously shown to promote the *in vivo* proliferation of adoptively transferred cultured T cells.[34] ATXBM T-deficient mice were used as hosts for these studies so that host L3T4$^+$ Th could not contribute to the outcome of therapy. Therapy with cyclophosphamide and purified Lyt-2$^+$ CTL alone again mediated only a small antitumor effect, but treatment with cyclophosphamide and CTL plus IL-2 cured 100% of mice (Table II). Thus, class I-restricted Lyt-2$^+$ CTL can be effective in ACIT if adequate helper factors necessary for *in vivo* proliferation are provided. These results have recently been confirmed with an FBL-specific Lyt-2$^+$ cytolytic T-cell clone—the therapeutic activity of the clone in ACIT was markedly augmented by daily administration of IL-2 following cell transfer.

5. Activation of Macrophages to a Tumoricidal State by Immune T Cells

The studies described above, as well as supporting results in other tumor therapy models,[10,35] have focused attention on mechanisms of tumor eradication in addition to direct T-cell-mediated cytolysis. The secretion of lymphokines by immune T cells can result in the activation of a variety of effector populations, including stimulation of B cells to produce antibodies that either directly lyse tumors or function in antibody-dependent cell-mediated cytotoxicity, activation of NK cells, induction of lymphokine-activated killers, and/or activation of macrophages to a tumoricidal state.[36-41] Analysis of the infiltrate in tumors undergoing rejection has suggested that, under certain settings, each of these tumoricidal mechanisms might contribute to eradication of FBL in our tumor model.[42-44] However, the infusion of FBL-specific antibodies has had little effect in tumor therapy,[45] and FBL is resistant to NK-mediated lysis *in vitro*. Studies in several tumor models have suggested that tumoricidal macrophages may be of critical importance to the outcome of tumor therapy.[44,46-48] Therefore, we have focused our efforts on the potential role of tumoricidal macrophages.

The induction of macrophages with tumoricidal activity requires that T cells secrete macrophage-activating factors (MAF), such as γ-IFN. Therefore, T cells immune to FBL were stimulated *in vitro* with FBL and the supernatants were harvested and tested for MAF activity. Since lymphokine production is not an exclusive property of L3T4$^+$ Th, the ability of Lyt-2$^+$ T cells, as well as unfractionated T cells and L3T4$^+$ T cells, to produce MAF was determined. The presence of MAF was defined as the ability of these supernatants to activate macrophages to lyse the relevant FBL target (Table III). Unfractionated spleen cells immune to FBL produced supernatant with MAF activity following *in vitro* stimulation with FBL, and depletion of T cells with anti-Thy-1.2 ablated production of MAF. The L3T4$^+$ subset (obtained by depleting the Lyt-2$^+$ population) also produced MAF following stimulation with FBL. By contrast, the Lyt-2$^+$ subset (obtained by depleting the L3T4$^+$ population) did not make significant quantities of MAF following stimulation with FBL alone. Since IL-2 has been shown to enhance lymphokine secretion by CTL clones,[49] we stimulated the Lyt-2$^+$ population with FBL in the presence of IL-2. Although stimulation with neither FBL alone nor IL-2 alone induced MAF secretion, Lyt-2$^+$ T cells stimulated with both FBL and IL-2 secreted large amounts of MAF.

Since γ-IFN is the only well-characterized murine MAF,[50] we assessed whether these T-cell subsets were producing γ-IFN. This was accomplished by utilizing Northern blots with mRNA prepared from stimulated T-cell subsets, and hybridizing to the blots with a cDNA probe for γ-IFN. Results analogous to those found in the functional MAF assay were obtained. Stimulation with FBL of L3T4$^+$

TABLE III
Production of MAF by FBL-Immune T Cells

Culture for generating MAF-containing supernatants[a]				MAF activity[c]
Responder	Preculture depletion	Stimulator	Addition to culture[b]	Percent cytolysis
$B6_{\alpha FBL}$	—	—	—	2
$B6_{\alpha FBL}$	—	FBL	—	15
$B6_{\alpha FBL}$	αThy-1.2 + C'	FBL	—	2
$B6_{\alpha FBL}$	αLyt-2.2 + C'	FBL	—	11
$B6_{\alpha FBL}$	αL3T4 + C'	FBL	—	2
$B6_{\alpha FBL}$	αL3T4 + C'	—	IL-2	3
$B6_{\alpha FBL}$	αL3T4 + C'	FBL	IL-2	24

[a] Supernatants were harvested after 20-hr culture of 10^7 responder cells with 5×10^5 irradiated FBL cells. Responder spleen cells from FBL-primed mice were used either untreated or after depletion of selected cell populations with antibody and complement.
[b] rIL-2 was added at the initiation of culture to selected wells at a final concentration of 350 U/ml.
[c] Macrophage monolayers were cultured with the supernatant for 24 hr, washed, and then incubated with labeled FBL targets to determine cytolytic activity.

T cells resulted in production of mRNA for γ-IFN, whereas no γ-IFN message was detected from Lyt-2$^+$ T cells stimulated with FBL alone. However, when Lyt-2$^+$ T cells were stimulated with FBL in the presence of exogenous IL-2, large amounts of mRNA for γ-IFN were detected. These results not only confirm that tumoricidal macrophages may be an important *in vivo* effector mechanism in ACIT of disseminated FBL with purified L3T4$^+$ Th, but suggest that at least part of the therapeutic efficacy observed following therapy with purified CTL and IL-2 may result from secretion of lymphokines by these immune T cells and activation of non-T-cell effector mechanisms such as tumoricidal macrophages.

This demonstration that both noncytolytic L3T4$^+$ T cells and cytolytic Lyt-2$^+$ T cell can mediate their *in vivo* antitumor activity by similar or overlapping mechanisms may explain some of the disparities that have been reported using different tumor models with regards to the effector T cells necessary to eliminate tumor cells *in vivo*. Since most tumors do not express class II antigens, it would be predicted that L3T4$^+$ T cells, which will require Ia$^+$ antigen-presenting cells for activation, would most likely be effective against tumors that shed tumor antigens and are either localized to sites in the reticuloendothelial system or associated with an inflammatory infiltrate. In these settings, L3T4$^+$ should be readily activated and capable of inducing the effector mechanisms necessary for tumor lysis. Alternatively, class I-restricted Lyt-2$^+$ T cells, which can directly recognize tumors, may prove to be more effective that L3T4$^+$ T cells and/or necessary for tumor elimination in settings in which presentation of tumor antigens by Ia$^+$ antigen-presenting cells is limiting. However, it may be incorrect to assume that activated Lyt-2$^+$ T cells can completely eradicate tumors by direct cytolysis. The data presented above demonstrate that Lyt-2⁺ T cells can induce tumoricidal macrophages, and it remains to be determined if the participation of such additional effector mechanisms is a mandatory component of tumor eradication by this subset.

6. Adoptive Therapy with Immune T Cells Expanded by *in Vitro* Culture

Immunotherapy with adoptively transferred T cells is predicated on the development of methodologies for generating adequate numbers of therapeutically active effector cells. For the ACIT studies discussed so far, lymphocytes were obtained from syngeneic hosts immunized to the tumor on the supposition that under appropriate conditions similar immune T cells could be isolated from the tumor-bearing host and expanded by *in vitro* culture. Although improved *in vitro* methods should make it increasingly possible to achieve this in selected patients, the use of *in vitro* expanded T cells may pose certain problems not evident with adoptively transferred fresh immune spleen cells. In particular, long-term culture of T cells has been shown in some instances to result in alterations of the cell surface that cause abnormalities in homing *in vivo*.[51] Therefore, we examined whether FBL-specific T cells could be expanded *in vitro* and used in specific adoptive therapy.

Spleen cells from B6 mice immune to FBL were activated by *in vitro* stimulation for 5 days with FBL, and then induced to proliferate *in vitro* by adding exogenous IL-2 to culture during repeated cycles of stimulation with FBL.[52] As a specificity control, spleen cells immune to EL4, a non-cross-reactive lymphoma also of B6 origin, were similarly activated and grown in culture. After 19 days of culture with IL-2 supplementation, cultured T cells had expanded numerically greater than sevenfold but retained specificity for the appropriate FBL or EL4 tumor target as reflected by *in vitro* cytotoxic reactivity. FBL-reactive cultured T cells exhibited dose-dependent efficacy in ACIT of FBL: treatment with cyclophosphamide alone extended median survival to day 25; treatment with cyclophosphamide plus 5×10^6 cultured T cells prolonged median survival to day 42 and cured 20% of mice; and treatment with cyclophosphamide plus 2×10^7 cultured T cells cured 80% of mice.[52] Cultured T cells retained specificity in ACIT, with FBL-reactive T cells only being effective in ACIT of disseminated FBL and EL4-reactive T cells only being effective in therapy of EL4.[52] Thus, T cells expanded by *in vitro* culture were effective in specific adoptive therapy for the treatment of disseminated tumors.

One limitation of this approach is the technical problem of expanding specifically immune T cells that may initially be present in very low frequency to large enough numbers for adoptive therapy. Since cultured T cells proliferate rapidly *in vitro* in response to exogenous IL-2 and mediate a dose-dependent effect in tumor therapy, we examined whether the *in vivo* administration of exogenous IL-2 to hosts after cell transfer could induce the *in vivo* proliferation and thereby augment the therapeutic efficacy of a relatively small dose of cultured T cells. For these studies, cultured donor T cells of B6 origin were transferred along with irradiated

FBL tumor antigen into congenic B6/Thy-1.1 hosts, and the donor Thy-1.2$^+$ T cells were enumerated following different schedules and doses of IL-2.[34] In hosts not receiving IL-2, most donor T cells died within 8 days of transfer. However, daily administration of IL-2 following cell transfer induced proliferation and accumulation of donor T cells in the peritoneal cavity, spleen, and lymph nodes. The growth of donor T cells was proportional to the dose of IL-2 infused, with doubling times of 12 hr achievable *in vivo*.[34] IL-2 pharmacokinetic studies have now been performed, and the results suggest that regimens which maintain a persistent detectable serum level (i.e., continuous infusion or intermittent depot injections) will be the most effective regimens for promoting growth of transferred T cells.[53] It should be emphasized that the doses of IL-2 which induced the proliferation of transferred cultured donor T cells were substantially below the dose that induces host toxicity or the dose necessary to induce LAK cells and no proliferative effect in normal (i.e., noncultured) host T cells was detectable.

The effect of IL-2, in doses which induce *in vivo* proliferation of transferred cultured T cells, on the efficacy of cultured T cells in therapy was examined.[54] B6 mice bearing disseminated FBL and treated on day 5 with cyclophosphamide alone or cyclophosphamide plus administration of exogenous IL-2 daily from days 5 to 10 had a median survival of 25 days. Treatment with cyclophosphamide plus 5×10^6 cultured T cells prolonged median survival to day 35. By contrast, therapy with cyclophosphamide plus 5×10^6 cultured T cells followed by administration of IL-2 on days 5 to 10 cured 75% of mice.[54] Thus, administration of exogenous IL-2 induced *in vivo* proliferation and augmented the therapeutic efficacy of transferred cultured T cells. Presumably, continuous infusion of IL-2 or administration of IL-2 more frequently and for longer periods would permit the use of even lower doses of cultured T cells in tumor therapy.

The long-term culture of some T-cell lines in the presence of supplemented exogenous IL-2 usually results in a dependency on exogenous IL-2 for survival and frequently with some loss of antigen specificity. Therefore, we attempted to circumvent these problems by adapting methods that have been described for the generation of antigen-driven T-cell lines without supplemented IL-2 to the generation of FBL-specific T-cell lines.[55] Spleen cells from FBL-primed mice were stimulated *in vitro* with FBL, and then were propagated by intermittent stimulation with FBL tumor and irradiated filler cells. Lines generated in this fashion were found to contain both Th and CTL reactive with FBL. Moreover, these FBL-specific T-cell lines produced adequate endogenous IL-2 upon stimulation with FBL to support *in vitro* proliferation and numerical expansion, were specifically cytotoxic to FBL targets, and could be rested *in vitro* without stimulation for greater than 1 month and still remained viable and antigen-specific.[56] After expansion of such an FBL-specific, antigen-driven B6 T-cell line for 62 days *in vitro*, the cells were tested in adoptive therapy of B6/Thy-1.1 hosts, and the fate of the cultured donor T cells after adoptive transfer was assessed.[56] These donor T cells proliferated rapidly *in vivo* in response to stimulation with FBL, and the growth rate could be augmented by the administration of exogenous IL-2 after cell transfer. Additionally, treatment of tumor-bearing hosts with cyclophosphamide and a cell dose of only 3.3×10^6 cultured T cells cured 100% of mice. Persistent donor T cells were detectable in the spleen, mesenteric and axillary lymph nodes of host mice 120 days after curative therapy, and these cells retained the capacity to specifically respond to FBL. Consistent with the observed capacity of exogenous IL-2 to increase the *in vivo* growth rate of these cultured cells, mice treated with cultured T cells and IL-2 contained a greater number of persistent FBL-reactive donor T cells at day 120 than mice not receiving IL-2.[56] Thus, FBL-specific T cells generated by these culture methods were effective in therapy of disseminated leukemia, proliferated *in vivo* in response to antigen, distributed widely in the host, responded to exogenous IL-2 *in vivo,* and persisted and retained specific immunologic function for long periods after adoptive transfer. The ability of these antigen-specific T-cell lines that have been expanded *in vitro* to function *in vivo* similar to noncultured immune T cells suggests that the generation of such T cells reactive to human tumors would have significant therapeutic potential.

7. Conclusions and Future Directions

The purpose of these studies has been to identify the requirements for effective adoptive therapy and to determine the mechanisms by which transferred T cells mediate tumor eradication. Additionally, we have tried to demonstrate the feasibility of using tumor-specific T cells that have been expanded by long-term *in vitro* culture for tumor therapy. Precisely how and when these principles can be applied to the treatment of cancer patients remain to be determined. At the present time, the major obstacle to human adoptive immunotherapy is a reproducible method for the identification and expansion of tumor-specific T cells. The existence of this problem does not necessarily imply that human tumors fail to express potentially immunogenic determinants. It should be recognized that, even with highly immunogenic transplantable animal tumors, T-cell reactivity to the tumor is rarely detectable in the host after the tumor has achieved an advanced state. Therefore, the failure to detect tumor reactivity in cancer patients under similar circumstances may reflect a low frequency of reactive T cells and immunosuppression induced by the advanced tumor that inhibits activation of effector cells. The further development of methods for isolating potential effector cells and/or activating such cells in limiting dilution even if suppressor cells are present, as done in other antigen systems, may prove useful for uncovering tumor reactivity. The technique of isolating tumor-infiltrating lymphocytes may permit detection of tumor reactivity that is not evident in peripheral blood T cells. Moreover, the continued development of *in vitro* methods for purifying antigens from the membranes of tumor cells and presenting them in potentially immunogenic matrices rather than on intact tumor cells, which may not be good stimulators due to issues such as low antigen density and release of suppressor or toxic molecules should enhance the ability to generate T cells reactive with tumors. The increased frequency with which retroviruses are being associated with human malignant disease and the findings of expression of abnormal oncogene products in some tumors support this bias that at least a fraction of human tumors will express determinants that can serve as targets for attack by specifically immune T cells. Thus, it is hoped that, with further elucidation of the immunobiology of T cell–tumor interactions and advances in culture technology, the therapeutic potential of specific adoptive immunotherapy for the treatment of human malignancies can be realized.

ACKNOWLEDGMENTS. Supported by Grants CA-30558 and CA-33084 from the National Cancer Institute, DHHS, and by Grant IM-304 from the American Cancer Society.

References

1. Zarling, J. M., and Bach, F. H., 1979, Continuous culture of T cells cytotoxic for autologous human leukemia cells, *Nature* **280**:685.
2. Lee, S. K., and Oliver, R. T. D., 1978, Autologous leukemia-specific T cell-mediated lymphocytotoxicity in patients with acute myelogenous leukemia, *J. Exp. Med.* **147**:912.
3. Vose, B. M., and Bonnard, G. D., 1982, Human tumor antigens defined by cytotoxic and proliferating responses of cultured lymphoid cells, *Nature* **296**:359.
4. Vose, B. M., and Bonnard, G. D., 1982, Specific cytotoxicity against autologous tumor and proliferative responses of human lymphocytes grown in IL 2, *Int. J. Cancer* **29**:33.
5. Mitsuya, H., Matis, L., Megson, M., Bunn, P., Murray, C., Mann, D., Gallo, R., and Broder, S., 1983, Generation of an HLA-restricted cytotoxic T-cell line reactive against cultured tumor cells from a patient infected with human T-cell leukemia/lymphoma virus, *J. Exp. Med.* **158**:994.
6. Vanky, F., Gorsky, T., Gorsky, Y., Masucci, M. G., and Klein, E., 1982, Lysis of tumor biopsy cells by autologous T lymphocytes activated in mixed cultures and propagated with T cell growth factor, *J. Exp. Med.* **155**:83.
7. Spitalny, G. L., and North, R. J., 1977, Subversion of host defense mechanisms by malignant tumors: An established tumor as a privileged site for bacterial growth, *J. Exp. Med.* **145**:1264.
8. Berendt, M. J., and North, R. J., 1980, T-cell-mediated suppression of antitumor immunity: An explanation for progressive growth of an immunogenic tumor, *J. Exp. Med.* **151**:59.
9. Dye, E. S., and North, R. J., 1981, T cell mediated immunosuppression as an obstacle to adoptive immunotherapy of the P815 mastocytoma and its metastases, *J. Exp. Med.* **154**:1033.
10. Fernandez-Cruz, E., Woda, B. A., and Feldman, J. D., 1980, Elimination of syngeneic sarcomas in rats by a subset of T lymphocytes, *J. Exp. Med.* **152**:823.
11. North, R. J., 1982, Cyclophosphamide-facilitated adoptive immunotherapy of an established tumor depends on elimination of tumor-induced suppressor T cells, *J. Exp. Med.* **55**:1063.
12. Perry, L. L., Benacerraf, B., McCluskey, R. T., and Greene, M. I., 1978, Enhanced syngeneic tumor destruction by *in vivo* inhibition of suppressor T cells using *anti*-IJ alloantiserum, *Am. J. Pathol.* **92**:491.
13. Eardley, D. D., Hugenberger, J., McVay-Boudreau, L., Shen, F. W., Gershon, R. K., and Cantor, H., 1978, Immunoregulatory circuits among T-cell sets. I. T-helper cells induce other T-cell sets to exert feedback inhibition, *J. Exp. Med.* **147**:1106.
14. Cantor, H., McVay-Boudreau, L., Hugenberger, J., Naidorf, K., Shen F. W., and Gershon, R. K., 1978, Immunoregulatory circuits among T-cell sets. II. Physiologic role of feedback inhibition *in vivo*: Absence in NZB mice, *J. Exp. Med.* **147**:1116.
15. Cantor, H., Hugenberger, J., McVay-Boudreau, L., Eardley, D. D., Kemp, J., Shen, F. W., and Gershon, R. K., 1978, Immunoregulatory circuits among T-cell sets: Identification of a subpopulation of T-helper cells that induces feedback inhibition, *J. Exp. Med.* **148**:871.
16. Treves, A. J., Cohen, I. R., and Feldman, M., 1975, Immunotherapy of lethal metastases by lymphocytes sensitized against tumor cell *in vitro*, *J. Natl. Cancer Inst.* **54**:777.
17. Greenberg, P. D., Cheever, M. A., and Fefer, A., 1980, Detection of early and delayed anti-tumor effects following curative adoptive chemoimmunotherapy of established leukemias, *Cancer Res.* **40**:4428.

18. Berenson, J. R., Einstein, A. B., and Fefer, A., 1975, Syngeneic adoptive immunotherapy of a Friend leukemia: Requirement for T cells, *J. Immunol.* **115**:234.
19. Cheever, M. A., Greenberg, P. D., and Fefer, A., 1980, Specificity of adoptive chemoimmunotherapy of established syngeneic tumors, *J. Immunol.* **125**:711.
20. Greenberg, P. D., Cheever, M. A., and Fefer, A., 1981, H-2 restriction of adoptive immunotherapy of advanced tumors, *J. Immunol.* **126**:2100.
21. Kolb, J. P., Poupon, M. F., Lespinats, G. M., Sabolovic, D., and Loisillier, F., 1977, Splenic modifications induced by cyclophosphamide in C3H/He, nude, and "B" mice, *J. Immunol.* **118**:1595.
22. Stockman, G. D., Heim, L. R., South, M. A., and Trentin, J. J., 1973, Differential effects of cyclophosphamide on the B and T cell compartments of adult mice, *J. Immunol.* **110**:277.
23. Zaleski, M., and Klein, J., 1978, Genetic control of the immune response to Thy-1 antigens, *Immunol. Rev.* **38**:120.
24. Greenberg, P. D., and Cheever, M. A., 1984, Treatment of disseminated leukemia with cyclophosphamide and immune cells: Tumor immunity reflects long-term persistence of tumor-specific donor T cells, *J. Immunol.* **133**:3401.
25. Greenberg, P. D., Kern, D. E., and Cheever, M. A., 1985, Therapy of disseminated murine leukemia with cyclophosphamide and immune Lyt 1^+2^- T cells: Tumor eradication does not require participation of cytotoxic T cells, *J. Exp. Med.* **161**:1122.
26. Kern, D. E., Klarnet, J. P., Jensen, M. C. V., and Greenberg, P. D., 1986, Requirement for recognition of class II molecules and processed tumor antigen for optimal generation of syngeneic tumor-specific class I-restricted CTL, *J. Immunol.* **136**:4303.
27. Greenberg, P. D., Cheever, M. A., and Fefer, A., 1981, Eradication of disseminated murine leukemia by chemoimmunotherapy with cyclophosphamide and adoptively transferred immune syngeneic Lyt 1^+2^- lymphocytes, *J. Exp. Med.* **154**:952.
28. Rosenstein, M. T., Eberlein, T. J., and Rosenberg, S. A., 1984, Adoptive immunotherapy of established syngeneic solid tumors: Role of T lymphoid subpopulations, *J. Immunol.* **132**:2117.
29. Evans, R., 1983, Combination therapy by using cyclophosphamide and tumor-sensitized lymphocytes: A possible mechanism of action. *J. Immunol.* **130**:2511.
30. Dailey, M. O., Pillemer, E., and Weissman, I. L., 1982, Protection against syngeneic lymphoma by a long-lived cytotoxic T-cell clone, *Proc. Natl. Acad. Sci. USA* **79**:5384.
31. Mills, C. D., and North, R. J., 1983, Expression of passively transferred immunity against an established tumor depends on generation of cytolytic T cells in recipients: Inhibition by suppressor T cells, *J. Exp. Med.* **157**:1448.
32. Mills, C. D., and North, R. J., 1985, Lyt 1^+2^- suppressor T cells inhibit the expression of passively transferred antitumor immunity by suppressing the generation of cytolytic T cells, *Transplantation* **39**:202.
33. Greenberg, P. D., 1986, Therapy of murine leukemia with cyclophosphamide and immune Lyt 2^+ T cells: Cytolytic T cells can mediate eradication of disseminated leukemia, *J. Immunol.* **136**:1917.
34. Cheever, M. A., Greenberg, P. D., Irle, C., Thompson, J. A., Urdal, D. L., Mochizuki, D. Y., Henney, C. S., and Gillis, S., 1984, Interleukin 2 administered *in vivo* induces the growth of cultured T cells *in vivo*, *J. Immunol.* **132**:2259.
35. Fujiwara, H., Fukuzawa, M., Yoshioka, T., Nakajima, H., and Hamaoka, T., 1984, The role of tumor-specific Lyt 1^+2^- T cells in eradicating tumor cells *in vivo*. 1. Lyt 1^+2^- T cells do not necessarily require recruitment of host's cytotoxic T cell precursors for implementation of *in vivo* immunity, *J. Immunol.* **133**:1671.
36. Badger, C. C., and Bernstein, I. D., 1983, Therapy of murine leukemia with monoclonal antibody against a normal differentiation antigen, *J. Exp. Med.* **157**:828.
37. Schulz, G., Staffileno, L. K., Reisfeld, R. A., and Dennert, G., 1985, Eradication of established human melanoma tumors in nude mice by antibody-directed effector cells, *J. Exp. Med.* **161**:1315.
38. Hanna, N., 1982, Role of natural killer cells in control of cancer metastases, *Cancer Metastasis Rev.* **1**:45.
39. Mazumder, A., and Rosenberg, S. A., 1984, Successful immunotherapy of natural killer resistant established pulmonary melanoma metastases by the intravenous adoptive transfer of syngeneic lymphocytes activated *in vitro* by IL 2, *J. Exp. Med.* **159**:495.

40. Russell, S. W., and McIntosh, A. T., 1977, Macrophages isolated from regressing Moloney sarcomas are more cytotoxic than those recovered from progressing sarcomas, *Nature* **268**:69.
41. Evans, R., 1982, Macrophages in neoplasms: New insights and implication in tumor immunobiology. *Cancer Metastasis Rev.* **1**:227.
42. Herberman, R. B., Holden, H. T., Varesio, L., Taniyama, T., Pucetti, P., Kirchner, H., Gerson, J., White, S., Keisari, Y., and Haskill, J. S., 1980, Immunologic reactivity of lymphoid cells in tumors, *Contemp. Top. Immunobiol.* **10**:61.
43. Key, M., and Haskill, J. S., 1981, Immunohistologic evidence for the role of antibody and macrophages in regression of the murine T1699 mammary adenocarcinoma, *Int. J. Cancer* **28**:225.
44. Evans, R., 1984, Phenotypes associated with tumor rejection mediated by cyclophosphamide and syngeneic tumor-sensitized T lymphocytes: Potential mechanisms of action, *Int. J. Cancer* **33**:381.
45. Fefer, A., 1969, Immunotherapy and chemotherapy of Moloney sarcoma virus-induced tumors in mice, *Cancer Res.* **29**:2177.
46. Montovani, A., 1978, Effects on *in vitro* tumor growth of murine macrophages isolated from sarcoma lines differing in immunogenicity and metastasizing capacity, *Int. J. Cancer* **22**:741.
47. Russell, S. W., Gillespie, G. Y., and Pace, J. L., 1980, Evidence for mononuclear phagocytes in solid neoplasms and appraisal of their nonspecific cytotoxic capabilities, *Contemp. Top. Immunobiol.* **10**:143.
48. Fidler, I. J., and Poste, G. 1981, Macrophage-mediated destruction of malignant tumor cells and new strategies for the therapy of metastatic disease, *Springer Semin. Immunopathol.* **5**:161.
49. Kelso, A., Glasebrook, A. L., Kanagawa, O., and Brunner, K. T., 1982, Production of macrophage-activating factor by T lymphocyte clones and correlation with other lymphokine activities, *J. Immunol.* **129**:550.
50. Pace, J. L., Russell, S. W., Torres, B. A., Johnson, H. M., and Gray, P. W., 1983, Recombinant mouse γ-interferon induces the priming step in macrophage activation for tumor cell killing, *J. Immunol.* **130**:2011.
51. Dailey, M. O., Fathman, C. G., Butcher, E. C., Pillemer, E., and Weissman, I., 1982, Abnormal migration of T lymphocyte clones, *J. Immunol.* **128**:2134.
52. Cheever, M. A., Greenberg, P. D., and Fefer, A., 1981, Specific adoptive therapy of established leukemia with syngeneic lymphocytes sequentially immunized *in vivo* and *in vitro* and non-specifically expanded by culture with interleukin 2, *J. Immunol.* **126**:1318.
53. Cheever, M. A., Thompson, J. A., Kern, D. E., and Greenberg, P. D., 1985, Interleukin 2 (IL 2) administered *in vivo:* Influence of IL 2 route and timing on T-cell growth, *J. Immunol.* **134**:3895.
54. Cheever, M. A., Greenberg, P. D., Fefer, A., and Gillis, S., 1982, Augmentation of the anti-tumor therapeutic efficacy of long-term cultured T lymphocytes by *in vivo* administration of purified interleukin 2, *J. Exp. Med.* **155**:968.
55. Fathman, C. G., and Fitch, F. W., 1984, Long-term culture of immunocompetent cells, in: *Fundamental Immunology* (W. E. Paul, ed.), Raven Press, New York, pp. 781–795.
56. Cheever, M. A., Britzmann-Thompson, D., Klarnet, J. P., and Greenberg, P. D., 1986, Antigen-driven long-term cultured T cells proliferate *in vivo*, distribute widely, mediate specific tumor therapy and persist long-term as functional memory T cells, *J. Exp. Med.* **163**:1100.

Index

A10, see T-cell hybridoma (A10)
A23187, 167–168
Abelson leukemia virus (A-MuLV), 123
　Abelson protein and, see Abelson protein
　erythroid cells and, interaction with, 134
　future directions, 138
　genome structure, 124
　hematopoietic cell interaction, 130–135
　infection and IL-3, 134–135
　target cells, analysis of, 133–134
　transformed lymphoid cells
　　differentiation markers expressed by, 131–132
　　Ig gene structure in, 132–133
　tumors induced by, 135
　viral genome expression, 124
Abelson protein
　function, 125
　structure, 124–125
　and transformation, 125–126; see also Transformation
abl gene, isolation of, 123; see also c-abl gene
ACIT, see Adoptive chemoimmunotherapy (ACIT)
Acute lymphocytic leukemia (ALL), cytogenetics of, 6–8
Acute myelogenous leukemia (AML), cytogenetics of, 10
Acute myelomonocytic leukemia (AMMOL), cytogenetics of, 10–11
Acute nonlymphocytic leukemias, cytogenetics of, 10–11
Acute promyelocytic leukemia (APL), cytogenetics of, 10
ADCC, see Antibody-dependent cellular cytotoxicity (ADCC)
Adenovirus, class 1 gene expression alterations and, 206

Adoptive chemoimmunotherapy (ACIT)
　described, 432
　and disseminated tumor elimination time period, 433–434
　efficacy in, 433
　T cells and
　　characteristics, 435–438
　　immune, 440
　　mediation by, 432–435
Adoptive therapy, see also Adoptive chemoimmunotherapy (ACIT)
　with immune T cells, 440–442
　of large tumors, 431–432
Adult T-cell leukemia (ATL), HTLV-1 involvement in, 187–188, 189–190
AED, see N-Iodoacetyl-N'-(5-sulfonic-1-naphthyl) ethylene diamine (AED)
AEV, see Avian erythroblastosis virus (AEV)
ALL, see Acute lymphocytic leukemia (ALL)
Allograft rejection, Lyt-1^+2^- T cell subset role in, 346
Amino acid sequence
　of EGF receptor, 95–97
　of erb-B-related proteins, 95–97
　of hCG, 395
AML, see Acute myelogenous leukemia (AML)
AMLR, see Autologous mixed lymphocyte reaction (AMLR)
AMMOL, see Acute myelomonocytic leukemia (AMMOL)
A-MuLV, see Abelson leukemia virus (A-MuLV)
Anchorage-independent growth, 75
Anti-asialo GM_1 serum
　large granular lymphocyte removal and, 262–264
　Nk depletion by, 264–265
Antibodies
　binding to cell-surface macromolecules, 72–73

Antibodies (*cont.*)
  of 212 cell line, 255–256
  monoclonal, *see* Monoclonal antibodies (MAbs)
Antibody-dependent cellular cytotoxicity
    (ADCC), 76, 77–78
Antigenic change, 317
  antigenicity and transformation comparison,
    320–322
  TSTA molecular analysis and, 318–320
  UV-induced and other tumors compared,
    317–318
Antigenicity
  and transformation, relationship between,
    320–322
  of tumors, 308
  UV-radiation induced, UV-induced suppressor
    T cell effect on, 321–322
Antigens, *see also* Antigenicity
  blood group A, biosynthetic pathway, 402
  helper T cells against, 357
  Ia, *see* Ia antigens
  immunogenic surface, 429–430
  on melanoma cells, 373, *see also* Melanoma
    differentiation, 374–377
  middle T (MTAg), 149–151
  M-MuLV
    cell surface expression of, 222–223, 227
    regulation of, 226, 227–228
  proteoglycan, on melanoma cells, 411, 414
  SV40, *see* SV40 tumor antigen
  tumor-specific versus tumor-associated, 279,
    see also Tumor-specific transplantation
    antigen (TSTA) system
  on UV tumors
    biochemically defined, 315–316
    changes in, 317–322, *see also* Antigenic
      change
    cytotoxic lymphocyte-defined, 312–313
    suppressor T lymphocyte-defined, 314–315
    transplantation-defined, 310–312
Anti-TsF antisera
  P815, effect on tumor growth, 282
  preparation of, 282–283
Antitumor effector T cells, 356–357
Antitumor immune response
  ACIT and, 434–435
  suppressor T cells as down-regulators of, 301–
    303
  T cell–T cell interaction and, enhanced
    induction, 359–361
APL, *see* Acute promyelocytic leukemia (APL)
Arginyl-glycyl-L-aspartyl-L-seryl (RGDS), and
    cell-surface receptors, 422–424
Asparagine-linked sugar chains, 387
  complex type, 390

Astrocytes, c-*src* gene product in, 152
ATL, *see* Adult T-cell leukemia (ATL)
Autoimmune diseases, Lyt-1^+2^- T-cell subset
    role in, 346
Autologous mixed lymphocyte reaction (AMLR),
    269
Avian erythroblastosis virus (AEV)
  genomic organization of different strains, 94
  v-*erb-B* gene of, 94–95
    implications, 103–106

B-cell tumors, cytogenetics of, 5–6
bcr/c-abl, in CBL, 137–138
Burkitt's lymphoma
  c-*myc* gene in
    deregulation, 29–30
    structure, 21–25
  cytogenetics of, 3–5
Bystander tumor growth inhibition, 342–343

C350 MAb, melanoma cells and, 376–377
Ca^{2+}, *see* Calcium (Ca^{2+})
c-*abl* gene, 136
  bcr/, product of, 137–138
  expression
    in chronic myelogenous leukemia (CML),
      137
    in normal cells, 136
  future directions, 138
  product of, 136–137
c-*abl* protooncogene, 123
Calcium (Ca^{2+})
  mobilization, and protein kinase C pathway,
    169–170
  and phorbol esters, synergism between, 167–
    168
cAMP second messenger system, and protein
    kinase C pathway, 168–169
Cancer
  aberrant *erb-B* protein expression in, 102–103
  cellular oncogene amplification in, 104
Carboxy-terminal region, of EGFR, 115
CAT, *see* Chloramphenicol acetyltransferase
    (CAT)
Cell attachment
  ganglioside role in, models for, 422–424
  melanoma and neuroblastoma, to extracellular
    matrix components, 415–422
212 cell line
  antibody plus complement lysis for, 255–
    256
  killing by allogeneic cytotoxic T lymphocytes
    (CTL), 256, 257
Cell-mediated lympholysis (CML), RadLV and,
    204–205

Cells, cancerous, aberrant erb-B protein expression in, 102–103
Cell-surface p185, down-modulation, by anti-p185 MAbs, 72–74
Cell-surface receptors
 features of, 60–61
 for fibronectin, gangliosides as, 408–410
 oncogene encoding, 60
  neu, 63–64
  v-erb-B, 62
  v-fms, 62–63
 RGDS and, 422–424
Cellular cytotoxicity, antibody-dependent (ADCC), 76
Cellular differentiation, c-myc gene role in, 27–28
Cellular homologue, of erb-B protein, 98–99
Cellular immune reactions, surface T-ag involvement, 235–236
Cellular processing, EGF-receptor complex fate in, 116
Cellular transformation, c-myc gene role in, 25–26
c-erb-B-2 gene, EGFR and, 99–100
c-erb-B-2 protein
 phosphorylation mechanism by C-kinase, 102, 103
 schematic illustration of, 98
 structure, 100–101
 and tyrosine kinase activity, 101
CF21 MAb, melanoma cells and, 376–377
CFA, see Complete Freund's adjuvant (CFA)
c-fms gene, 86
 and FeLV, recombination between, 87
 future directions, 87–88
 regulatory control region, alterations to, 88
c-fms protooncogene product
 CSF-1 and, 85–86
 and its ligand, 83–84
c-H-ras expression, NK sensitivity mediated by, 248–255
Chloramphenicol acetyltransferase (CAT) activities
 LTR directed, 193–194
 with SV40 promoter, 197–198
 pLTR-, deletion mutations in, 196
Choriocarcinoma, hCG produced in, sugar chains of, 392–401
Chromosome alterations, see also Chromosome translocations
 in human leukemia and lymphoma, 11–12
 in human solid tumors, 12–14
Chromosome translocations
 Burkitt's lymphoma, 3–5
 c-myc gene breakpoint in, 23–25

Chromosome translocations (cont.)
 in leukemia
  acute lymphocytic (ALL), 6–8
  acute nonlymphocytic, 10–11
  nonlymphocytic, 9–10
 other B-cell lymphomas, 5–6
 T-cell tumors, 8–9
Chronic granulocytic leukemia, tumor progression in, 2
Chronic myelogenous leukemia (CML), 123
 c-abl gene expression in, 137
CHX, see Cycloheximide (CHX)
C-kinase, erb-B protein phosphorylation by, 101–102, 103
Clones
 evolution of, 2
 H-2 cosmid, 211
 K19 and K11, cytolytic T lymphocytes (CTL), 237–238
 NK, see NK clones
 from T-lymphocytes, protooncogene expression in, 48–49
CML, see Cell-mediated lympholysis (CML); Chronic myelogenous leukemia (CML)
c-myc gene, 21
 in Burkitt's lymphoma
  deregulation, 29–30
  structure, 21–25
 cellular differentiation, role in, 27–28
 cellular transformation, role in, 25–26
 expression regulation of, 29
 function, 25
 mitogenic response and, role in, 26–27
 promoter, 33
 similar genes, 28
 transcriptional regulation of, 30–33
c-myc protein, structure and function of, 28
Colony-stimulating factor-1 (CSF-1)
 binding by v-fms gene, 84–86
 gene, 86
 macrophage, 84
 receptor (CSF-IR), features of, 60–61
Complement lysis, of 212 cell line, 255–256
Complete Freund's adjuvant (CFA), 286, 288–289
Complex type sugar chains, 388–389
c-onc gene, analyses of, 123
c-onc protooncogenes, v-onc genes versus, 145
Concomitant immunity
 decay of, 297
 associated with suppressor T cell generation, 298–301
 progressive tumor and, unsuccessful response to, 296–298
 and secondary tumors, 298

Concomitant immunity (cont.)
T cells and, 297–298
down-regulation, 301–303
triggering of, 297
CSF-1, see Colony-stimulating factor (CSF-1)
CSF-1 gene, 86
c-src gene products, cell-type-specific expression of, 151–153
CTB6-4A, protooncogene expression in, 48–49
CTL, see Cytolytic T lymphocytes (CTL); Cytotoxic T lymphocytes (CTL)
Cycloheximide (CHX), and protooncogenes
expression regulation, 49–54
levels in PBL, 44–45
Cytogenetics
of acute nonlymphocytic leukemias, 10–11
of lymphomas, see Lymphomas, cytogenetics of
of malignant melanoma, 12–13
of neuroblastoma, 13–14
and oncogenes, 2–3
of small cell carcinoma, 14
of T-cell tumors, 8–9
of tumor progression, 1–2
Cytokines, production by LGL, 273
Cytolysis
of NK clones, 253–254
of ras-transfected cell lines, 251–252
Cytolytic T lymphocytes (CTL), see also Cytotoxic T lymphocytes (CTL)
clones K19 and K11, 237–238
disseminated leukemia and, 436–438
and immunity mediation, 296
RadLV and, 205–206
-reactive sites
fine mapping of, 237–238
localization, 236–237
surface T-ag and, 235–236
SV40-specific, 236
Cytoplasmic domain, of epidermal growth factor receptor (EGFR), 113–116
Cytotoxic T lymphocytes (CTL), see also Cytolytic T lymphocytes (CTL)
antihapten T-cell responses and, 357–358
envelope glycoprotein recognition, 223–224
gag gene recognition by, 224–225
generation
in cellular immune response, 221
in NK-depleted spleen cells, defective, 264–265
in NK-depleted spleen cells, restoration of, 265–267
restored by IFN and IL-2, 268, 269, 270
212 killing by, 256, 257
and Lyt-1^+2^- T cells, 332, 334–335
overriding by TsC, 281

Cytotoxic T lymphocytes (CTL) (cont.)
and P815 specificity, 290–291
response to M-MuLV:MSV complex, 222
target recognition structures, M-MuLV antigens as, 222–225
and TATA heterogeneity, 359–361
UV-induced tumor antigens and, 312–313
Cytotoxicity, cellular, antibody-dependent (ADCC), 76

Delayed-type hypersensitivity (DTH) response
haptens and, 357–358
and tumor protection correlation, 335–336
Diacylglycerols
and phorbol ester relationship
in protein kinase C activation, 165–167
structural analogy, 174–175
and protein kinase C activation, 158–159
Differentiation
antigens, melanoma and melanocytes, 374–377
cellular, c-myc gene role in, 27–28
Diffusion chamber
construction and implantation of, 337
double, 339–340
tumor growth inhibition in
by Lyt-1^+2^- T cells, 338–339
mechanism, 341
specificity, 338
time course, 337
Disseminated leukemia, recognition and eradication of, T-cell responsibility, 435–438
DNA
in Burkitt's lymphomas, 22–23
cancer induction and, 59
class I, RadLV tumorigenesis and, 209
H-2, RadLV association and, 207, 209
repetitive sequences in gene expression regulation, 216–217
replication suppression in SMLR, 272
and SV40 infection, 231–232
synthesis in T-lymphocytes, kinetics of, 47–48
Double diffusion chamber, 339–340
mechanism of tumor growth inhibition, 341
DTH, see Delayed-type hypersensitivity (DTH) response

Effector cells, NK cells and
characteristics, 254
spleen cells as, 252
Effector T cells, assays used to measure production of, 303
EGFR, see Epidermal growth factor receptor (EGFR)

INDEX 451

Envelope glycoprotein, cytotoxic T lymphocytes (CTL) recognition of, 223–224
Epidermal growth factor (EGF), 111
  erb-B encoding of, 98–99
  ^{125}I labeled, 116
Epidermal growth factor receptor (EGFR)
  aberrant erb-B protein expression and, 102–103
  amino acid sequence, 95–97
  cell physiology, 115–116
  and c-erb-B-2 gene, 99–100
  complex, in cellular processing, 116
  cytoplasmic, 113–116
  detection of, 111
  extracellular, 113
  features of, 60–61
  isolation of, 112–113
  mechanisms of action, 117
  schematic illustration, 98
  structural features, 113, 114
  and TGF-α, 111–112
  and v-erb-B gene, 93
erb-B gene
  c-, EGFR and, 99–10
  v-, see v-erb-B gene
erb-B protein, see also c-erb-B protein
  aberrant expression in cancerous cells, 102–103
  amino acid sequence of, 95–97
  cellular homologue of, 98–99
  phosphorylation of, 101–102, 103
  schematic illustration, 98
Erythroblastosis, transforming ability of erb-B gene of AEV, 94–95
Erythroid cells, A-MuLV interaction with, 134
Extracellular domain, of epidermal growth factor receptor (EGFR), 113
Extracellular matrix components, cell attachment inhibition to, 415–422

FBL leukemia
  ACIT model and, 432–435
  T-cell nature and, 435–438
Feline leukemia virus (FeLV)
  and c-fms gene, recombination between, 87
  and c-fms protooncogene sequences, 82
Feline sarcoma virus (SM-FeSV), v-fms oncogene associated with 62, 81–83
  transformation, 84–86
FeLV, see Feline leukemia virus (FeLV)
Fibroblasts
  altered sensitivity to NK cells, methylcholanthrene (MCA) and, 257
  c-src gene product in, 151–152
  EGFR's in, 115

Fibroblasts (cont.)
  enhanced NK killing of, 245–248
  growth events of, 56
  infection by M-MuLV:MSV complex, 226–227
  mitogens for, 161–162
  protooncogene role in, 55
  transformation
    region I role in, 126–127
    region IV lethal effect in, 127
Fibronectin, gangliosides as cell receptors for, 408–410
Fine mapping studies, of cytolytic T lymphocyte (CTL) reactive sites, 237–238
fms gene, see c-fms gene; v-fms gene
Focal adhesion plaques, of melanoma cells, GD_2 and GD_3 in, 410–414

gag gene
  of Moloney murine leukemia virus (M-MuLV), 223
  products, cytotoxic T lymphocyte (CTL) recognition of, 224–225
Ganglio-N-tetraosylceramide (asialo GM_1)
  LGL removal and, 262
  LGL removal and, see also Anti-asialo GM_1
Gangliosides
  in cell attachment
    melanoma, 420–421
    models, 422–424
  as cell surface receptors for fibronectin, 408–410
  composition of, 410
  GD_2 and GD_3, melanoma cells and, 410–414
  MAbs directed to, 415–422
  and tumor cell–substratum interactions, 408
  perspectives, 424–426
Gene expression
  class I
    alterations of, 203–207
    flanking viral sequences alteration of, 215–217
    RadLV effects on, 203–205
    regulation by viral and class I promoter interactions, 210
γ-Glutamyl transpeptidase (γ-GT), and hepatoma, 390–392
Glycoproteins, see also Human chorionic gonadotropin (hCG)
  envelope, cytotoxic T lymphocytes (CTL) recognition of, 223–224
  malignant changes of sugar chains, see Sugar chains
  as tumor markers, 385
Glycosyl transferases, sugar chains and, 401–403

GM-CSF, *see* Granulocyte-macrophage colony stimulating factor (GM-CSF)
Goldberg-Hogness box, 210
gp 120v-*fms* polyprotein, 82
gp 140v-*fms* polyprotein, 82
Granulocyte-macrophage colony stimulating factor (GM-CSF), 88
Growth factors, *see also* Individually named growth factors
  protein kinase C and
    activation by, 160–164
    receptor transmodulation, 162–163
  protooncogenes and, 40, 57
γ-GT, *see* γ-Glutamyl transpeptidase (γ-GT)

H-2 cosmid clones, 211
H genes, virus integration adjacent to, 212–215
Hapten immunotherapy, for BCL$_1$
  and MDP helpers, 366–368
  protocol, 365–366
  and TNP helpers, 366
Hapten-reactive helper model, 358–359
Haptens, and anti-TATA induction, 357–358
hCG, *see* Human chorionic gonadotropin (hCG)
Helper antigen system, 357–358
Helper T cells
  against antigens, 358–359
  Lyt-1$^+$2$^-$ T cells as, 334–335
Hematopoiesis, in mouse, 130–131
Hematopoietic cell interaction, A-MuLV, 130–135
Hepatoma, γ-GTs and, 390–392
Heterogeneity, of responses, after protein kinase C activation, 172–174
Host responses, to tumors, enhancement of, 430
[^{3}H]-PDBu binding sites, phorbol esters and, diacylglycerol inhibition and, 166
HTLV-1, *see* Human T-cell leukemia virus type 1 (HLTV-1)
Human chorionic gonadotropin (hCG)
  amino acid sequences of, 395
  unusual sugar chains in, 392–401
  structure, 396
Human T-cell leukemia virus type 1 (HLTV-1)
  ATL and, 187–188, 189–190
  facts about, 188
  genomic structure, 188–189
  leukemogenesis mechanism of, 190–191
  pX and, *see also pX*
    gene expression, 192–193
    sequence, 191–192
Hydatidiform mole, hCG sugar chains in, 400
HZ-FeSVII, 135–136

Ia antigens
  IL-2 production and, 268
  on melanoma and melanocytes, 377–378
  SMLR and, 273–274
IFN, *see* Interferon (IFN)
IGF-1, *see* Insulinlike growth factor-1 receptor (IGF-1)
IgG$_3$ MAb
  melanoma cells and, 374–375
  melanoma treatment and, 380–381
IL-2, *see* Interleukin-2 (IL-2)
IL-3, *see* Lymphokine IL-3
Immune complexes, v-*fms* oncogene product and, 82–83
Immune T cells
  adoptive therapy with, 440–442
  macrophage activation to tumoricidal state by, 438–440
  mediation by, ACIT and, 432–435
Immunity
  concomitant, *see* Concomitant immunity
  generation and loss during immunogenic tumor growth, 301
  and T lymphocyte mediation, 296
  in tumor mass *in situ*, 344–345
Immunization, tumor immunogenicity revealed by, 295–296
Immunofluorescent staining
  with anti-p185 MAbs, 73
  of gangliosides GD$_2$ and GD$_3$, in melanoma cells, 410–414
Immunogenicity, 295–296
Immunoprecipitation, of p21, *ras* protein synthesis and, 250
Immunosurveillance, SV40 tumor antigen in, 238
Inositol-1,4,5-triphosphate (IP$_3$), PI turnover and, 158
Insulin receptor, features of, 60–61
Insulinlike growth factor-1 receptor (IGF-1), features of, 60–61
Interferon (IFN)
  defective cytotoxic T lymphocyte (CTL) generation restoration by, 268, 269, 270
  production by purified LGL, 267–268
Interleukin-2 (IL-2)
  defective cytotoxic T lymphocyte (CTL) generation restoration by, 268, 269, 270
  effect on protooncogene levels in PBL, 44, 46–47
  production by purified LGL, 267–268
  protein kinase C and, 163–164
  and signal transduction mechanisms, 47–49

N-Iodoacetyl-N'-(5-sulfonic-1-naphthyl) ethylene diamine (AED), 358
IP_3, see Inositol-1,4,5-triphosphate (IP_3)

Large granular lymphocytes (LGL)
  in allogeneic T-cell response, 274–275
  autoregulation of, 272–273
  cytokine production by, 273
  depletion and enrichment of, 262–264
  IL-2 and IFN production by, 267–268
  and NK activity, 261, 271–272
LDLR, see Low-density lipoprotein receptor (LDLR)
Lectin, see Phytohemagglutinin (PHA)
Leukemia
  chromosome alterations in, 11–12, see also Chromosome translocations
  chronic granulocytic, tumor progression in, 2
  chronic myelogenous (CML), see Chronic myelogenous leukemia (CML)
  disseminated, 435–438
  FBL, see FBL leukemia
  T cell–T cell interaction and, 365–368
Leukemogenesis
  etiological agent of, 189–190
  pX function involvement in, 198–199
  trans-acting viral function for, 190–191
Ligand
  of c-fms protooncogene product, 83–84
  for protein kinase C, structure–function relationship, 174–175
L-myc gene, 28
Long transposable elements, in class I gene expression alteration, 215–216
Low-density lipoprotein receptor (LDLR), features of, 60–61
LTR
  mediated by pX, trans-activation of, 193–194
  $p40^x$ as trans-activator, 194–195
Ly genes, virus integration adjacent to, 212–215
Lymphocytes, see also T lymphocytes
  cytolytic T (CTL), see Cytolytic T lymphocytes (CTL)
  cytotoxic T (CTL), see Cytotoxic T lymphocytes (CTL)
  tumor-infiltrating, see Tumor-infiltrating lymphoid cells (TILC)
Lymphocytic tumors, chromosome translocations in, 3–5
Lymphoid cells, mitogens for, 163–164
Lymphoid transformation
  A-MuLV, Ig gene structure and, 132–133
  region I role in, 129

Lymphoid transformation (cont.)
  region III role in, 129–130
  region IV and, 127–128
  region IV role in, 127–128
Lymphokine IL-3, A-MuLV infection and, 134–135
Lympholysis, cell-mediated (CML), 204–205
Lymphomas, cytogenetics of
  Burkitt's, 3–5; see also Burkitt's lymphoma
  other B-cell, 5–6
  T-cell, 8–9
$Lyt-1^+2^-$ T cells
  allograft rejection and autoimmune diseases, role in, 346
  antitumor effect of, and nonspecific tumoricidal effectors collaboration, 336–342
  and DTH response, 335–336
  as helper T cells, 334–335
  and in vivo tumor immunity, 361–362, 363
  TATA recognition by, 345–346
  tumor eradication in vivo, role in, 332–333
  tumor growth inhibition in diffusion chamber by, 338–339

Macrophage
  growth factor for, 84
  transformation model systems involving, 88
Macrophage-activating factor (MAF)
  immune T cells and, 438–440
  production
    by FBL-immune T cells, 439
    T-cell subsets responsible for, 340–341
Major histocompatibility complex (MHC)
  antihapten T-cell response and, 358
  class I, requirements in cytotoxic (CTL) recognition, enhanced expression, 226–227
  class I, requirements in cytotoxic T lymphocyte (CTL) recognition, 225–227
  regulation, 226
  restricted, 225–226
  class II, diffusion chamber culture system and, 341–342
  viral sequences in, 210–212
Malignant melanoma, cytogenetics of, 12–13
MCA, see Methylcholanthrene (MCA)
MDP, see Muramyl dipeptide (MDP)
Melanocytes, see also Melanoma
  antigen analysis of, 375–376
  Ia antigens on, 377–378
Melanoma
  antigen systems on, 373
  categories of, 376

Melanoma (*cont.*)
  cell attachment to extracellular matrix
    components, 415–422
  cell heterogeneity, 378–380
  focal adhesion plaque of, GD_2 and GD_3 in,
    410–414
    MAbs directed, 415–422
  genetic material loss, 379
  genetic studies, 379
  Ia antigens on, 377–378
  malignant, cytogenetics of, 12–13
  mouse MAb detecting GD_3 and, 380–381
  phenotypes, 376
Metabolic processing, of EGFR, 116
Methylation, of class I genes, RadLV
    transformation and, 207–209
Methylcholanthrene (MCA)
  and fibroblast sensitivity to NK cells, 257
  -induced sarcoma, class 1 gene expression
    alterations and, 206
MHC, *see* Major histocompatibility complex
    (MHC)
Middle T antigen (MTAg), 149–151
Mitogenic lectin, *see* Phytohemagglutinin (PHA)
Mitogenic response, c-*myc* gene role in, 26–27
Mitogens
  for fibroblasts, 161–162
  for lymphoid cells, 163–164
M-MuLV, *see* Moloney murine leukemia virus (M-MuLV)
Moloney murine leukemia virus (M-MuLV)
  antigen regulation by, 226
  cell surface expression of, 222–223
  fibroblasts infection by, 226–227
  MHC-restricted recognition of, 225–226
  Moloney sarcoma virus (M-MuLV:MSV)
    complex, 222
Moloney sarcoma virus (MSV)
  antigen regulation and, 226
  fibroblast infection and, 226–227
  and M-MuLV, 222
Monoclonal antibodies (MAbs)
  anchorage-independent growth inhibited by,
    75
  antiganglioside, effect on melanoma cell
    attachment, 418, 420
  anti-p185, 72
  and cell-surface p185, down-modulation of,
    72–74
  effect on tumorigenic growth, 77, 78
  hybridomas secreting, selection method, 71
  immunofluorescent staining with, 73
  immunologic antitumor effects of, 76–77
  in vitro cytotoxic effects of, 76
  C350, melanoma cells and, 376–377

Monoclonal antibodies (MAbs) (*cont.*)
  CF21, melanoma cells and, 376–377
  ganglioside antigen directed, perspectives on,
    424–426
  GD_3
    melanoma cells and, 374–375
    melanoma treatment and, 380–381
  high-molecular-weight glycoproteins reaction
    with, 375
  preparation of, anti-TsF antisera and, 282–283
  reactive with *neu* oncogene product, 71–72
  TA99, melanoma cells and, 376–377
  and tumor cell–substratum interactions, 407–
    408
Mono[^{125}I]iodotyrosine, EGF and, 116
mRNA
  class I, RadLV transformation and, 207–209
  in c-*myc* gene
    cellular differentiation, 27–28
    and c-*myc* correlation, 28
    and expression regulation, 29–33
    levels, 26–27
  c-*src* gene product in, 152
  erb-B-related, 100
  protooncogene, accumulation over time, 41–43
  in T-lymphocytes, 48–49
MSV, *see* Moloney sarcoma virus (MSV)
MTAg, *see* Middle T antigen (MTAg)
Muramyl dipeptide (MDP)
  haptenic, structure of, 367
  in leukemia immunotherapy, 366–368
Murine tumors, immunogenicity of, 280
Mycobacterium tuberculosis, haptens and, 358

Natural killer (NK) cells
  abrogation of activity after anti-asialo GM_1
    injection, 262, 263
  in allogeneic T-cell response, 274–275
  asialo GM_1^+ cell fraction analysis, 264
  characterization, 245, 246–247
  clones, *see* NK clones
  fibroblasts, enhanced killing of, 245–248
  murine, *see* Large granular lymphocytes (LGL)
  regulatory features of, 271–274
  sensitivity, 244, 245–246
    c-H-*ras* expression mediation of, 248–255
    v-*fps* oncogene and, 248
  and spleen cells, *see* Spleen cells, NK-depleted
  susceptibility, spleen cells as effectors, 252
  in vitro studies with, 243–244
    results, 244–257
Neoplasia
  karyotypic alterations in, generalizations
    about, 1
  resistance to, 203–207

Neoplasms, immune responsiveness to, 280–281
neu oncogene, 70–71
neu oncogene product
  monoclonal antibodies reactive with, 71–72
  as receptor, 63–64
  transformed NIH 3T3 cell line, see NIH 3T3 cell line
Neural tissue, c-src gene product in, 152
Neuroblastoma
  cell attachment to extracellular matrix components, 415–422
  cytogenetics of, 13–14
Neurons, c-src gene product in, 152
NIH 3T3 cell line, anti-p185 MAbs and, 72
  down-modulation, 72–74
  effect on tumorigenic growth, 77, 78
  transformed phenotype reversion, 74–76
4-Nitroquinoline oxide (4-NQO), tumor induction by, 317
NK clones
  cytolysis of, 253–254
  phenotype of, 253
N-myc gene, 28
Nonlymphocytic leukemias, cytogenetics of, 9–10
4-NQO, see 4-Nitroquinoline oxide (4-NQO)
Nuclear matrix, protooncogene association with, 55–56

Oligosaccharides, from hCG
  gel permeation chromatographies, 397
  structures, 397–398, 399–400
Oncogenes
  cytogenetics and, 2–3
  -induced transformation, phorbol ester enhancement of, 165
  PI turnover induction by, 164–165
  products
    neu, 63–64
    phosphorylation of, 165
    v-erb-B, 62
    v-fms, 62–63
  retroviral, see Protooncogenes
  viral and nonviral, 59–60
Oncogenesis
  DNA role in induction of, 59
  viral and nonviral oncogenes in, 59–60

P815 mastocytoma
  immune response to, manipulation of, 285–290
  mastocytoma, 280–281
  and T-cell hybridoma (A10), 283–285
  TsC development and, 281
  tumor, resistance to, 288–289
  tumor growth, anti-TsF antisera effect on, 282–283

PBL, see Peripheral blood lymphocytes (PBL)
[^{3}H]-PDBu binding sites, phorbol esters and, 160
PDGF, see Platelet-derived growth factor (PDGF)
Peripheral blood lymphocytes (PBL),
  protooncogene expression in, 40–41
  mitogen-stimulated, 41–43
  mRNA accumulation time course, 41–43
  PHA-stimulated, 44–47
Peripheral blood platelets, c-src gene product in, 152–153
PHA, see Phytohemagglutinin (PHA)
Phenotype, transformed, reversion by p185 monoclonal antibodies, 74–76
Phorbol esters
  biological effects of, 159
  and calcium, synergism between, 167–168
  as carcinogenesis promoters, 157–159
  80,000-dalton protein phosphorylation and, 160–161
  and diacylglycerol relationship
    in protein kinase C activation, 165–167
    structural analogy, 174–175
  mitogenesis induction by, 159–160
  oncogene-induced transformation enhancement by, 165
  protein kinase C down-regulation and, 160
  and signal transduction mechanisms, 47–49
Phosphatidylinositol (PI) turnover, 158
  induction by oncogenes, 164–165
[^{32}P]phosphoric acid, v-fms oncogene product and, 83
Phosphorylation
  of 80,000-dalton protein, 160–161
  of erb-B protein, 101–102, 103
  of oncogene product, 165
  and substrate specificity, 170–171
  tyrosine, TPA and, 101–102
Phytohemagglutinin (PHA)
  PBL stimulated by, protooncogene expression in, 44–47
  and signal transduction mechanisms, 47–49
PI turnover, see Phosphatidylinositol (PI) turnover
PI-4,5-diphosphate (PIP$_2$), PI turnover and, 158
PIP$_2$, see PI-4,5-diphosphate (PIP$_2$)
Plasmacytoma cells, in mouse, c-myc gene deregulation in, 29–30
Plasmic pMTPX expression, 194–195
Platelet-derived growth factor (PDGF)
  protooncogenes and, 40, 55
  receptor (PGDFR)
    features, 60–61
    and v-sis gene, 93
Polyoma virus middle tumor antigen, pp60^{c-src} tyrosine kinase activity and, 149–151

Polyprotein
  gp 120v-*fms*, 82
  gp 140v-*fms*, 82
Pro-B cell, 130
Progressive tumor, concomitant immunity as unsuccessful response to, 296–298
Protein kinase C, 158–159
  calcium mobilization and, 169–170
  cAMP second messenger system and, 168–169
  down-regulation following phorbol ester treatment, 160
  growth factors and
    activation, 160–164
    receptor modulation, 162–163
    and Interleukin-2 (IL-2), 163–164
  ligand structure–function relationships, modeling studies of, 174–175
  oncogene product phosphorylation by, 165
  and oncogenes, interaction between, 164–165
  phorbol ester and diacylglycerol relationship, 165–167
  response heterogeneity mechanisms after activation, 172–174
Proteins
  erb-B-related, amino acid sequence of, 95–97
  pX sequence and, 191–192
Protooncogenes
  expression
    in normal cells, 40
    regulation in CHX studies, 49–54
  identified, 39
  products, 39–40; *see also* Oncogenes
    c-*fms*, *see* c-*fms* protooncogene product
    products of, protein, 55–56
    regulation in human PBL, 40–41, *see also* Peripheral blood lymphocytes (PBL), protooncogene expression in
  and signal transduction mechanisms, 47–49
Psoralen and UV-A (PUVA), tumor induction by, 317
p40^x, *trans*-activation of LTR by, 194–195
PUVA, *see* Psoralen and UV-A (PUVA)
pX
  function, involvement in leukemogenesis, 198–199
  gene expression, mechanism of, 192–193
  LTR mediated by, *trans*-activation, 193–194
  sequence in HLTV-1, 191–192

RadLV
  and cell-mediated lympholysis (CML) response, 204–205
  and cytolytic T lymphocytes (CTL), 205–206
  effects on class 1 gene expression, 203–205

RadLV (*cont.*)
  transformation, states associated with, 207–209
  tumorigenesis, and class I DNA, 209
ras Expression, Northern blot analysis, 249
Receptors, cell-surface, *see* Cell-surface receptors
Resistance, to neoplasia, 203–207
Responder T cells, and SMLR inhibition, 269, 271
Restriction-fragment-length polymorphisms (RFLPs), 209
  melanoma cells and, 379–380
Retinoblastoma, cytogenetics of, 14
Retroviruses, 187–188
  in *H* gene polymorphism, 214
  HTLV-1, *see* Human T-cell leukemia virus type 1 (HTLV-1)
  tissue specificity of, 193–194
RFLPs, *see* Restriction-fragment-length polymorphisms (RFLPs)
RGDS, *see* Arginyl-glycyl-L-aspartyl-L-seryl

Sarcoma, methylcholanthrene (MCA)-induced, 206
Short interspersed repeats, in class I gene expression alteration, 215
Signal transduction
  mechanisms in normal T-lymphocytes, 47–49
  protein kinase C pathway, 168–170
SILR, *see* Suppressor-induced late regression (SILR) phenomenon
Skin cancer
  phorbol esters in promotion of, 157–159
  UV radiation induced, *see* UV tumor system
Small cell carcinoma, cytogenetics of, 14
SM-FeSV, *see* Feline sarcoma virus (SM-FeSV)
SMLR, *see* Syngeneic mixed lymphocyte reaction (SMLR)
Solid tumors
  chromosome alterations in, 12–14
  T cell–T cell interaction and, 362–365
Spleen cells, NK-depleted
  cytotoxic T lymphocyte (CTL) generation in
    defective, 264–265
    restoration of, 265–267
    restoration of, by IL-2 or IFN, 268, 269, 270
  cytotoxic T lymphocyte (CTL) response of, 266
  and IL-2 and IFN production, 267–268
src gene family, *see also* c-*src* gene products; v-*src* gene products
  v-*erb-B* gene as member, 95–98
Substrate specificity, phosphorylation and, 170–171

Sugar chains
  asparagine-linked, 387
    complex type, 390
  and blood group A antigens, 401–402
  complex type, 388–389
  of glycoproteins, 385
    structural rules, 386–387
  and glycosyl transferases, 401–403
  of γ-GTs
    from kidney, 394
    from liver, 393, 394
  in hCG, 392–401
Suppression, see also Suppressor T cells;
    Suppressor T lymphocytes
  passive transfer of, 299–300
  tumor specificity and, 300
Suppressor T cells
  as adoptive therapy obstacle, 431–432
  assays used to measure production of, 303
  and concomitant immunity decay, 298–301
    role in, 300–301
  as down-regulators of antitumor immune
    response, 301–303
Suppressor T lymphocytes
  UV-induced tumor antigens and, 314–315
  UV-radiation induced, effect on UV-induced
    antigenic variants, 321–322
Suppressor-induced late regression (SILR)
    phenomenon, 286–287
SV40 tumor antigen (T-ag)
  cell surface localization of, 232–233
  cellular immune reactions and, 235–236
  cytolytic T lymphocyte (CTL)-reactive sites
    fine mapping of, 237–238
    localization of, 236–237
  fine mapping studies, 237–238
  in immunosurveillance, 238
  multifunctional nature of, 231–232
  nuclear and surface forms, comparative
    analysis, 234
  and SV40-specific cytolytic T lymphocyte
    (CTL), 236
  in TSTA activity, 232
Syngeneic mixed lymphocyte reaction (SMLR)
  DNA replication suppression in, 272
  inhibition of, 269, 271
  and syngeneic Ia antigenic determinants, 273–274

T cells, see also Helper T cells; Suppressor T cells
  antitumor effector, increasing number of, 356–357
  immune, see Immune T cells

T cells (cont.)
  in recognition and eradication of disseminated
    leukemia, 435–438
T cell–T cell interaction
  model for, 357
  tumor-specific immunity by, 358–362
    in leukemia system, 365–368
    in solid tumor system, 362–365
T suppressor cells, see also Suppressor T cells;
    Suppressor T lymphocytes
  and T suppressor factors in tumor models, 281
T suppressor factors (TsF)
  anti-P815, effect on P815 tumor growth, 282–283
  immunization with, 289
  in tumor models, 281
TA99 MAb, melanoma cells and, 376–377
TAA, see Tumor-associated antigens (TAA)
T-ag, see SV40 tumor antigen (T-ag)
Target cells, A-MuLV, analysis of, 133
TATA, see Tumor-associated transplantation
    antigens (TATA)
T-cell hybridoma (A10)
  and P815 immune response manipulation,
    285–290
  P815-specific TsF, 283
    biochemical characterization of, 284–285
    and cytotoxic T lymphocyte (CTL)
      generation suppression, 283–284
T-cell response
  antihapten, 357–358
  LGL in, 274–275
  for tumor eradication, 434–435
T-cell tumors, cytogenetics of, 8–9
Tetradecanoate-phorbol-13-acetate (TPA)
  and signal transduction mechanisms, 47–49
  in tyrosine phosphorylation increase
    prevention, 101–102
TFR, see Transferrin (TFR)
TILC, see Tumor-infiltrating lymphoid cells
    (TILC)
T-lymphocytes, see also Responder T cells
  cloned (CTB6-4A), protooncogene expression
    in, 48–49
  growth events of, 56
  immunity mediated by, 296
  and in vivo immune protection, 361–362
  protooncogene
    mRNA accumulation, 48–49
    role, 55
  signal transduction mechanisms in, 47–49
TNP, see Trinitrophenol (TNP)
TPA, see Tetradecanoate-phorbol-13-acetate
    (TPA)

trans-activation of LTR
  elements responsible for, 196–198
  mediated by pX, 193–194
  by $p40^x$, 194–195
Transcription
  of class I gene expression, 215–216
  of c-myc gene, 30
    control regions near, 32
    protein encoding by, 33
    rate studies, 31–32
    and erb-B-related genes, 100
Transferrin (TFR), protooncogenes and, 41–43
Transformation
  and Abelson protein regions, 125–126
    region I, fibroblasts and, 126–127
    region II, 126
    region IV, lethal effect in fibroblasts, 127
  antigenicity and, relationship between, 320–322
  lymphoid, 127–130; see also Lymphoid transformation
  model systems, 87–88
  oncogene-induced, phorbol ester enhancement of, 165
  RadLV, 207–209
  related properties, zinc induction, 250
  by v-src protein, 146
Transformed phenotype, reversion by anti-p185 MAbs, 74–76
Transforming growth factor, type (TGF-α), see Epidermal growth factor (EGF)
Translocation, see Chromosome translocations
Transmodulation, of growth factor receptors by protein kinase Cf, 162–163
Trinitrophenol (TNP)
  212 cell line and, 256–257
  haptens and, 358
  immunotherapy
    in MCA-induced autochthonous fibrosarcoma model, 363–365
    protocol for, 362, 363
    solid tumor regression, 362, 364
  -reactive helper model, 358–359
Tryptic digestion, of EGFR, 115
TSA, see Tumor-specific antigens (TSA)
TsC, see T-suppressor cells (TsC)
TsF, see T-suppressor factors (TsF)
TSTA, see Tumor-specific transplantation antigen (TSTA) system
Tumor cell–substratum interactions, 407–408
  gangliosides and, 408
  perspectives in, 424–426
Tumor eradication in vivo
  and bystander tumor growth inhibition, 342–343

Tumor eradication in vivo (cont.)
  irrelevant tumor cells, mechanism for, 343
  Lyt-1^+2^- T cells role in, 332–333
    allograft rejections and autoimmune diseases, 346
    mechanisms underlying, 334–342
  by tumor-infiltrating lymphoid cells, 347, 348
Tumor immunogenicity, revealed by immunization, 295–296
Tumor markers, defined, 385
Tumor mass in situ, antitumor immunity, 344–345
Tumor models, T suppressor cells and factors in, 281
Tumor progression, cytogenetics of, 1–2
Tumor protection, and DTH response correlation, 335–336
Tumor-associated antigens (TAA), tumor-specific antigens (TSA) versus, 279
Tumor-associated transplantation antigens (TATA)
  and antitumor effector T cells, 356–357
  evidence for, 355–356
  heterogeneity of, 359–361
  immune responses to, 331–332
  and Lyt-1^+2^- T cells, 332, 334–335
  and Lyt-1^+2^- T cells, recognition by, 345–346
Tumoricidal effectors, nonspecific, and antitumor effect of Lyt-1^+2^- T cells, collaboration between, 336–342
Tumorigenesis
  and neu-transformed cells, growth of, 77, 78
  RadLV, and class I DNA, 209
Tumor-infiltrating lymphoid cells (TILC)
  and antitumor immunity, 344–345
  cell populations constituting, 344, 345
  tumor eradication in vivo by, 347, 348
Tumors
  adoptive therapy of, 431–432
  A-MuLV induced, 135
  antigenic changes in, 317–318
  cell modification, 357–358
  elimination of, kinetics, 433, see also Tumor eradication in vivo
  host response to, enhancement of, 430
  immunogenic surface antigens and, 429–430
  lymphocytic, chromosome translocations in, 3–5
  markers for, defined, 385
  M-MuLV induced, see Moloney murine leukemia virus (M-MuLV)
  murine, immunogenicity of, 280
  progressive, 296–298
  solid, see Solid tumors
  UV, see UV tumor system

Tumor-specific antigens (TSA), versus tumor-associated antigens (TAA), 279
Tumor-specific immunotherapy models, T cell–T cell interaction in establishment of, 362–368
Tumor-specific T cells, *in vitro* expansion before transfer, 440–442
Tumor-specific transplantation antigen (TSTA) system
  molecular analysis of, 318–320
  nuclear T-ag involvement in, 232
  origin and function of, 307–308
  purification using biochemical approaches, 315–316
  tumor-associated antigens and, 279
  in UV tumor system, 310–311
Turnover, of EGFR, 116
Tyrosine kinase
  and c-*erb*-B-2 protein, activity, 101
  for c-*src* and v-*src* gene products, activity compared, 148–149
  pp60^{c-src}, activation by polyoma virus middle tumor antigen, 149–151

UV radiation, tumors induced by, *see* UV tumor system
UV-1591 tumor cells, TSTA molecular analysis on, 318–320
UV tumor system, 308–310
  antigens in
    biochemically defined, 315–316
    changes, origin of, 317–322
    cytotoxic T lymphocyte (CTL)-defined, 312–313
    suppressor T lymphocyte-defined, 314–315
    transplantation-defined, 310–312
  1591 cells, TSTA molecular analysis on, 318–320
  properties of, 309–310
  uniqueness of, and other tumors compared, 317–318

v-*erb*-B gene
  of avian erythroblastosis virus, 94–95
  implications, 103–106

v-*erb*-B gene (*cont.*)
  epidermal growth factor receptor (EGFR) gene and, 93
  as member of *src* gene family, 95–98
  protein related to, *see* erb-B protein
v-*erb*-B oncogene product, 62
v-*fms* gene, CSF-1 transformation by, 84–86
v-*fms* oncogene product, 62–63
  feline sarcoma virus and, 62, 81–83
v-*fps* oncogene
  and NK sensitivity, 248
  and v-Ki-*ras* oncogene, similarities between, 247
Viral replication, *pX* proteins and, 198–199
Viral sequences
  class I gene expression alteration by, 215–217
  in MHC, 210–212
Viruses
  and class I promoter, interactions between, 210
  integration adjacent to *H* and *Ly* genes, 212–215
v-Ki-*ras* oncogene
  fibroblasts transfected with, enhanced NK killing, 245–248
  Southern blot analysis of, 247
  and v-*fps* oncogene, similarities between, 247
  YAC and, competitive inhibition, 246, 247
v-*onc* genes, versus c-*onc* protooncogenes, 145
v-*sis* gene, PGDFR and, 93
v-*src* gene products, and c-*src* compared
  function, 147–148
  structure, 146–147
  tyrosine kinase activity, 148–149
v-*src* protein, transformation by, 146

X5563 tumor cells, growth inhibition of, 340

YAC, and v-Ki-*ras* oncogene, competitive inhibition, 246, 247

Zinc induction, of transformation-related properties, 250